S0-BFD-413

SHORTWAVE
RADIO LISTENING
WITH THE EXPERTS

Edited by
Gerry L. Dexter

Howard W. Sams & Co.
A Division of Macmillan, Inc.
4300 West 62nd Street, Indianapolis, IN 46268 USA

NOTICE TO LISTENERS

As this book went to press, a bill concerning the privacy of electronic communications was working its way through Congress. At this point, it is not known exactly how this legislation, if passed, will affect certain forms of radio listening. The Association of North American Radio Clubs' Government Affairs Liaison is following the bill's progress and working to see that it has minimal effect upon those who monitor the airwaves as a hobby activity. Readers interested in more information may contact this committee in care of Robert Horvitz, 1634 15th Street NW, Washington, DC 20009.

Those who monitor non-broadcast, non-amateur communications should also be aware of certain provisions of the Communications Act of 1934 which make it illegal to repeat the content of such two-way transmissions.

Additionally, certain states have laws against monitoring some types of VHF/UHF transmissions on scanner radios—usually involving the listening to law enforcement communications on an in-car scanner. Readers are advised to check for any applicable laws which may be in effect in their state of residence.

GERRY L. DEXTER

© 1986 by Thomas P. Harrington

FIRST EDITION
FIRST PRINTING—1986

All rights reserved. No part of this book shall be reproduced, stored in a retrieval system, or transmitted by any means, electronic, mechanical, photocopying, recording, or otherwise, without without written from the publisher. No patent liability is assumed with respect to the use of the information contained herein. While every precaution has been taken in the preparation of this book, the publisher assumes no responsibility for errors or omissions. Neither is any liability assumed for damages resulting from the use of the information contained herein.

International Standard Book Number: 0-672-22519-0
Library of Congress Catalog Card Number: 86-61479

Satellite and Amateur Radio Series Editor: *Thomas P. Harrington*
Edited by: *Frank N. Speights*
Illustrated by:*William D. Basham and William R. Hurley*
Designed by: *T.R. Emrick*
Cover art by: *Dean Glascock*

Printed in the United States of America

CONTENTS

PREFACE

Thirty years ago, Indonesia was an uncharted land from a DX standpoint. Music heard on a Venezuelan station was described as "Latin American," all Chinese stations were "Radio Peking," all Russian stations were "Radio Moscow," and the use of single-sideband reception techniques to clear up a signal transmitted in amplitude modulation was unheard of. "Grayline" might have been thought to be a transportation company, and "krongkong" considered as a monster in some grade-D Japanese horror film.

If the opportunity should ever arise, there is a very revealing test you can make. Read a shortwave club bulletin from the 1950s and compare it to one from the 1980s. Ignore the fact that one bulletin will be offset printed while the other was ground out on a mimeograph machine. Look at the content. When you've completed the test, you may agree with yours truly on this point: the remarkable technical advances made in equipment over the past several years grow pale in comparison to the advances made in the general knowledge of shortwave listening and DX'ing.

Today's DX'er knows far, far more about the intricacies of DX'ing and listening than did his counterpart 30 years ago, assuming equal experience levels. And, a newcomer, still getting his wrist in shape, comes off the starting blocks much faster than used to be the case.

Scanners did'nt exist then. And, FM and TV were barely out of the cradle. The ordinary frequency monitor shrugged his shoulders at a radioteletype signal, realizing it was beyond his ken (assuming he could recognize it), and everybody sent International Reply Coupons (IRCs) with their reception reports.

Equipment innovations came from those commercial concerns with the financial wherewithal to develop and improve upon the technical side of reception. The new learning knowledge came largely from the hobbyists themselves —each contributing new learning, new techniques, new approaches through their own research, experimentation, and station loggings.

The advances in knowledge made by enthusiastic listeners and DX'ers is far more worthy of a salute than the advances in technology, welcome though such advances are to all of us. One expects science to march along, bringing us new wonders on a regular basis. Indeed, we've almost come to expect it as something to which we are entitled.

Praise, however, is better handed to those individual listeners or groups who, working together, have blazed new trails, thanks to their deep interest and involvement in the fascinating hobby of shortwave listening. Every person who gets involved in DX'ing of any kind owes a debt to those who were not satisfied with the status quo, who believed and then proved that there was more to Indonesia than the overseas service from Jakarta, and that a indices could mean something to the average listener.

When you think about it, the bottom line in successful DX'ing isn't fancy equipment, it's knowledge. Digital readout, selectable sideband, super-sensitivity, and all kinds of filters and outboard gadgets are all nice, and they certainly do contribute a great deal to one's listening successes, but a person with the necessary knowledge can make an old klunker of a receiver perform miracles which someone lacking in knowledge cannot duplicate with the most expensive, state-of-the-art equipment.

The amount of information available on the subject of listening and DX'ing has multiplied severalfold over the years as well—from general frequency lists to club-produced phamplets covering the more esoteric areas of DX'ing. One of those "books on wheels" mobile library units, which are available in many areas, could very likely be filled with publications relating to listening and DX'ing. A few years ago, however, the available works on those subjects might have taken up little more than a single shelf.

Why add yet another volume to the library? There are several good reasons. The knowledgeable and experienced DX'er, even if specializing in just one DX'ing mode, rarely is equally strong in all facets of his or her band speciality. *Radio Listening with the Experts* offers the chance to fill in some gaps. For the newer listener, the book provides what we feel is the key to gaining this information faster than could be done otherwise and thereby permits advancing in the hobby more quickly. No matter which side of the experience line you find yourself standing on, this handbook will also provide a superb introduction to other forms of monitoring and DX'ing.

The authors featured here represent some of the brightest stars in the DX'ing game—the equivalent of million-dollar-a-year professional atheletes. All have either pioneered the research and development of their particular areas or have made significant contributions of knowledge and techniques in their chosen specialities. More than that, they've agreed to share their knowledge and experience with you. I know many of the authors in this book personally and the rest by name and reputation. I can attest to their qualifications. You can rely upon their information and advice.

There are subjects yet to be treated and explored (after all, DX'ing is as big as the world itself) and we hope to deal with other areas in the future. Meantime, what you hold here is the combined experience and learning of inquiring minds who looked at a given phase of the hobby and thought, "There must be more to it than that," and proceeded to prove it.

There's plenty to digest here and it all comes right from the "horse's mouth." Read on!

GERRY L. DEXTER

SETTING UP YOUR SHACK

Fred Osterman and Larry Cunningham

Communications is more than a hobby for **Fred Osterman**. He is the manager of Universal Short-wave Radio (a division of Universal Amateur Radio, Inc.) of Reynolds-burg, Ohio, where he has developed the Universal Bulletin Board and Information Exchange (UBIX) —first computer-based bulletin board designed expressly for communications enthusiasts.

For Fred Osterman, it all began on the medium waves 20 years ago. Today, he is a recognized expert in the utilities, especially RTTY, CW, and FAX. His articles on RTTY and other utility-related subjects have appeared in most of the commercial and non-profit shortwave publications. He is the author of the *Shortwave Log* and the *DX'ers Directory*, both published by Universal Shortwave.

Fred's other interests include astronomy, cats (he has four!), computers, and amateur radio (he holds General Class call N8EKU). Fred and his wife, Barbara, live near Columbus, Ohio.

It takes a special kind of DX'er to haunt the tropical shortwave broadcast bands. **Larry Cunningham** is one of those special types.

Larry has also been a pioneer in utilizing microcomputers in the short-wave hobby field. Several of his programs are now part of the Association of North American Radio Clubs (ANARC) computer library. He has also been involved in the development of UBIX (Universal Bulletin Board and Information Exchange).

Larry works for the State of Ohio (in the personnel field). He and his wife, Colleen, (and two more cats!) live near Columbus, Ohio.

The most important factor in setting up a radio shack, in many cases, is not the "radio" but rather the "shack." Radio listening must be done in an environment which is, at the same time, both comfortable and conducive to concentration. Research supports the idea that one's environment directly affects one's attitude and degree of success. This certainly applies to the surroundings you create for yourself during your listening. Creating a pleasant and functional environment can make DX'ing more enjoyable, and more productive.

General Environment

Location

The first decision that needs to be made in setting up a radio shack is which room of the house is to be used. In many cases, because of space limitations and other factors, you may have little choice in the matter. The room should be isolated enough to permit listening at any time of day without disturbing other family members. The room should offer easy access for antenna leads and a short access to an earth ground. The room should be as far as possible from television sets, microwave ovens, or any other appliance that emits radio frequency interference (RFI). Basements usually offer one of the best locations to build a "radio" room, presuming that it can be kept warm and dry.

Temperature

A ventilated space is vital, especially if you have tube-type equipment. You'd be amazed at how a couple of Hammarlund (tube) receivers can raise the temperature of a small room!

If you are restricted to a small room or a section of the basement and you smoke, you should also consider an exhaust vent. Constant smoke can damage electronic equipment. Potentiometers get prematurely scratchy, wafer switches can get noisy, etc. Computer equipment (especially disk drives) is also very susceptible to smoke damage. A small room air conditioner (4000 to 6000 BTUs) is ideal for maintaining the desired temperature and air quality.

If you elect to utilize the basement area, you may want to consider a dehumidifier and a supplementary source of heat to protect you and your equipment from dampness.

Lighting

Lighting is subjective to some extent; however, a few rules do apply. Lighting should not originate from behind or on top of the equipment. It is desirable to have control of the lighting, but beware of light dimmers. Many light dimmers produce a horrendous amount of interference (RFI). For the same reason, fluorescent lights must also be avoided.

Although it may sound strange, a lamp with a red light bulb throws a very pleasing soft light that is often ideal for late-night listening. It provides the necessary light, but doesn't tire the eyes.

Electrical Power

As you progress in the hobby, you will be surprised how your need for additional electrical outlets will grow! Even a modest listening post will require four outlets (receiver, tape recorder, clock, lamp, etc.). If you add a computer and perhaps some RTTY equipment, your needs can quickly increase to over 10 AC outlets.

Ideally, the listening post should be on a separate 110-V AC main circuit. This will insure sufficient capacity, and will diminish potential interference from household appliances and other induced sources.

Interference is a growing problem for listeners. Every appliance that uses a motor, a switch, or a microprocessor can create interference. Some of this noise is radiating through the air and some of it radiates through the AC power lines. A line filter, such as the Drake LF-2 or LF-6, may reduce some of the noise coming through the lines. Many of these line filters will also protect your equipment from line transients. Line transients are those voltage spikes that are caused by electrical storms or problems in the power lines. Solid-state radios (and computers) can easily be damaged by these rapid voltage fluctuations. Line filters are a small investment for both reduced noise and equipment protection.

Newer homes use 12-gauge wire and 15-ampere circuit breakers for standard wiring. This should provide sufficient capacity for most listening posts. Fortunately, the solid-state receivers of today draw much less current than their tube-type predecessors. The average tube-type unit draws 350 watts; the average solid-state unit draws only 70 watts. It is still a good idea, however, to add up the current requirements of all your equipment and check it against the capacity of the outlet you're using. And, owners of older homes *beware!* Some older homes use 14-gauge wire, or aluminum wire. Just because you find a 20-amp fuse in the fuse box for your room's outlets doesn't mean this is the correct value. The previous owner may have run out of the correct (lesser) value! The ground system in each outlet should be tested for correct hookup. You can also inspect the mechanical connection at the main ground system, which is located near the main service entrance. If there are any doubts, you should consult an electrician!

Many DX'ers use a "master" switch to turn off the entire shack (except the clocks) when they are not listening. A good master switch will disconnect both AC lines (hot and neutral) from the wall. This offers the *best* protection from line transients. Alpha Delta Communications has built filtering, line protection, and a master switch into one unit as shown in Fig. 1-1.

**Fig. 1-1. The
Alpha Delta
Model MACC-4
master
controller.**

Furnishings for the Shack

The choice of furniture used in the listening room can make a big difference, both in terms of comfort and success in DX'ing. The principal furniture required is a chair and a table (or a desk) on which your radio gear will be arranged. The key point to remember is that shortwave listening is a "stationary" hobby, and if you are uncomfortable when you sit down to listen, you will not find yourself spending many hours at the radio chasing DX! Choose a chair that is comfortable and from which your radio gear is within easy reach. However, it is best not to choose a chair that is *too* comfortable, since late-night sessions at the radio could end up being merely a long nap as you fall asleep at your set! Because of this, a hard-backed chair rather than a soft living-room chair may be the best choice.

The Desk

Several points should be considered when selecting a radio desk. You need space for your radio and its associated accessories, *and* sufficient open space for notepads, station schedules, and other needed materials. Most radio listeners make notes on what they hear, and there is nothing so frustrating as not having enough space for comfortable writing. Another factor to consider is the *height* of the radio desk or table. The ideal desk height for most people is 30 inches. The height of the desk, in combination with the type and height of your chair, should permit easy operation of the radio controls and easy note taking. A desk or table whose surface is too high off the floor will produce unpleasant results in the long run.

An "L"- or "U"-shaped desk will bring more surface area within reach. Desks such as these (Fig. 1-2) can often be purchased used at office supply centers or at garage sales. Some DX'ers, who are skilled with wood, are able to construct their own custom desks.

Fig. 1-2. Fred Osterman uses a common secretarial "L"-shaped desk to keep equipment within easy reach in his shack.

Because of space limitations and other factors, many listeners use a system of shelves placed at the rear of the desk (Fig. 1-3). A vertical arrangement of your radio gear does wonders in opening up extra space for other needs. Beware, however, of placing often-used equipment too high on a shelf. The inconvenience of having to reach too far will cause the equipment to rapidly fall into disuse.

Fig. 1-3. Larry Cunningham uses shelves for maximum space utilization at his listening post.

While the particular arrangement you choose for your equipment is largely a matter of personal taste, there remain several agreed upon "do's" and "don'ts." In placing your equipment on your table or desk, safety should be the first consideration. In today's world of solid-state devices, problems of heat buildup are generally not as critical as they were in the days of vacuum-tube technology. However, even solid-state gear can overheat and be damaged if sufficient ventilation is not provided. Avoid any arrangement in which the ventilation holes in equipment are blocked or impeded. Do not stack other equipment, papers, books, etc., on top of equipment that has ventilation holes on top. If the piece of equipment has ventilation holes on the bottom, avoid the temptation of shoving papers or books underneath it, thus blocking needed air circulation. Lastly, when deciding on the arrangement, consider what access you will have to a good RF *ground* for your equipment. The subject of grounding is covered in detail elsewhere, but keep in mind that the grounding cable needs to be as short as possible for greatest safety. Try to place equipment which needs to be grounded as close as possible to your ground wire.

Telephones

A telephone in the "shack" is very useful. If you are fortunate enough to be in touch with other listeners in your area, a nearby phone will allow you the convenience of sharing DX tips and information *while* you are listening. Leaving the room at 04:59:50 to answer the phone, when you have waited for 29 minutes for an ID, is no fun. Again, having the phone within reach is very helpful!

You may want to add a "flasher" to your phone so it will flash a light instead of ringing a bell during incoming calls. This is nice for those late-night DX tips from fellow listeners.

A phone near the radio is also nice if you have a computer hookup capable of telephone communications. You can use your computer to access radio-related bulletin board systems (like UBIX), while remaining at your listening post.

Recorders

After the radio and antenna, the most important accessory for the listening post is a tape recorder. Given the tendency of stations to fade, and with the some-times far-from-ideal reception conditions experienced on the shortwave bands, a tape recorder often makes the difference between a tentative logging and a confirmed catch. Running a recorder while listening enables you to re-listen to station IDs, or review program segments missed by an interruption. Many listeners record their favorite programs or "catches" so that they can enjoy or share them at a later date.

Most current shortwave sets have a "RECORD" output jack that is used to feed a tape recorder. The signal that comes from this output is at a constant level. It is independent of the receiver's volume or AF gain control. If your receiver does not have such a jack, the headphone output jack can be used for connecting the recorder. In this case, however, when you change the audio level on the receiver, you will also change the signal level going to the recorder. In some cases, you may find that the signal level going through the "RECORD" jack on your receiver is not sufficient to provide a good recording. Again you may be forced to use the headphone jack. If you wish to use the headphone jack for listening *and* recording, you should obtain a "headphone splitter." This device splits the jack output to provide two outputs.

Several points should be considered when choosing a tape recorder for your listening activities. The two principal choices are reel-to-reel and cassette recorders. Cassette recorders are compact and usually less expensive than reel-to-reel types. Cassette tapes are inexpensive, easy to find, and easy to use. Reel-to-reel tape provides longer recording times and is easier to edit. Also, some listeners feel that reel-to-reel decks offer better-quality recordings.

Shortwave signals are monophonic, so either a mono or stereo recorder may be used. However, there are several ways in which you can utilize a stereo tape recorder. You may wish to keep track of the exact time at which a logging or ID (identification) takes place. One way to do this is to use one channel (track) to record time signals (WWV, CHU, etc.) from a second receiver. During playback, you can use your balance control to listen to either the station itself, the "time track," or both. You may also use the "extra" stereo track to record your own verbal notes via a microphone.

Headphones

A good set of headphones is a must for the well-equipped shack. Other family members may not have the same "appreciation" for the "sounds" of shortwave that you have. Headphones will also insulate the listener from external noise. Most listeners will also agree that you can hear a weak signal better when using headphones!

In choosing headphones, make sure that you select a comfortable set that will not cause irritation over long periods of time. High-fidelity headphones are not desirable as they tend to overemphasize the bass and treble portions of the audio. So-called "communications" headphones with good midrange response produce the best results.

Most listeners use monophonic headphones. However, if your setup involves use of diversity reception (discussed later), you will need a pair of stereo headphones. Be sure the set you buy has an impedance rating that is compatible with your receiver.

Station Clock

Another valuable accessory is an accurate clock. In the world of international radio, most timekeeping is done using Universal Coordinated Time, or UTC (formerly know as Greenwich Mean Time, or GMT). Station schedules are published in this 24-hour time format. A 24-hour clock set to UTC (GMT) is very desirable. Many modern receivers have this feature built in. A variety of 24-

hour desk and wall clocks (both digital and dial) are available. One example is illustrated in Fig. 1-4.

Fig. 1-4. The Benjamin Michael LCD 24-hour digital clock.

Records, References, and the Log Book

Apart from the equipment required for radio listening, information sources are also important for the serious listener. Proper reference material, near at hand, will enhance the listening hobby immensely. Several reference books are nearly mandatory if you are to find your way around the world of radio. For this reason, a bookcase or, at least, a shelf dedicated to books and located near your radio desk is highly recommended.

Through radio listening, you will be discovering and exploring many areas of the world which may be unfamiliar. For this reason, a good current world atlas is an important item to have. In addition, a world map (such as the ARRL great circle map) mounted on the wall near your radio desk can quickly provide important needed information, as well as adding an "international" flavor to the hobby. Many listeners use map pins or dots to indicate countries heard and/or verified.

Another important source of information is an up-to-date guide to station frequencies and schedules. There are two principal forms in which such information is available. First, there are books on the subject that are available through your local radio store or by mail order. The better books are published frequently in order to keep up with the changing shortwave frequency spectrum. Each book generally specializes in a particular area of the listening hobby. The best guide to shortwave broadcast stations is the *World Radio TV Handbook*, often referred to as the *WRTH* or *WRTVH*. This annual publication is considered the "bible" for shortwave broadcast listeners. Similar books are available for the nonbroadcast shortwave listener. Every utility listener should have the *Guide to Utility Stations*. Every RTTY listener will want to keep a copy of the *List of Worldwide RTTY Stations* handy.

Shortwave club journals provide a rich source of information on shortwave stations and their schedules. Since these journals are published monthly and the information contained in them is very timely, the serious listener would do well to join several shortwave organizations which cater to his or her area of listening interests.

Many of the radio stations that you will hear do not broadcast in your native language; therefore, dictionaries and language guides can be a valuable addition to your listening post. The most frequently heard languages are English, Spanish, French, and Portuguese. Dictionaries for those languages will help increase the number of stations you can identify and enjoy.

Keeping records on what you hear, and on what you would *like* to hear, is a necessary part of one's listening activity. Many beginners to the hobby start out by simply scanning the radio bands, taking in all the exciting stations they discover without keeping records on what they hear. However, as you progress in the hobby, some systematic method of keeping track of what you have heard becomes a great asset. One method of accomplishing this is through the use of a simple log book. A log book should include, at minimum, the UTC date, UTC time, frequency, station name, reception quality, and some details of what you hear for each station you log. This can be done using a loose-leaf or spiral-bound notebook, using 3 × 5 cards, or by using more sophisticated methods, such as an entry into a home computer. The use of a card system or home computer has advantages over a "chronological" log book in that, by using one of these methods, you can sort your loggings by country of origin, time heard, frequency, etc.

If you are the type of listener who gets the greatest enjoyment out of logging previously unheard stations, you may want to consider keeping records on stations you would *like* to hear. Based on information from DX programs and shortwave bulletins, you will soon accumulate a long list of stations you would like to log. A systematic method of keeping records will enable you to better spend the time you have for pursuing these elusive stations, rather than just randomly cruising the shortwave spectrum in search of new finds. The time spent maintaining such records is rewarded by the focused searches they make possible. These records, like the loggings of stations heard, can be kept on a nearby bulletin board, chalkboard, card file, or on a home computer.

Equipment Needs

Receiver Positioning

Most listeners agree that the receiver you are using should be slightly elevated and/or tilted. Building a small riser for your rig is an excellent investment in time and effort. The riser will bring the front panel dials (or display) closer to eye level, and will also provide a nice storage space for your accessories and your log. The riser should not be higher than 5 inches, though. You need to be able to rest your elbow on the desk and still reach the tuning knob. Fig. 1-5 shows a typical arrangement.

Fig. 1-5. Fred Osterman utilizes risers for efficient equipment placement in his shack.

Antenna Switching

There is no "ultimate" antenna. Therefore, most DX'ers use more than one antenna. For maximum efficiency, these antennas are cut and tuned for the DX'er's favorite bands.

Switching antenna connectors manually is cumbersome, so one of the first additions to most listening posts is an antenna switch. You can find special antenna switches which select from 2, 3, 4, 5, or even 6 antennas. Buy a switch to meet your present *and* future antenna switching needs. All antenna switches are not created equal. Avoid antenna switches that utilize slide or rocker switches. Don't forget that even the slightest amount of resistance (as in a poor switch) can cancel your reception.

The Daiwa Company makes some very high quality 2- and 4-position switches (Fig. 1-6). These are the coax type that maintain the impedance of the feed line. The most common is the 50-ohm feed line with standard fittings properly attached to the coax line ends. Barker & Williamson Co. also makes a wide variety of switches. One that is of special interest is the Model 590 G pictured in Fig. 1-7. This is a 5-position switch that grounds all unused antennas.

Equipment Grounding

There are many good reasons to ground your antenna and your equipment. In the unlikely event of a direct lightning strike, a good grounding system will help direct the strike to ground and away from your equipment and house. In the more likely situation of static buildup during an electrical storm (or high winds), a grounding system will discharge the electricity to ground. Even a small static buildup in an antenna can destroy or desensitize the "front ends" of today's modern solid-state receivers. Grounding the equipment will also prevent the

Fig. 1-6. Daiwa 2- and 4- position coax switches.

Fig. 1-7. The B & W Model 590 G switch allows selection of five antennas and grounds all nonused antennas.

likelihood of the operator receiving a shock from a defective piece of equipment. Lastly, grounding will improve reception. A good *RF* ground will enhance shortwave reception—especially below 5 MHz. A good RF (radio-frequency) ground is a must for long-wave (LW) and medium-wave (MW) DX'ing.

There are two types of grounds—an RF ground and an AC (electrical) ground. An RF ground is always an AC ground, but an AC ground is not always an RF ground. Water pipes (metal, not plastic) and house-wiring ground leads are good AC grounds, but not good RF grounds. RF grounds are direct connections to the earth by means of heavy copper wire that is attached to an 8-foot ground rod. The length of the copper wire should not exceed 100 feet.

To ground your antenna properly, you need a lightning arrestor, some ground wire, and a ground rod. For many years, listeners used the standard "blitz plug" lightning arrestor (shown on the left in Fig. 1-8). This device has an internal air gap that will discharge large voltages across the gap and pass them harmlessly to ground. Recently, a new generation of improved lightning arrestors has become available. One is shown at the right in Fig. 1-8. Instead of an "air gap," new arrestors utilize a hermetically sealed ceramic tube of gas. The gas tube has a more controlled and predictable "break-down" voltage, affording much better protection for sensitive solid-state equipment.

Fig. 1-8. A standard "air gap" arrestor (left) and an advanced "gas-tube" arrestor (right) made by Transi-Trap™.

The lightning arrestor should be attached to the coaxial lead-in cable before it enters the house. The arrestor should be connected to the ground rod by a short length of #10-gauge (or thicker) solid wire. The ground wire should be kept short and have no sharp bends. The grounding rod should be at least 5 feet in length.

It is important to ground the receiving equipment as well. Ideally, you should connect the chassis of each piece of equipment to a heavy ground wire and run the wire to a good ground rod. The ground wire should be as short and straight as possible. If a ground rod is not practical, you may have to connect your ground wire to a cold water pipe (not a gas pipe!). See Chapter 2 for further information on antennas.

Antenna Tuners

An antenna tuner may or may not be required for your radio. In some situations, an antenna tuner can provide a better impedance match between the radio and the antenna. If you utilize a tuner, it should be connected ahead of any preamps or preselectors. (See Chapter 2 for more on antennas.)

Preselectors

Like antenna tuners, preselectors and preamplifiers are valuable in some situations and not in others. A preselector or preamp will not cure an intrinsically poor antenna or radio. If these devices are used, they should be connected between the tuner and the receiver's antenna-input connection.

Audio Filters

An audio filter is like a sophisticated tone control designed for use with a shortwave receiver. There are many types available, and they all serve to tailor the audio output of the receiver to make it more readable and the listening more pleasurable. Audio filters are used to eliminate high-pitched whistles, or heterodynes, and to accentuate certain portions of the audio signal, such as the voice of an announcer. They can also be valuable in digging out certain utility signals, such as Morse code and radioteletype. It must be emphasized that audio filters act only on the *audio output* of a receiver, and they have no effect on the receiver's actual sensitivity or selectivity. Considerable practice is required to use an audio filter effectively.

Audio filters are usually designed to be connected to the headphone jack of the receiver. Then, your headphone or, in some cases, an external speaker is connected to the output jack of the audio filter. Some audio filters come equipped with their own built-in speaker, but these speakers are generally not as good as an external speaker or a headphone.

External Speakers

Most shortwave receivers come equipped with an internal speaker. In many cases, these speakers are quite adequate for general radio listening. In other cases, they are the weakest part of an otherwise excellent radio. Many built-in speakers don't face the listener. For this reason, external speakers are available for attachment to the radio. Kenwood and Yaesu manufacture external speakers containing simple built-in audio filters. Some receivers provide a jack specifically designed for connecting an external speaker, and attaching an external speaker to this jack will usually deactivate the receiver's own internal speaker.

Audio Mixers

If you have more than one receiver and/or recorder, you may consider constructing some sort of audio mixer. This device is also handy if you are feeding other devices that require an audio input, such as an RTTY or FAX unit. Mixers can be simple or complex, and may or may not include provision for matching audio levels. Perhaps the simplest device is a switch used to select between two receivers (Fig. 1-9).

Fig. 1-9. A sample mixer consisting of a switch to select one of two receivers.

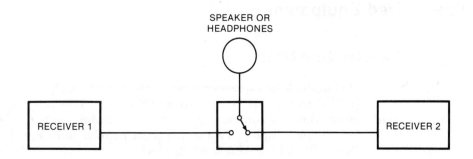

If you have the luxury of two receivers, you may wish to try some other things as well. If you have separate antennas on your two receivers, you can try diversity reception. If your antennas are significantly separated from each other, they will pick up the incoming signal in a slightly different manner. You may wish to blend (or mix) the audio from each, using the circuit shown in Fig. 1-10.

Fig. 1-10. A mixer which allows the blending (mixing) of signals from two receivers into one speaker or into headphones.

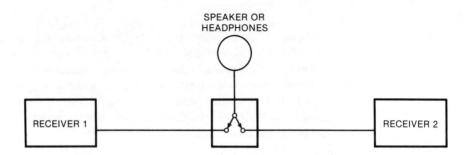

Another variation to this can be accomplished if you have stereo headphones. The idea here is to feed the output of one receiver into the left earpiece and the output of the other receiver into the right earpiece, as shown in Fig. 1-11. You will be surprised at the unusual effects this will produce.

Fig. 1-11. An arrangement which allows signals from two receivers to be fed to stereo headphones.

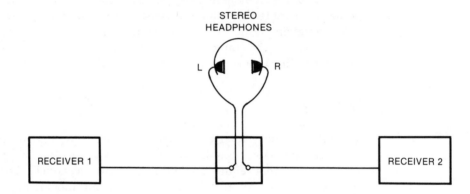

Specialized Equipment

Radioteletype Units

One of the least expensive ways of entering the exciting world of radioteletype (RTTY) monitoring is with an RTTY reader. This self-contained device takes audio from any communications receiver and presents readable text on a "billboard-type" display. Two of the most popular units are the Kantronics Mini-Reader™ and the AEA MBA Reader™.

You will want to keep the reader close to the radio, with the three connecting wires (power, audio in, and monitor out) as short as possible. Readers do not work well standing up or lying down; therefore, the best alternative is to set it at an angle. Kantronics supplies a very nice angled holder with its mini-reader (Fig. 1-12). The AEA Company offers a similar unit as an option ($4.00). This is a good investment as it holds the unit at a pleasing angle.

Fig. 1-12. The
Kantronics
Mini-Reader on
an angled
stand.

With the increased popularity of microcomputers, there has been a proliferation of special interfaces that attach to these computers to copy RTTY and Morse code. They usually consist of two parts: the interface, itself, and the software (a ROM cartridge or diskette). Careful thought should be taken when incorporating a computer interface into the shack.

You will want to have the display (monitor or TV) close to the receiver. Your eyes will be constantly traveling between the radio and the display. Also, the keyboard must be *very* handy. When DX'ing RTTY, you may be scrutinizing up to 20 or 30 signals per hour as you tune the band. You will be testing the signal under several parameters which are controlled on both the computer keyboard and the computer interface. To test one signal for just the three most common speeds, and the three most common shifts, you will have to do the following:

- Try 60 WPM 170 Hz shift—normal
- Try 60 WPM 425 Hz shift—normal
- Try 60 WPM 850 Hz shift—normal
- Try 66 WPM 170 Hz shift—normal
- Try 66 WPM 425 Hz shift—normal
- Try 66 WPM 850 Hz shift—normal
- Try 100 WPM 170 Hz shift—normal
- Try 100 WPM 425 Hz shift—normal
- Try 100 WPM 850 Hz shift—normal

- Try same—reverse
- Try same—reverse
- Try same—reverse
- Try same—reverse
- Try same—reverse
- Try same—reverse
- Try same—reverse
- Try same—reverse
- Try same—reverse

Needless to say an efficient placement of computer, interface, and receiver is required! The best orientation for a right-handed listener is shown in Fig. 1-13.

While it is desirable to have the computer very close to the radio for operator convenience, this can present some problems. All computers emit radio interference, and some computers are worse than others. The following table lists the shielding level of several popular computers used for RTTY reception:

Fig. 1-13.
Orientation of
equipment for a
right-handed
listener.

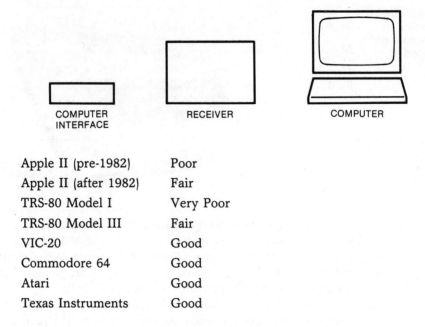

Apple II (pre-1982)	Poor
Apple II (after 1982)	Fair
TRS-80 Model I	Very Poor
TRS-80 Model III	Fair
VIC-20	Good
Commodore 64	Good
Atari	Good
Texas Instruments	Good

Adding peripheral devices (printers, disk drives, etc.) to any of these will usually aggravate the problem. If you do "hear" your computer in your receiver, you may try some of the following solutions. Buy an AC line filter for your radio and/or your computer. The Drake LF-2 or LF-6 line filters often attenuate the interference significantly. Another solution is to buy several large toroidal coils (Amidon FT-240-61 or equivalent) and wind the equipment AC cords through the coil, as shown in Fig. 1-14.

Fig. 1-14. An AC
line cord
wrapped
through a
toroidal coil to
reduce
interference.

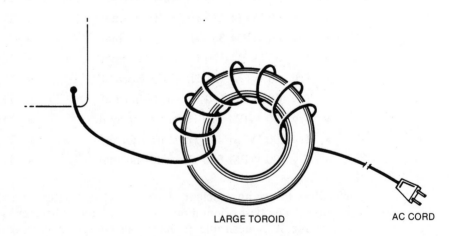

Dedicated RTTY units are self-contained, free-standing devices. They require audio from the receiver and then display the output to a monitor (and/or printer). They offer dramatically improved performance over readers or computer interfaces. They usually offer additional features beyond interfaces or readers, and are not keen sources of RFI.

Some dedicated units, such as the Info-Tech models (M-200E, M-200F, and M-600A) will actually determine the speed for you, if it is unknown. However, you will still be making numerous adjustments to your RTTY unit as you encounter and examine RTTY signals. Once again, the display and the RTTY unit should be as close to the receiver as practical. The configurations in Fig. 1-15 are ideal for a right-handed DX'er.

Fig. 1-15. Setup of listening post with an RTTY unit.

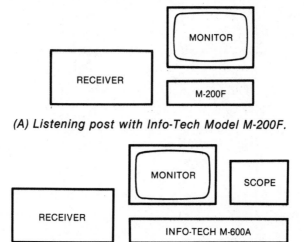

(A) Listening post with Info-Tech Model M-200F.

(B) Listening post with Info-Tech Model M-600A and an oscilloscope.

2

ANTENNAS FOR SHORTWAVE RECEPTION

Thomas P. Harrington

Tom Harrington (W8OMV) has been in radio communications for many years and has actively monitored most regions of the radio spectrum. He flew in the United States Army Air Corps during World War II, and during this period became interested in shortwave propaganda broadcasts in Europe. Later, through Amateur Radioteletype, Tom became interested in all types of radioteletype services and world news RTTY. When he is not engaged in his electronics and radio business, this interest in RTTY fills much of his time.

Tom Harrington is a Life Member of the American Radio Relay League, the Quarter Century Wireless Association, and the Experimental Aircraft Association.

Many shortwave listeners spend many hours and considerable money on their shortwave receivers, while spending only minutes and a few dollars on their shortwave antenna. As we all know, receiver performance is directly related to the effectiveness of our antenna. The ideal shortwave shack, or "setup," is like a chain and thus is only as strong as its weakest link. The first link in our shortwave chain is, of course, the antenna that picks up that small amount of electromagnetic energy that comprises the shortwave signal.

A good efficient shortwave antenna is easy to build and costs only a few dollars. *Take the time to erect the best antenna that you can put into service;* you will never regret the time spent on this important key to your total shortwave enjoyment, as a good antenna is as equally important to good shortwave reception as a good receiver.

Long-Wire Antennas

One of the simplest and least understood of the receiving antennas is the random long-wire antenna. This efficient antenna is just a length of wire, from 30- to 80-feet long, which is erected in a straight line on a horizontal plane, with an insulated lead connected into your shack; and then to your shortwave receiver. The long-wire antenna performs best when erected from 20 to 50 feet off the ground, in a spot clear of trees, buildings, and other obstructions, and mounted away from power lines and other areas that could induce electrical noise into your antenna and on into your receiver (Fig. 2-1). Both ends of the long-wire antenna carry suitable insulators, and the actual antenna wire can be any suitable wire, insulated or bare. The lead-in wire should be insulated to keep it from shorting to other obstructions, such as parts of the house, window frames, metal screens, etc.

The random long-wire antenna, when used with a simple antenna tuner, will cover a frequency range from 1.6 MHz to 30 MHz. Most of the modern solid-state shortwave receivers of today cover a wide range of frequencies; therefore, we need an antenna that will afford good operation from the low end (1.6 MHz) to the highest end of the receiver's frequency range (30 MHz). The long-wire antenna fills this need for extreme frequency ranges, requiring only a quick peaking of the knobs of a simple antenna tuner. The receiver's "S" meter, or *signal strength meter*, is read for maximum signal peaking; just tune for maximum signal reading on the "S" meter. This maximum tuning takes but a second or two and results in a strong, clear reception. This peaking should cover each of the shortwave bands and thus need not be redone until a major frequency shift is made, or a band change takes place.

Antenna Tuners

The antenna tuner is a simple passive device (nonpowered) that will allow you to maximize both the pickup and transfer of the signal from the antenna to your receiver.

Fig. 2-1. A
typical long-
wire antenna.

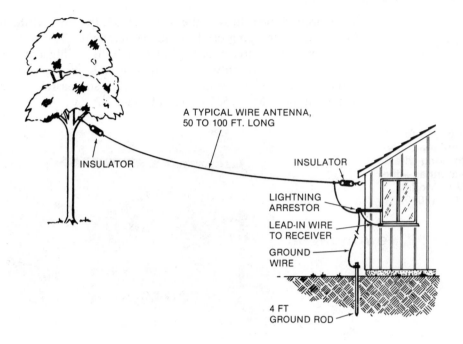

A TYPICAL WIRE ANTENNA,
50 TO 100 FT. LONG

INSULATOR

INSULATOR

LIGHTNING
ARRESTOR

LEAD-IN WIRE
TO RECEIVER

GROUND
WIRE

4 FT
GROUND ROD

Basically, the tuner will do two things:

1. Through the use of an inductor, which is a wound and tapped coil, it will allow you to match the resonant length of your antenna to most of the shortwave bands in use. This section of the antenna tuner allows the shortwave signals to be absorbed more readily to a resonant length; therefore, the signal efficiency or pickup is much greater, and will result in an increased signal strength to the receiver.

2. The remaining section of the antenna tuner is made up of a variable condenser, which allows you to match the incoming signal to the impedance of your receiver. Most modern solid-state shortwave receivers have a 50-ohm coaxial connector for the antenna input.

The wire lead-in from the antenna will go into the antenna tuner as a single wire and, then, will be sent to the receiver through a short length of RG-58U (a 50-ohm coaxial cable) to maintain the proper match. This simple cost-effective all-band antenna has been my favorite type of antenna for many years and always with exceptional results.

Variations of the Long-Wire Antenna

Many times, I have had to resort to other simple forms of the long-wire antenna. Recently, I lived in an apartment for a year while building a new home. During this period, I taped about 50 feet of fine wire, obtained at a local Radio Shack store, around the ceiling of the room. This is the fine, white, insulated wire used for the wire-wrapping of electronic components. The wire was placed high up on the wall in the area near the junction of the ceiling and the wall; the wire was held in place (invisibly) with small ¾-inch pieces of ordinary transparent Scotch®* tape. This antenna was used very successfully with the MFJ Antenna Tuner Number MFJ-16010 shown in Fig. 2-2.

Fig. 2-2. An antenna tuner for long-wire antennas—the MJF Model MJF-16010.

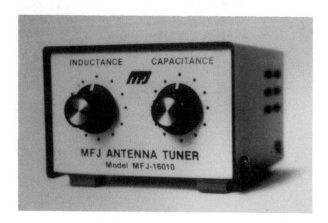

Tuned Dipoles

The tuned dipole is a basic half-wave antenna that is cut to a specific frequency of the band of frequencies in the HF spectrum (Fig. 2-3). Dipoles and multidipoles are used by many SWL'ers whose principal activity is confined to a single shortwave band or bands. More than one dipole can be combined in the dipole system (see Fig. 2-4). Dipole length is calculated by the following simple formula:

$$L = \frac{234}{F}$$

where
L = length in feet for each half of the dipole,
F = center frequency of the band in MHz.

For example, the center frequency of the 25-meter broadcast band (11.6 MHz to 12.0 MHz) is 11.8 MHz.

$$234 \div 11.8 \text{ MHz} = 19.83 \text{ feet}$$

$$0.83 \text{ of a foot } (12 \times 0.83) = 9.96 \text{ inches, or } 10 \text{ inches}$$

---

* Scotch is a registered trademark of the 3M Company.

Therefore, 19 feet plus 10 inches is the length for each half of a dipole for the 25-meter band. This is illustrated in Fig. 2-4. Frequencies and dipole lengths (singularly and overall) for various bands are given in Table 2-1.

Fig. 2-3. A single-band dipole cut for a given frequency.

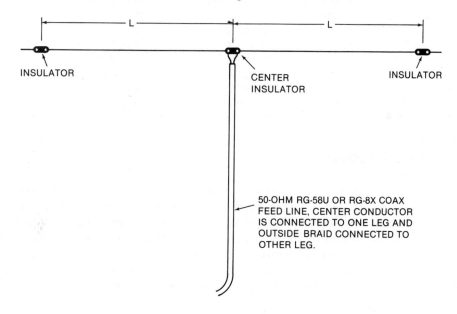

INSULATOR

CENTER INSULATOR

INSULATOR

50-OHM RG-58U OR RG-8X COAX FEED LINE, CENTER CONDUCTOR IS CONNECTED TO ONE LEG AND OUTSIDE BRAID CONNECTED TO OTHER LEG.

Fig. 2-4. Multiband dipoles fed with a single 50-ohm coax feed line (not to scale).

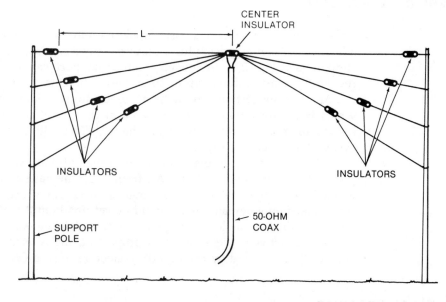

CENTER INSULATOR

INSULATORS

INSULATORS

SUPPORT POLE

50-OHM COAX

The single dipole, or a series of multidipoles, will have a directivity factor broadside to the length of the dipole; therefore, a cut or tuned dipole can be erected to favor, within a preferred broadcast band, a preferred set of stations or a particular area of the world.

Table 2-1. Dipole Antenna Dimensions	Band	Frequencies (MHz)	Each Side of Dipole (L)	Overall Dipole Length
	11 Meter	25.4 to 26.1	9 ft. 1 in.	18 ft. 2 in.
	13 Meter	21.45 to 21.75	10 ft. 10 in.	21 ft. 8 in.
	16 Meter	17.7 to 17.9	13 ft. 2 in.	26 ft. 4 in.
	25 Meter	11.6 to 12.0	19 ft. 10 in.	39 ft. 8 in.
	31 Meter	9.2 to 9.7	24 ft. 9 in.	49 ft. 6 in.
	41 Meter	7.1 to 7.4	32 ft. 3 in.	64 ft. 6 in.
	49 Meter	5.9 to 6.4	38 ft. 6 in.	77 ft. 0 in.
	60 Meter	4.75 to 5.0	48 ft. 0 in.	96 ft. 0 in.
	90 Meter	3.2 to 3.4	70 ft. 11 in.	141 ft. 10 in.

Trapped Dipoles

Several antenna manufacturers are now producing all-band shortwave antennas based on the trapped dipole principle. These antennas have proven themselves in several years of actual use. My recent experiments with several of these antennas indicate very satisfactory performance (see Fig. 2-5).

These trap-antenna systems do not require any type of band switching or the use of antenna tuners, which makes them cost-effective and practical for all-band SWL DX'ing, as they cover all of the active SWL broadcast bands.

Inverted-V Antenna

The inverted-V antenna is a basic dipole mounted in a V position. This has the advantage of saving space, plus it requires only a single center pole for the main support (see Fig. 2-6). A low-cost center support can be easily made using regular TV mast sections, up to 25-to 30-foot heights.

Another suggestion would be to erect an additional half-wave antenna, which is cut for another favorite band, at the top of the support. Mount this second antenna at a 180° angle from the first. This will form a very good guying system for the center support.

It is also possible to place other shorter-length half-wave dipoles inside the area of the other antennas. An SWL friend of mine has six carefully cut dipoles erected on a single 30-foot-high TV mast system. The shorter antenna (which is the higher-frequency antenna) should be set up about 3 feet below the longest antenna (which is the lower-frequency antenna). There should not be too much interaction between the antennas if proper spacing is used. This system works very well, as attested by my friend's collection of rare QSL cards.

Vertical Antennas

The vertical antenna is omnidirectional; that is, it will receive signals equally well from all directions. However, it has one serious drawback—the vertical

Fig. 2-5. An all-band shortwave receiving antenna with trap circuits to electrically separate the antenna segments. The overall length is 43 feet.

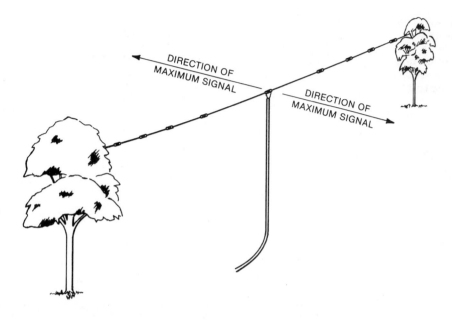

DIRECTION OF MAXIMUM SIGNAL

DIRECTION OF MAXIMUM SIGNAL

(A) Side view.

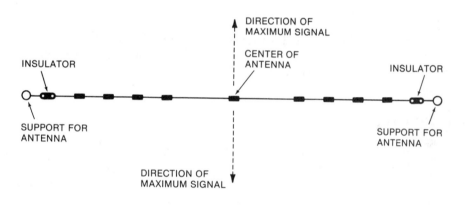

DIRECTION OF MAXIMUM SIGNAL

CENTER OF ANTENNA

INSULATOR

INSULATOR

SUPPORT FOR ANTENNA

SUPPORT FOR ANTENNA

DIRECTION OF MAXIMUM SIGNAL

(B) Top view.

antenna is very responsive to noise pickup, both man-made and natural or atmospheric. Most verticals seem to have high background-noise factors which tend to cover up weak signals.

An ideal height for the average vertical antenna would be 25 feet; also, the antenna must be installed clear of buildings, trees, and other obstacles in order to receive well in all directions.

A new 25-foot fiberglass flagpole appeared in the backyard of an acquaintance who lives in a highly restricted subdivision which forbids outside antennas of any type. Close examination of the flagpole disclosed three carefully cut, 3-band, vertical antennas constructed of wire and running up the sides of the

Fig. 2-6. The inverted-V antenna requires one center support. Multiband trap antennas can be used in the inverted-V configuration with good results.

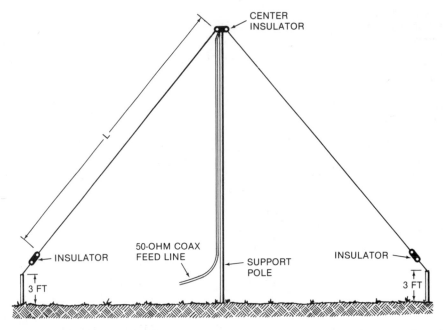

flagpole. The 50-ohm coax feed line was carefully buried underground and it resurfaced next to his radio-shack window. This was a neat trick that worked extremely well because the subdivision also had all the power lines and telephone lines buried—not too many in our hobby have perfect radio locations. Buried power lines and ground-mounted transformers are big factors in reducing power-line noise figures.

Outlaw Antennas

Property restrictions, space limitations, landlords, and sometimes our own family will take a heavy toll on our antenna endeavors. Let's face it, we all do not have enough space to erect our dream antenna farm. There is a real world out there and we must live in this world.

When there are total restrictions and other detrimental factors, we simply have to put our brains to work! At this point, here are some of the alternatives to the problem of no antenna:

1. Camouflaged vertical as flagpole.
2. Very thin, almost invisible wires up in trees.
3. Thin wires around the side of the house.
4. Thin wires dropped out of your shack window and taken up by a fishing reel when not in use.
5. Use of a B & W TV window antenna; remove it when not in use.
6. Thin wire around the ceiling of your shack.

7. Concealed antenna system in attic or crawl space.

8. Multiband amateur vertical antenna mounted on your car. Run the coax to the shack window. No law about antennas on cars.

9. Antenna mounted under eaves of house.

10. Use of metal gutters, downspouts, ungrounded metal rails, etc.

I am sure you can figure out ways to have several good working SWL antennas to use, no matter what the restrictions or limitations are. THINK! THINK! Then, do your work at night or very early in the morning.

An Alternate Antenna System

When all else fails and, for some reason, you cannot come up with a good outside antenna, and you do not feel a makeshift indoor antenna will do, there is one other good possibility left! This would be one of the newer indoor active antennas that are available from most shortwave dealers. This system is a completely self-contained active device, with electronic sections to tune all the major shortwave bands. The units have an attached whip antenna plus a tunable preselector with ample amplification to deliver a good signal to the receiver. The indoor unit sits on your desk—ready to listen to the world. Unique circuitry minimizes intermodulation, and provides added RF selectivity. These units can also be used as a preselector for external or outside antennas. Most active antenna systems cover frequencies of 300 kHz to 30 MHz in 5 bands. The unit is simple to hook up and use.

Several special active antenna systems are built as remote active antennas. They have a 54-inch remote whip which mounts outdoors away from electrical noise for maximum signal and minimum noise pickup. The units can be mounted anywhere—atop the house, on balconies, apartments, mobile homes, and even on board ships. The Sony, MFJ, and McKay Dymek companies manufacture these units for the SWL trade. One is illustrated in Fig. 2-7.

Ground Systems

A simple and effective RF ground system is shown in Fig. 2-8. It consists of a steel copper-clad ground rod, 6 to 8 feet long, driven into moist ground. The clamp must be positive clamping to obtain a proper ground-wire connection. The ground wire must be heavy and as short as possible. In dry climates and dry periods of the year, a generous watering of the ground-rod area will do wonders for your ground system.

Lacking the ability to install your own RF ground system dictates that you must seek a substitute ground system that might exist in your home. While these are not perfect RF grounds, many times they are better than no ground

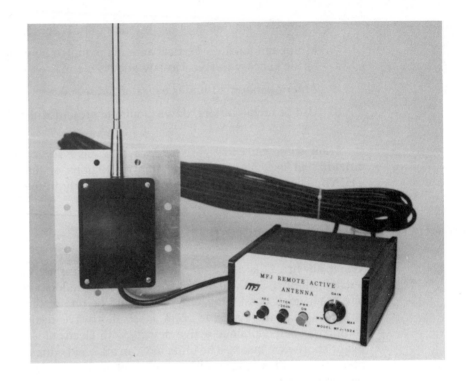

Fig. 2-7. The MJF-1024 remote active shortwave antenna unit. The whip mounts outdoors in the clear and as high as possible.

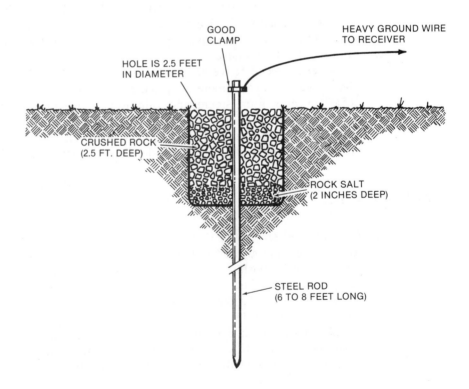

Fig. 2-8. A good shortwave RF ground system. It should be placed as near to the receiver as possible.

system. Some of these grounds would be plumbing pipes (metal of course), metal drainage pipes, etc., plus other known, grounded, metal objects.

I do not advise the use of your AC power-outlet ground wire as a ground, as this can be a source of induced noise from the many other devices using the same common ground system. I also do not recommend the use of the phone company's ground, as this ground also can be very noisy. For some reason, I have seen small voltages emanate from these so-called telephone grounds. In other words, I do not consider either of these two grounding systems a good RF ground system and I would not recommend that they be used!

Lightning and Atmospheric Static

With a proper RF ground system, your receiver and other radio equipment can be protected with one of the newer lightning arrestor gas-discharge units. See Chapter 1 for full information on the arrestor units and their sources of supply.

Fig. 2-9. The shortwave DX'er's dream antenna. A log-periodic rotary beam system having continuous coverage from 1.6 to 30 MHz and giving a typical 6- to 8-dB gain. This antenna belongs to Ben Bissman III of Mansfield, Ohio.

Summary

We have just covered some of the basic shortwave antenna systems here. There are several complete books on the subject that are worthy of future study. We do not feel that there is one ultimate shortwave antenna, and as you progress in your SWL DX'ing, you will soon be able to determine which type of antenna, or combination of antennas, you will want to build.

Most DX'ers will first choose their favorite broadcast bands and areas of interest. At that time, it is possible by further study and experimentation to select antennas that fill your wants and needs. Tune in the world! Fig. 2-9 illustrates the "Dream Antenna," a log-periodic rotary beam system belonging to Ben Bissman III of Mansfield, Ohio. It has a continuous coverage from 1.6 to 30 MHz, and a typical gain of 6 to 8 dB.

PURCHASING THE RIGHT SHORTWAVE RECEIVER

Lawrence Magne

Lawrence Magne, who has, since 1971, played a major role in the publication of *World Radio TV Handbook* at Billboard Magazine, is now editor of the highly successful *Radio Database International* series of shortwave publications.

Larry, a Trinidadian by birth but now an American, has long been active in computer database design and shortwave broadcast frequency management. For the past 15 years, he has provided articles for a wide variety of radio, travel, and technical publications. Larry is also widely known for his two regular monthly features on Radio Canada International's "Shortwave Listener's Digest" and "Canada a la Carte" programs. A recent RCI survey found Larry's monthly equipment review feature to be far and away the most popular feature of "Shortwave Listener's Digest"—itself, the most widely listened-to SWL program in North America.

Since 1981, Larry has been president of International Broadcasting Services, Ltd., headquartered in Penn's Park, Pennsylvania. His wife, Jane, is a molecular biologist at the University City Science Center in Philadelphia. During rare free moments, they often can be found tilling the fields or repairing the barn at their Revolutionary-era farm in rural Bucks County.

What was shortwave listening once like? Was it, as some remember, the clear voice of Winston Churchill rallying his nation over interference-free shortwave channels? Or was it a nightmare of ear-wrenching squeals and roars? Actually, it was either, or both, depending, as in the elephant parable, on your perspective, ... and your radio.

If your memory goes back far enough, you may recall that exciting period, a half century back, when the medium was aborning. London's BBC Empire Service was beginning to span the globe, while the Axis powers and Moscow's Comintern discovered in the shortwaves the ultimate vehicle for international propaganda. Signals were relatively few and transmitter powers weak, but spectrum space was more than ample and jamming all but unknown. Shortwave listening was a pleasurable pastime for individuals and families alike, with the emphasis in receiver design being placed on sensitivity and "good audio." Sound had to be undistorted, even if it was bassy, to mask the static that tended to intrude upon the relatively weak signals. Tight selectivity (although unnecessary, given the few stations then on the air) was avoided to allow for the pleasant wideband audio so often associated with shortwave's early days.

The post-World War II years changed all that. Shortwave stations flourished during and after the War to the point where, by the early 1950s, President Truman was moved to proclaim congestion within the shortwave spectrum to be of serious concern. Older unselective "wideband" radios became increasingly painful to listen to, as the desired signals became more subject to interference due to the vast growth in the number and power of shortwave transmitters —and the Soviet bloc jammers—on the air. About this time, television, and then FM radio, came into their own. With these media providing the "carrot" and degraded shortwave reception providing the "stick," shortwave listening faded from public visibility to languish in a 30-year trough.

Receiver Design Little Changed for Decades

During the period of shortwave's rise and fall—from roughly 1930 to 1980, shortwave receiver design underwent relatively few changes. The basic single-conversion superheterodyne circuit with an logarithmic analog frequency readout was the workhorse. Such radios ranged in size from the hefty "boat anchor" shortwave specialty receivers to large wooden-cabinet home radios sporting a shortwave "band" to tiny transistorized portable sets. Early sets (tube-type, of course) tended to need regular attention, as the alignment drifted with tube age and the wax-coated capacitors (with paper dielectrics) petered out. Tubes needed an occasional replacement, while every now and then, a resistor would "fry."

Later tube-type sets incorporated refinements, including capacitors with an epoxy coating in place of wax and with Mylar dielectrics in place of paper. Some of the finest examples of refinement in tube-type communications receiver design were found in the various "R-4" series models produced by the American firm of R. L. Drake (which no longer produces nonprofessional shortwave receivers). The R-4B is still considered to be one of the better shortwave receivers for shortwave broadcast DX'ing. The more recent and selective R-4C, as modified by Sherwood Engineering, is still considered to be among the finest "competition-grade" receivers for use by amateur radio enthusiasts ("Hams").

Of course, the tube-type receiver has long since ceased to be produced. Some secondhand tube-type models, if inexpensive enough and in good operating condition, continue to provide commendable results, provided you are able to keep them properly maintained. In this regard, certain makes are preferable to others. The R. L. Drake Company continues to provide service and all parts—even for its earliest models, which go back a quarter century. At the other extreme, Hammarlund went out of business years ago.

International Broadcasting Services, Ltd., in its custom databank of shortwave receiver reviews, rates—aside from the aforementioned Drake R4 series—the Collins 51S-1, the military surplus R-390 and related models, the Racal RA17, and the Hammarlund HQ-180/HQ-180A among the better models. You can check advertisements in amateur radio and SWL club and commercial publications for information on equipment availability, but the best bet is to attend "Hamfests" (Amateur Radio flea markets) in your area. These days, a tube-type "boat anchor" (this excludes the Drake models, which are relatively light and compact) almost has to be purchased locally unless you wish to wrestle with the nightmare of having such a unit crated and shipped.

The Mixed Blessing of the Transistor

The introduction of the transistor and the transistorized radio (Fig. 3-1) may have resulted in radio being brought to nearly every village in the world, but it did little to gladden the hearts of tube-tested shortwave veterans. Even today, many transistorized shortwave radios suffer from inadequate dynamic range as compared to comparable tube-type designs.

What is dynamic range? We'll get to that shortly, but suffice for now that in this important regard, transistors tend to be at a disadvantage as compared to tubes. Still, the transistor has brought about three major improvements: improved reliability, reduced power consumption, and miniaturization. Ma Bell's little solid-state creation thus terminated the era of the "boat-anchor radio."

Dr. Wadley's Loop Ushers in New Era

The first major advance in transistorized shortwave portable design appeared in the early 1970s, not in Japan or the United States or Germany, but in the unlikely country of South Africa. To understand this development, we have to go back to England some 20 years earlier. At that time, Dr. Wadley, while working on Racal's RA17 design team, conceived what has come to be known as the *Wadley-loop principle*. The Wadley loop allowed not only for considerable receiver stability (important for reception of "single-sideband" signals used for

Fig. 3-1. The Sony Model ICF-2002 "pocket-book size" receiver (Courtesy Sony Corp.).

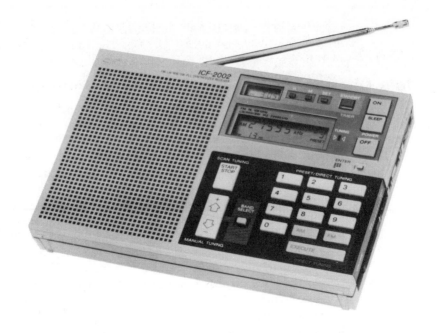

point-to-point shortwave radio communication), but also for relatively accurate linear analog frequency readout. This meant the listener could tune directly to any desired frequency, rather than having to fumble about trying to tune in a station "by ear." Dr. Wadley eventually migrated to South Africa where, in conjunction with the South African electrical firm of Barlow's, he designed a transistorized shortwave portable radio utilizing a revised and much lower-cost version of his beloved loop circuit. The multiple-conversion portable set, Model XCR-30, performed well and was straightforward to operate.

Although intended primarily for the domestic South African market (until relatively recently, South Africa had no television, and most of the "white" domestic radio channels were on shortwave), the Barlow-Wadley XCR-30 had a stunning impact, worldwide, on the future of shortwave radio design. Japanese manufacturers pored over Dr. Wadley's revolutionary achievement, and then began producing portable and table model shortwave radios containing linear analog frequency readout and multiple conversion features. Some, such as the Sony ICF-5900 and the Panasonic RF-2200, utilized "Rube Goldberg" schemes to provide linear readout without using the complex Wadley-loop circuit. Other models, such as the Yaesu FRG-7, used Wadley-type circuits that were improvements on the pioneering XCR-30 design. In due course, sophisticated digital phase-locked loop (PLL) circuits emerged. PLL circuitry is now commonplace, not only in shortwave radio design, but also in AM, FM, and TV receiver design.

The reaction to these new receivers in Japan was overwhelming. A shortwave "craze" arose among male Japanese adolescents, which prompted Japanese manufacturers to try to market these radios abroad. To their surprise, the reaction in the U. S. and Western Europe was positive, but not among teenagers.

Rather, adults curious about world events and foreign cultures found shortwave to be a complement to the domestic news and entertainment media.

Models Differ Widely in Design and Performance

Does this mean that most new shortwave receivers, just as most new television sets, are generally identical? Hardly. In most "developed" countries, the television set is a popularly understood item. Were a manufacturer to attempt to market a distinctly inferior or overpriced model, it would be widely perceived as such and would fail. Shortwave broadcasting, on the other hand, is a little-understood, almost arcane, medium; most individuals considering the purchase of a shortwave radio haven't any idea of what to look for. Not surprisingly therefore that the market is open to all manner of abuse.

CAVEAT EMPTOR!

There is really no other term but abuse to describe the marketing of old-technology shortwave radios by credit card companies, airline and magazine direct-marketing operations, and some retail radio outlets. Most of these "OEM nightmares" (OEM standing for "original equipment manufacture") are straight from the technology of the 1960s and come with frequency guides that may be years out of date. The results for the hapless buyer are predictable: dreadful noisy reception with no indication to show where or what one is tuning. The vendor makes his fast buck, and the buyer, stung and disappointed, abandons shortwave listening, perhaps for life.

This is why it is so essential to understand what to look for in a shortwave radio, and which sources to trust for advice and information on current models.

How to Evaluate a Shortwave Receiver

What do you look for in a shortwave receiver? The two obvious starting points are price and portability. Although the price situation is fluid and varies greatly from country to country and year to year, a coarse rule of thumb within the United States is that shortwave radios costing under $75.00 rarely provide satisfactory performance except under the most choice reception circumstances. In the $75–$125 range, very roughly, can be found a number of portable units that will provide acceptable performance for listening to major shortwave broadcasters. From about $125–$300 lies a fertile ground for the listener who is seeking good performance in a shortwave portable set. From about $300–$1000 are found a number of communications-type table model receivers, some of which will provide enviable performance for the shortwave enthusiast interested in hearing weaker, as well as stronger, stations (Fig. 3-2).

Above $1000? With the demise of the superb Drake R-7/R-7A, radios in this price stratosphere now tend to be divided into two categories: models for military-intelligence/commercial-maritime use, and "glitzy" consumer models de-

Fig. 3-2. The ICOM Model IC-R71A receiver (Courtesy ICOM America, Inc.).

signed to impress those whose financial good fortune is not equaled by their understanding of radio value. However, as with any other generalization, this has its exceptions. Sony's costly CRF-1 is the best portable we've tested to date, whereas Drake's R4245 professional communications receiver flies in the face of military/commercial receiver design by being manually flexible for serious DX applications.

Portable or Table Model?

The line between portables and table models has blurred with advances in solid-state circuit development. Now, the primary determinant of whether you obtain a portable or a table model radio is price. Under $400, nearly all models are portable units, whereas, the range above that is largely table model territory. The reason is not difficult to grasp: more sophisticated (and costly) receivers are designed for much more than run-of-the-mill shortwave listening. Instead, they tend to be used primarily by "Hams" and professionals for point-to-point communication in any of a number of specialized transmission modes. Few of these uses call for portability, so the emphasis is on performance and ease of repair. The compactness so prized in portables thus becomes a liability.

However, there are a number of "quasi-portables" that perform as well as communications receivers, yet are self-contained—if a bit cumbersome—for certain field portable applications (Fig. 3-3). At the other end of the portability spectrum is the miniature or compact portable set (Fig. 3-4), which is ideally suited to the needs of the airline traveler.

Once the questions of price and portability are resolved, the more difficult issue of performance rears its head. Alas, there is little one can tell from sales literature that will translate to assurance that a given model will perform in the desired fashion. However, there are a number of basic guideposts used to separate an acceptable model from a Klutzrundfunkempfanger.

Accurate Frequency Readout a "Must"

Will your dream receiver allow you to find a station readily?

Fig. 3-3. The
Sony Model
ICF-6800W
receiver
(Courtesy Sony
Corp.).

Fig. 3-4. The
Uniden Model
CR-2021
receiver
(Courtesy
Uniden
Corporation of
America).

Imagine what it would be like to tune a television set having no precise channel indicator. This is essentially the case with nearly all old-technology shortwave radios, which may have as little as one number to indicate a given "megahertz" or "meter" band. With such sets, you may be able to tune in up to 100 separate radio channels in the vicinity of that solitary numerical marking. You can imagine the frustration when you are trying to find a favorite newscast at a specific hour!

Shortwave channel spacing is at every five kilohertz (5 kHz). Better shortwave radios provide a digital frequency readout to the nearest kilohertz to tell you precisely where you are tuning. An inexpensive compromise arrangement involves a "bandspreaded" analog frequency readout, but this technique is fading away as digital-electronics circuitry reduces in cost.

The same digital technology that provides a digital frequency readout also makes it possible to include such tuning aids as *programmable channel memories*, *push-button frequency slewing*, and *keypad frequency entry*. Whether these are of interest is a matter of individual taste, but keep in mind that most radios that include push-button ("up" and "down") frequency slewing do not include a tuning knob.

With a digital frequency readout, you can tune directly to a specific broadcast. You can also get a good idea of what is coming in should you be tuning about and come across a station you can't immediately identify from the program content. However, to do this, you need a "map" of the shortwave spectrum to tell you what's on when and on what frequencies. This is easier said than done, as international broadcasters make basic changes in their schedules every 6 months, and sometimes more often, so keeping abreast of the thousands of stations on the air is somewhat more complicated than keeping track of which TV stations operate on what channels.

This need for a shortwave directory was the genesis of *Radio Database International*, an up-to-the-minute shortwave listening publication that appears in three editions throughout the year, as well as the venerable *World Radio TV Handbook*. The *Database* is a quick-access guide, whereas the *Handbook* is a reference tome, so active shortwave listeners tend to keep both on hand.

Which Bands Are Covered?

The shortwave spectrum is broken up into various segments, or bands, some of which are reserved for the exclusive use of broadcasters. There are no less than 18 such shortwave broadcasting bands, with the higher bands (15–26.1 MHz) tending to provide the best daytime reception, and the lower ones (2.3–14 MHz) coming to life with the approach of darkness. In general, a radio that covers the entire shortwave spectrum from 2.3–26.1 MHz (Fig. 3-5) is preferable to one that covers only part of the shortwave spectrum, as your listening choices are maximized. If you listen only during the evening, the higher coverage may seem to be irrelevant. However, even at night, some distant parts of the world come in only on the higher bands. Many persons new to shortwave listening also discover that they enjoy daytime listening, as well. For example, the most interesting English and French language programming—from France—is heard in eastern North America on the higher bands in the morning and early afternoon.

Fig. 3-5. The Kenwood Model R-600 receiver (Courtesy Trio-Kenwood Communications).

How to Defeat Shortwave's Howls and Squeals

A factor called "multiple conversion" is used to help reduce the presence of "false" radio signals which cause unnecessary interference to the station you're trying to hear. There is such congestion within the shortwave spectrum that every good radio must be rigorously designed to eliminate almost every vestige of such "false" signals, i.e., those generated within the receiver itself as a result of design weaknesses or shortcuts. So look for such indications as "double conversion" or "triple conversion" in the description of any receiver you're considering. Another way to spot this design enhancement is to look in the specifications for two or more "IF" or "intermediate frequency" values for *shortwave* reception. Two such frequencies mean the set is a double-conversion unit, three frequencies, triple conversion, and so forth. However, more is not necessarily better; what you should avoid is the single-conversion design.

Image-rejection specifications are about as meaningful as the 55-mile-per-hour speed limit in west Texas. Measurement can be made anywhere in the tuning range of the receiver and, on inexpensive double-conversion radios, a decent image-rejection figure (implicitly for the second IF) may not take into account image-rejection problems introduced by the cheap first IF. Too, the image frequency (double the value of the IF frequency) may be such that image interferences are coming from other shortwave broadcasting bands, where signals—and thus image interferences—are unusually strong. For example, in the Panasonic RF-2800/RF-2900, a 2-MHz IF produces 4-MHz images. As a result, powerful images from the 7-MHz (41 meter) International Broadcasting Band virtually swamp the weak signals found in the 3-MHz (90 meter) Tropical Band.

Nevertheless, image-rejection specifications can give a rough clue as to what you can expect. In general, image rejection over about −60 dB is good, under −40 dB awful. If you're paying dearly for the "Receiver of a Lifetime," look for −72 dB, or greater, rejection.

Shortwave broadcasters are often crammed together like cattle en route to Kansas City. The ability to separate the station you wish to hear from the rest of the herd is arguably the single most-important feature in shortwave receiver design. This ability is called "selectivity" but, alas, it is often not ascertainable from the promotion-minded literature put out by manufacturers.

Selectivity has three major aspects: *bandwidth*, *ultimate rejection*, and *shape factor*. Ideally, a radio "boxes off" the desired signal from competing signals on adjacent frequencies. In reality, it's something of a shoving match, with weak stations alongside strong ones, and so forth. Guess which ones come out on top! Obviously, it's not simply a matter of separating stations; it's also a question of being able to fight off the heavyweights who are leaning against the smaller and weaker station you're trying to tune in.

"Bandwidth" commonly defines the amount of radio signal that will be let in when the station you wish to hear is either of comparable strength or stronger than the competing stations on adjacent channels. Not surprisingly, given 5-kHz channel spacing, bandwidths of between 4–6 kHz are common on shortwave receivers. Given that conditions of reception, including interference, vary greatly from situation to situation, better shortwave radios include those with more than one bandwidth. To take an exceptional case, the recently discontinued Drake R-7 series had up to five separate voice bandwidths, which is one reason this model continues to be so highly prized by serious listeners, monitors, and DX'ers, even though it is complex to operate.

Bandwidth can also be expressed as a given number, or as plus/minus of half that number. For example, a ± 2.5-kHz bandwidth is simply another way of expressing a 5-kHz bandwidth. Either means of expressing bandwidth measurement is acceptable; just be sure to watch for the presence of the "+/−" signs!

"Ultimate rejection" refers to the maximum signal depth in which a receiver can select a desired signal against a relatively powerful signal alongside. As signal strength is measured in decibels (dB), ultimate rejection is also expressed in dB. Good receivers possess an ultimate rejection of at least −60 dB, with −72 dB (or better) being typical of most excellent communications receivers. Modest performers go no further down the scale than −50 dB, and the clunkers even less that that.

"Shape factor" is one of the most important measurements, not only of selectivity, but of receiver performance as a whole. Shape factor is the ratio between the bandwidth, traditionally measured at −6 (or −3) dB and the equivalent measurement at −60 (or −66) dB. A good shape factor provides the best defense against adjacent powerful signals muscling their way into your desired station's turf. Ideally, a shape factor should be 1:1. Given present technology, however, the best that is available—very rarely, and at high cost—is 1:1.3. A good rule of thumb is that a shape factor (−6/−60 dB) equal to or less than 1:2 is excellent, 1:2–1:3 good, 1:3–1:4 fair, and anything higher poor.

The importance of shape factor vis-à-vis bandwidth is found when you are attempting to reduce powerful adjacent-channel interference to a relatively weak desired signal. A narrower bandwidth reduces interference, of course, but, at the same time, it degrades the audio fidelity of the desired signal. However, by using a filter of the same bandwidth, but having improved skirt selectivity, fidelity is not downgraded (in fact, audio tends to become less muffled), yet interference is reduced. A filter with poor skirt selectivity and having a bandwidth of 4 kHz thus may produce inferior audio and also let in more interference than a filter with excellent skirt selectivity and having a bandwidth of 6 kHz.

One cautionary note: some receivers provide shape factor information at −6 and −50 dB. Not only does this give an inflated figure (obviously, any filter is narrower at −50 dB than −60 dB), it also tips you off that the ultimate rejection of the receiver is almost certainly little better than −50 dB. For casual listening purposes, a −50 dB ultimate rejection is adequate. However, for reception of weak signals alongside powerful signals, a −50 dB ultimate rejection serves as a yellow caution flag that the receiver is not designed for serious DX'ing purposes. If such a receiver is inexpensive, fine. But $500 receivers with such pedestrian performance are overpriced.

In summation, then, the best receivers will have two or more voice bandwidths (between about 2–6 kHz) from which to choose, an ultimate rejection of at least −60 dB, and a shape factor of 1:3 or less.

"Dynamic range" refers to a receiver's ability to receive weak signals with the presence of strong signals in the vicinity, usually elsewhere in the same band, whereas "selectivity" is limited to the ability to minimize interference to the desired signal from any adjacent or nearly adjacent signals.

The most common symptom of inadequate dynamic range is *overloading*, in which a station not on the same frequency as the station you're trying to hear will "ghost in" with other signals throughout much or all of the band. The symptom, but not the problem, can be alleviated by reducing signal input to the circuitry via shortening the antenna, switching in an attenuator circuit, or by turning down an RF gain control.

Inadequate dynamic range has been a curse of transistorized shortwave radios. Whereas even simple tube-type sets could handle strong signal inputs without overloading, only the best solid-state models have performed comparably. Thankfully, this is improving, but, as of now, the vast majority of transistorized portable radios, and, to a lesser extent, many table model sets, are limited in their ability to handle weak signals in the presence of the types of powerhouse signals that are common on today's shortwave broadcasting bands.

Dynamic range can be measured in various ways to produce results that can be deceptive to the consumer and flattering to the manufacturer. For this reason, dynamic range and related figures (e.g., third-order intercept point) can be misleading, even in those exceptional instances when manufacturers publish them in the specifications.

In order to alleviate this shortcoming, International Broadcasting Services, in conjunction with Sherwood Engineering, performs various dynamic-range laboratory measurements for its comprehensive reports published in each Summer/Autumn edition of the *Radio Database International*. These measurements now allow the potential buyer to evaluate beforehand whether the dynamic range of a specific model is likely to be adequate for the purposes to which the radio will be put.

Sensitivity Is Important in Tropical and High Bands

In the old days, when weak stations were the norm, sensitivity was a prized aspect of receiver performance. Now, given the powerful transmitters in use, its importance has become secondary to selectivity and certain other factors.

The problem with ascertaining sensitivity is that manufacturers' published measurements provide little valid guidance. Various measuring standards can be used; usually, those which provide the most flattering results are chosen by the manufacturer to provide the specifications for ads and brochures.

Probably the most practical way to ascertain sensitivity is to try two or more models at the same time on weaker stations. Listen not only to whether a given station is more audible on one model that another, keep your ear perked also for *hiss* mixing in with the radio signal—a sign of a mediocre signal-to-noise ratio. Remember, sensitivity tends to vary from band to band, so be sure to check out the bands that are of most interest for your listening purposes. At the same time, keep in mind that in the evening, the 6-MHz (49 meter) band is so chockablock with powerful signals that high sensitivity is almost counterproductive. On the other hand, daytime signals—in the higher bands, of course—tend to be relatively weak, as are the signals of the unusually interesting low-powered domestic stations in the "Tropical Bands" below 5.73 MHz.

Be sure to test for approximate sensitivity, with each model being compared under the same circumstances. They should be as close as possible to those you expect to encounter. For example, if you plan to use a portable set on battery power most of the time, check for sensitivity using batteries. With certain models, sensitivity rises once the set is connected to the AC main power. (Indeed, if you select a set that is insensitive with batteries, but sensitive with AC power, be careful to ensure that the power lines are quiet where you plan to do most of your listening. If not, that increased sensitivity, which results from the coupling of the receiver to the power-supply network, will bring with it unwanted noises from the electrical system.)

Another rough "on-site" test that you can perform at a store to determine receiver performance is the one for *stability*. A highly unstable receiver will require periodic retuning to maintain good reception, whereas a stable receiver will rest "spot on" for hours on end. If you are interested in receiving shortwave broadcasters only, simply tune in the shortwave radio to a powerful station (preferably on the highest band that is functioning at the time you are listening) and leave the set untouched for a half hour or so. Your ears will tell you if the station is drifting, but, after the half hour is up, see how much, if any, you have to retune. Obviously, a drifty set should be avoided.

Special Characteristics for Single-Sideband Reception

Thankfully, almost all the new-technology shortwave radios—even the less costly models—are reasonably stable for listening to shortwave broadcasts. Broadcasting signals consist of a "carrier" and two "sidebands." These three-component signals are used for broadcasting as they can be well received on easy-to-tune low-cost radios. However, by eliminating the carrier and one of the two (redundant) sidebands, the transmitter can put all its energies into the single remaining sideband. Thus, half as much energy will deliver the same transmitter power. Also, a single-sideband signal takes up only half as much radio-spectrum space—another, and very important, form of conservation. Consequently, almost every type of user (except broadcasters) of the radio spectrum operates voice communications exclusively in the single-sideband mode or one of its variants, such as *independent sideband (ISB)*.

The problem is that to receive single-sideband (SSB) signals properly, you should have a *product detector*, a *beat-frequency oscillator (BFO)*, and an exceptionally *stable receiver* (Fig. 3-6). The product detector demodulates the SSB signals with minimal distortion. The BFO provides a locally generated carrier in lieu of the carrier not transmitted, since high stability is needed to keep the local carrier of the radio properly tuned vis-à-vis the sideband so that the voices and music sound natural (the audio equivalent of tuning a TV tint control for lifelike flesh tones).

Fig. 3-6. The Japan Radio Model NRD-515 receiver (Courtesy Japan Radio Co.).

If you intend to listen to signals in the SSB mode, knowing the stability of the receiver is very important. Although "Hams" and others operate in the SSB mode, the most common application for shortwave listening is the reception of what are known as "feeder" signals. These are shortwave signals in SSB that are fed from a primary location to a remote relay or rediffusion facility. For example, within the USSR, SSB feeders are used to feed transmitter sites scattered throughout the country. Also, some SSB signals from broadcasters are intended to be heard by the general public. For example, SSB is sometimes used by the Australian Broadcasting Corporation to reach listeners in the West Australian outback, whereas some broadcasters, such as Radio Sweden, air their domestic service relays to diplomatic posts and mariners via SSB.

Schedules and other information on these broadcaster SSB signals are included in each edition of the *Radio Database International*, as it has been found that certain broadcasts are better received in North America and other parts of the world via SSB than via the usual broadcasting channels. For example, Radio Australia sometimes comes into eastern North America very well via SSB, but is rarely more than tolerable on the announced normal channels. Similarly, the jammed services of Radio Free Europe/Radio Liberty, the BBC, the Voice of America, and the Deutsche Welle can be heard—free from jamming—by tuning to their SSB and ISB feeder channels. Although, in principle, there is no real difference between receiving SSB or ISB signals, good ISB reception requires at least one relatively narrow bandwidth (e.g., around 2.0–2.5 kHz) with good skirt selectivity.

Reducing Interference, Fading, and Distortion

A novel use for a receiver with exceptional SSB reception qualities is the receiving of ordinary double-sideband broadcasting signals as though they were SSB. This tuning technique, known as *exalted-carrier selectable sideband (ECSS)*, allows the listener to choose the quieter of the broadcast's two sidebands. For example, if Canada (on 15325 kHz) is interfered with by AFRTS (on 15330 kHz), Canada's upper sideband is likely to be seriously degraded, whereas its lower sideband may be sufficiently far enough away from AFRTS that it comes in free from interference (provided, of course, that 15320 kHz does not contain a significant source of interference). So, in this instance, the receiver would be carefully tuned to the 15325-kHz lower sideband.

Not only can ECSS reception reduce or eliminate interference, it can also replace the fade-prone transmitted carrier with the fade-free carrier (BFO) generated by your receiver. So the impact of fading is reduced, especially the distortion resulting from "selective" fading. In fact, ECSS is such a powerful technique for improving the fidelity of shortwave (and medium-wave AM) reception that manufacturers are designing and beginning to produce receivers that automatically *phase*, or *fine tune*, the BFO with the received signal. As of 1984, this process is only beginning to appear in the marketplace, but during the life of this book, the movement in this direction should accelerate, especially if certain forms of "AM stereo" become widely accepted. These forms of AM stereo also benefit from *exalted carrier reception*, albeit of both sidebands.

Clues to Quality of Construction

There is no way to ensure that a given model of receiver will have a relatively long and trouble-free life unless it has been in use for a number of years. However, there are some clues to help separate the nags from the thoroughbreds.

Simplicity in design and construction reduces the number of opportunities for problems to develop. This is one reason that the discontinued Yaesu FRG-7 performed relatively reliably. However, even complex designs can be laid out in a systematic and straightforward manner to accomplish the same end. For example, receivers with only a few circuit boards, each having gold-plated interface contacts, tend to have a higher MTBF (mean time between failure) than those receivers of similar design which have numerous boards that use tinned contacts and a "spaghetti bowl" of coaxial cables running from board to board. They're also easier to repair.

Component quality is another barometer of MTBF. For example, epoxy circuit boards are nearly immune from cracking. "Mil spec" components, manufactured to the rigorous specifications of NATO military services, are almost invariably less failure-prone than ordinary components. Components with higher-than-necessary load-handling ability also help ensure long life. With digital circuits, some redundancy helps ensure continuity should one portion of a chip fail. Finally, chassis and front-panel construction help determine a receiver's resistance to physical abuse. The copper-clad steel chassis of the discontinued Drake R-4B, with riveted and threaded screw holes, is obviously superior to that of Drake's new model R-7A—also discontinued—which uses a stamped aluminum chassis and self-tapping screws. Similarly, the cast aluminum chassis of the Racal RA6719/GM is nigh indestructable. A cast-metal front panel also helps limit torquing and its resulting equipment damage.

Although some of these indicators of quality are not readily apparent to the buyer, they are considered when equipment evaluations are performed at International Broadcasting Services. Additionally, MTBF is in the specifications published by the manufacturers of most of the professional-quality models targeted for the military/intelligence community.

Should Receivers Be Modified?

A number of shortwave specialty firms in the United States, Canada, Western Europe, and Australia provide modifications to "stock" receivers so that one or more aspects of performance can be upgraded. Typically, an existing low-quality bandwidth filter is replaced by a filter of higher quality so as to improve the shape factor.

Such modifications can go a long way in improving receiver performance if the receiver is basically well-designed, but performs below snuff in only one or two respects. However, a mediocre receiver with countless modifications invariably performs peculiarly, as the inherent shortcomings of the receiver keep cropping up to intrude upon the improvements. Generally, then, improvements make the most sense when they are well-engineered and straightforward, flowing in concert with the design concepts of the factory receiver.

Where to Buy?

Ordinary shortwave portable sets sold by such consumer firms as Sony and Panasonic are often obtained through local dealers or mail-order houses, much as are other small appliances. The only disadvantage to treating such sets as ordinary appliances is that the consumer receives very little dependable point-of-purchase guidance from general retail outlets.

Another problem (or opportunity, depending on your viewpoint) is the presence of the so-called "gray market." In the gray market, radios are obtained by dealers for low cost from unofficial overseas sources, rather than via the established domestic distributor. The radios may or may not have identical features and performance as compared with the usual market versions. However, in most instances, they do not carry the factory warranty, which is the responsibility of the domestic distributor and which, in effect, is included in the price paid for the normal version. For example, it took three months and cost $74.00 to have Sony of America replace two clear plastic windows on the cabinet of a recently purchased $180 gray-market Sony ICF 7600D; an ICF-2002 presumably would have been repaired at no cost under Sony of America's one-year warranty. Early in 1984, the Federal Communications Commission began a "crackdown" on the so-called gray-market dealers, conducting a number of visits to several stores in an attempt to close off this practice. It remains to be seen whether the FCC will continue to maintain a firm antigray-market policy or whether enforcement efforts will slacken in the future.

Table-model communications receivers are marketed somewhat differently from the popular shortwave portable sets. For one thing, most manufacturers of communications receivers are primarily manufacturers of amateur radio equipment. Their dealers are almost invariably amateur radio stores and mail-order houses. As amateur radio operators tend to be technically savvy, the caliber of product, sales, and service is generally better than that of consumer-related electronics firms. On the other hand, most amateur radio outlets are used to selling equipment that is of interest to "Hams," not shortwave radio enthusiasts, so the sales personnel may not be so knowledgeable as they would be were you seeking, say, a transceiver.

The "shortwave specialty firm" has sprung up in response to the growing demand for helpful sales-and-service oriented to the shortwave listener. In some cases, such as with Electronic Equipment Bank near Washington, DC, and Universal Amateur Radio near Columbus, Ohio, there exist retail amateur/shortwave radio vendors that have facilities especially established to allow the shortwave listener to make his own receiver choice. If such a store is in your vicinity and has available the various models you wish to consider, this is an excellent way to select a receiver. Each set is connected to an antenna so that the customer can operate it himself and reach his own judgement as to which model provides the optimum interface of price and performance. Otherwise, most shortwave specialty firms are mail-order businesses.

As of now, no shortwave specialty firms are known to carry gray-market merchandise. However, some are authorized factory-service facilities, whereas others are not. This can make a difference in how repairs are handled, especially when modified receivers are purchased. Although such modifications can audibly improve receiver performance, they may invalidate the factory warranty if they aren't performed by a factory-authorized firm. On the other hand, some firms that do equipment modifications, such as Radio West of Escondido, California, will provide their own warranty in lieu of that of the manufacturer.

With a few models of shortwave receivers, quality control at times has been shaky. For this reason, at least some dealers routinely inspect and realign all receivers before they are sold. This is not an academic consideration for, in 1983, an estimated 2000 faulty units of one highly regarded model reportedly were sold, mostly in Europe, before the factory took steps to resolve the problem. One reason the problem was less prevalent in North America was the relative preponderance of shortwave specialty dealers that perform pre-sale inspection and alignment. While European consumers were complaining publicly about a "poorly designed model," Americans were generally pleased because unbeknownst to them, their dealers had been adjusting or returning an unusually high proportion of the sets received from the factory's U.S. distributor.

Unbiased Equipment Reviews Essential

As with high-fidelity loudspeakers, the differences among various models of shortwave receivers are so great that a decision to purchase should be made only after the prospective buyer has had a period of time in which to put various models "through the paces." But, unlike speakers, shortwave radios are not always amenable to valid in-store testing. With the exception of those specialty retailers, such as mentioned earlier, most radio stores lack proper antennas, are constructed of concrete and steel (which absorb shortwave radio signals), and are awash in electrical noises, such as those from fluorescent lights and business equipment. Additionally, most store hours do not conform to the hours in which the consumer will be listening to shortwave broadcasts. Given that propagation varies considerably from one time of the day to another, this just about rules out the practicality of checking out a shortwave radio in most stores.

Fortunately, this does not leave the consumer at the mercy of sales material. Since 1978, International Broadcasting Services, Ltd. has been commissioned by the *World Radio TV Handbook* to provide laboratory and user tests of shortwave receiving equipment. These tests are now a regular feature of the *Handbook*. Additionally, in 1984, International Broadcasting Services inaugurated advanced laboratory and field tests and evaluations, which are published each June in the Summer/Autumn edition of the shortwave listening guide, *Radio Database International*.

Equally unbiased are summary reviews aired by Radio Canada International and Radio Netherlands. Similar opinions on equipment are found in the *World Radio TV Handbook Newsletter* and in Radio Netherlands' free *Receiver Shopping List.*

International Broadcasting Services also makes available, upon request, individual printed reviews of any of the dozens of specific models of most shortwave receivers—from inexpensive portable sets to the exotic professional models that have been on the market in recent years, plus a number of models that go back to the vacuum-tube era. This allows prospective buyers to concentrate on those models that are of most interest to them. These reviews are now available to consumers, as well as professionals, at nominal cost plus postage.

So, if you choose carefully, you can obtain a state-of-the-art shortwave receiver that provides a caliber of reception that until recently was a thing of the past—before the shortwave spectrum became congested with competing signals. Unlike those pleasant old sets of the 1930s, the new ones are compact, reliable, and let you to know exactly where you're tuned. Thanks to digital chip technology, they are affordable, too.

So, take the time to read up on the current crop of receivers. If you do your "homework," you'll be rewarded with years of listening pleasure, and can bury most shortwave squeaks and squawks for good.

Publications

Radio Database International, the *World Radio TV Handbook*, the *World Radio TV Handbook Newsletter*, and the *Receiver Shopping List* all provide current addresses for distributors and shortwave specialty firms. These publications can be reached at:

- Radio Data Base International
  International Broadcasting Services, Ltd.
  P. O. Box 300
  Penn's Park, PA 18943 USA
  Tel: (215) 794-8252

- World Radio TV Handbook
  World Radio TV Handbook Newsletter
  Billboard Publications, Inc.
  1515 Broadway
  New York, NY 10036 USA
  Tel: (212) 764-7300

- "Media Network"
  Radio Netherlands
  P. O. Box 222
  1200 JG Hilversum, Holland

FUNDAMENTALS OF SHORTWAVE RADIO PROPAGATION

David D. Meisel, Ph.D.

David Meisel became interested in radio at the age of 12, when he constructed his first "cat's whiskers" crystal radio. This was followed by a lifetime affair with astronomy and radio propagation.

To attempt to list David's accomplishments and other professional activities would, quite literally, take several pages in this book. He holds degrees in physics and astronomy, has done research or worked under grants from the National Science Foundation, National Aeronautics and Space Administration, and the State University of New York. He has written some 56 articles on astronomy, propagation, and associated subjects.

David Meisel's working career has seen him in various positions as assistant, lecturer, instructor, assistant professor, and professor at a number of universities and observatories, as well as at the Goddard Space Flight Center. Currently, he is Professor of Astronomy at the State University of New York in Geneseo.

Somehow, David still finds time for shortwave listening.

Introduction

One of the aspects of shortwave radio most often taken for granted is the fact that properly directed signals are often able to circumnavigate the globe rather than simply escaping out into space. While many radio enthusiasts can get satisfactory results without a clear understanding of this behavior, a good grasp of the basics is often helpful, whether you are trying to pick up that new or rare DX broadcast, or simply trying to improve daily reception of broadcasts beamed specifically to your area. Although the subject matter we will discuss here is usually considered to be a matter of radio engineering, it is best to remember that information from many other scientific and technical areas must be used in making various operational broadcasting and reception decisions.

Although the advent of the digital receiver has removed a major obstacle to the enjoyment of shortwave radio by providing accurate frequency tuning, the novice is soon bewildered when the previously reliable reception of a desired station suddenly deteriorates or ceases entirely. When this event is not related to some alteration in local conditions—equipment failure, interference from "local" stations, or interference from local electrical equipment—there is little that can be done to get immediate relief. However, as we will indicate later, the majority of these nonlocal disruptions are only temporary and, with experience, the radio enthusiast can learn to recognize their signs and, thus, not be led into looking for a "local" solution to the problem when, in fact, there is none.

What Are Radio Waves?

Radio waves are simply one form of what are known as *electromagnetic (E-M)* waves (Fig. 4-1). Visible light, X rays, ultraviolet light, and microwaves are other examples of E-M waves. In free space, these waves do not stand still, but "propagate" at the speed of light (300,000 kilometers per second). Every wave has a certain distance between its peaks called its *wavelength*; the rapidity with which the wave vibrates is called its *frequency*. It is possible to arrange E-M waves according to their wavelengths or frequencies into an *E-M spectrum*. (The frequency dials of nondigital radios are simply graphical displays of sections of the E-M spectrum.)

Fig. 4-1.
Regions of the electromagnetic spectrum.

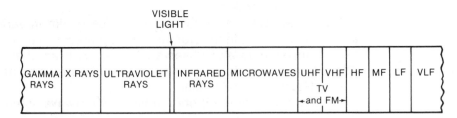

E-M waves can be made to vibrate along a specific direction (in a plane perpendicular to the direction of travel) and thereby display the property of *polarization* (Fig. 4-2). Radio waves are most often produced by the motion of electrons in metallic structures called *antennas*. When the antennas are long rods, the waves produced are said to be *linearly polarized*. When the antennas are made in the form of loops, the radio waves produced are said to be *circularly polarized*. Reception of radio waves is accomplished most efficiently by antennas whose polarization characteristics match those of the incoming waves. While antenna properties are said to be "reciprocal" (structures of optimum dimensions can be used for either transmission or reception), a transmitting antenna must be built to withstand high power levels that would literally burn up the sometimes fragile wires used for receiving antennas. As radio waves move from transmitting antenna to receiving antenna, they can be influenced by various objects and conditions along the way. A description of the conditions that affect the way a wave travels is called the *mode of propagation*.

Fig. 4-2. Wave polarizations.

LINEAR POLARIZATION

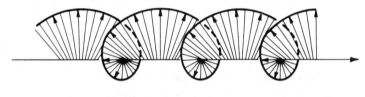

CIRCULAR POLARIZATION

Because waves of different frequencies are affected differently by prevailing conditions, each main type of radio-wave propagation is usually described separately. When a radio wave leaves the vicinity of the transmitting antenna, it spreads out in all directions. Some of the energy stays near the ground in a *ground wave* while some of the energy heads skyward in the so-called *sky wave*. Near the receiving antenna, the ground is able to reflect incoming waves and thereby alter the conditions of reception considerably. This is particularly true for those waves whose direction of vibration is parallel to the ground itself (horizontal polarization).

Shortwaves are those radio waves whose frequencies are between 2-million and 30-million times per second. The proper way to refer to frequency is through the unit, hertz (abbreviated Hz and named after Heinrich Hertz, who first experimentally demonstrated the existence of E-M waves). Thus, the shortwave region is considered to go from 2-million hertz (or two megahertz = 2 MHz) to 30-million hertz (or 30 megahertz = 30 MHz). Using the relationship

$$\text{Wavelength (meters)} = \frac{300,000}{\text{Frequency (kHz)}}$$

to calculate wavelengths shows that the name "shortwave" applies to a range of 150 meters to 10 meters, which by human standards is hardly short! (Radio engineers often prefer the names "high frequency" and "decimeter" as being more accurate descriptions.) High-frequency (HF) waves that are directed along the ground are not able to "cling" to the curved surface of the earth for very long, so most shortwave radio propagation involves sky-wave behavior.

Using Sky Waves for Radio Communication

When an HF wave leaves the transmitting antenna as a sky wave, it travels through the earth's lower atmosphere essentially in a straight line. For reasons to be discussed later, the earth's upper atmosphere becomes electrified with free electrons and other ions, starting at a height of about 60 kilometers. When a sky wave encounters such a charged region, the speed of the wave is altered by an amount which depends on the wave frequency, its angle of incidence on the layer, and the local electron density. If the electron density of the upper atmosphere increases along the wave path, the signal trajectory begins to curve. If the angle of incidence is steep enough and/or the electron density is high enough, this bending becomes extreme and the direction of the sky wave is changed by an angle sufficient enough to send the wave heading back toward the ground. Although this behavior is properly called the *refraction* of radio waves, it is commonly referred to as *reflection*. If conditions are favorable, the returning sky wave is RE-reflected by the ground and heads skyward again. Under favorable circumstances, this process can repeat many times. With high-power transmitters, the original waves can even be made to travel completely around the world. (This, in fact, can sometimes be heard on the shortwave bands as "round-the-world" echoes. In some cases of reported "round-the-world" propagation, however, the reported delays are much too long to be simple global circumnavigation and, therefore, must be due to some other mechanism.)

For a given electron density:

1. The amount of radio-wave bending is greater for lower frequencies than for higher ones.

2. There is a maximum frequency to the wave that can be returned to earth if the sky wave is projected straight up.

3. If a sky wave is projected into the charged region at an oblique angle, the wave that can be deflected will always have a higher maximum frequency than the wave which can be reflected in a vertical projection.

4. As a radio wave travels through an electrified layer where there is a sufficient amount of atmospheric pressure (in the lower levels of the upper atmosphere, for example), a loss of signal strength will occur. This is often called sky-wave *absorption*. In general, the higher the frequency used and the higher the take-off angle of the wave, the lower will be the sky-wave absorption.

5. Radio waves traversing an electrified layer surrounded by a magnetic field are split into two separate components. One component is called the *ordinary ray* and the other is called the *extraordinary ray* (Fig. 4-3). These two rays propagate in somewhat different ways. For HF frequencies, the extraordinary rays are reflected from "lower" levels in the atmosphere than are the ordinary rays. The extraordinary rays also suffer "higher" degrees of absorption than do the ordinary rays.

**Fig. 4-3.
Sky-wave
deflection by an
electrified layer
having a
magnetic field.**

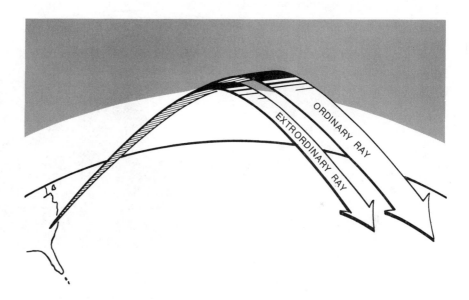

By appropriately setting the operating frequency and antenna take-off angle, the radio broadcaster can, in principle, maximize the signal received in a specific target area. As rules-of-thumb, these simple guidelines indicate that for local coverage, a broadcaster should use the lowest-possible frequency and the highest-possible signal take-off angle (vertical). Likewise, a broadcaster interested in worldwide reception should use low take-off angles and the highest-possible frequencies. Unfortunately, there are a number of reasons why this approach can fail to achieve perfect results. To be able to explain these reasons clearly requires some description of the behavior of the electrified regions that do the "reflecting."

Structure of the Earth's Radio-Wave Deflecting Region

The electrified region of an upper atmosphere is generally called an *ionosphere*, but electrons are usually the most numerous ions present. In sky-wave propagation, the electron density is indeed the most important factor influencing the propagation of the waves through the region.

In the case of the earth, the part of the ionosphere that most drastically affects HF waves occurs between 60 and 1200 kilometers (km) above the ground. A considerable amount of research has indicated that the earth's ionosphere usually consists of several separate regions or layers. It is customary to denote these layers/regions by letter designations, as shown in Fig. 4-4.

Fig. 4-4. Regions of the earth's ionosphere.

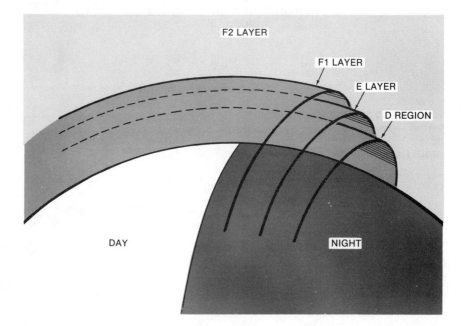

F2 LAYER
F1 LAYER
E LAYER
D REGION
DAY
NIGHT

- **D Region**—A region of relatively high atmospheric density but weak electron density at heights from 60 to 100 km. This region is extremely weak on the nightside of the earth. Most of the radio-wave absorption observed for sky-wave propagation occurs here and is greatest during the daylight hours.

- **E Layer**—The lowest normal ionospheric layer; usually located at heights between 100 and 140 km. The normal E layer is weak at night. Some daytime radio-wave absorption can occur in this layer.

- **F Layers**—One or more layers located above 140 km. During the daytime, there are often two F layers—the F1 layer peaks around 200 km while the stronger F2 layer peaks at an average height of 320 km. At night, the F1 layer disappears, leaving a single F2 layer at about 250 km high.

In general, conditions in the D region and the E layer determine the lowest usable frequency (LUF) that can be used on a given path while the F-layer conditions determine the maximum usable frequency (MUF).

Since the motion of charged particles is always affected by the presence of a magnetic field, it should come as no surprise that the earth's magnetic field (which makes magnetic compass needles point north and south) has a profound effect on the ionosphere—particularly on the F layer. (In recognition of this, the region of the upper atmosphere located above the F layer is often called the *magnetosphere*.) The *geomagnetic field* not only influences the structure and behavior of the earth's ionosphere, but, because the very motions of the electrons themselves are disturbed, there are characteristic magnetic effects observed in sky-wave radio propagation.

The earth's magnetic field is not symmetrical around the globe. First, the magnetic poles are not coincident with the geographic poles. This means that the magnetic equator is inclined at an angle of about 15 degrees with respect to the geographic equator. Secondly, the line connecting the two magnetic poles is not on a diameter, but a chord of the earth. This means that there can be a significant difference between the magnetic field strength at two opposing (or *conjugate*) points on either side of the magnetic equator. (The geomagnetic field is weakest in an area between Africa and South America called the *South Atlantic Anomaly*.)

For the purposes of explanation of shortwave radio propagation, it is convenient to divide the earth into three separate domains:

- **Polar Regions**—The magnetic field is strongest here with the lines of force nearly vertical (like those shown for bar magnets with iron filings spilled over them). Geomagnetic propagation effects are greatest in these regions and, because of the long seasons of sunlight and darkness, the structure and behavior of the polar ionosphere is highly variable with extreme and complicated summer/winter contrasts.

- **Mid-latitude Regions**—The magnetic field is average in strength here and the lines of force come out of (or go into) the ground at oblique angles. The mid-latitude ionospheric structure is fairly well understood and shows distinct seasonal changes.

- **Equatorial Regions**—The magnetic field is weakest here and the lines of force are parallel to the ground. Geomagnetic propagation effects are least in these regions and, together with the nearly constant cycles of day and night, enable the equatorial ionosphere to be extremely stable and reliable and show little seasonal change.

The ultimate source of the ionosphere's charge production energy is our star —the sun. It has been shown by many years of scientific research that solar high-energy radiation (X rays, ultraviolet light, and cosmic ray particles) controls the structure and strength of the earth's ionosphere.

Structure of a Typical Star—Our Sun

The sun (Fig. 4-5) is the nearest star to the earth and much scientific research time has been spent studying its behavior. In many respects, our sun is considered to be of "average" size, mass, and temperature. The part of the sun "visible" from the earth can be divided into three separate regions:

Fig. 4-5. Solar features and regions.

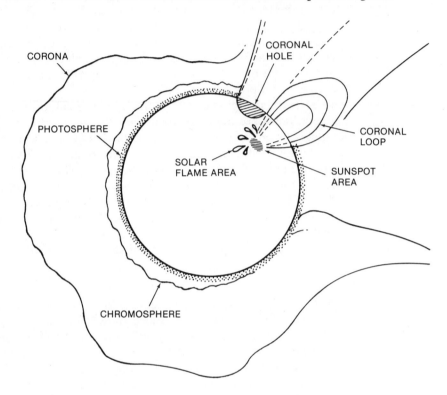

- **Photosphere**—This is the "surface" part of the sun from which most of the "visible" solar radiation appears to come.
- **Chromosphere**—A region in the sun's atmosphere above the photosphere. The lower part of the chromosphere is cooler than the photosphere while the top of the chromosphere is about three times hotter than the photosphere. The chromosphere is a source of solar ultraviolet light as well as microwave energy.
- **Corona**—This is the upper region of the sun's atmosphere, above the chromosphere. The temperature in the corona reaches TEN times that in the photosphere. The corona of the sun is a source of ultraviolet light and X rays, as well as HF and VHF radio noise.

These three regions have been observed to undergo cyclical changes called *solar activity*. (Recent astronomical research indicates that some other stars, which are like the sun, also undergo similar changes.) The main features that indicate the prevailing degree of solar activity are:

- **Sunspots**—These are features in the solar photosphere that appear darker and, therefore, cooler than the surrounding areas. Counts of sunspots provide a simple measurement of the amount of solar activity that is present.

- **Solar Flares**—These are explosive features in the chromosphere that last for several minutes and emit pulses of high-energy radiation—ultraviolet light, X rays, and charged particles (solar cosmic rays). Regions that produce solar flares are hotter than normal for the chromosphere and, as a consequence, emit larger than normal amounts of microwave energy. Measurements of this solar microwave energy (called simply *Solar Flux* values) provide a much smoother and a more reliable measurement of solar activity than do sunspot counts. (However, because sunspots have been observed since the 1600s, they provide a much longer historical baseline.)

- **Coronal Loops**—These are semipermanent, high-temperature features in the solar corona above the sunspot and solar-flare areas. Coronal loops emit copious amounts of X rays and ultraviolet light.

- **Coronal Holes**—These are "dark" areas located between the loop features in the corona. Research with the Skylab spacecraft established that the coronal holes were the regions where high-speed solar-particle streams emerged to form the so-called *solar wind*.

Since the changing pattern of solar activity was first indicated by the behavior and amount of sunspot activity, the *activity cycle* is often referred to simply as the *sunspot cycle* (Fig. 4-6). Although the sunspot areas themselves do not emit the high-energy radiation to which the earth's ionosphere is sensitive, the solar-flare areas and coronal loops that usually occur near sunspots do emit pronounced amounts of X rays and ultraviolet light which, in turn, do produce ionospheric changes. In this way, a statistical (but not directly causitive) relationship between the number of sunspots and the state of the ionosphere (above specific locations on the earth) results. The term, *quiet sun*, is used to refer to those features of the sun not directly affected by solar activity, while the term, *active sun*, is used to denote those aspects which directly reflect the presence of solar activity.

Variations of HF Radio-Wave Propagation

As seen from the earth, solar activity goes through a considerable amount of variation. Although the fluctuations are basically *random* in character, mathematical analysis shows that the average level of solar activity contains some elements of cyclical behavior. But, because the random variations are so large,

**Fig. 4-6.
Schematic of a
typical sunspot
cycle.**

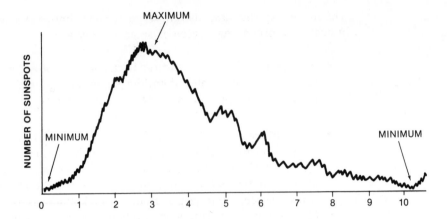

solar-activity trends are very difficult to predict with acceptable scientific accuracy. Since the earth's ionosphere responds to solar-activity changes as well as having random elements of its own, HF radio propagation is always subject to random variations and, therefore, also lacks real predictability. However, certain cyclical factors have been noted from studies of past behavior.

- **Day/Night Variations** (diurnal)—These have already been discussed earlier. The D, E, and F1 layers are essentially daytime features. When the sunlight is gone, the charged particles recombine into neutral ones. This happens more rapidly at low altitudes than at high altitudes and thus the F2 layer does not disappear at night, *but* it is lower and weaker than during the daytime. The day/night transition takes several hours so, in many locations, the "twilight" period offers the best DX reception conditions. (The D-region/E-layer absorption is still fairly low and the F layers are able to reflect reasonably high frequencies.) The listener who wants steady reception, however, will often be disappointed by the rapidly changing conditions found near the time of local sunrise and sunset.

- **Seasonal Variations** (summer–equinox–winter)—At the equinoxes, the ionosphere is placed reasonably symmetrically between the northern and southern hemispheres. In summer, the mid-latitude D, E, and F1 layers are stronger than in winter while, for the F2 layer, the opposite is true. In winter, the daytime F2 layer is lower and more dense than the summer daytime. The nighttime F2 layer, however, is weaker in winter and stronger in summer. This seemingly contradictory behavior occurs because the electrons of the F2 layer on the summer (hotter) hemisphere boil off and get channelled by the earth's magnetic field to the winter (cooler) F2 layer. At the equinox periods, the electrons migrate at roughly equal rates between the two hemispheres. At night, the F2 region reverts to a more normal pattern—higher density in summer than in winter. Thus, there is a greater contrast in HF reception for the winter conditions than for the summer and equinox periods.

- **Monthly Variations** (27-day)—Because the sun's period of rotation as seen from the earth is roughly 27 terrestrial days, solar-activity levels have a tendency to repeat with roughly the same period. (Actually, because the sun is not solid, different parts of it spin at different speeds. This means that the monthly patterns can have repetition periods that range from 26 to 30 days. However, the standard period is often taken as 27.26 days.) Since the ionosphere is sensitive to solar activity, radio-reception quality and reliability tends to exhibit a similar 27-day pattern of variation.

- **Solar-Cycle Variations** (11 year)—Past studies of sunspot numbers (activity) establish that solar activity generally goes through long-term variations averaging 11–12 years in length. (Because the sunspots themselves are magnetic and their magnetic polarities switch when a new cycle begins, the solar-activity cycle is considered to be magnetically controlled by some mechanism deep within the interior of the sun.) At the beginning of a new solar cycle, activity rises rapidly, with the peak coming 3 or 4 years from the start. During the remaining 8 or 9 years, activity levels decline erratically until a minimum level is reached again. The sensitivity of the ionosphere to solar activity produces a long-term, solar-cycle (11–12 year) pattern in HF-reception conditions upon which the other variations (daily, monthly, seasonal, and random) are superimposed.

In general, HF radio-wave reception is considered to be better at the maxima (peaks) of the sunspot cycle than at the minima (valleys). This is because, at the times of sunspot maxima, higher levels of ultraviolet light (to which the F2 layer is most sensitive) produce a strengthening of its "reflectivity." This improvement is not as spectacular as it might at first seem because D-layer absorption and multireflective "shielding" by the E and F1 layers of the F2 layer are also high at sunspot maximum. An even more-important degradation that peaks at sunspot maximum comes from solar flares (Fig. 4-7) and their consequent random *disruption* of HF reception. Solar-flare effects include:

- **Shortwave Fadeouts (SWFs)**—These occur on paths crossing the daylight side of the earth. They usually last only for tens of minutes, but if several flares are going off at once, a net effect of several hours can seem to occur. These fadeouts are the result of X-ray enhanced absorption in the D region and occur coincident in time with the solar flare itself.

- **Polar Blackouts**—These may last several days in the polar regions and are concentrated around the geomagnetic poles. They are now known to be caused by solar particles (mainly protons and electrons) ejected by flares or caused when coronal loops or filaments disappear. There is often a delay of one or two days between the solar event and the corresponding blackout. The strongest blackouts are often accompanied by visible *aurora* displays. HF reception for paths going through the polar regions may be completely disrupted or, at least, subject to extremely rapid (sometimes called *auroral*) fading.

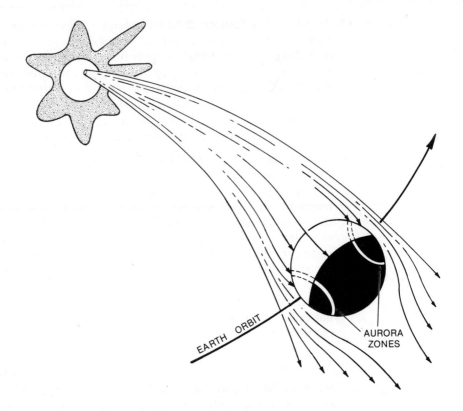

Fig. 4-7. Solar-flare bombardment of the earth.

EARTH ORBIT

AURORA ZONES

- **Magnetic Storms**—Even if the solar-flare particle spray does not hit the earth directly, but only passes nearby, the magnetic field carried along by the stream of charges is able to interact with the geomagnetic field and thereby disturb HF reception with increased fading and absorption. In some cases, magnetic storms produce effects that are not simply restricted to the polar regions. The zone of maximum geomagnetic effect is in a roughly circular band concentric with the geomagnetic poles. These bands are about 10 degrees of latitude thick and are located (under average conditions) some 15 degrees from the magnetic poles. Since aurora are seen very well from these parts of the earth's poles, these areas are often called the "auroral zones." The maximum effect on HF signals during polar blackouts and geomagnetic storms occurs for paths that are tangent (i.e., going east-west rather than north-south) to the auroral zone and when the F layers are much lower than at sunspot maximum. Of course, geomagnetic-storm effects are generally more severe on transmissions crossing the polar and mid-latitude regions. During geomagnetic storms, not only is D-region/E-layer attenuation of signal much greater, but the "reflectivity" of the F layers is often decreased considerably. This narrows the range of frequencies that are available to give satisfactory reception. (Some geomagnetic disturbances are so strong that the MUF and LUF overlap and nothing is heard on any HF bands!)

At sunspot minimum, all ionospheric layers are weakened. In addition, the coronal holes (mentioned earlier as sources of continuous particle streaming) cover more of the sun and become a significant source of reception disturbance. Like the flare-produced streams mentioned earlier, the coronal-hole streams that come near (or hit) the earth also produce magnetic storms. Since individual coronal holes are quite long lasting, their accompanying magnetic storms happen each time the sun's rotation carries them past the earth. Coronal-hole-related magnetic storms tend to show a repeating 27-day pattern and are often called *recurrent magnetic storms*. While recurrent magnetic storms are rarely as strong as flare-induced ones, they may last for several days and (since the reflective layers are in a generally weakened state) seem quite disruptive to HF reception.

Each solar cycle has its own peculiarities. Some, such as those that peaked in 1957 and 1980, were considered strong while those that peaked in 1804 and 1816 were considered weak.

There are some shortwave propagation anomalies that do not depend directly on solar activity. One of these is called the *sporadic-E* phenomenon. Sporadic-E (abbreviated Es) appears on a worldwide basis, but its form and severity is subject to wide variations. In some areas, such as Japan and southern Italy, it is a fairly permanent feature. In others, such as North America, it is essentially a summertime occurrence. Most Es is apparently associated with intense wind shearing and turbulence in the E layer of the ionosphere. Some forms of sporadic-E are able to reflect even very-high-frequency (VHF) signals (30 to 150 MHz) for several thousand kilometers. For the most part, Es is detrimental to HF radio broadcasting because it is able to "cut off" transmissions that otherwise would be able to penetrate the normal E layer and be reflected by the F layers. When strong (*blanketing*) Es occurs, "local" shortwave transmissions come in with unusual strength, and the more distant signals coming down from the F layer just bounce off the Es region from above and are never able to penetrate to the ground.

A second phenomenon not directly related to solar activity is called *spread-F*. Spread-F occurs mainly over the equatorial regions and seems to be related to the development of intense turbulence in the F region. This turbulence is called *scintillation* and is observed on VHF transmissions to and from communication satellites. On the HF bands, spread-F is most commonly observed on paths that traverse the equator in a north-south direction. It produces flutter-fading not unlike that heard on polar paths during magnetic disturbances, but spread-F is often strongest when there are no magnetic storms in progress. The turbulence associated with spread-F is often able to scatter VHF signals as high as 50–75 MHz between stations located several thousand kilometers on either side of the equator.

Shortwave radio-broadcast engineers and frequency planners go to a considerable amount of trouble trying to get strong signals into specific target areas. They try to take diurnal, seasonal, and solar-cycle variations into account and, when broadcast schedules are published, they often reflect these considerations. The usual broadcaster response to reception problems at particular target areas is to simply put more power into higher-gain antennas and use a wider selection of frequencies. It is often the case that the prime listening times in a target area correspond to a less-than-favorable time for direct reception. If satisfactory service cannot be obtained by way of direct transmissions, the wealthier HF broadcasting organizations reroute their transmissions over paths that are more reliable using relay transmitters placed at more favorable locations relative to the target areas. Many organizations also have a corps of technical monitors in the target areas to keep them informed of changing reception conditions. Most of the larger shortwave broadcast organizations also have available graphical or digital computer models of the ionosphere as aids to their planning efforts. In some countries, radio propagation information used for military purposes is available to the civilian broadcast organizations. In many countries, there are civilian scientific organizations collecting data on solar activity and these data are available to broadcasters, but not necessarily to the shortwave listeners.

While the smaller broadcast organizations do not have the resources available to do extensive propagation analyses, their engineers are often radio amateurs who have had a fair amount of operational experience on the HF bands. Over the years, amateur radio operators (Hams) have developed a good deal of expertise in propagation matters and their publications reflect this. In some countries, amateur radio operators exchange propagation information over the air and make contacts to discuss propagation on a regular basis. Hams have tried over the years to systematize their loggings into regular prediction systems with some apparent success.

Of course, there is little that any broadcaster or listener can do when ionospheric conditions are extremely poor. However, listeners often can minimize the effects of ionospheric disturbance with effective antenna design and planning. Here are some points to consider: (1) Because of geomagnetic effects, HF transmissions rarely follow a great circle path, although that approximation is often made. (2) During geomagnetic disturbances, signals can arrive from virtually any compass heading. This is particularly true for high-power transmissions, which may "bounce" into a target area from the side of the beam (*sidescatter*). (3) Thus, it may be desirable to have available for trial-and-error selection, several antennas with different directional sensitivities.

Although the most-frequently used HF broadcast antennas emit mostly horizontally polarized waves, reflection by and transmission through the ionospheric layers produces both horizontally and vertically polarized signals at the receiving end of a path. On east-west paths, the ordinary ray is mainly vertically polarized. On north-south paths, the ordinary ray is mainly horizontally polarized. Since the ordinary ray is least affected during geomagnetic disturbances, proper selection of receiving polarization may produce improved results.

At any given time, the highest frequencies received often come in at extremely low angles. This is particularly true at times of sunspot minimum. Because of ground reflectivity, this means ALL receiving antennas, regardless of polarization sensitivity, should be placed as far off the ground as possible. For those vertical-antenna designs that require near-ground mounting, improved low-angle performance is obtained by adding a large number of horizontal wires (*radials*) that are at least ¼-wavelength long to the *ground* connection of the receiver.

Shortwave listeners who keep extensive logs of their results are eventually able to acquire the needed experience and will learn to recognize the normal seasonal, diurnal, and solar-cycle variations as they occur. They are also often able to correctly identify various abnormal propagation conditions simply from the quality of received transmissions. As experienced DX'ers know, some of the most interesting listening occurs during periods of ionospheric disturbance when signals beamed directly to their geographical areas are relatively weak. It is often quite useful to be able to recognize when these disturbed periods are (or about to be) occurring.

Improving Your Propagation Awareness

On the assumption that this essay has heightened your awareness of some of the propagation factors controlling HF broadcasting, we conclude with some suggestions for further investigation.

Most shortwave listeners have the equipment necessary to compile a useful technical reception log. Usually it is a matter of selecting several representative paths for daily monitoring. For example, on the east coast of North America, nonrelay broadcasts of NHK (Radio Japan) provide excellent indicators of the reception conditions on a transpolar path, and broadcasts of Radio R.S.A. provide an indication of both polar and transequatorial conditions, while reception of Radio Australia provides an excellent probe of the transequatorial conditions.

Once suitable probe transmissions have been selected, and assuming that solar activity (and hence, ionospheric conditions) repeat after about 27 days, one should be able to predict both favorable and unfavorable reception circumstances for the next 4 weeks. (The maintenance and updating of your prediction system would seem to be an easy task for a home computer. It could also keep track of your hits and misses!)

For those shortwave listeners wanting to attempt to correlate their reception results directly with professionally obtained indices of solar and geomagnetic activity, the procedure may be more difficult to carry out because none of the modern solar-activity and geomagnetic-activity indices are readily available worldwide nor can they be readily compiled on a real-time basis by nonprofessional observers.

Shortwave listeners in North America—at least at the time of writing of this chapter—can obtain values of the solar flux and estimates of geomagnetic activity (called the "A-index" and "Boulder K-index") over station WWV at 18 minutes past the hour. Several fairly elaborate prediction systems using these indices have been set up and some are used extensively by radio amateurs in the United States. More details on such systems can be found in several monthly Ham radio magazines.

RECOGNIZING LANGUAGES

Richard E. Wood, Ph.D.

Richard E. Wood, globe-trotting DX'er and linguist, began his career in 1957, in England. As part of a worldwide college teaching career in languages and linguistics, he has lived and DX'ed in Saudi Arabia, New England, New York, and the southern United States.

President of the International Radio Club of America, Richard specializes in worldwide DX on the medium-wave band and in searching for nonEnglish-speaking stations, particularly in Latin America and the Arab world. He is a member of ten DX clubs in six countries and a longtime contributor to Radio Sweden International's "Sweden Calling DX'ers" and other DX programs.

Richard Wood is now a businessman in Hilo, capital of the Big Island of Hawaii, which some consider to be the finest DX location in the United States.

You Don't Have to Speak Them, Just Learn to Identify Them

Linguists estimate there are about 4000 languages in today's world. What constitutes a language? The definition is more political than scientific. Someone said, "A language is a dialect with an army and a navy." By any neutral linguistic measurement, for example, Cantonese is more different from Standard Chinese than, say, Norwegian is from Swedish. But the government of China (whether in Beijing or in Taipei) insists that Cantonese is a "dialect" of Chinese, whereas Sweden and Norway are two separate countries and, thus, have two separate languages. (Actually, Norway has two official languages, Bokmäl and Nynorsk, but you won't hear much of the latter one on shortwave so we won't discuss it here.)

How many of these languages can be heard on shortwave? Less than one tenth of the world's total. And, mostly the more important ones, having the largest number of speakers. But not always—Icelandic, with a mere quarter-million speakers, can be heard, and so can Kiribertese (formerly Gilbertese) with a lot fewer than that.

What's the language with the largest number of native speakers that *isn't* heard on shortwave? It seems to be Javanese, which is estimated to be the language of the thirty million citizens of the island of Java, site of the Indonesian capital of Jakarta. The reason is clear. As far back as 1926, Indonesian nationalists voted to make Bahasa Indonesia, which is derived from Malay, the national language and, ever since independence in 1945, all other languages have been discouraged. Most Javanese now speak or understand Indonesian, and foreign stations get through to them just as well in Indonesian, so there is no point in broadcasting in Javanese.

This isn't the case with other minority languages in the Asian countries, where minority ethnic groups are at loggerheads with their government, especially when they are predominantly Christian but the government is based on a nonChristian majority group. Christian missionary stations broadcast in these minority languages, such as Karen and Kachin, and beam them to those tribes in Burma from Christian stations in the Philippines. The same applies in Africa, where the partly Christian Oromo people of Eritrea are fighting the Marxist Ethiopian régime of Mengistu Haile Mariam. The missionary station in the Seychelles, FEBA, broadcasts to them, and the Oromo identification, listed along with the others in Table 5-1, was monitored from those broadcasts.

Table 5-1.
Station
Identifications
in Various
Languages

	Europe and European Ex-colonies
Albanian:	"Ju flet..."
Belorussian:	"Havoryts..."
Bulgarian:	"Govori...", "Tuk e..."
Czech and Slovak:	"Volá..."; listen also for "Rozhlas" (broadcasting) and "vysilaní" (broadcast)
Danish:	"De lytter til..." (you are listening to)
Dutch:	"Hier is...", "Dit is...", "U luistert naar..." (you are listening to); "U bent afgestemd op..." (you are tuned to)

Table 5-1 (cont.)
Station
Identifications
in Various
Languages

Europe and European Ex-colonies (cont.)

Esperanto:	"Parolas..." (speaking)
Estonian:	"Siin..."
Faroese:	"Útvarp..." (broadcasting)
Finnish:	"Täällä..."
French:	"Ici...", "Vous écoutez...", "Vous êtes à l'écoute de..." (you are listening to). "Broadcasting" is "Radiodiffusion".
German:	"Hier ist...", "Sie hören..."; "Radio" is also "Rundfunk".
Greek:	"Edhó..."; "Broadcasting" is "Radhiofonia".
Hungarian:	"Itt..."
Icelandic:	"Útvaro..." (broadcasting)
Italian:	"Qui...", "Questa è...", "Ascoltate...", "Trasmette..."
Latvian:	"Rūna..."
Lithuanian:	"Kalba..."
Norwegian:	"Dette er..."; "Broadcasting" os "kringkasting".
Polish:	"Tu...", "Tu mówi", "Mówi"; (sounds just like English "movie").
Portuguese:	"Aqui...", "Fala...", "Sintonizam..." (you are tuned to); "Escutam..." (you are listening to). *Note:* "Fala..." and "Escutam..." are used mainly in Portugal and ex-Portuguese Africa. "Aqui..." is used in Brazil, Portugal, and Africa.
Romanian:	"Aici..." (one syllable; sounds like the name of the letter h as pronounced by a Cockney or an Australian).
Russian:	"Govorit..." (stress is on last syllable, like English eat); "Vy slushaete..." (you are listening to). Listen also for "peredacha" (broadcast).
Spanish:	"Aquí...", "Esta es...", "Transmite...", "Sintonizan..." (you are tuned to); "Están sintonizando con...", "Están en sintonía con..." (you are in tune with); "Escuchan...", "Habla" (speaking). *Note:* "Habla" is used mostly in Spain and in Equatorial Guinea (ex-Spanish Guinea), not in Latin America.
Serbo-Croatian:	"Ovdje...", "Ovde..."
Ukrainian:	"Hovoryt..."

Africa

Afar:	"Ah..." (sounds like the German exclamation "ach" but not so throaty).
Afrikaans:	"Dit is...", "U luister naar...". *Note:* This language is descended from Dutch and sounds like it. Thus, identifications also sound similar.
Amharic:	"Yith ye...new"
Berber:	"Dahab..."
Malagasy:	"...ity" (follows station name)
Oromo:	"Kun... t" (t is added after station name)
Setswana:	"Ke..."
Somali:	"Halkani waa..."
Swahili:	"Hii ni..." (this is...), used in ex-British colonies, e.g., Kenya, Tanzania, Uganda; "Hapa..." (here), a translation of French "ici" used only in the ex-Belgian colonies of Zaïre, Burundi, and Rwanda. Listen also for "Sauti ya..." (voice of).
Swazi:	"Lona ngu..."

Table 5-1 (cont.)
Station
Identifications
in Various
Languages

Middle East

Arabic:	"Huna..." (here is), almost always used; rarely "Hadha..." or "Hadihi..."; English or French "Radio" is rarely used. Listen for "Idhaa(t)...", pronounced "Izaa(t)..." in Egypt and Sudan.
Armenian:	"...khosum" (follows station name, but sometimes goes before it)
Azerbaijani:	"Danishyr..."
Dari:	"Inja..." or "Inja...ast" (station name goes between the two words); this is a kind of Persian spoken in Afganistan and ID is the same.
Farsi (Persian):	"Inja... ast"; a very long a, like English "jaw".
Georgian:	"Laparakobs...", Hebrew "Kol..." (means "voice," the beginning of station names.) No word corresponding to "this is" or "here is" is used.
Pushtu:	"Da...dai" (station name goes in the middle.)
Turkish:	"Burasi..."

Asia

Bangla (Bengali):	Name of station is given with no introductory words.
Burmese:	"Thima..."
Cantonese:	"...(po) tientoi", "...deentoy", "...tin t'oi" (there is no official English transcription)
Chinese:	"...bo diantai (Standard Chinese, formerly Mandarin); spelling on Taiwan: "...po tien tai" (follows station name).
Ohivehi (Maldivian):	"Mee..."
Hindi:	"Yeh...", "Yeh...hai" (in the latter, station name is sandwiched between the two words).
Indonesian:	"Inilah..." or, rarely, "Disini...". Watch also for "Suara..." (voice).
Japanese:	"Kochirawa...desu" (name between as for Hindi. Sometimes only "Kochirawa")
Khmer (Kampuchean):	for "Withayu" meaning "radio")
Korean:	"Yeogineun...". Watch also for frequent "...nida" at end of sentences.
Lao:	"Thini..." (same as Khmer and Thai)
Malaysian (Malay):	"Inilah..." (same as Indonesian. "Suara" for "voice" is also the same)
Monogolian:	"...yarsh baina" (follows station name)
Nepali:	"Yo...ho" (sandwiched around station name)
Sinhala:	"Me..." (cf. Ohivehi, above)
Tamil:	As for Bengali, the station name alone is stated.
Urdu:	"Yeh..." "Yeh...hai" (same as Hindi)
Vietnamese:	"Dà la..."

Pacific

Most stations speak English or French, but let's look at a few Polynesian and Melanesian languages:

Kiribertese:	"Alo..."
Niuean:	"Ko e..."
Tahitian:	"O..."
Tok Pisin:	(Neo-Melanesian; formerly Pidgin English.) "Yupela woklong harim..." (cf. English "you-fellow walk 'long hear 'im")
Tongan:	"Ko e..." (same as Niuean)

Table 5-1 (cont.)
Station
Identifications
in Various
Languages

The Americas

Like the Pacific, the Americas are now dominated by languages of European origin—English, Spanish, Portuguese, French, and Dutch, in that order. Only a few Amerindian languages are heard, and most such stations give identifications in European languages, mainly Spanish or English. But three Creole (part-European, part-African) languages can be heard, two of them from Radio Netherlands.

Haitian Creole:	"Du écouté..." (you are listening to)
Papiamentu:	"Bo ta skucha..." (as for Haitian)
Sranan Tongo:	"Joe arki" (pronounced "you"—joe is a Dutch-based spelling for you, and arki comes from obsolete English "hark'ee, hark ye")

An easy way to learn to recognize languages is to tune in to the Voice of America broadcasts, since an announcer always introduces them in English first. Fig. 5-1 shows a schedule from an issue of the VOA's magazine *Voice*. This might be your only chance to see this listing. The VOA refuses to send it to American listeners, or to respond to U.S. inquiries about it.

You can learn to identify all these languages. It takes time and practice, of course. You can probably identify some of them already—perhaps French, German, Spanish, and Italian. Build on that base, and add more "tongues" belonging to the same language family. For example, if you can identify Spanish, learn to recognize Portuguese. They're really very different. And once you can recognize German, you can add others in the Germanic group, like Dutch and Swedish. The first nonEuropean languages should come easily, too. Arabic is heard very widely on the shortwave bands and is easy to recognize. Chinese and Japanese have similar writing systems, but in shortwave listening you're concerned with the spoken language, and you'll very soon realize that spoken Chinese and Japanese have nothing in common. It's much more of a challenge to tell Chinese apart from, say, Vietnamese (which has a completely different writing system, using a modified form of our own Latin alphabet), Thai, Lao, Burmese, and the other tone languages.

Even if you can't put your finger on the exact identity of the language, it's a big step in the right direction if you can pin it down to a continent or a cultural area. Most U.S. DX'ers seem to enjoy chasing signals from Third World countries on the tropical bands—60, 90, and 120 meters. When you hear African stations, one of the first steps you can take is to decide whether it's a station in a former British colony or a former French or Belgian possession. How can you tell? Listen to the names of countries in the newscasts. If it's a newscast from Nigeria, Ghana, Botswana, Lesotho, or some other ex-British colony in Black Africa, country names will mostly be in English; they'll say "America," "Lebanon," "Soviet Union," and so on. But if the station is in Senegal, Ivory Coast, Upper Volta, Zaïre, or another of the host of officially French-speaking African countries, it will be "Amérique," "Liban," or "Union Soviétique." The same applies to technical and political words and to the pronunciation of the word "radio."

OTHER LANGUAGE BROADCASTS

(GMT) FREQUENCY Asterisk (*) indicates medium wave

MIDDLE EAST

ARABIC (0400-0600) 15195, 5965, 1260; **(1700-2100)** 15305, 7205, 1260*
GREEK (1800-1830) 15195, 11915, 9735, 5965
PERSIAN (DARI) (0230-0300) 11805, 9700, 9565, 6050; **(1715-1800)** 15435, 11740, 9680, 7280
PERSIAN (FARSI) (0300-0430) 11805, 9750, 9700, 6060; **(1800-2000)** 11835, 9680, 7280, 6150
TURKISH (0330-0400) 15195, 9530, 7130, 6080, 1260*; **(1630-1700)** 15305, 15195, 9735, 7205, 5965, 1260*

USSR AND PARTS OF EASTERN EUROPE

ARMENIAN (0200-0215) 9615, 9530, 7170, 7120, 6105, 1260*; **(1500-1600)** 21520, 17855, 15415, 15195, 11805, 9670
AZERBAIJANI (1500-1530) 15435, 11845, 9565, 7280,
ESTONIAN (0245-0300) 9530, 7170, 7120, 6080; **(1430-1500 and 1530-1600)** 15270, 11960, 11865, 9735
GEORGIAN (0215-0230) 9615, 9530, 7170, 7120, 6105, 1260*; **(1400-1500)** 21520, 17855, 15415, 15195, 11760, 9670
LATVIAN (0300-0315) 9530, 7170, 7120, 6080; **(1400-1430)** 15270, 11960, 11865, 9735; **(1600-1630)** 21650, 15270, 11865
LITHUANIAN (0315-0330) 9530, 7170, 7120, 6080; **(1500-1530)** 15270, 11960, 11865, 9735; **(1700-1730)** 21630, 15280, 11865,
RUSSIAN (0200-0500) 11760, 9770, 7270, 7105, 6160, 6150, 6090, 6025, 6020, 1197*
   EAST ASIATIC USSR **(0800-1100)** 21625, 17865, 17740, 15430, 15410, 15325, 11965, 11930
   CENTRAL ASIATIC USSR (1200-1400) 25880, 21570, 21540, 21520, 21500, 17865, 17855, 15280, 15235, 15225, 15120, 11740
   EUROPEAN USSR (1500-2300) 15415, 15235, 11960, 11855, 11835, 11805, 11710, 9690, 9670, 9660, 9625, 9585, 9530, 7280, 7270, 6140, 6095
UKRAINIAN (0200-0400) 11850, 9760, 7190, 6180, 6125, 792; **(1600-1800)** 17855, 15415, 9660, 7245, 6150
UZBEK (0000-0100) 11945, 9690, 9615, 7270, 5985; **(1400-1500)** 25920, 25880, 21540, 21500, 17800, 15235, 15225

SOUTH ASIA

BENGALI (0130-0230) 21630, 17820, 17785; **(1600-1700)** 17850, 15185, 11965, 1575*
HINDI (0030-0100) 11810, 9635, 6020; **(1530-1630)** 21680, 17800, 15435, 11845, 9680, 7280
PASHTO (0130-0230) 11805, 9700, 9565, 6050; **(1630-1715)** 15435, 11740, 9680, 7280
URDU (0100-0130) 21630, 17830, 11810, 9635, 6020; **(1400-1500)** 21610, 15300, 11805, 9565

AFRICA

AMHARIC (1800-1830) 15135, 11740, 9620
FRENCH (0500-0630 Mon.-Sat.) 15240, 11890, 11875, 11850, 9565, 7265, 6180, 6020; **(1830-2230)** 21470, 17800, 17640, 15315, 15195, 9605, 7135
HAUSA (1600-1630) 25800, 17740, 15320, 11760
PORTUGUESE (1730-1830) 25800, 21610, 17740, 17705, 15330, 15320, 11715
SWAHILI (1630-1730) 25800, 21660, 21610, 17740, 17705, 15330, 15320

LATIN AMERICA

SPANISH (1130-1400) 21610, 21580, 21490, 17885, 17830, 15265, 15205, 15195, 11890, 9525, 1180*; **(0000-0300)** 17780, 17710, 15400, 15375, 11895, 9720, 9670, 7400, 6190, 1180*
PORTUGUESE (1000-1100) 21490, 17830, 15195, 11715; **(2300-2400)** 17775, 15600, 15240, 9670

EUROPE

ALBANIAN (1600-1630) 15195, 9735, 5965; **(1900-1930)** 15280, 9650, 5965
BULGARIAN (0400-0415) 9615, 9530, 7220, 7130, 6080, 3980, 1197*; **(0645-0700)** 11710, 9615, 7210, 6130; **(1830-1900)** 15245, 11915, 9735, 5965; **(2000-2030)** 15280, 9650, 5965
CZECHOSLOVAK (0500-0515) 11710, 9615, 9530, 7220, 6080, 3980, 1197*; **(0600-0615)** 11710, 9615, 7210, 6130; (1630-1700) 21630, 15280, 11865, 3980, 1197*; **(2000-2100)** 7180, 6150, 6060, 3980, 1197*
HUNGARIAN (0515-0530) 11710, 9615, 9530, 7220, 6080, 3980, 1197*; **(0615-0630)** 11710, 9615, 7210, 6130; **(1730-1900)** 15280, 11790, 7130, 6060, 3980, 1197*; **(2100-2130)** 7180, 6150, 6060, 3980
POLISH (0500-0530) 9605, 7130, 6160, 5955; **(0530-0545)** 11710, 9615, 9605, 9530, 7220, 7130, 6160, 6080, 5955, 3980, 1197*; **(0545-0630)** 9635, 9605, 7130, 6160, 5955, 1197*; **(0630-0645)** 11710, 9635, 9615, 7210, 7130, 6160, 6130, 5955, 1197*; (0645-0700) 9635, 7130, 6160, 5955, 1197*; **(1900-2000)** 15245, 11915, 7180, 6160, 6060, 3980, 1197*; **(2000-2130)** 15245, 11915, 11845, 6160; **(2100-2130)** 1197; **(2130-2200)** 15245, 11915, 11845, 6160, 6150, 6060, 3980, 1197; **(2200-2400)** 11845, 11915, 9770, 6160, 6150, 6060, 3980, 1197
PORTUGUESE (2130-2230) 15400, 9580, 7120, 6130,
ROMANIAN (0430-0445) 9615, 9530, 7220, 7130, 6080, 792*; **(1700-1800)** 15195, 9735, 5965, 792*; **(1930-2000)** 15280, 9650, 5965, 792*
SERBO-CROAT (0445-0500) 9615, 9530, 7220, 7130, 6080, 792*; **(2030-2130)** 15280, 9650, 5965, 792*
SLOVENE (0415-0430) 9615, 9530, 7220, 7130, 6080, 3980, 1197*; **(0545-0600)** 11710, 9615, 9530, 7220, 6080, 3980, 1197*

NORTH AFRICA

ARABIC (0500-0600) 11840, 9700, 6125, 6090; **(0730-0800)** 15245, 11875, 11775, 9580, 9565, 6180, 6150, 6020; **(1800-2200)** 17740, 15235, 6015

FAR EAST, SOUTHEAST ASIA, OCEANIA

BURMESE (1400-1500) 15250, 11930, 9630, 1575*
CHINESE (2200-0100) 11930, 9545, 7210, 6130; **(1000-1600)** 17765, 17740, 15410, 11965, 9555, 7285, 6185
INDONESIAN (2200-2330) 17780, 15155, 11805; **(1100-1200)** 15250, 11930, 9630; **(1400-1500)** 15105, 9730, 6030
KHMER (2200-2230) 11780, 9630, 7275, 6015, 1575*; **(1330-1400)** 15365, 11895, 9620, 1575*; **(1500-1530)** 15160, 9545, 6030
KOREAN (2130-2200) 17780, 15215, 11925, 9545, 6110; **(2330-2400)** 17780, 15155, 11805; **(1330-1400)** 15250, 9725, 6125, 6030
LAO (1200-1230 and 1500-1530) 15250, 11930, 9630, 9620, 1575*
THAI (2330-2400) 17810, 15215, 11780
VIETNAMESE (2230-2330) 11780, 9630, 7275; **(1230-1330 and 1530-1630)** 15250, 11930, 9630, 9620, 1143*

Fig. 5-1. (cont.)

PROGRAM SCHEDULES

EAST ASIA (GMT) FREQUENCY Asterisk (*) indicates medium wave

Sunday	Monday - Friday	Saturday

Northeast Asia **(2200-0100)** 11760, 15290, 17740, 17820; **(2200-2400)** 26000. *Southeast Asia* **(2200-0100)** 9770, 15185. *Indonesia* **(2200-0100)** 11760. *Oceania* **(2200-0100)** 11760, 15290, 17740.

Sunday	Monday - Friday	Saturday
2200 News	**2200 News**	**2200 News**
2210 VOA Morning	2210 Newsline	2210 VOA Morning
2230 Special English News & Features	2230 Special English News & Features	2230 Special English News & Features
2245 VOA Morning	2245 VOA Morning	2245 VOA Morning
2300 News	**2300 News**	**2300 News**
2310 VOA Morning	2310 Newsline	2310 VOA Morning
0000 News	2330 VOA Morning	**0000 News**
0010 VOA Morning	**0000 News**	0010 VOA Morning
0030 Special English News & Features	0010 Newsline	0030 Special English News & Features
0045 VOA Morning	0030 Special English News & Features	0045 VOA Morning
	0045 VOA Morning	

Northeast Asia **(1100-1330)** 11715; **(1100-1500)** 7230, 9760, 15425. *Southeast Asia* **(1100-1500)** 9760, 15160; **(1130-1200, 1230-1300)** 1575*. *Indonesia Only* **(1100-1400)** 6110. *Oceania* **(1100-1330)** 11715; **(1100-1400)** 6110.

Sunday	Monday - Friday	Saturday
1100 News	**1100 News**	**1100 News**
1110 New Horizons	1110 Newsline	1110 This Week
1130 Issues in the News	1130 Music USA	1130 Press Conference USA
1200 News	**1200 News**	**1200 News**
1210 Critic's Choice	1210 Focus	1210 American Viewpoints
1230 Special English News & Features	1230 Special English News & Features	1230 Special Engligh News & Features
1300 News	**1300 News**	**1300 News**
1310 International Viewpoints	1310 Newsline	1310 Weekend
1330 Studio One	1330 Magazine Show	**1400 News & Editorial**
1400 News & Editorial	**1400 News & Editorial**	1415 Music USA Jazz
1415 The Concert Hall	1415 Music USA Jazz	

MIDDLE EAST (GMT) FREQUENCY Asterisk (*) indicates medium wave

Sunday	Monday - Friday	Saturday

(0600-0800) 1260*, 5965, 7325, 15185; **(0300-0600)** 7200; **(0400-0800)** 11925; **(0400-0700)** 15205, **(500-0700)** 9770, **(0700-0800)** 9760.

Sunday	Monday - Friday	Saturday
0400 News	**0400 News & Newsline**	**0400 News**
0410 VOA Morning	0430 VOA Morning	0410 VOA Morning
0500 News	**0500 News & Newsline**	**0500 News**
0510 VOA Morning	0530 VOA Morning	0510 VOA Morning
0600 News	**0600 News & Newsline**	**0600 News**
0610 VOA Morning	0630 VOA Morning	0610 VOA Morning
0700 News	**0700 News & Newsline**	**0700 News**
0710 VOA Morning	0730 VOA Morning	0710 VOA Morning

(1500-1630,2100-2200) 1260*; **(1500-1700)** 15205; **(1500-1800)** 15260; **(1500-2200)** 9700; **(1700-1830)** 11760; **(1700-2000)** 9760; **(1700-2200)** 6040; **(2100-2200)** 7205.

Sunday	Monday - Friday	Saturday
1500 News	**1500 News**	**1500 News**
1510 New Horizons	1510 Newsline	1510 This Week
1530 Special English News & Features	1530 Special English News & Features	1530 Special English News & Features
1600 News	**1600 News**	**1600 News**
1610 International Viewpoints	1610 Focus	1610 American Viewpoints
1630 Music USA Standards	1630 Music USA	1630 Press Conference USA
1700 News	**1700 News**	**1700 News**
1710 Critic's Choice	1710 Newsline	1710 This Week
1730 Special English News & Features	1730 Special English News & Features	1730 Special English News & Features
1800 News	**1800 News**	**1800 News**
1810 Sunday Report	1810 Focus	1810 Weekend
1830 Issues in the News	1830 Music USA	**1900 News**
1900 News	**1900 News**	1910 American Viewpoints
1910 International Viewpoints	1910 Newsline	1930 Press Conference USA
1930 Music USA Standards	1930 Magazine Show	**2000 News & Editorial**
2000 News & Editorial	**2000 News & Editorial**	2015 Music USA Jazz
2015 The Concert Hall	2015 Music USA Jazz	**2100 News**
2100 News	**2100 News**	2110 Weekend
2110 New Horizons	2110 World Report	
2130 Studio One		

Shortwave Frequency/Wavelength Conversions

Meters 11 13 16 19 25 31 41 49 75

MHz 26 21 17 15 11 9 7 6 4

PROGRAM SCHEDULES

For program schedules in any of VOA's 42 languages write to the language service of your choice, VOA, Washington, D.C. 20547, U.S.A., or see post office box listings on page 19.

17

Muslim Countries

Here's another cultural area you can identify, even if different languages are used. Some words will always be the same. Look at the listings of the station IDs for Hindi and Urdu in Table 5-1. They're both the same. India, broadcasting in Hindi, says "Ye Akashvani hai" (Akashvani, the Hindi name for All India Radio, means "the voice from the heavens"). And Pakistan, in Urdu, says "Yeh Radio Pakistan hai." (By the way, since Pakistan is a former British colony, it pronounces "Radio" with the same first vowel as in English; not with the "continental" vowel found in French or in Asian countries like Iran which have never been British colonies.) But listen to what Radio Pakistan says just before it says "Yeh Radio Pakistan hai." It greets the listener with "As-salaamu aleikum." They call that Urdu, but it comes from the Arabic; it's the Muslim greeting "Peace be with you," and it is found in all the languages of the traditionally Muslim peoples. That includes Hausa in Nigeria and Somali, Fulani in Niger and Upper Volta, and even into the historically Muslim republics of the officially atheist Soviet Union. In fact, if you send enough reception reports, you can join the "Salaam Aleikom" (that's their spelling) Club of Radio Tashkent, the English-speaking station in the Soviet Central Asian republic of Uzbekistan. But Hindi, in mainly Hindu India, which generally sounds like Urdu, doesn't say "As-salasmu aleikum." It says "Na mastai."

Most of the Muslim languages take their technical and cultural vocabulary from the Arabic. So you will recognize the same words in many different languages. Often heard in Somali, for example, is the Arabic word for "program"—*barnamaj* (the stress is on the middle syllable). Broadcasts in Farsi from Iran, or in Dari from Afghanistan, often refer to the Soviet Union. Its name in those two countries is "Ittihad Shuravi" (stress on the final syllable of each word) and "Ittihad" is Arabic for "Union." Another word heard in far-flung Muslim languages is "Jumhuriya" meaning "republic." In Egypt, it's pronounced "Gumhuriya." In Farsi (Iran), it's "Jumhuri," while in Swahili (East Africa), "Jamhuri." In a country like Pakistan, which gets its advanced vocabulary from both English and Arabic, you can hear both "Jumhuriya" and "Republic." Also, all Muslim countries broadcast recitations from the Holy Qur'an in its original Arabic. There are no official translations of the Qur'an, which must be recited in the original. So, Radio Kaduna in Muslim northern Nigeria will announce the Qur'an chapter and verse in its local language, Hausa, and then the recitation itself will be in Arabic (but in a recognizably northern Nigerian style, which is easy to tell apart from the style used in, say, Saudi Arabia).

African Languages

We've already discussed the basic division of the Black African continent into British Commonwealth and Francophone (French-speaking) nations. To these we can add the Lusophone (Portuguese-speaking) countries of Angola, Mozambique, Cape Verde (now active on 3930 variable), Guinea Bissau, and São Tomé e Príncipe (which has recently reactivated 4807 kHz). And did you know that there's an official Spanish-speaking Black African country? It's Equatorial Guinea, with two active stations which broadcast about half the time in Castillian Spanish and the rest of the time in African languages; they're on 4924 and 6250 kHz.

Besides English and French, there's another two-way division of languages in Black Africa that cuts across the English/French divider. It's between Bantu and nonBantu languages. The Bantu family is the most important African language grouping, and it contains the single most important Black African language, Swahili. It's easy to recognize Swahili. Like most African languages, and especially most Bantu languages, it sounds rhythmic and melodious—not strident or warlike, but musical and very relaxed. In other words, it sounds quite like Italian among the European languages. It has just about the same vowels as Italian or Spanish, and, exactly as in Italian, every word ends in a vowel—*a, e, i, o,* or *u*. Swahili originated among the Muslims of the coast of what is now Kenya and Tanzania, and so it contains all the Arabic words mentioned above. The Swahili for "language" is *lugha*, from the Arabic, which contains a typically Arabic throaty *gh* (made deeper in the throat than any consonant in English). But all foreign words are Africanized so that they sound musical and contain an alternation of vowels and consonants—"Swahili" is *Kishwahili*, "English" is *Kiingerezi*, "German" is *Kijeremani*, and so on.

Swahili is different from most other Bantu languages in two ways. First, it contains many Arabic words; "book," for example, in Swahili is "kitabu," from the Arabic "kitab." Other Bantu languages didn't have the early cultural contacts with the Arabs, so their word for "book" usually comes from English, "buku." The second difference is that Swahili isn't a tone language. Most tribal Bantu languages—Lingala, Kinyarwanda, Zulu, Siswati and all the others—change the meaning of words as the tone goes up or down. (By the way, this is the principle of Africa's talking drums, which is why talking drum messages cannot be sent in Swahili, but only in the tribal tonal languages.) In other words, they are like Chinese. Think of English "mm-hmm" or "uh-huh," or however we can spell these grunts or inarticulate sounds we use. They go up and down, they can mean "yes" or "no," or they can express agreement, skepticism, or disagreement. If you hear such tones in an African language, it isn't Swahili. And if you hear an African language which has a lot of such ups and downs in tones (sometimes called *tone terracing*), and which doesn't have an even relaxed tone, with lots of harmonious word-final vowels, you are probably hearing one of the nonBantu languages of West Africa—the region stretching from Senegal to Cameroon. Those languages are among the world's most difficult languages to copy—much more difficult than Swahili—but, luckily, all stations in those countries broadcast also in English, French, or Portuguese, so you can wait for them to switch to those languages.

Fig. 5-2 shows a sample of Amharic, the official language of Ethiopia and the traditional language of the Ethiopian Orthodox Church. Fig. 5-2 is taken from the monthly program schedule of Vatican Radio. Identification is "Yih Radio

Vaticana tabia new'' and you can try for it at 1500–1515 UTC on 17730 (best in North America), 15120, and 11810 kHz.

Fig. 5-2. Sample pages from the Vatican Radio program schedule showing examples of the Amharic language (Courtesy Vatican Radio).

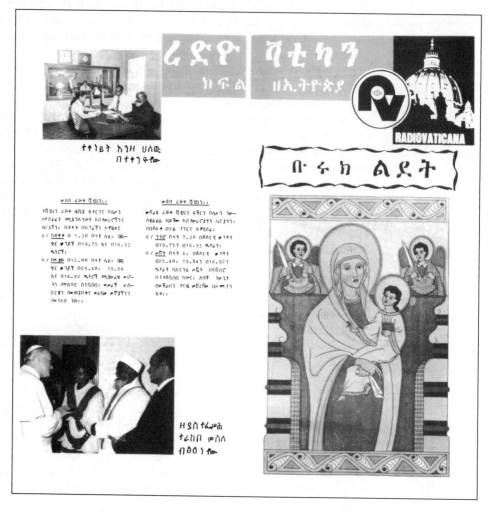

Asian Languages

All languages in a broad cultural area tend to influence each other and come to sound a little bit alike. Something about the quality of the voice—the intonation and rhythm—sounds similar, and languages borrow from each other even if, historically, they belong to different families and different cultural traditions. Thus, the Far East has a group of languages that bear a vague resemblance to each other, even if the details are very different. It's only in this vague superficial way that Chinese, Japanese, Korean, Tagalog (in the Philippines), and others in East Asia sound vaguely similar.

When we consider specifics, the first division we can make in the Far East is between tone languages and nontonal languages—the same division we made between Swahili (nontonal) and the other Bantu languages of eastern and southern Africa (tonal). Nontonal languages are on the island and peninsular fringe of Asia; tone languages are in China and countries south and southwest of it. And, though they might still be difficult, nontonal languages are easier for an English speaker to learn or, at least, recognize. Why? Because English isn't a tone language, with the marginal exception of the "mm-hmm" and "uh-huh" sounds mentioned earlier.

The division between tonal and nontonal languages also corresponds to a division between monosyllabic and polysyllabic. Again, polysyllabic is easier for an English speaker to handle, because English is, well, a polysyllabic language—words can consist of more than one syllable. Classical Chinese is the classic monosyllabic language; every syllable can stand alone and has a meaning by itself. This isn't true in English, although English has actually moved closer to being a monosyllabic language than any other European language. It's easy to make up simple sentences in English where every word has one syllable, but it is almost impossible in, say, Spanish or Italian. Modern Chinese has compounds, and when you hear a Chinese announcer identify a station as "Zhongyang renmin guangbo diantai," or "This is the Central People's Broadcasting Station" (in Beijing), you may not be able to tell that it is a monosyllabic language. But listen to the distinct tone on every syllable. If the tone changes, the meaning changes. This doesn't happen in Korean, Japanese, Tagalog, or Malay/Indonesian. The Japanese identification begins "Kochirawa..." and ends "...desu." In fact, most Japanese sentences end with "...desu," which means something like "is" or the English verb "to be." The same thing in Korean. Identification begins "Yeogineun" and the great majority of sentences end in "...nida" which corresponds to Japanese "...desu." You can't break these words up into separately meaningful syllables, and the tone is insignificant. So..., do you still think you can't tell the difference between Chinese and Japanese? With these hints, and some practice in listening, you should be able to.

Indonesian

Let's go down to Southeast Asia. There we reach the paradise of the DX'ers, Indonesia. That sprawling republic, a former Dutch colony, is neck and neck with Brazil for the number of operational shortwave broadcasting stations. A year or two ago, there was a scare, a rumor, that all the stations would be transferred to the medium waves (the standard AM broadcast band), except for Suara Indonesia, the Voice of Indonesia, the external service from Jakarta, but the fact is that only the many commercial stations are supposed to be confined to the MW band, and some smaller official stations are also moving there. Meanwhile, the number of shortwave stations, operated by Radio Republik Indonesia, and by the regional and local administrations, and by the armed forces, continues to grow. Indonesian is an important language for the DX'er to recognize and, if possible, to write reception reports in—with the aid of the report forms or report models produced by the Australian Radio DX Club and other clubs. And we're lucky; Indonesian is the easiest Asian language for Easterners, perhaps the easiest nonWestern language of all. That's the reason for its great success in spreading as the official language (and daily spoken language) of the more than 10,000 islands where, until recently, it had quite a few native speakers—a success story which is the envy of governments in countries like India and the Philippines, where a stiff resistance has been met in spreading their designated national language—from speakers of other languages who are proud of their own tongues.

So, let's look more closely at Indonesian. It isn't a tone language, and it isn't monosyllabic. It's wide open to cultural influences from the waves of invaders, missionaries, and other cultures that have swept through Indonesia in the past and been absorbed by it. Thus, we find hundreds of Arabic words in Indonesian —the greeting "Salamat" (health), often found paired with a word for the time of day, e.g., "Selamat pagi" or "good day," "Selamat sore" or "good evening." Watch for these greetings at the sign-on and sign-off times of Indonesian stations. Another Arab-derived word often heard in station IDs is "Khusus" meaning "special." Next, let's examine the influence of Sanskrit and the other languages of India. This gave Indonesian words like "guru" (a teacher, e.g., a schoolteacher). The first Europeans to reach the East were the Portuguese. They left words like "sekolah" for "school," "gereja" (church), and "sapatu" (shoe). Note also the Indonesian word for "two"—dua, which resembles the Portuguese "duas" (feminine; the masculine is "dois"), but this may be a coincidence. Then came the Dutch, who gave works such as "Republik" and "listrik" (electricity).

These European-derived words are fairly easy to recognize. The whole tone and sound pattern of Indonesian has been influenced by the Dutch, so that if you hear an Asian language which still sounds rather Western, or, specifically, rather Dutch, it's Indonesian you're hearing. Probably the easiest guide of all is to listen for the reduplicative plurals of nouns. In Indonesian you don't add s as you usually do in English, you say the noun twice. So, "wavelengths" is "gelombang-gelombang." This is quite long and clear and easy to recognize. These words used to be written once, followed by the number "2"; thus, the old spelling was "gelombang2," which was neat and creative. Now, they are written twice, which seems wasteful of both time and energy, a poor policy for a developing country. "Sir" or "Mr." is "Saudara," and "listeners" are addressed as "Saudara-Saudara penengar." There are many music-request shows on RRI regional stations, so "Saudara" will be heard before the names of the listeners; women are addressed as "Nyonya" (Miss). Watch for the prefixes "ber-," "per-," "ke-," and others.

Iberian Languages
Spanish

The Spanish or Castillian language is official in 19 Latin American republics, Puerto Rico, Spain, and Equatorial Guinea (Africa). It is also one of three official languages in the Philippines, but it is in decline there and is never heard in shortwave broadcasts (although the indigenous Filipino languages have borrowed thousands of Spanish words, including the Spanish numerals).

For listeners in North America, Spanish is second only to English in its importance on the radio dial, and, on the tropical bands, it may be more important than English, since the Latin American stations are closer and are audible for longer periods, while the African stations where English is strongest (along with French) are further away and audible mainly in two peak periods—the North American late afternoon, and the period around North American midnight (dawn in Africa).

Two pages of the "Letterbox" feature printed in Vatican Radio's monthly bulletin (sent free to listeners) are given in Fig. 5-3 and show six languages. Can you identify them? From top left, they are: German, Polish, Serbo-Croatian (Yugoslavia), Portuguese, Spanish, and Esperanto. Could you recognize them if you heard them?

Table 5-1 shows some typical ID announcements from Spanish-speaking stations, and they are easy to follow. It might be useful to learn the Spanish alphabet, as call letters, which are in deep decline in other parts of the world, e.g., Europe and Africa, are still widely used in most parts of Latin America, especially where there is a multiplicity of commercial broadcasters—which means all the Spanish-speaking countries except Cuba and Nicaragua. Even in Nicaragua, the Sandinista régime has recently shown its appreciation of the importance of call letters by "revolutionizing" that aspect, too, of Nicaraguan life, by changing the prefix of every station in the country from the old *YN-* to the new *HT-*, and adopting a new pattern of one subsequent letter (indicating the province) and three following numerals (specifying the frequency). Spanish letters are mostly similar to their English names, but be careful with *h* ("ache"), *w* ("double u"; in some countries "doble v"), *j* ("jota", like English "hota"), *z* ("zeta"), *x* ("equis," the former name of a station in Managua, Nicaragua: Estación Equis, "Station X", the personal station of the Somoza family, which is now Radio Sandino), and *y* ("i griega," but in Venezuela, etc., "ye").

Spanish as spoken by professional announcers is pretty standardized and unified throughout the Spanish-speaking world. But there are some regional differences which exist even among top professionals. In Spain, *z* (in all positions) and *c* (before the vowels *e* and *i*) is pronounced like English *th* in "thin." This "castizo" (pure, Castillian) pronunciation is simply never heard from a Latin American speaker. Also, in Spain, the *s* is pronounced higher up toward the palate, toward the roof of the mouth. It produces a narrow hissing sound similar to caricatures of a toothless person speaking in English. All broadcasts from Spain have this distinctive Castillian accent, except for some from Radio Exterior de Espana beamed to Latin America, which seem sometimes to use Latin American announcers, perhaps in Madrid for training or on an exchange. The Spanish of Spain sounds brisk and crisp, not relaxed, and very different from both Portuguese and Italian.

Argentina, in the nineteenth century, received many immigrants from Italy, as well as from parts of Spain, such as Galicia on the Portuguese border, far

Fig. 5-3. Two
pages of the
Vatican Radio
monthly bulletin
showing
examples of six
of the
languages used
(Courtesy
Vatican Radio).

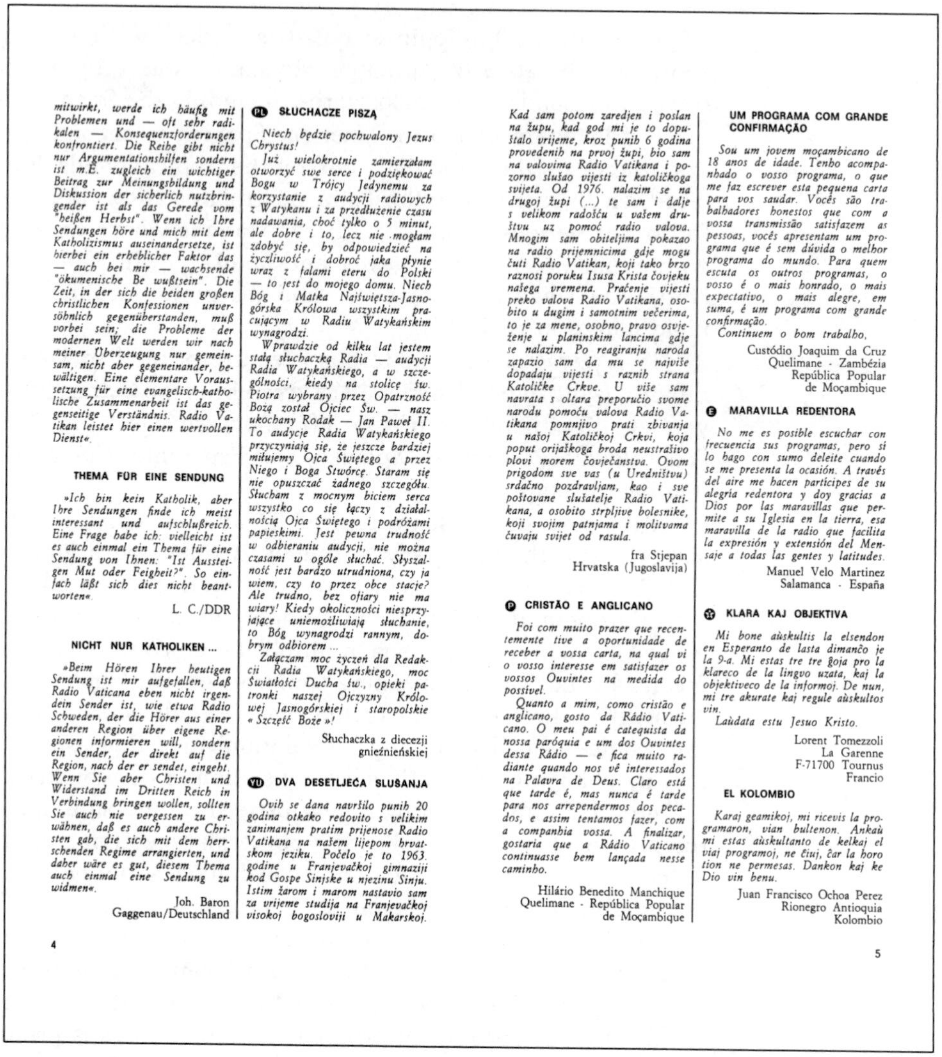

mitwirkt, werde ich häufig mit Problemen und — oft sehr radikalen — Konsequenzforderungen konfrontiert. Die Reihe gibt nicht nur Argumentationshilfen sondern ist m.E. zugleich ein wichtiger Beitrag zur Meinungsbildung und Diskussion der sicherlich nutzbringender sein; die Probleme der "heißen Herbst". Wenn ich Ihre Sendungen höre und mich mit dem Katholizismus auseinandersetze, ist hierbei ein erheblicher Faktor das — auch bei mir — wachsende "ökumenische Be wußtsein". Die Zeit, in der sich die beiden großen christlichen Konfessionen unversöhnlich gegenüberstanden, muß vorbei sein; die Probleme der modernen Welt werden wir nach meiner Überzeugung nur gemeinsam, nicht aber gegeneinander, bewältigen. Eine elementare Voraussetzung für eine evangelisch-katholische Zusammenarbeit ist das gegenseitige Verständnis. Radio Vatikan leistet hier einen wertvollen Dienst.«

THEMA FÜR EINE SENDUNG

»Ich bin kein Katholik, aber Ihre Sendungen finde ich meist interessant und aufschlußreich. Eine Frage habe ich: vielleicht ist es auch einmal ein Thema für eine Sendung von Ihnen: "Ist Aussteigen Mut oder Feigheit?". So einfach läßt sich dies nicht beantworten«.

L. C./DDR

NICHT NUR KATHOLIKEN ...

»Beim Hören Ihrer heutigen Sendung ist mir aufgefallen, daß Radio Vaticana eben nicht irgendein Sender ist; wie etwa Radio Schweden, der die Hörer aus einer anderen Region über eigene Regionen informieren will, sondern ein Sender, der direkt auf die Region, nach der er sendet, eingeht. Wenn Sie aber Christen und Widerstand im Dritten Reich in Verbindung bringen wollen, sollten Sie auch mit vergessen zu erwähnen, daß es auch andere Christen gab, die sich mit dem herrschenden Regime arrangierten, und daher wäre es gut, diesem Thema auch einmal eine Sendung zu widmen«.

Joh. Baron
Gaggenau/Deutschland

SŁUCHACZE PISZĄ

Niech będzie pochwalony Jezus Chrystus!

Już wielokrotnie zamierzałam otworzyć swe serce i podziękować Bogu w Trójcy Jedynemu za korzystanie z audycji radiowych w Watykanu i za przedłużenie czasu nadawania, choć tylko o 5 minut, ale dobre i to, lecz nie mogłam zdobyć się, by odpowiedzieć na życzliwość i dobroć jaka płynie wraz z falami eteru do Polski — to jest do mojego domu. Niech Bóg i Matka Najświętsza-Jasnogórska Królowa wszystkim pracującym w Radiu Watykańskim wynagrodzi.

Wprawdzie od kilku lat jestem stałą słuchaczką Radia — audycji Radia Watykańskiego, a w szczególności, kiedy na stolicę św. Piotra wybrany przez Opatrzność Bożą został Ojciec Św. — nasz ukochany Rodak — Jan Paweł II. To audycje Radia Watykańskiego przyczyniają się, że jeszcze bardziej miłujemy Ojca Świętego a przez Niego i Boga Stwórcę. Staram się nie opuszczać żadnego szczegółu. Słucham z mocnym biciem serca wszystkiego co się łączy z działalnością Ojca Świętego i podróżami papieskimi. Jest pewna trudność w odbieraniu audycji, nie można czasami w ogóle słuchać. Słyszałność jest bardzo utrudniona, czy ja wiem, czy to przez obce stacje? Ale trudno, bez ofiary nie ma wiary! Kiedy okoliczności niesprzyjające uniemożliwiają słuchanie, to Bóg wynagrodzi rannym, dobrym odbiorem ...

Załączam moc życzeń dla Redakcji Radia Watykańskiego, moc Światłości Ducha św., opieki patronki naszej Ojczyzny Królowej Jasnogórskiej i staropolskie « Szczęść Boże »!

Słuchaczka z diecezji gnieźnieńskiej

DVA DESETLJEĆA SLUŠANJA

Ovih se dana navršilo punih 20 godina otkako redovito s velikim zanimanjem pratim prijenose Radio Vatikana na našem lijepom hrvatskom jeziku. Počelo je to 1963. godine u Franjevačkoj gimnaziji kod Gospe Sinjske u njezinu Sinju. Istim žarom i marom nastavio sam za vrijeme studija na Franjevačkoj visokoj bogosloviji u Makarskoj.

Kad sam potom zaređen i poslan na župu, kad god mi je to dopuštalo vrijeme, kroz punih 6 godina provedenih na prvoj župi, bio sam na valovima Radio Vatikana i pozorno slušao vijesti iz katoličkoga svijeta. Od 1976. nalazim se na drugoj župi (...) te sam i dalje s velikom radošću u vašem društvu uz pomoć radio valova. Mnogim sam obiteljima pokazao na radio prijemnicima gdje mogu čuti Radio Vatikan, koji tako brzo raznosi poruku Isusa Krista čovjeku našega vremena. Praćenje vijesti preko valova Radio Vatikana, osobito u dugim i samotnim večerima, to je za mene, osobno, pravo osvježenje u planinskim lancima gdje se nalazim. Po reagiranju naroda zapazio sam da mu se najviše dopadaju vijesti s raznih strana Katoličke Crkve. U više sam navrata s oltara preporučio svome narodu pozorno valova Radio Vatikana pomnijšo prati zbivanja u našoj Katoličkoj Crkvi, koja poput orijaškega broda neustrašivo plovi morem čovječanstva. Ovom prigodom sve vas (u Uredništvu) srdačno pozdravljam, kao i sve poštovane slušatelje Radio Vatikana, a osobito strpljive bolesnike, koji svojim patnjama i molitvama čuvaju svijet od rasula.

fra Stjepan
Hrvatska (Jugoslavija)

CRISTÃO E ANGLICANO

Foi com muito prazer que recentemente tive a oportunidade de receber a vossa carta, na qual vi o vosso interesse em satisfazer os vossos Ouvintes na medida do possível.

Quanto a mim, como cristão e anglicano, gosto da Rádio Vaticano. O meu pai é catequista da nossa paróquia e um dos Ouvintes dessa Rádio — e fica muito radiante quando nos vê interessados na Palavra de Deus. Claro está que tarde é, mas nunca é tarde para nos arrependermos dos pecados, e assim tentamos fazer, com a companhia vossa. A finalizar, gostaria que a Rádio Vaticano continuasse bem lançada nesse caminho.

Hilário Benedito Manchique
Quelimane · República Popular
de Moçambique

UM PROGRAMA COM GRANDE CONFIRMAÇÃO

Sou um jovem moçambicano de 18 anos de idade. Tenho acompanhado o vosso programa, o que me faz escrever esta pequena carta para vos saudar. Vocês são trabalhadores honestos que com a vossa transmissão satisfazem as pessoas, vocês apresentam um programa que é sem dúvida o melhor programa do mundo. Para quem escuta os outros programas, o vosso é o mais honrado, o mais expectativo, o mais alegre, em suma, é um programa com grande confirmação.
Continuem o bom trabalho.

Custódio Joaquim da Cruz
Quelimane · Zambézia
República Popular
de Moçambique

MARAVILLA REDENTORA

No me es posible escuchar con frecuencia sus programas, pero sí lo hago con sumo deleite cuando se me presenta la ocasión. A través del aire me hacen partícipes de su alegría redentora y doy gracias a Dios por las maravillas que permite a su Iglesia en la tierra, esa maravilla de la radio que facilita la expresión y extensión del Mensaje a todas las gentes y latitudes.

Manuel Velo Martínez
Salamanca · España

KLARA KAJ OBJEKTIVA

Mi bone aŭskultis la elsendon en Esperanto de stara dimanĉo je la 9-a. Mi estas tre tre ĝoja por la klareco de la lingvo uzata, kaj la objektiveco de la informoj. De nun, mi tre akurate kaj regule aŭskultos vin.

Laŭdata estu Jesuo Kristo.

Lorent Tomezzoli
La Garenne
F-71700 Tournus
Francio

EL KOLOMBIO

Karaj geamikoj, mi ricevis la programaron, vian bultenon. Ankaŭ mi estas aŭskultanto de kelkaj el viaj programoj, ne ĉiuj, ĉar la horo tion ne permesas. Dankon kaj ke Dio vin benu.

Juan Francisco Ochoa Perez
Rionegro Antioquia
Kolombio

4   5

---

from Madrid. As a result, its accent contains some features which distinguish it both from Castillian and from what might be called general Latin American Spanish. There is more relaxed, slightly "sing-song" tone, with some lengthening of stressed vowels, which places Argentine Spanish a little closer to Italian. But this kind of Spanish is more likely to be heard from sportscasters and, especially, in interviews with nonprofessional speakers; not so much in newscasts. One Argentine, or general "Southern Cone," feature which is often heard in newscasts, is the pronunciation of Spanish *y* as *zh*, the sound of the last consonant in English "garage" or the first one in "azure." This sound is also heard in words spelled in Spanish with an *ll*, so it would begin the first and third words in the phrase "yo me llamo" (I am called).

Finally, the Spanish of the Caribbean—including Cuba, Puerto Rico, Venezuela, Panama, and the coast (but not the interior) of Colombia—has a tendency to drop the -s at the end of the words. This makes it very difficult for, say, English speakers to tell the plural of nouns. This is definitely a nonstandard feature, and professional radio announcers avoid it. So listen to it mostly in sportscasts and interviews. It is more likely to be heard from a country like Cuba, where class barriers have been upset, than from a more traditional Caribbean society like the Dominican Republic. Spanish "señoras y señores" (ladies and gentlemen) would, in this pronunciation, become something like "señora y señore."

Portuguese

Anyone who says he can't tell Spanish from Portuguese is probably a real beginner in the SWL'ing hobby. The two don't sound very much alike at all. In fact, in its overall sound pattern, Portuguese has been compared to Russian, not to Spanish. DX'ers have reported being fooled by a weak signal on a frequency like 4825 or 4780 kHz—they thought it was a Brazilian or an Angolan speaking in Portuguese, but it turned out to be a Soviet regional speaking in Russian! Spanish and Portuguese are similar in their origin and in the spelling of many words, but the sound patterns are very different. For example, Spanish has only five vowels (the same as Swahili), and they are neither short nor long. Also, they aren't nasal (made through the nose). That's really the simplest and best system of identification, and it's part of the reason why Spanish has remained remarkably unified wherever it is spoken in 23 countries. We might note in passing that the international language, Esperanto, adopted exactly the same vowel system as Spanish—and, yet, if you hear Esperanto over Radio Polonia or, maybe, Radio Beijing or SRI Switzerland, you aren't likely to mistake it for Spanish, because of the Slavic and Germanic influences that make Esperanto sound nonHispanic.

Getting back to Portuguese, we find a language rich in linguistic subtleties and hard to pin down or describe—just the opposite of Spanish. In Spanish, whether or not a vowel is under stress (accent), it sounds pretty much the same. Not in Portuguese; its quality changes. For example, in Spanish "kilómetro" (kilometer), the vowel is pretty much the same, whether in the accented syllable (ó) or the unaccented one (o). In Portuguese "quilómetro," the first ó is something like the English "aw," but the second, unstressed o, is like the English "oo." So, the sounds of Portuguese depend upon where they are in the word and what comes next to them. In Spanish, every vowel is clear and precise; even if spoken fast, the language sounds distinct and the station identification is usually easy. Portuguese is a sweetly flowing language. The listener gets a general impression of it, but is hard pressed to fish out the details. It has been described as "Spanish spoken with marbles in the mouth," or "Spanish spoken by a Russian," or "halfway between Spanish and French." It's a subtle language because it has countervailing patterns fighting each other. Like Spanish or English, it has work-stress; i.e., one syllable in each word carries more stress than the other syllables. But it also has sentence-stress, and that means that the syllable which carries the stress in a word may not carry it in the sentence. In fact, the stressed syllable in a sentence may be one which is unstressed in a word. Thus, Portuguese, especially that of Portugal and Africa, is tantilizingly hard to grasp.

Luckily for the North American DX'er, the main source of Portuguese-speaking DX stations is in Brazil, and Brazilian Portuguese is a little easier to follow. Some people claim it is closer to Spanish, but that isn't really true, except in a few points, and only if one is thinking of the Spanish of the Southern Cone. The rhythm of Brazilian Portuguese seems to have been influenced by the tone languages of West Africa and it also resembles that of the Brazilian samba. An English speaker might call it "sing-song" and see a resemblance to Swedish and Norwegian—the closest things to tone languages found in Europe. One thing that makes Brazilian Portuguese easier to follow is that unstressed vowels, which are often completely lost in Lisbon Portuguese, are still heard in Brazilian —though they are not what the spelling might make you think they are. Take "noite" meaning "night." In Portugal, Angola, and Mozambique, it's a single syllable; in English, we might write it "noyt." In Brazil, the t is affected by the following vowel (compare English "nation" or "consortium"), and the e becomes like the English -ee, so it sounds like "noy-chee." In this way, it's more like the Spanish noche, so claims of a similarity between Brazilian and Spanish are true on this point. The kind of consonant change found when Brazilians pronounce "noite" is also found with d when it comes before the vowels e and i; so, in Brazilian Portuguese, "rádio" comes out like "hah-djioo." In Brazilian, but not Lisbon Portuguese, r at the beginning of a word has passed through a French-type throaty (uvular, as linguists call it) r to a sound similar to the English h. Now you see why Portuguese is considered "difficult" and why some DX'ers avoid Brazilian stations, even though they will certainly provide more different catches and different QSL verifications than any other country—even the word "radio" is hard to recognize at first!

Let's conclude our summary of Portuguese by saying that it's like English in many ways, but not in ways that make it easier for an English-speaking person to understand! How many vowels does Spanish have? Easy…, it has five! How many vowel sounds does English have? Well, ah, hmm…. Linguists don't agree, and it varies with dialect and region. Maybe thirteen, perhaps fourteen, maybe more, maybe less. Frustrating! Well, Portuguese is the same way; it has more than a dozen vowels, and the long vowels have different qualities from the short ones, and the stressed from the unstressed. And, unlike English, Portuguese has lots of nasal vowels. Still, Portuguese is an important language; it is the official language in countries of four continents and is spoken by well over twice as many native speakers as French. Above all, Brazilians wish their linguistic identity to be recognized and don't take kindly to being mistaken for Spanish speakers.

French

Most of us are familiar with French. Like Portuguese, it has many nasal sounds, and its sound system is a subtle complex one, quite different from that of any other language. Most European languages have words with one syllable stressed, the rest unstressed. French isn't like that. It has sentence stress—the whole sentence has a pattern, but not the individual words. (As we saw earlier for Portuguese, that Iberian language has both patterns, crisscrossing each other.) To the English speaker, this makes French words sound as if they are stressed on the last syllable, but this isn't technically true. The spelling system of French is very complex and involves many silent letters—some of which are actually pronounced at times, through the rules of liaison (linking one word with the following word in a sentence). With Spanish (or Indonesian), it's possible to copy words which one doesn't know very well, or may never have heard before. In French, this is impossible. You have to know French very well before you can try to take diction; i.e., copy a radio signal in it.

More so than English, Spanish, or Portuguese, French is a highly standardized, centralized language. The speech of the educated residents of Paris is the standard everywhere, and the French used by, for example, newscasters on radio stations in Francophone Africa is much closer to the European standard. These speakers are really "black Frenchmen." For example, as I type these lines, Radiodiffusion Nationale du Tchad (in the Chadian capital, N'djamena) has just come back on the air on 4904.5 kHz, after being knocked off the air a year ago in that country's civil war (which included Libyan intervention). On its very first day observed back on the air, Radio Tchad was using excellent Parisian French. The country is in chaos, wracked by civil war and drought, partially occupied and annexed by Libya, but the linguistic standards have not slipped; the two Chadian announcers heard on the 1900 UTC (Coordinated Universal Time) newscast could easily have been hired on a French domestic station in Paris. (A Nigerian or Ghanaian announcer, speaking in English, would have no chance of being considered for use on the domestic services of the BBC in London.)

So, a distinctively African French will more likely be heard on nonnews programs, such as sports and interviews, folklore shows, drama, etc. There are only two important points about such African French, and they are the same as those for Canadian French—the nasal vowels sound different. In particular, the vowels normally written *on* and *en* move toward the third nasal vowel, *in*. And, the *r* is pronounced with a trilling of the tip of the tongue (as in Spanish, or in Scottish English), not with the *uvula*, the little flap of tissue at the back of the mouth which sounds like gargling. In Canadian French, the dental consonants undergo a shift similar to the one described earlier for Portuguese. Before *e* and *i*, the consonants *d* and *t* are changed to *dz* and *ts*; thus, "dis" (say, tell me) sounds more like "dzee" and "tigre" (tiger) sounds like "tsigre." Finally, in Canada, the very frequent French vowel *i*, which in Paris (and Africa) sounds like the English *ee*, takes on the same sound as the English short *i*. So, *Afrique*, which sounds in Parisian French like the English "a freak," sounds in Québec French like "a frick." On Radio Canada International (on shortwave), these traditional Québécois pronunciations can be heard mostly on interviews, especially on the excellent RCI DX program "Allô DX", when Québec DX'ers, radio engineers, and others who are not professional announcers, educators, or actors are interviewed. And AM and FM DX'ers can often catch Québec French, especially from the small rural stations; on the big-city stations, something closer to the Parisian standard dominates.

Russian

Let's close the discussion with Russian, a Slavic language. It isn't a very important language for DX'ers, as most stations that broadcast in Russian also use other languages, and many of them accept reception reports in English. But there are some American DX'ers who specialize in Soviet regional DX and who write reception reports in Russian, either using a report form or writing in longhand. One problem with this is that Soviet official broadcasters are suspicious of those who write in Russian or who listen to Soviet domestic programming. Some refuse to verify domestic program reception and will admit only that the external services can be heard.

Anyway, let's look briefly at it. It's written in the Cyrillic alphabet, so we'll just use a transliteration here. Transliteration is normally done by the Library of Congress system, but that doesn't give a very realistic idea of the sound of the language. Like English spelling, Russian spelling is rather old-fashioned and doesn't reflect changes in the sound of words which have taken place over the centuries. For example, the genitive ending of most masculine and neuter adjectives is, in the Cyrillic original and the Library of Congress translation, *-ogo*. But the real sound is *-ovo*. The same with the usual word that introduces station IDs: "Govorit." That's the spelling. But, in Russian (as in English, or Portuguese, but not Spanish), when *o* isn't stressed, its quality is reduced. It becomes like *a* (or *uh*), so the word sounds more like "gava-REET." It means "speaking," as in "Govorit Moskva" (Moscow speaking), and, again, since "Moskva" has stress on the last vowel, it sounds like "mask-VA."

Russian is a pleasant, melodious language that is easy to recognize. Nothing sounds much like it, except the Slavic languages genetically closest to it, especially Ukrainian. The more distant Slavic languages, like Polish, Bulgarian, and Czech, have different sound patterns, though they can still be identified as Slavic through the very common adjective endings of *-ski* (*-sky*), *-ska* (*-skaya*), *-sko*, and by words which all or most Slavic languages have, like *rabota* (work), *slovo* (word), *Bog* (God), and so on. Russian words often heard in broadcasts include *gazeta* (newspaper), *Pravda* ("truth," the name of a newspaper), *da* (yes), *net* (pronounced "nyet," for "no"), and *tovarishch* (comrade). Many stations sign on with the phrase "Dobroe utro, tovarishchi," or "Good morning, comrades."

Radio Sweden International, a typical medium-sized international broadcaster, uses the seven languages shown in Fig. 5-4. Its broadcasts in Russian ("po-russki") are not subject to jamming—Radio Liberty, Deutsche Welle, and other broadcasters are not so lucky.

We hope this brief introduction to the endlessly fascinating subject of language recognition in shortwave radio will get you started developing skills in the field. As Larry Magne never fails to say when he finishes his incomparable equipment reviews on Radio Canada International's "SWL Digest,"—"Bonne écoute", good listening!

Fig. 5-4. An example of Radio Sweden International's programming schedule (Courtesy Radio Sweden International).

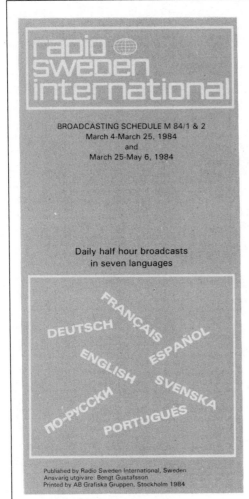

radio ⊕ sweden international

BROADCASTING SCHEDULE M 84/1 & 2
March 4-March 25, 1984
and
March 25-May 6, 1984

Daily half hour broadcasts
in seven languages

FRANÇAIS
DEUTSCH
ESPAÑOL
ENGLISH
SVENSKA
ПО-РУССКИ
PORTUGUÊS

Published by Radio Sweden International, Sweden
Ansvarig utgivare: Bengt Gustafsson
Printed by AB Grafiska Gruppen, Stockholm 1984

Schedule March 4 - March 26, 1984

Time UTC GMT	Frequency kHz	Wavelength meters	Primary target (Also heard elsewhere)	Beam direction	Transmitter
SVENSKA					
0100	11705	25,6	Latin America	235°	H2
''	9695	30,9	Latin America	275°	H1
0200	9695	30,9	North America	320°	H1
0330—0830[1]	6065	49,5	Europe, Africa	210°	K
0330—0430[1]	9605	31,2	Europe	180°	H1
0430—0830[1]	15390	19,5	Africa	180°	H1
1000	9630	31,1	Europe, Africa	210°	K
''	17820	16,8	Aust. N. Zealand	70°	H2
''	21690	13,8	Middle East	145°	H1
1130 P1	9630	31,2	Europe, Africa	210°	K
''	21610	13,9	Middle East	145°	H2
1300	15190	19,7	East Asia	55°	H1
''	21690	13,8	Europe, Africa	210°	K
1430	17860	16,8	North America	305°	H1
''	17850	16,8	South Asia	85°	K
1700 P1	1179	254	Europe	non dir	H3
'' P1	6065	49,5	Europe, Africa	210°	K
'' P1	17710	16,9	North America	305°	H2
1900	1179	254	Europe	non dir	H3
''	6065	49,5	Europe, Africa	210°	K
''	15240	19,7	Africa	165°	H1
2130	1179	254	Europe	non dir	H3
''	6065	49,5	Europe, Africa	210°	H1
2330	1179	254	Europe	non dir	H3
''	11705	25,6	Latin America	275°	H2
''	9695	30,9	North America	290°	H1
ENGLISH					
0230	9695	30,9	North America	320°	H1
''	11705	25,6	North America	320°	H2
1100	9630	31,1	Europe, Africa	210°	K
''	17860	16,8	Aust. N. Zealand	70°	H1
1230	15190	19,7	East Asia	55°	H1
''	21690	13,8	Africa	180°	K
1400	17850	16,8	South Asia	85°	K
''	17860	16,8	North America	305°	H1
1600	6065	49,5	Europe, Africa	210°	K
''	15330	19,6	South Asia	85°	H1
1830	1179	254	Europe	non dir	H3
''	6065	49,5	Europe, Africa	210°	K
''	15240	19,7	Africa	165°	H1
2100	1179	254	Europe	non dir	H3
''	11845	25,3	Africa	180°	H1
''	11955	25,1	Middle East	145°	K
2300	1179	254	Europe	non dir	H3
''	9695	30,9	North America	290°	H1
''	11710	25,6	North America	290°	H2
PORTUGUÊS					
0030	9695	30,9	Latin America	235°	H1
''	11705	25,6	Latin America	235°	H2
0200	11705	25,6	Latin America	235°	H2
1730	1179	254	Europe	non dir	H3
''	6065	49,5	Europe, Africa	235°	K
''	15240	19,7	Africa	165°	H1
2200	11705	25,6	Europe, Africa	210°	K

[1] Riksprogrammet/Swedish Home Service, 0330-0830
vard. P3 0330-0430; P1 0430-0830,
lörd. P3 0330-0525, 0715-0830; P1 0525-0715,
sönd. P3 0330-0630; P1 0630-0830

PRINTED MATERIALS FOR THE DX'ER

John C. Herkimer

Shake hands with **John Herkimer** and you will experience a case of "instant like." John is a printer by profession, supervising the design and composition department of a large printing firm in his home town of Caledonia, New York. He has turned his graphic design and printing abilities to advantage in his radio activities as well.

"Herk" began DX'ing in 1968 and, like so many other shortwave broadcast DX'ers, soon developed a special fondness for the *musica folk-lorica* of the Andes. A former editor of the "Shortwave Center" section in *Frendx*, the bulletin of the North America Shortwave Association, John still serves on the NASWA Executive Council.

John and his wife, Mary Ann, have two children.

If a budding DX'er stays in the hobby long enough—and takes it seriously—there will be a time when the need or desire for printed personalized items will arise. The incentive will often come from seeing a fellow hobbyist's letterhead, report form, prepared card, or customized report insert. For others, it is the realization that successful QSL'ing is often a case of "attention getting" and special-looking reports may provide the needed edge. And, for the DX'er whose recordkeeping and logging methods are haphazard and inefficient, and an investment in a home computer is a long way off, an organized system utilizing printed aids can streamline that failing approach.

If you have never considered investing in printing, you're not alone. For most persons, the idea of spending money (and time) on items such as letterheads is superfluous; function should supercede appearance. But consider this: the past few years have seen an increased awareness by businesses of the importance of printing and graphics. They've discovered that their long-neglected corporate identity—their graphics—tells the customer that there is more to their company than what meets the eye. Applying that idea to DX'ing and the salesmanship of QSL'ing, it is easy to see that printing is an opportunity to gain an advantage!

Another common complaint is of a lack of "artistic" ability. While it's true that artistic talents are beneficial, I've found, in observing the printed items used by DX'ers over the years, that they lack imagination more than talent. The paucity of unique, creative, printed materials is a clear signal that the area has great potential for an imaginative DX'er. Instead of worrying about talent, concentrate on knowing exactly what you want and on being able to convey that to your printer, who is probably not a DX'er.

Hopefully, this chapter will help you tap your imagination so that you may utilize printing to make your hobby more enjoyable and rewarding.

Letterheads

How often have you received a verification from a Latin American station that had a dazzling letterhead design? Hard to forget, isn't it? And you've likely displayed it prominently in your collection, right? Then, it should be no secret that many of the top DX'ers use personalized letterheads when sending reception reports. When you consider the importance of making your report stand out from the rest, a distinctive letterhead (Fig. 6-1) may be just the trick in getting that stubborn Peruvian gerente to answer your report. And, like those companies who are just waking up to the importance of graphics, your letterhead is an excellent investment. Arriving at your own letterhead design isn't as difficult as it may sound.

You'll need to present your printer with some sort of copy anyway, so you may as well start now with a few sketches set on standard 8½ × 11-inch paper. First, you'll want your name and address, including country, to appear prominently. Spell it all out, without abbreviations. Next, no letterhead would be complete without some sort of artwork or logo—one that immediately calls

Fig. 6-1.
Examples of
letterheads.

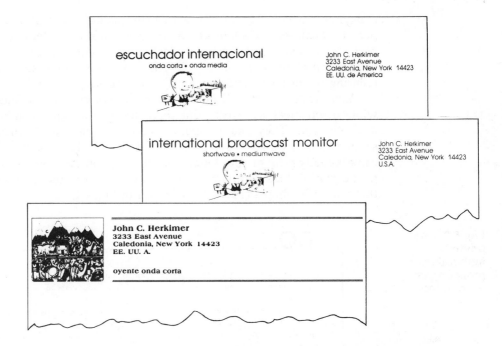

attention to your report. The best sources for such pieces of artwork are hobby magazines and amateur radio publications.

A globe with headphones? If your targets are Andean, perhaps a piece of artwork typical of the locale? Maybe a steamy, tropical scene for those Indonesian reports? Whatever you decide on, the art should be good-quality, high-contrast black and white art copy, if possible. If you can't find anything you like, most print shops will have books of black and white art in a variety of topics. However, since they are used primarily for advertising, there is usually little for the DX'er.

A phrase or slogan would complete the letterhead design. Maybe "oyente onda corta" for those Latin reports? "International Broadcast Monitor" or "Listening to the World" might better help define your interests.

Perhaps you want to feature a picture of yourself sitting at the dials on your letterhead, instead of artwork. In past years, printers were only interested in using quality black-and-white photographs for reproduction. Luckily, these days, color prints work fine provided they are clear and sharp. Of course, your photo will have to be converted into a halftone print (a method of reproducing a continuous tone picture by offset lithography), and you can expect to pay about $15.00 for this service. If you have ever had to make up copies of pictures to send with reports, you'll agree that a printed picture is certainly cheaper when you consider quantity.

Once you're satisfied with a design, it's time to visit a printer. Most communities have printing facilities that range from the large commercial printing companies (which deal mostly with high-quality advertising) to local newspapers to franchised "quick print" operations. It's at the latter that you'll find your best value, since quick printers deal mostly with walk-in clientele and can handle just about any kind of job. And, pricewise, they are very competitive, tailoring the order to suit your budget if necessary.

Your rough draft will now have to be converted into quality black-and-white mechanical art. Most print shops have some sort of in-house artist who can help you. They charge about $14.00 to create a letterhead design from scratch. If you want to save that fee, you can try to present the printer with a camera-ready print of your letterhead. In fact, you can save from 20–25% on the total cost of just about any printing job if you prepare the art yourself. Some tips that one print shop suggested are shown in Fig. 6-2.

Fig. 6-2. One printer's instructions for preparing copy for printing.

DO

Do use BLACK ink

Do use a very white paper not easily seen through (opaque)

Do use a carbon film or Mylar ribbon if one is available for your machine, or black India ink if doing artwork

Do keep the degree of darkness (density) of your lines and type as consistent as possible

Do use rubber cement or a wax adhesive for all paste-ups

Do keep your original if you think you may need future reprints. Avoid giving us a copy to work from Even our copies won't reproduce as well as a properly-prepared original

Do use a liquid ocver-up correction fluid (such as Wite-Out) to correct typing errors

Do allow a generous clear margin around the edges of your pages. At least 3/8" of clear space is necessary on all sides — but 1/2" is preferable

Do keep your layout/design clean, neat, and as open as possible

Do minimize paste-up and use white paper similar to the background

Do check copyrights. We cannot accept legal responsibility for reproducing copyrighted material

DON'T

Don't use blue or pastel colored pencils or inks. They don't reproduce well

Don't use pencil or ball-point pen — for the same reason

Don't use onion skin, erasable bond, tracing paper, or colored papers

Don't use typewriters that type unevenly or have broken or filled-in letters

Don't run copy all the way to the edges, top, bottom, or sides

Don't use a copy as an original. A little quality is lost from each generation of a copy-from-a-copy-from-a-copy-from-a-copy ... etc.

Don't over-type corrections until they are darker than the rest of the type. It shows

Don't allow dirt to get into your original. As a general rule, what you see is what you get

Don't use large black or solid areas in your layout/design if you can avoid it. Printing ink dries slowly, especially when applied over a large area

Don't use different colors on the same original

Don't allow rubber cement under the paste-up to bleed to the edges where it will collect dirt like a magnet. Use a rubber cement "pickup" to clean up around cemented pieces

Don't use newspaper pictures or color photographs. They reproduce poorly. Most black-and-white photos will reproduce well, but remember they add time and expense to your project

Don't shade areas with pencil, crayon, or paints

Don't use transparent tape, or masking tape, to hold items on your paper

Don't use staples. Staple holes will show on the copies as black spots

Now, your only task will be to select a type style from those offered by the printer. The advent of phototypesetting has made greater varieties of typeface available for small printers, and the styles are both conservative and contemporary, depending on your tastes. While the quality will not be as good as typeset, copy can be prepared using just a typewriter and a pen (Fig. 6-3).

Fig. 6-3. Examples of the use of typewriter and pens to prepare copy for printing.

Typewritten Copy.

The best reproduction of typewritten copy comes when you use an electric typewriter with black carbon ribbon. Cloth ribbon can be used, but carbon is preferred. Copy typed on a manual typewriter is acceptable but usually is irregular and difficult to reproduce evenly and sharply. Especially when using a cloth ribbon, be sure to clean your typewriter keys and use a new, very black ribbon.

> For good, sharp, quality repro-
> ductions, keep your original copy
> clean, crisp and legible.

This is a sample of copy typed on a manual typewriter with a cloth ribbon.

> For good, sharp, quality repro-
> ductions, keep your original copy
> clean, crisp and legible.

This is a sample of copy typed on an electric typewriter with a carbon ribbon.

Marking Pens.

You will get very good reproduction when using marking pens with felt or nylon wick points. Black is the best color to use. Red is acceptable. Do not use blue.

For good, sharp, quality

Example of marking pen writing.

Ball Point Pens.

Again, the rule is use black or red ink for best reproduction. Generally speaking, ball point pens make good original copy. Remember, do not use blue ink.

For good, sharp, quality

Example of ball-point pen writing.

Once the art has been prepared, it's time to make some choices. Paper? Your printer will offer you a wide variety of textured colored papers, usually with matching envelopes, but you can just forget about all of them. They are too heavy! With high international airmail rates, it doesn't take much paper to put you over the limit. My advice is to limit your paper choice to paper that is not heavier than 16-pound stock, or less if possible. But, there are limits to how light you can go, too. One well-known DX'er uses a 10-pound Eagle-A stock that most printers won't touch. That kind of weight is hard to control in the press and takes extra care. The choices in the 16-pound and under range are excellent though and, as an added bonus, the cheaper bond papers come in a variety of colors: green, salmon, cherry, canary, blue, goldenrod, and, of course, white. With colored papers still difficult to obtain in the developing countries, these are a good inexpensive choice for reception reports. The quantity you order will depend on your needs, but be sure to get plenty of blank sheets for those two-page reports.

Since the question of envelopes will eventually come up, I would say that most DX'ers will find little use for imprinted envelopes unless their needs are away from the practice of writing reception reports. Some printers will imprint standard airmail envelopes, but even that extra investment is questionable.

The last choice you'll have to make is the ink color(s); black is considered the standard. If you want a color other than black, you can expect to pay an extra $6.00. Two or more colors? Well, there's no question that a multicolor letterhead is distinctive, but the price changes dramatically, adding $20.00 for each color change. Let your budget decide.

A sample of the current quick-print charges is given in Fig. 6-4. Your printer will show you any of the paper samples you might be interested in; keep in mind, however, that prices will, of course, vary. The prices given in Fig. 6-4 indicate printing charges only. Artwork is extra.

Fig. 6-4. One printer's sample price list.

8½" x 11" (Letter Size)

PAPER	No. of COPIES 1-25	50	100	200	250	300	400	500	600	700	800	900	1,000	1,500	2,000	5,000	Addt'l 1,000's
20# Bond	4^{75}	5^{25}	6^{25}	8^{50}	9^{75}	11^{25}	13^{25}	16^{25}	18^{00}	21^{00}	23^{00}	25^{00}	27^{48}	38^{00}	48^{00}	101^{00}	23^{00}
60# Offset	7^{50}	8^{50}	10^{00}	14^{75}	17^{00}	19^{00}	23^{00}	27^{00}	32^{00}	36^{00}	41^{00}	45^{00}	49^{00}	64^{00}	92^{00}	221^{00}	43^{00}
Rippletone	8^{00}	9^{50}	12^{00}	16^{75}	19^{00}	22^{00}	27^{00}	32^{00}	37^{00}	43^{00}	48^{00}	53^{00}	58^{00}	84^{00}	110^{00}	266^{00}	46^{00}
Classic Laid	9^{00}	10^{25}	16^{00}	24^{00}	28^{00}	32^{00}	40^{00}	48^{00}	56^{00}	65^{00}	73^{00}	81^{00}	89^{00}	130^{00}	172^{00}	—	51^{00}
67#	8^{50}	10^{00}	13^{00}	18^{00}	20^{50}	23^{00}	29^{00}	34^{00}	40^{00}	46^{00}	51^{00}	57^{00}	62^{00}	88^{50}	115^{00}	—	49^{00}
110# or 70#	10^{50}	13^{00}	18^{00}	28^{00}	31^{50}	35^{00}	37^{00}	43^{00}	50^{00}	58^{00}	65^{00}	72^{00}	79^{00}	PER QUOTE			58^{00}
2nd Side	4^{75}	5^{25}	6^{25}	8^{00}	9^{25}	11^{00}	12^{50}	15^{00}	16^{00}	17^{00}	21^{00}	22^{00}	23^{00}	33^{00}	41^{00}	86^{00}	21^{00}

8½" x 14" (Legal Size)

Photo Copies 10, 15, 25, cents

PAPER	No. of COPIES 1-25	50	100	200	250	300	400	500	600	700	800	900	1,000	1,500	2,000	5,000	Addt'l 1,000's
20# Bond	7^{50}	8^{25}	9^{50}	13^{00}	15^{00}	17^{00}	21^{00}	25^{95}	29^{00}	33^{00}	36^{00}	40^{00}	44^{00}	46^{00}	60^{00}	144^{00}	28^{00}
2nd Side	7^{00}	7^{50}	8^{50}	10^{50}	11^{50}	13^{00}	15^{00}	17^{50}	20^{50}	22^{00}	24^{00}	27^{00}	28^{00}	45^{00}	57^{00}	130^{00}	20^{00}

11" x 17" (and Misc Sizes, Including Setup)

Colored Stock - add 10%

PAPER	No. of COPIES 1-25	50	100	200	250	300	400	500	600	700	800	900	1,000	1,500	2,000	5,000	Addt'l 1,000's
60# Offset	15^{00}	18^{00}	23^{00}	28^{00}	30^{00}	32^{00}	34^{00}	39^{00}	46^{00}	52^{00}	59^{00}	65^{00}	71^{00}	102^{00}	136^{00}	321^{00}	59^{00}
67# Bristol	18^{00}	21^{00}	26^{00}	32^{00}	38^{00}	44^{00}	53^{00}	64^{00}	75^{00}	86^{00}	97^{00}	109^{00}	120^{00}	176^{00}	232^{00}	570^{00}	98^{00}
2nd Side	15^{00}	18^{00}	23^{00}	25^{00}	28^{00}	30^{00}	32^{00}	35^{00}	40^{00}	43^{00}	51^{00}	58^{00}	64^{00}	90^{00}	115^{00}	—	50^{00}
Gov't Postcards	13^{50}	14^{00}	16^{00}	17^{00}	17^{50}	18^{00}	22^{00}	25^{00}	28^{00}	30^{00}	32^{00}	34^{00}	36^{00}	54^{00}	71^{00}	176^{00}	35^{00}

T.C.S.* for multiple-originals on 20# Bond

No. of COPIES	No. of ORIGINALS 8-10	11-25	26-50	51-75	76-100	101-125	126-200
25	—	—	2^{20}	2^{10}	2^{00}	1^{90}	1^{80}
50	3^{75}	3^{15}	3^{65}	2^{45}	2^{00}	2^{35}	2^{25}
75	4^{59}	4^{40}	3^{50}	3^{41}	3^{21}	3^{03}	2^{89}
100	6^{00}	5^{40}	4^{90}	4^{80}	4^{70}	4^{60}	4^{40}
200	7^{80}	7^{20}	6^{70}	6^{60}	6^{50}	6^{40}	6^{30}
300	9^{10}	8^{50}	7^{65}	7^{15}	6^{85}	6^{65}	6^{50}
400	10^{00}	9^{50}	8^{50}	8^{10}	7^{70}	7^{50}	7^{00}

ADDITIONAL CHARGES
Collating: add ½¢ per sheet; $7.50 minimum
14-Inch Paper: add ⅓¢ per sheet
3-Hole Paper: add 1¢ per sheet $5.00 Minimum
Colored Papers: add 10%
*T.C.S. stands for Total Copy Service. Our investment in automation saves your cash!

Report Forms

If you're strapped timewise but still enjoy QSL'ing, report forms (Fig. 6-5) can provide the answer. The stigma attached to a "fill in the blanks" approach to reporting is well known—they're impersonal and too brief to allow sufficient reporting of details. They're a compromise.

I've kept a collection of report forms through the years that I received from DX'ers, radio clubs, broadcast stations, and some forms that were offered commercially. All, unfortunately, had serious shortcomings because of design problems: poor layout, skimpy room for reporting details, information missing, etc. While these types of forms will never replace the personal reception report,

**Fig. 6-5.
Examples of
report forms.**

RECEPTION REPORT

DATE

FROM

TO:

DEAR SIRS:

I AM PLEASED TO SUBMIT

DATE OF RECEPTION

PROGRAM DET

QUALITY OF

SIGNAL STRENGTH

RECEPTION CONDIT

MY RECEIVER IS A

A VERIFICATION
APPRECIATED IF

ADDITIONAL CO

Reception Report

Date:

From:

To:

Dear Sir/Madam:

I am pleased to submit this report of reception of your station

Date of Reception	Time of Reception		Frequency	Wavelength
		G.M.T.	kHz	Meters

Program Details I am listing below the details of the program I heard:

Quality of Reception

Signal Strength	Interference	Noise/Static	Fading	Overall Merit	S	I	N	P	O

Reception conditions were

My receiver is a _____ Interference, if any, was from

A verification of this reception report, either by QSL-card or letter, would be greatly appreciated if this report is found to be correct and useful. _____ My antenna is a

Additional comments:

Sincerely,

it *is* possible to design a report form that is both useful to the station and provides a sense of completeness to the DX'er. The trick is getting all that information on a standard 8½ × 11-inch sheet of paper!

The most common type of form in use today is the do-it-yourself approach. The DX'er types a reception report as he would normally, and leaves blanks where the details are to be entered. The copy is then copied on a duplicator as needed. This certainly works and is economical; plus, the quality of copiers has improved tremendously over the years. But the obvious drawback is the unprofessional appearance.

Designing a report form is actually quite easy and will mean utilizing the fundamentals of business-form design. Will you be filling in the blanks on the report form with a typewriter? Then, you'll want the form to follow standard typewriter spacing, usually at 1/6-inch × 1/10-inch increments. Will you enter the data by hand? Then be sure that you have left enough room to compensate for your handwriting. Does the form flow in a logical manner like a reception report? Is there enough room for detailed program notes? All of these questions will need answering. And, yes, allow sufficient room for program details as this is the single biggest problem most report forms have. The space allotted for program notes is somehow always squeezed in the middle. Start there first and work in both directions.

If you think stations don't appreciate report forms, you're wrong. Large broadcasters appreciate the brevity of the form and it helps speed up the checking process, since the volume of mail they usually receive takes a great deal of time to read. In fact, both Radio Bucharest and HCJB complimented me on the design shown. Their claim was that it was simple and easy to read.

Approaching the printer with your report-form copy is about the same as for a letterhead, except that due to the amount of text, the report form will cost more to typeset. You can figure about $25.00 for one of the forms shown in Fig. 6-5, and $15.00 to have 500 copies printed on 16-lb. stock. Make sure that your copy is clear and legible. The terms we use in DX'ing won't mean much to the typesetter and you may wind up with a form loaded with typos.

Report Cards

If creating a form on an 8½ × 11-inch sheet of paper is a challenge, then trying to squeeze one onto a postcard is nearly impossible. Unless you were interested only in the very large broadcasters, a report card makes little sense. There is just no way you can give a station an accurate detailed report on a 3 × 5 card and, yet, cards of this type have been used and are commercially available as well. If you insist, a card can be printed on an index stock for about $20.00 for 500 cards. Your typesetting costs should run in the area of $7.00.

There is one approach to a report card that is practical, however. One experienced DX'er, to save time, created a card in Spanish for reports. Along with the card, he enclosed a detailed form letter describing his hobby interests and background. His reports were of such high quality that the card actually worked to his advantage. For most, however, I simply cannot recommend it.

Prepared Cards

If you're not using prepared cards (Fig. 6-6) with your reception reports to the small outlets, you're probably missing out on some worthwhile *veries*. When you consider that a lot of regional stations have no idea what a ''verie'' is or what you want, a prepared card makes good sense.

Fig. 6-6.
Examples of
prepared QSL
cards.

A prepared card is simply a QSL card that you design and fill out ahead of time with all the needed details. Enclosed with your report, the station can merely sign and/or stamp the card and return it to you. Your reception is verified and the station has fulfilled your wish. As an added bonus, the station will often include a letter of their own, hopefully on their stationary.

I've always felt that prepared cards should be more than a quickly typed index-card approach. The verie will be on display in your collection, so it makes sense that you want it to be attractive and aesthetically appealing. On the other hand, you are possibly instructing the station on what a verie is and yours may serve as their example. Consider this story: Well-known U.S. DX'er, Ralph Perry, designed a prepared card some years ago for use with reports to Latin America. Now affectionately dubbed the ''Kenosha Card,'' his prepared card has served as a model for a number of Latin American stations; most recently, Radio Los Andes in Péru.

In designing a prepared card, it's important to include all the important details, such as time, date, frequency, power, and, above all, the verie text which states that your report is being confirmed as correct. Pick a paper stock that's heavy. Some stations will return your card and nothing else, and that long mail trip back to your listening post won't be too kind to lightweight paper. A popular paper is cover stock, which is a highly textured paper used for promotional folders and the like. It's expensive and your printer is required to buy a minimum quantity from his distributor. The result is that the printer will always have a small supply on hand that he cannot use. It's your chance to walk away with some expensive paper at an attractive price.

For my own prepared cards, I used a leather-finished 90-lb. cover stock (Leatherette), and each card was designed around a specific Latin American country. The typesetting cost $15.00, and I had 100 of each printed at $30.00. The size was standard postal card size of 3½ inch × 5½ inch.

Custom Report Inserts

You're now using your newly printed letterheads and prepared cards and furiously preparing your reception reports. The next question is: "How can the DX'er take this a step further?" The answer is a customized report insert—an extra printed item that will be an incentive for the station to take notice and answer.

The most common forms of report inserts are well-known and too often used: stamps, decals, postcards, travel brochures, etc. Latin America, alone, must be well-stocked with listener's humble offerings. Their impact has diminished from overuse.

As the name implies, a custom report insert is a unique personalized item that you use to supplement your report. An imaginative DX'er can utilize printing to design an insert that any station personnel will notice. Plus, it can help cut time in preparing lengthy reception reports.

The personal data that we include with our reports (family information, local history, hobby background, etc.) can be incorporated into a separate printed pamphlet. Along with your photo, you can include a map of your area, scenes of local interest, and an in-depth explanation of our hobby. In fact, an excellent example is the Spanish pamphlet created by the National Radio Club (available from the NRC Publications Center, P.O. Box 164, Mannsville, NY). It explains, in detail, the hobby to confused station managers and even uses actual veries for examples. Typesetting such an item would be rather expensive, costing perhaps as much as $100, and the extra weight could create added problems in postage costs. But, my own experiments in this area have yielded extraordinary results.

Some DX'ers have used cards similar to the ones that Ham operators use. The glossy stock often used on such cards will require a special type of printing process. Ham publications, such as CQ and QST, frequently advertise QSL printers.

A popular item these days is the pressure-sensitive stock (peel apart) that can be printed into stickers or bumper stickers. Die-cut circles are also available for specialized business labels and a creative DX'er can find a good use for them.

Since there is no such thing as a standard custom insert, the limits are boundless. It's a matter of asking yourself, "What will a station notice and tempt it to respond with a verie?" Only recently, a well-known DX'er suggested an insert design that was unlike anything I'd ever seen before. It illustrated the imagination that separates a skilled veteran from the novice.

Logs and Recordkeeping

Unless you've invested in home computing, you'll still need a way to keep records and logs (Fig. 6-7). And, for most of us, we learn too late that our methods are inefficient.

Fig. 6-7. Examples of log and recordkeeping forms.

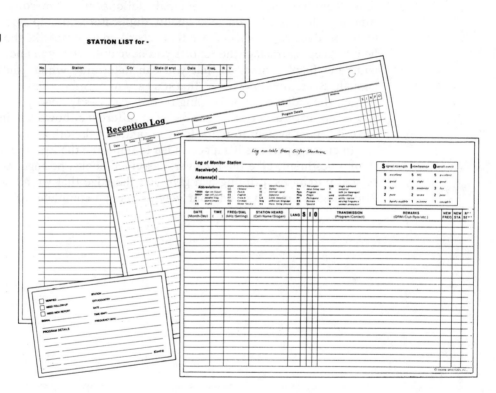

Logs fall into two categories: the day-to-day logs from casual listening and band scanning, and the more detailed logs of those stations that you record and send reception reports to. Printing can handle both. Log books and log sheets have been around for years. Gilfer Shortwave (P.O. Box 239, Park Ridge, NJ 07656) has log sheets available that are printed on easy-eye green stock. Some DX'ers even use ARRL logbooks which are designed for amateur radio use. I designed my log sheets and loosely based them on previous formats. The type-setting cost was about $14.00, while the cost of 500 sheets on white 16-lb. bond was roughly $18.00.

The other aspect of log keeping—records of stations I've sent reports to and their program details—was another matter. I first started out using small stenographer's notebooks for these logs. It worked well for awhile but the problem arose that the data could not be reorganized without detaching the pages from the books. I had read about index cards and so I took that a step further.

The cards I designed and printed were 3″ × 5″, with text on both the face and back. I can file them anyway I want—by country, region, state, etc.—and the style of paper (110-lb. index) comes in a variety of colors so they can be color-coded to suit your filing system. The data recorded at the top right of the card instantly tells me when it's time to send a follow-up report or when a new report is due. The typesetting was $7.00 and 500 cards cost $15.00.

Another form I use helps me determine how many stations I have from those countries and where regional stations are numerous, such as Latin America. Here is where a good-quality duplicator will help since you obviously won't need too many sheets. Another record, in a format like the one above, will help in keeping track of the reports you have out, what you enclosed with them, and can "flag" a time, so that you can determine when you've waited long enough. Remember, however, that it's easy to become a slave to recordkeeping. In fact, the tedious job of keeping records can easily keep you from the dials if you're not careful. Printing can make a difference, however, in allowing your recordkeeping system to work for you, and not the other way around.

In closing, perhaps it's a good idea to reinforce the idea that quality printing is an investment. Like the receiver, cassettes, books, postage, and other hobby-related purchases, printing can give you the edge that we all sometimes need. And that's when you're on your way to being a complete DX'er!

7

TAPE RECORDING AND SHORTWAVE

Jerry Berg

Like a number of other listeners who began their careers in the 1950s, **Jerry Berg** credits his interest to having built a one-tube radio from an experimenter's kit back in 1956. He has been an active and avid shortwave DX'er and listener ever since. His library of recorded excerpts from broadcasts of shortwave stations runs into the multihundreds. Those who know him believe his library of quips to be nearly as large.

Jerry has written a number of articles on various shortwave-related subjects and takes an active interest in hobby affairs. He is a loggings editor for *Frendx*, the monthly bulletin of the North American Shortwave Association, and also serves on that club's Executive Council.

Jerry lives in Lexington, Massachusetts, with his wife and two children. He is an attorney and the court administrator for the Massachusetts District Court.

Many hobbyists find that, besides their receiver and antenna, a tape recorder is the most useful piece of equipment they can own. It isn't long before even the novice shortwave listener discovers the advantages of being able to replay what has been heard over the air; many DX'ers find the tape recorder an essential tool for serious DX'ing.

A tape recorder can be used for several purposes in the radio-listening hobby:

- Programs can be recorded and preserved for later use.
- Using a timer, programs can be recorded with the receiver unattended and then played back at a later, more convenient time. (This is known as "program shifting" in the home-video recording industry).
- Recordings can be played over and over again, enabling the DX'er to explore the recording in detail in pursuit of an identification or more precise program details.
- Station IDs can be recorded and cataloged, making an interesting collection that is both entertaining to listen to and useful as proof of reception.
- You can send taped reports to stations, showing exactly how they were heard at your location.

All of these aspects of shortwave recording are discussed below.

Choosing a Recorder

The first step in selecting a tape recorder for shortwave purposes is deciding exactly what you want to use it for. This is important, since not every recorder will have all of the features necessary to achieve all of the purposes previously outlined.

For example, if you want to physically manipulate the tape and put IDs or programs from different stations onto different tapes, you will probably want a reel-to-reel ("open reel") recorder. Cassette recordings are difficult to edit unless you have two recorders, or a cassette recorder with two decks, and can re-record particular segments from one to the other. If you like the idea of having your recorder running all the time while you are listening, either so you won't have to remember to turn it on or so you won't miss something before you have a chance to start the recorder, you will probably want a unit that offers a long playing time. If you plan to use your recorder for deciphering hard-to-nail-down IDs, it will help to have a unit with tone controls to help you make the most of a noisy QRM-ridden channel. And, of course, these and other factors must be considered in the context of what is usually the most important factor of all: cost.

To help you decide what you want, let's review the various recorder features that you should be aware of, and assess them in terms of how you plan to use your recorder.

Open Reel vs. Cassette

The most fundamental decision is whether to go open reel or cassette. Open-reel recorders utilize reels of tape, usually 7 inches in diameter. The tape is ¼-inch wide and comes in varying thicknesses (see the section on "Recording Time"). You place the supply reel on the left side of the recorder, thread the tape through the various heads, rollers, and guides, and then insert the end of the tape into the "take up" reel on the right. Unlike the old days, when the tape had to be physically snaked around heads and rollers, most modern open-reel recorders have straight-line threading, i.e., the heads and rollers are located in a straight line and the tape can just be dropped into a slot in the housing that covers these components. Thus, threading open-reel recorders is usually a simple matter.

While open-reel recording was once the familiar standard, it has been replaced almost universally by cassette recorders. There is no need to dwell on the properties of audio cassettes, as most everyone is familiar with them. Contained within the plastic cassette housing is the supply reel, the take-up reel, and the tape, which is about one-seventh of an inch wide. The tape passes over a pressure pad that is designed to press the tape against the recording or playback head. Although users still differ in their opinions of the relative fidelity of cassette recording vs. open-reel recording, there is no disputing that the audio capability of the modern cassette recorder is a technological wonder. The cassette format was originally developed by Phillips® Ltd. for dictation purposes where low-fidelity sound reproduction was more than adequate. With the technology available in the early 1960s, there just wasn't enough tape surface to achieve a decent dynamic range, or a signal-to-noise ratio, or enough speed to leave room for high-frequency response. We have come a very long way since those early days, however.

What cassette tapes have is convenience, and this has led to their overwhelming popularity in today's audio market. They are easier to handle than 7-inch reels, and have permitted the development of a whole generation of portable cassette machines, many of which are not much larger than a book, and some (the Sony WM-10 "Walkman" player, for example) only nominally larger than the cassette itself. (If you doubt the ability to get good sound quality from a cassette recorder, listen to the WM-10.)

For the shortwave hobbyist, the decision to use open-reel tapes or cassette tapes will probably hinge on two factors: the need to edit the tape and the cost. Open-reel recording facilitates the physical editing of a segment of a longer recording. The author, for example, keeps a collection of Latin American station IDs, snipping out the desired ID from the original open-reel recording and splicing it onto a master tape containing recordings from that particular country. This is virtually impossible with cassettes.

Cost may well be the deciding factor, however, even if tape editing is desired. With few exceptions, open-reel machines have become basically semiprofessional pieces of equipment. They are designed for the really serious recordist and, consequently, they are very expensive. The sky is the limit as to price, with most units being $1000 or more. And, with few exceptions, they are tape decks only, requiring a separate amplifier and speaker. Cassette units, on the other hand, come in many sizes and configurations, from decks to hand-held portable units, and they are readily available on the consumer market in a wide variety of styles and prices.

Another possible drawback to open-reel recording, from the shortwave hobbyist's standpoint, is that most of the open-reel decks are designed for stereo use, i.e., more than one channel can be recorded on the tape at the same time. This is accomplished by using more than one "track" on the tape, with each track having its own recording head. There are two popular track configurations in open-reel stereo recording—the quarter-track format, where you record on two tracks in one direction and then turn the tape over and record on two other tracks in the opposite direction, and the half-track format, where you record on two tracks in one direction only. In the quarter-track format, the recording heads are about one fourth the tape width in size; in the half-track format, they are about one half the tape width in size.

There are some who feel that a stereo format is useful for shortwave recording because it permits you to record different material on each of the two channels at the same time. You might, for example, record a station on one track and WWV on the other, giving you a ready time reference for later playback; or you could record different programs from two receivers at once; or you could record a station on the first track while making oral "notes" of reception on the second track. Although there may be instances where such flexibility is desirable, in most instances, a single track is probably all that is really needed. On a stereo unit, this means using one of the two tracks and turning the other off (since nothing is gained by recording the same material on two tracks). Thus, unless you are an open-reel stereo music fan, stereo capability on an open-reel recorder may be of little real benefit in most shortwave situations. What may be more desirable is half-track monaural recording, where you record on a single track (half the width of the tape) in one direction, and then turn the tape over and record on the other track in the opposite direction.

Most roads lead to cassette units in the open reel vs. cassette debate. Cassette recorders come in both stereo and monaural formats, and they are relatively inexpensive.

Open-Reel Recorders

If the need to edit the tape is paramount, however, and if you want a new open-reel recorder in a mono format and are willing to pay the price, a few such units are available. One is the Tandberg Model 1521 (Fig. 7-1), a monaural open-reel recorder that has a built-in 10-watt amplifier and speaker. The unit accommodates up to 7-inch reels and operates at 1⅞, 3¾, and 7½ inches per second (ips). Tandberg is an old and well-known Norwegian company specializing in the audio-visual, educational, and professional markets more than in the home stereo market. While not generally available off the shelf, the Model 1521 can be obtained by Tandberg's dealers from Tandberg's American headquarters in New York.* The current retail price is approximately $700.

Fig. 7-1. An open-reel recorder (Courtesy Tandberg of America, Inc.).

Another open-reel monaural recorder that is available (although at the suggested retail price of $1300, it will be out of the price range of most of us) is the Uher Report Monitor AV.** The AV is the most recent generation of the Uher Report Monitor tape recorder series, which has been in use by radio reporting teams around the world for many years. It is smaller than the Tandberg and is portable. It will accept a maximum reel size of only 5 inches, but it operates at a fourth speed of $^{15}/_{16}$ ips for extra running time (frequency response at that speed: 25–6000 Hz). The AV also comes in more expensive 2-track stereo and 4-track stereo versions.

Recording Time

Recording time, or the length of time that you can record without having to either change tapes or turn the tape over, is a function of the recording speed plus the quantity and thickness of the tape on the reel. Recording speed also affects the quality of the recording.

Generally speaking, open-reel recording offers a longer recording time than does cassette recording. Open-reel recorders typically operate at three speeds: 1⅞, 3¾, and 7½ ips. (Some recorders will also run at $^{15}/_{16}$ or 15 ips.) Open-reel tape also comes in several thicknesses: 1.5 mils, 1 mil, and 0.5 mil (1 mil is 0.001 inch thick). The thicker the tape, the stronger it is, but even the thin 0.5-mil "long-playing" tapes are easy to handle. A 7-inch reel will hold as much as 3600 feet of 0.5-mil tape. This yields a running time of 96 minutes (1½ hours plus) at 7½ ips, 192 minutes (3 hours plus) at 3¾ ips, and 384 minutes (6 hours plus) at

---

* Tandberg of America, Inc., Labriola Court, Armonk, NY 10504
** Uher equipment is manufactured in Germany. It is distributed in the United States by Walter Odemer Co., Inc., 1516 W. Magnolia Boulevard, Burbank, CA 91506.

1⅞ ips. If the reel is then turned over, the running time is doubled. Of course, increased running time and slower speed are obtained at some expense to fidelity. On the Tandberg Model 1521, for example, frequency response is 40–18,000 Hz at 7½ ips, 40–13,000 Hz at 3¾ ips, and 40–7000 Hz at 1⅞ ips.

Cassette tape generally comes in C-30, C-60, C-90, and C-120 cassette sizes, meaning that the cassette has a *double-sided* running time of 30, 60, 90, or 120 minutes. Single-sided running times are half that, of course, and this represents a considerably shorter running time than what is available with open-reel recorders. Moreover, the combination of the fragility of C-120 cassettes and the automatic design of cassette decks makes the use of C-120 cassettes somewhat risky.

One means of getting the longer playing time of open-reel recording with the convenience of cassettes may be by means of a specially designed long-playing cassette recorder. One firm (undoubtedly there are others), Extendo-Tape-Systems,* sells several portable cassette machines that have been modified to provide extra-long running time. These devices permit the recording of as much as 5 hours on *one* side of a C-120 cassette, and as much as 7½ to 10 hours on one side of a C-180 cassette. The author does not know firsthand how these devices perform in terms of audio quality, tape handling, and location finding on playback, but they are available and might warrant consideration. Prices are in the $125–300 range.

Counters

To be useful for shortwave purposes, a recorder should have either a mechanical or an LED (light-emitting diode) counter. Most recorders have such a feature, but care should be taken to ensure that the counter is visible from a distance of a few feet. Some mechanical counters have reached such a state of miniaturization that it is a challenge to the eyesight of mere mortals. Four-digit counters are preferred over three-digit counters, as they provide a more accurate reading for a particular point on the tape. Some recorders also have handy electronic *tab markers* or *cueing* devices that put an electronic "mark" on the tape when activated. The mark is inaudible at playback but can be detected as a short "beep" when the unit is in cue mode (fast forward or reverse, combined with "play").

Tone Controls

For years, shortwave-receiver manufacturers paid no attention to the need for tone controls. Many of the new generation of shortwave receivers have tone controls, however, and they are very useful in improving the intelligibility of a shortwave signal. The same is true of tone controls on tape recorders. They can be helpful in highlighting or de-emphasizing certain audio frequencies when trying to identify a station from a recorded ID. Using the tone control, high

---

* Extendo-Tape-Systems, P.O. Box 1600 LC, Temple Terrace, Fl 33687

frequencies can be highlighted to the audibility of spoken material, or the low frequencies can be emphasized to reduce the effect of static bursts and other noise.

Lower-priced cassette recorders often lack tone controls, so check carefully. Ideally, the recorder should have separate base and treble controls, thereby giving better control over the full spectrum of audio frequencies. Try the controls before you buy to see if they are truly sensitive and if they make a difference.

It should be noted that much the same effect, with greater range and versatility, can be obtained by using an audio filter (on playback) between the recorder output and the speaker (assuming you are using an external speaker). A number of manufacturers produce such devices for shortwave receivers. They add extra cost to the playback process, however, so all other things being equal, choose the recorder that is already equipped with tone controls.

Other Features

Among the other features to look for in a recorder are:

1. The recorder should have a jack or terminals for recording directly from a receiver or from some other device besides a microphone. Sometimes the microphone jack can double for both uses (see a later discussion on this).

2. Try to obtain a recorder with a microphone input jack that will accommodate a switchable microphone; i.e., one that will start and stop the recording process by using a switch on the microphone. This is handy, for instance, when speaking into the microphone, such as when preparing a taped reception report.

3. There should be an output jack for an external speaker, as you may want to connect one for better audio quality.

4. The recorder should have a pause control in order to interrupt the recording process, temporarily, without leaving the record mode.

5. There should be a record-level indicator. Some units, particularly the portable cassette machines, adjust the audio input level automatically and, therefore, do not have recording-level indicators. For recorders that do have indicators, meters are preferred over other types of indicators, as they provide greater precision in monitoring the recording levels.

6. An off-tape monitor is also a useful feature. This allows you to listen, by way of a headset, to what was recorded an instant before. It is a useful way of ensuring that your machine is in fact recording and that the recording level is correct.

All things considered, what should you buy? If you are able to make a major investment, an open-reel recorder is probably the preferred choice, with monaural capability probably being more practical than stereo. You may want to keep an eye on your local "want ad" magazine, as these units sometimes become available on the used-equipment market. However, financial considerations and ease of availability will probably cause most people to opt for a cassette tape unit. Heed the advice outlined earlier and then follow the usual rule for such purchases: get the best you can afford.

Using the Recorder

Let's look at some of the shortwave-related uses of a tape recorder.

Preserving Programs

There are many shortwave program items that you might want to preserve, depending on your taste. The following are some ideas that might interest you.

MUSIC

If you like a particular type of national music, you might be able to build a good collection if there is a well-heard station that plays what you like. (Don't expect hi-fi quality on shortwave, however.) Some examples are: the Ecuadorian music that is played in the morning over the HCJB domestic service, the Portuguese Fado music over Radio Portugal, and the African rhythms over Africa Number One.

SPECIAL PROGRAMS

Drama, news, entertainment, and various feature programs are common fare on the BBC and some of the other major broadcasters. Some special programs may also be hobby-related. I am still pleased to have in my tape collection a 1960 recording of veteran DX'er Ken Boord's organ recital that was aired in a special HCJB Easter broadcast, and, also, the last broadcast of Keith Glover's "Listener's Mailbag" over Radio Australia. Alas, however, much of the programming on shortwave is of limited interest—particularly the rather heavy propaganda-laden output of some of the more powerful international broadcasters. But, when there *is* something interesting on the air, it's nice to know that you can preserve it for later playback.

MAILBAG PROGRAMS

If you think your letter is going to be read over the air, or your question answered, record it. It will be a novelty listening to it years from now.

DX PROGRAMS

We have all run into the problem of not having accurately jotted down a tip broadcast over a DX program; you couldn't copy the frequencies quickly enough, the telephone rang, etc. It is a good idea to record these programs so that you can replay an item you might have missed.

Program Shifting

If you like to listen to your favorite shortwave programs on your time schedule rather than the station's, you might want to record programs, during your absence, for later playback at a more convenient time. With some extra effort, this can be accomplished.

TIMERS

In order to record while you are away from your receiver, you must use one or more timers to turn your equipment on and off. (Some receivers have a built-in clock timer to turn the receiver on and off at designated times.) However, this technique can only be used if your receiver does not drift. If it does drift, you will find that after a time your set will be off frequency and this will most likely result in a disappointing recording. In such a case, you may want to tune in the correct frequency after the receiver has fully warmed up and then leave the receiver on until after the recording is made. This is illustrative of one of the basic problems with program shifting: the inability to provide the fine tuning that is often necessary on even the best shortwave signal in order to obtain the best audio transmission. If there will be someone at home when the recording is made, show him or her how to adjust the tuning dial for the best audio when the program begins.

You will also need a timer for the recorder, and you must remember to put the recorder in the record mode before leaving. Typically, the timer is placed between the recorder and the AC power outlet and serves the function of providing power to the unit at a particular time. Therefore, the recorder must already be in the *record* mode when the power is applied. Some timers, however, are designed to be connected to a remote-control jack on the recorder. Make sure the timer is compatible with your recorder's design.

CONNECTING THE UNITS

The final step is to connect the receiver and recorder together and adjust the recording level. The connection can be made either by a cable connected to the speaker output of the receiver or by connection to a separate recording output. As explained at the end of this chapter, the former will require an adjustment of the receiver's audio gain while the latter connection will not. In either case, however, the recording level on the recorder will have to be preset (in the absence of an automatic recording-level capability). You will have to learn, by trial and error, just how your recorder should be set for a signal of a particular strength.

If you record from the speaker connections of the receiver, the speaker will, of course, be "live" and will be emitting audio during the recording process. If you are connected to a separate recording output terminal on your receiver, a silent recording can be made (other people in the room will not be bothered).

FEATURES

Timers are available in many price ranges, and generally will differ in terms of the number of devices the unit will control, the precision with which the "on" and "off" functions can be set, how far ahead you can pre-set the starting time, whether the unit also functions as a regular clock, whether the unit will automatically turn itself "off" after a given period of time (if you forget to activate the "off" function), and whether the timer can be overridden without disconnecting it.

Among the many firms selling timers are the Radio Shack® stores, Radio West (3417 Purer Road, Escondido, CA 92025), and ESE (142 Sierra Street, El Segundo, CA 90245). Both of ESE's ES 1296 and ES 1372 presettable tape timers are more expensive and semiprofessional types of unit. Some units, such as those manufactured by J.C. Labs (P.O. Box 183, Wales, WI 53183) and Capri Electronics (Rte. 1-M, Cannon, GA 30520), activate the recorder only when a signal is being received.

Taped Reception Reports

One of the most interesting uses of the tape recorder is in the preparation of taped reports to send to stations. Written reception reports describe the programming and the quality of reception, but a taped report permits the station to join you in front of your receiver and experience the reception exactly as you did. Most stations are happy to get a taped recording (Fig. 7-2).

Hobbyists differ in their opinions as to whether taped reports are really of much use to stations, and some question their value to the listener who is seeking a QSL. In 1980, the European DX Council conducted a survey of international broadcasters on various aspects of shortwave listening. Of the 40 stations surveyed (which accept taped reports), 23 stations found them more valuable than written reports and 17 found them less valuable.

The author is of the opinion that, as a general rule, the larger and more powerful the station, the less genuine interest it will have in taped reports. There are several reasons for this. First, many of the large international broadcasters make use of professional or semiprofessional monitors, who participate in organized arrangements for monitoring the station's signal on a continuous basis. Second, the larger stations typically receive a great many written reports, and taped reports add only marginally to this body of reception information. Third, and perhaps most important, taped reports are much more difficult for the station to handle. Playback equipment with a track configuration that is compatible to that of the listener's is necessary, and it takes much longer to listen to a tape, even a short one, than it does to process a written report.

Of course, a tape recording does provide the most definitive proof of reception, and so the listener who wants to eliminate all doubt in the mind of the broadcaster might naturally favor a tape. But the best advice, for the larger well-established international stations, is to stick to written reports unless you know (from the *World Radio TV Handbook*, for instance, or from an item in a club bulletin) that the station is interested in taped reports.

Taped reports should be seriously considered in other instances, however. One is in the case of a new station. A new station has no transmitting history on which to determine how it is being received, and, in such a case, a tape is indeed

Fig. 7-2. The hoped-for reply showing a station's appreciation of receiving a tape of their broadcast.

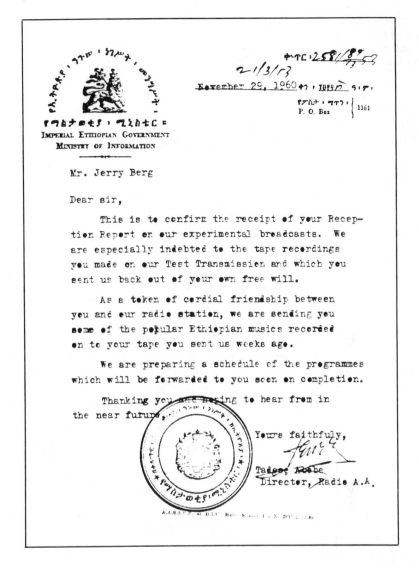

worth a thousand words. This is also likely to be the time when the station's QSL policy is the most favorable to listeners. A tape could set your report apart from the bulk of written reports, and may well bring a more personalized reply (Fig. 7-3). Of course, your's will not be the only taped report received, and, if you are not in the initial wave, you may find your report consigned to a heap of tapes destined for later reply as time allows. On the balance, however, this is a good time to send a tape.

Another good time to think about a taped report is when you have heard a rather unusual station and want to both prove your reception beyond all doubt and have the station share fully in the novelty of the reception. One of the best examples in this category is the reception of the European pirate stations. Operating at very low power, they are nonetheless heard in the United States from

Fig. 7-3. A reply
from a station
that just came
on the air.

Trans World Radio

P. O. Box 3518, Agana, Guam 96910 / Phone: 344-4198 / Cable: VOTAN

September 12, 1977

Paul E. Freed, President

Dear Mr. Berg:

Every now and then we get a letter and report like yours and feel compelled
to do more than just send a QSL card and a schedule. It surely was a
thrill to get your very welcome and thorough letter and the tape. I immediately
copied the tape after listening to it and gave it to the man who handles our
propagation and program coordination, Dave Fisher. It was quite interesting
how you happened to be tuning in and caught us at that particular time. With
all those years of DXing behind you, I suppose you have lots of pleasant
experiences similar to this one.

You mentioned Joan Mial and the DX card you received in 1961. Joan Mial
and her husband Bill have been in Monte Carlo until just about a year or so
ago, and now they are heading for Hong Kong where Trans World Radio will
be setting up a production studio and coordination center for our Asian
programming. They were due to arrive there this past week, but have
been delayed in the States due to the illness of a daughter.

You will see from our program schedule that most of our programming
is aimed to the Far East. We are beaming from the northern direction of
Japan southward to almost a southwest direction of Indonesia. Our equipment
consists of two Harris 100 kilowatt transmitters, and we will be able to
broadcast in four international bands of 9, 11, 15, and 17 megahertz. Each
transmitter feeds into a TCI antenna, Model 611, which is a 4-band, 4-element
curtain antenna. Each antenna has three patterns, and is slewable plus or
minus 30 degrees. This in a sense gives us six distinct patterns from which
we can direct our signal in what you might call a searchlight fashion to cover
the various areas as indicated on our schedule.

International Headquarters · Chatham, New Jersey
Transmitting Locations . Monte Carlo . Bonaire . Swaziland . Cyprus . Guam

time to time, and they are most interested in knowing how their signal is
received. They seem particularly receptive to taped reports, often returning
them with a personal message or with a studio-quality recording of their broadcast. Use of tape in such cases is helpful, however, only if reception was good
enough so that the station can definitely recognize the recording as their station.

Fig. 7-3 (cont.)

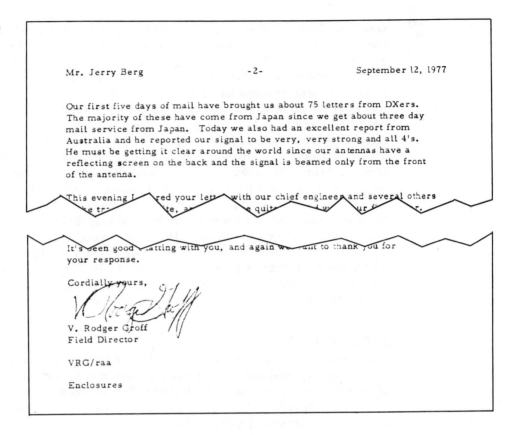

Mr. Jerry Berg   -2-   September 12, 1977

Our first five days of mail have brought us about 75 letters from DXers. The majority of these have come from Japan since we get about three day mail service from Japan. Today we also had an excellent report from Australia and he reported our signal to be very, very strong and all 4's. He must be getting it clear around the world since our antennas have a reflecting screen on the back and the signal is beamed only from the front of the antenna.

This evening I ____red your lett____ with our chief engineer and several others ____ tr____te, a____ ____e quit____ ____ w____ ____ur f____ r.

It's ____een good ____atting with you, and again w____ ____ant to thank you for your response.

Cordially yours,

V. Rodger Groff
Field Director

VRG/raa

Enclosures

If you could barely make out the signal yourself, the station will probably not have the patience to listen and will not do as well as you did in deciphering the recording. In such a case, you are better off with a written report.

Another category of station where a taped report may be desirable—indeed, required—is where the reception was very brief or when some other aspect of reception makes it difficult to convincingly describe what you heard. The author encountered an example of this when he ran across a weak, but audible, signal from the seldom-heard Polish Pathfinders Station. Reception lasted only two minutes before being blocked by another station. During that time, PPS was playing its interval signal, without voice identification. After confirming that this interval signal did indeed belong to the Polish Pathfinders Station, I sent them a tape of my reception, with apologies for the very limited nature of the "program details." A QSL was received, but, without the tape, I probably would not have been able to verify them.

There is one area where submission of a taped report becomes more problematical. That is when the listener has heard a station which he believes to be a particular broadcaster but is not absolutely certain. There may be a temptation to send the station a recording and ask them whether it was their station or not. This is different from sending a tape as evidence of the reception of a signal which you have satisfied yourself as originating from a particular station. Where you have not identified the station, there are no hard and fast rules. Certainly you ought not send the tape just as "a shot in the dark." You should have some concrete reasons for believing that it is a particular station, such as language, program format, propagation, a familiar anthem, etc., before reporting it. The reception should also be of sufficient quality so that the broadcaster can clearly hear the reception on the tape. A recording of a weak, completely unidentified signal should not be sent in the hope that the station will somehow be able to divine its signal and verify your report. If voices or music are audible, and if you have good reason to believe that it is a particular station, it would seem permissible to send the tape and request confirmation, provided that you inform the station that you have not positively identified them. This approach should only be used in those rare instances where it is unlikely that further monitoring will yield a more positive identification. However, if you have heard the station only once and have not tried for it again, continue listening until you have exhausted all reasonable means of identifying it fully. Only if you are unable to get an identification based on extended listening, but are satisfied based on all the circumstances that it is a particular station, should you send a tape.

TAPE PREPARATION

Once you decide to send a tape, preparation of the tape is relatively straightforward. It is suggested that you send your report on a cassette tape, since cassette players are in use virtually everywhere. The European DX Council (EDXC) survey of international broadcasters revealed that cassette tapes are acceptable by the great majority of such stations.

If you wish to send an open-reel tape, record it at the "professional" speed of 7½ ips, which was indicated to be the most acceptable open-reel speed in the EDXC study (though many stations can also accommodate other speeds). Put the open-reel recording tape onto a 2½-or 3-inch reel (available at most audio stores); don't wrap the tape around a pencil, or stuff it into an envelope, or use some other novel method. Consider also whether you want to keep a copy of the recording, either for your own collection or for possible future use as a follow-up. Mark the cassette with your name and address and the date, time, and frequency of the recorded reception. If you are sending a 3-inch reel, put this information, plus the speed at which the recording was made, on a piece of leader tape at the beginning of the recording.

TAPE LENGTH

How long a recording should you send? In the EDXC survey, 16 stations indicated that they prefer a series of 3-minute program extracts, made on different frequencies during a particular time period on the same day. Fourteen stations indicated that they prefer 3-minute extracts made at the same time, on the same frequency, on consecutive days, and only six stations indicated that they prefer a complete uninterrupted recording of approximately ½ hour. The implication is that the main interest of the major international broadcasters is in the quality of reception, and that this can be determined with one or more brief recorded segments. So, in these cases, keep it short.

In other cases, for instance where you are reporting an unusual reception to a seldom-heard station, or where you are sending a report to a new station, you may want to send a longer recording. If the reception varied over the listening period, include several segments recorded at various times, so the station will get an idea of how reception changed over time. If you have a long recording with station IDs located at various points, you might want to include the "best" IDs. You probably should not send a recording longer than 30 minutes, unless there is some special reason to do so.

ADDITIONS TO THE TAPE

Should you talk on the tape? This adds a personal touch and, if you are fluent in the station's language, is useful in describing what the station will be hearing on the tape. It also permits you to highlight certain portions of the tape. Perhaps the best rule to follow is: the greater the likely novelty to the station of learning of your reception, the more reason there is to talk on the tape. Keep your comments brief, however.

Even if you send a tape, you should still send a complete reception report, including a written description of the program details, just as if you were not sending a tape at all. This will aid the station in reviewing the tape, and it will also be useful for you to have a complete written report to use for follow-up purposes, if that becomes necessary.

MAILING TAPES

Although taped reports are more expensive to mail because of their added weight, they should always be sent airmail. The tape should be enclosed with the written report and not sent separately. A tape and written report sent separately will probably arrive at different times, and may be directed to different persons, and this will lessen the chance of a reply.

SUMMARY

Sending taped reports is not without risk. Though the author has no concrete evidence, it is felt that there is a greater likelihood that taped reports go astray in the international mails. Also, a station may not have the time or inclination to play your tape, and your report may languish in the files even though you have enclosed a written report. On the other hand, taped reports have one special advantage, and that is that the station may return the tape to you with some interesting material over-recorded on it. You should not expect the tape to be returned, however, as most will not be. Some stations do return tapes, however, but some do so only on request. The best policy is probably to either say nothing about it in your report or, else, request that the tape be over-recorded if the station wishes to return it (noting at the same time that you would be happy to have the station keep the recording if it wishes).

Tapes that are returned can be most interesting. Sometimes, they contain a station's national music, or a studio-quality dub of the station's ID or programming, or perhaps a recording of a mailbag program where your report was read (a favorite of the European pirates). My favorite is a return tape that I received nearly 25 years ago from Emissora da Guine in (then) Portuguese Guinea, with the announcer sending me greetings, chatting about the station, and playing Portuguese music. Such are the things of which great DX experiences are made.

Using a Tape to Identify Stations or Improve Program Details

There are many instances when you listen to a station, and hear what sounds like an identification, but you are unable to tell exactly which station you are hearing. You may hear the announcer give the station name, but you cannot quite identify it. This is a particular problem with the Latin American stations, where the language, a weak signal, and the noisy tropical-band reception will often conspire to cause you to miss the ID the first time around.

The best technique is to continue listening until the reception is over, keeping the recording running and noting the tape counter number of any IDs. If you still cannot identify the station, go back and play and then replay the IDs. If you are not successful the first time, don't give up. Put the tape aside for a few days and then go back to it. I am always amazed at how different a station can sound after being away from it awhile. I have had instances where an ID that I had missed completely becomes apparent when listening to it later. Alas, I have also had cases where I was unable to distinguish an ID, later, that I was sure I heard before, so it can work both ways. Taping can only serve to improve the accuracy of your DX'ing, however.

A similar use for tape recorders in DX'ing is in the review of program details when preparing a written reception report. DX'ers should always strive to include in their reports those details that will best satisfy a station that it was actually heard. One of the best items of program information is the verbatim text of an ID or station promotion, and the recorder is extremely helpful in this area. It will be impossible to copy the text verbatim while you are listening, particularly if the station is not broadcasting in English. If you note the time of the ID and the tape counter number, you can go back over the tape as many times as necessary until you have pieced together part or all of the ID text. This technique is useful for other program details as well, such as the names of products, artists, songs, programs, etc. These are among the best details to include in your report.

A Tape Library

Most DX'ers who want something tangible from their radio listening will seek QSLs. Sending reception reports may not appeal to everyone, however. Particularly, in the case of DX stations, broadcasts may be almost exclusively in a foreign language and feature programming which, to the untrained ear, offers no clues to a station's location. In the absence of some knowledge of the language, advertisement strings and other announcements are likewise usually difficult to decipher. In addition, the highly erratic verification policies of some stations means that obtaining QSLs will often require an unusual perseverance. Reports often must be written in another language, making the already difficult problem of providing accurate and complete program details even more complicated, and repeated follow-ups may be necessary. And, the cost of reporting, both in terms of time and money, is well known to everyone.

One way to have some tangible results for your DX efforts and still have time left for other pursuits, and, maybe, even for the wife and kids, is to start a library of taped station identifications. Using a reel-to-reel recorder is best, as it permits the editing and assembling of the separate taped segments into long-playing tapes. It is a good idea to keep the recorder operating at all times while you are listening. Even if you are just "tuning around," turn it on—you never know when you might come across something interesting. This is a good rule for any DX session, but it is especially important when you are seeking IDs.

It is best to make the recordings at sign-on or sign-off times, whenever possible, since that is when the most complete ID can be heard. The taped segments might range from one to several minutes. Try to include part of the national anthem or the tuning signal, plus some of the music played. Then, place the recording onto a tape containing the recordings from that particular region. When the tape is full, prepare an index for it.

This can be an interesting approach to shortwave listening. When the DX doldrums set in, it is enjoyable to listen to a tape and relive past DX experiences. Also, the prospect of hearing additional stations and keeping your tape library up-to-date will cause you to develop a continuing familiarity with the various shortwave bands.

Although a tape recording is evidence—indeed, proof—of reception, it is not the same as a QSL and does not and should not "count" as a verification for any purpose. For many individuals, however, taping may offer an attractive alternative to the time-consuming and expensive practice of sending reception reports.

Hooking Up

The cardinal rule in recording from a receiver is to never use a microphone. Just placing a microphone in front of the speaker will act against producing a good quality recording, since it will introduce ambient noise onto the tape and will be affected by the relative qualities of both the speaker and microphone. Recordings made with a microphone will not faithfully reproduce the signal and are definitely not suitable for taped reports.

The only suitable method for recording from a receiver is to connect a cable between an audio source at the receiver and the input of the recorder. This will permit the recorder to take its audio directly from the receiver and be totally unaffected by room noise, talking, etc. This produces the best quality recording.

There are generally two readily available audio sources on the receiver. One is the speaker or headphone jack (or terminals) that is controlled by the receiver's audio gain (volume) control. You can generally connect to this source by way of a "Y" connector or other appropriate fitting. (If you connect your recorder to the speaker connection, make sure that insertion of the headphones into the headphone jack does not automatically mute the speaker, as this will also mute the audio to the recorder.)

The disadvantage of using this connection point is that it requires that there be an audio signal at the connecting point. If you use the speaker connection, the speaker must be in use and the volume must be sufficient to drive the recorder. This could create problems if you are recording when others want to sleep. If you use the headphone jack, it may be that, in order to drive the recorder, you will have to increase the volume to a level too high for comfortable headphone listening. Another problem is that, by using a jack that is connected to the volume control, both the receiver volume control and the recording-level control on the recorder must be adjusted. If you set the volume control too low and the recording level too high, or vice versa, the quality of the recording will suffer. Although experience will teach you the best settings for these controls, the interplay between them introduces an extra inconvenience to the recording process.

Many receivers solve this problem by providing a special recording output jack. This jack is independent of the receiver volume control; that is, a constant-level audio signal is present at this jack without regard to the setting of the volume control. Indeed, the volume can be turned all the way down and the constant-level audio signal will still be present at the receiver output. By connecting a cable between this output and the recorder input, the recording level can be controlled solely by the recording-level control on the tape recorder. This simplifies the recording process and is the best way of recording from the receiver. It has one disadvantage; if you have a device like an audio filter connected between your receiver and the speaker (or headphones), a recording made from the recording jack will, of course, record only the unprocessed audio signal.

A final comment: Receiver recording outputs are often of a high-impedance type, while tape recorder inputs are often low-impedance inputs. Connecting a high-impedance output to a low-impedance input will produce a highly distorted audio and a completely useless recording. This can easily be remedied by purchasing an inexpensive signal-reducing jack/plug and placing it in the recording cable between the receiver and the recorder. This will compensate for the impedance mismatch.

8

THE VERIFICATION GAME

Henry Lazarus

Every time that **Henry Lazarus** and I sit down to discuss QSLs and building our collections, I become more and more convinced that he is the most enthusiastic, determined, and avid devotee of this particular phase of DX'ing that I have ever met. And, I'm no slouch in that department.

The first QSL arrived in Henry's mailbox in 1956, although he didn't really get active in collecting until 1974. Latin America is Henry's prime area of interest, and it is quite like him to take on the most difficult challenge. That the techniques, attitudes, and approaches outlined here do indeed work is attested to by the remarkable record of success he has received in just over a decade of QSL'ing.

Henry holds a law degree, and lives in New Orleans with his wife and two "receivers." His other interests include ancient history, collecting historic newspapers, and bicycling.

What Is a QSL?

A QSL may be defined as a card, letter, or other document with which a radio station certifies that, in its judgment or opinion, a particular shortwave listener has heard its transmission. This judgment or opinion is based upon evidence supplied to the radio station by the shortwave listener, with the evidence consisting either of a description of the programs heard by the listener, or an open-reel or cassette tape recording of the broadcast, or both.

Whether a particular card or letter qualifies as a QSL depends entirely upon the language contained therein. What DX'ers look for is a "verification statement" or "verification language." The most common verification statement is "We verify (or "confirm") your reception of our station" (Fig. 8-1) or "We verify (or "confirm") your reception report". Another term frequently used as a verification statement is "Your report is correct" (Fig. 8-2). An additional satisfactory phrase is "The details in your reception report correspond with our program log." Similarly, a station may acknowledge that a recording sent by the DX'er was indeed a recording of the station's broadcast. Still another verification statement is "You heard our station." Also, if a card is entitled "QSL," "Verification of Reception", or other similar term, it properly qualifies as a QSL.

The previous list of verification statements is not, of course, intended to be all-inclusive but illustrates the type of language that a DX'er should look for when he receives a reply from a station. Nor can it be said with authority, either in this chapter or by anyone, just what language constitutes a verification statement, or what card or letter constitutes a QSL. The DX'er soon finds that many cards and letters from stations do not contain precise certifications that the listener has heard such stations. While it would, of course, be desirable to

Fig. 8-1. Most African stations have their own printed QSL cards, many of which are quite colorful. This card from Radiodiffusion Nationale du Burundi, La Voix de la Révolution, the government-operated station in Bujumbura, Burundi, confirms a reception.

LA VOIX DE LA REVOLUTION DU BURUNDI
B. P. 1900 BUJUMBURA

Monsieur,

Nous avons le plaisir de vous confirmer la réception du rapport d'écoute Q.S.L. que vous avez eu l'amabilité de nous faire parvenir.

Nous joignons à cette carte-réponse nos vifs remerciements.
La Voix de la Revolution émet actuellement sur les fréquences suivantes : 1100 KHZ-3300 KHZ et 6140 KHZ, aux heures suivantes :

de 5 H 30' à 8 H   de 12 H à 15 H
et de 17 H à 23 H   heures locales = GMT + 2 h

Bonne écoute

receive technically pure verification statements from every station that replies to a reception report, practically speaking, this is impossible. Consequently, the DX'er is forced to relax his standards to a certain extent and accept as a QSL any card or letter which, in his opinion, evidences an intent on the part of the station to confirm that he heard its signal (Fig. 8-3). Such a judgment is quite subjective and depends in large part on just what satisfies a particular DX'er. This topic will be discussed again later in this chapter in the section on "prepared cards."

While the standards by which the language in a particular card or letter are judged may be relaxed by a DX'er in order to cope with the realities of the hobby, this should not change the definition of a QSL, as set forth above. Technically speaking, *a QSL is a certification by a station that a listener has heard its signal,* and it is important to keep this in mind so as to avoid confusing a QSL with other hobby-related items, such as pennants or program schedules, as well as that evidence of the reception of a station which was not received from the station itself (such as a DX'er's own logs, or the tape recordings of stations he has heard).

Thus, the need to avoid confusing QSLs and hobby-related items is quite important, because a realization of the precise nature of a QSL helps clarify a dispute which has existed in the hobby for some time: *whether a QSL is proof of reception.* This dispute perhaps is derived, at least in part, from the fact that it is common to refer to a QSL as a "verification of reception," and this has possibly led some people to think of a QSL as an *actual verification* or *proof of reception.* This misconception has given rise to a contrary argument, in which it is contended that a QSL is not really a verification or proof of reception, because some stations are not careful about checking the evidence submitted in a reception report. Consequently, QSLs from such stations may have been sent in response to reception reports having insufficient program details or other satisfactory evidence of reception. Also, there is always the possibility of an unscrupulous person knowingly sending a station a false report of reception, based perhaps on program details he has seen reported elsewhere.

This contention has generated an even more basic dispute: assuming the correctness of the premise that a QSL does not truly verify or prove the reception of a particular station, is there really any reason to obtain one? In this connection, it might be argued that a better proof of verification of reception is a tape recording of the station that was made by the listener. Pursuing this line of reasoning, it might also be said that a QSL is nothing more than a souvenir from a radio station, thus implying that obtaining a QSL is the same as receiving a pennant or a calendar from such a station.

When viewed in its proper perspective, this argument really misses the point, and the issue which it raises is not a real one. The contention runs afoul of the basic definition of a QSL as set forth earlier. A QSL is not a proof of reception and is not intended as such. Rather, it is nothing more than the statement of an opinion by the station. Obviously, it is impossible for a station to actually verify or prove that a DX'er has heard its signal. Instead, the station can only give its opinion that the evidence submitted by a listener adequately supports his claim of having heard its broadcast.

Fig. 8-2. A
South American
QSL card,
which was
received from
Radio El Mundo
in Guayaquil,
Ecuador.

Furthermore, the argument is fallacious because of its misplaced emphasis on the false issue of whether or not a person has successfully proved his reception of a station. The truth is that the reception of a station's signal is really not susceptible of absolute proof by anyone without sophisticated methods, such as the ability to take direction-finding bearings. Therefore, it is impossible for the DX'er, himself, to prove to any third person, including a radio station, that he has heard a particular transmission. The evidence he gives may have been falsified: program details could have been copied from someone else's report, or a recording could have been duplicated from a tape made during another's reception of the station. There is no way that a third person to whom such evidence has been submitted can know with certainty whether it is genuine or falsified.

Moreover, it is sometimes impossible for a DX'er even to prove to himself that he actually has heard a particular station from a particular location. The DX'er may know that he is on a certain frequency and that he has heard a broadcast in which a voice said that the transmission originated from a certain station at a certain location, but he cannot really be sure that he is actually hearing that station from its announced location. Consider, for example, the radio station commonly termed as a "black clandestine," which is a station pretending to be another station (this sometimes occurs during wars and civil wars). Also, unlicensed revolutionary stations (sometimes called "clandestine stations") sometimes intentionally make false statements about the location of their transmitters in order to avoid detection by unfriendly forces or agencies. Again, confusion can be caused by the common practice of certain large shortwave stations, who use relay stations located in other countries and even cause some of their programs to be broadcast over transmitters belonging to other stations. A person unfamiliar with Radio Moscow's practice of relaying some programs over the transmitters of Radio Habana Cuba would probably believe

Fig. 8-2 (cont.)

Mr. Gerry L. Dexter
West Bend, Wisconsin

Estimado Señor:

Agradecemos a usted por el reportaje hecho
de nuestra Emisora el 8 de Julio de 1.966, en la frecuencia
de los 4.750 Kc., en el que nos indica una Sintonía Correcta.
Es un placer para nosotros ser escuchados en los Estados Uni-
dos. Esperamos que los programas ofrecidos sean de vuestro
agrado.
Hasta tener el gusto de recibir nuevos reportajes de Ud. en
la frecuencia de los 4.750 Kc. Onda Corta y quedando sumamen-
te agradecidos, nos suscribimos

Muy atentamente,

Guayaquil, Julio 14 de 1.966

GABRIEL VERGARA JIMENEZ
Gerente "R E M"

that he was really listening to a transmission from the Soviet Union when he was actually hearing a broadcast from a Radio Habana Cuba transmitter. Similar confusion could also result from the many transmissions of the Voice of America, Deutsche Welle (the Voice of Germany), and the British Broadcasting Corporation, who sometimes cause their programs to be transmitted either from relay stations in other countries or from transmitters belonging to or operated by other stations.

Still another conceivable source of uncertainty is a deliberate hoax. Various people have access to shortwave transmitters and they sometimes transmit without government licenses. A practical joker with a shortwave transmitter could conceivably broadcast fake transmissions allegedly emanating from a rare, seldom-heard station over the listed frequency of such a station. If the person putting together the programming and doing the announcing were particularly clever and a good actor as well, he could deceive numerous DX'ers into thinking that they were really logging an extremely rare station. While this may seem far-fetched, I can point to at least one occasion where there was concern over such a possibility. In early 1980, the Falkland Islands Broadcasting Station, which at that time was transmitting on the difficult frequency of 2370 kHz, was being fairly widely reported in the United States. I logged the station during this period and heard British Broadcasting Corporation news programs, followed by a local announcement, and a very clear station identification immediately before the station signed off at 0130 Greenwich Mean Time with the anthem "God Save The Queen." A few days later, I read a report in a shortwave club bulletin to the effect that this signal might not have originated from the Falkland Islands, but instead from an unlicensed station in the United States, which, as a hoax, may have been transmitting fake Falkland Islands Broadcasting Station programs on 2370 kHz at the same time as the real station was reportedly transmit-

Fig. 8-3. Some of the regional government broadcasters in Indonesia have their own printed QSL cards. This card was issued by the Radio Republik Indonesia station in Surakarta.

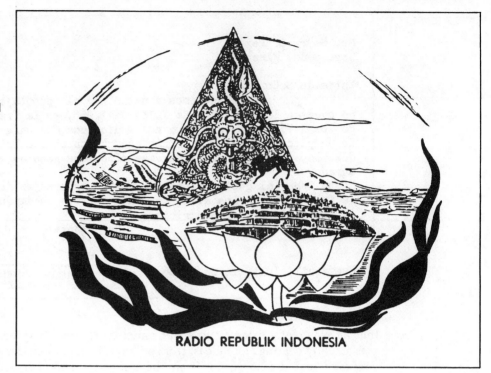

RADIO REPUBLIK INDONESIA

Dear ....*Sir*................

DIREKTORAT RADIO
R. R. I.
SURAKARTA
DEPARTEMEN PENERANGAN R.I.

We gratefully acknowledge your report on

station : ....*YDG*........

frequency : ....*4932 kc/s*........

date : ....*21. 10. 1972*........

time : ....*11.30 — 11.45 GMT*........

Remarks :   Director,

Sangat gembira telah menerima laporan dari odr dan masih mengharapkan laporan lagi yg lebih terperinci dan secara periodik terutama pd musim² kering dan hujan. terima kasih

Radio Republik Indonesia
Kep. Bag. tehnik   *(SOEWONDO)*

ting. I was naturally concerned about this report. How did I know that the BBC programming I had heard had not been taped by a private individual in the United States and then transmitted on 2370 kHz as a hoax? Furthermore, how did I know that the announcer claiming to be speaking from the Falkland Islands was not really someone using a Scottish or British accent and broadcasting from an unlicensed station in the United States as a joke? In truth, I could not know. There was absolutely no way for me, with the receiver and equipment I was using, to know for certain whether I had heard the real Falkland Islands Broadcasting Station or a hoax.

We must therefore dispense with the preoccupation of proving one's reception of a station. Such proof is a practical impossibility, and this preoccupation serves only to obscure the true nature, value, and purpose of a QSL. Thus, it is unfair to say that a QSL is of no value because it is not absolute proof of reception. True, a QSL is not proof of reception, but neither is it a mere souvenir, because it constitutes a certain type of certification: it is an official expression of opinion by the station that its transmission was heard by a particular DX'er. The station can give no more than that, since absolute proof is impossible.

A QSL is in the nature of, and quite analogous to, various certifications, judgments, and opinions which society often recognizes as official determinations of certain facts insusceptible of absolute proof. For example, when persons present their dispute to a court, each side submits witnesses and other items of evidence in support of its claim. A judge or jury then renders a decision or a verdict based on his or its opinion of the truth after an examination of the evidence. This decision is not absolute proof, but it is accepted by law and by society because unquestionable proof is not possible. A judge's decision or a jury's verdict is a legally binding determination, even though the judge or the jury may have been careless in considering the evidence, in the same way as a radio station might have been remiss in failing to carefully check the program details in a reception report. Similarly, how can society really prove whether a person accused of a crime is guilty or innocent? It cannot. A jury must give its verdict that, in its opinion, based on the evidence submitted in the case, the accused person is either innocent or guilty. Actually, the verdict may be incorrect and an innocent person might be convicted or a guilty person acquitted. In like manner, universities confer degrees based on the opinions of their faculty members as to the achievements of each student, even though a student might have cheated in taking written examinations or possibly paid others to take such tests or write essays for him. Likewise, boards and commissions routinely grant or deny occupational licenses based on their opinions of people's ability, knowledge, or expertise in particular fields, which opinions are not necessarily infallible. Numerous awards for achievements in all types of endeavors are given each year based on the opinions of various persons and boards. A QSL is similar to these various certifications in that it is not really proof of reception but an official expression of opinion as to the sufficiency of the evidence presented by a DX'er in his reception report. While it would be unreasonable to argue that such an opinion constitutes an absolute proof of reception, it still is entitled to the same type of recognition as society grants to the numerous certifications discussed above.

Thus, when we dispose of the confusion surrounding the false issue of whether or not a QSL is a proof of reception and recognize the QSL for what it really is, we can see that it is quite valuable. The QSL is a certification of reception which is truly unique, for it is given by the one source that is most qualified to make such a determination and render such an opinion: the station itself.

The Uniqueness of QSL Collecting

There are many hobbies involving collecting. For years it has been popular to collect coins, stamps, and other items. However, the collecting of QSLs is unique, for it is one of the few hobbies in which the item collected cannot be purchased. While a person may buy a desired coin or stamp by paying the right price at a store or to another collector, a QSL can be obtained only in a certain way: a DX'er must hear a station (not always an easy task), and then convince that station to send him its certification that he has heard its transmission.

There are several attractive aspects of QSL collecting. The hobby is quite exotic, due to the fact that most shortwave stations are in other countries, including little-known nations with which an average person would have no contact at all during his lifetime. For example, how many people, other than QSL collectors, have had correspondence with people in such countries as Brunei, the Cook Islands, Kampuchea, Kiribati, or Upper Volta? Furthermore, the smaller stations are often located in towns and cities which are completely unknown to most people. The average person in the United States would have no idea whatsoever where Juanjui (Peru) or Mendi (Papua New Guinea) are located, or even what they are. Yet, there are shortwave stations in these towns, and quite a few DX'ers have corresponded with these stations and received QSLs and personal letters from them.

QSL collecting also features many side benefits, such as foreign stamps and numerous enclosures which may be received from radio stations, including program schedules, magazines, pennants, postcards, tourist brochures, key chains, and even books. Sometimes DX'ers receive extensive literature from government-run stations anxious to present their views to persons in other countries, as well as from unlicensed revolutionary ("clandestine") stations seeking recognition of the causes they are promoting. Some persons at the smaller stations write DX'ers long, friendly, personal letters describing in detail the cities, towns, and regions in which they are located. These side benefits add to the fascination of the QSL collecting hobby.

The Reception Report

Having discussed the nature of the QSL and some of the more interesting aspects of the hobby of QSL collecting, let us now turn to the mechanics of obtaining a QSL. Obviously, the first step is to log a shortwave station. A DX'er should monitor a station as long as reasonably possible so as to hear the most program details and station identifications. This enables him to provide the station with as much evidence of reception as possible. In this connection, it might be advisable for the DX'er to tape his loggings, as this makes it possible to later recheck and more fully describe the program details in a transmission.

Generally speaking, a reception report should at least include a statement of the time, date, and frequency of reception, a description of the quality of reception, a discussion of the programs heard by the DX'er, and a description of the type of receiver and antenna used to log the station.

When writing a reception report on the transmission of a station broadcasting an international service, as well as when reporting to most of the stations located outside the Western Hemisphere, the time should be stated in *Coordinated Universal Time* (also called *Greenwich Mean Time*). On the other hand, reception reports directed to the smaller stations, particularly those in Indonesia and in Central and South America, should be stated in accordance with the local time of the countries in which such stations are located. Whenever there is any doubt about the correct time zone, this should be pointed out and Greenwich Mean Time should be put in parentheses. If there is any uncertainty about the local time of a Brazilian station (Brazil has several time zones), it might be a good idea to put the time of Brasilia (the capital city) in parentheses. It is also good practice to list both the local time and Greenwich Mean Time when writing to stations in Papua New Guinea.

When describing the quality of reception, a listener should discuss such items as signal strength, readability, and interference from other stations (identifying, if possible, the interfering stations). When reporting to a large international broadcaster, it is often best to use the "SINPO" Code. This code uses a scale of 1 (poor) to 5 (excellent) and covers five points: signal strength, interference, noise (static), propagation (fading), and overall merit. Thus, for example, one might describe reception of a station as follows: S-4 (good signal strength), I-4 (only slight interference), N-3 (moderate static), P-4 (slight fading) and O-4 (overall merit of reception was good). This information can also be written as SINPO 43444. On the other hand, when writing to the smaller stations, it is probably better to simply describe the quality of reception in as clear language as possible.

The most important part of the reception report is the description of the programs which the DX'er heard during the transmission. This is a vital part of the reception report, for it provides the evidence upon which the station must base its determination of whether the listener really heard its broadcast. It is very important for a DX'er to describe the programs in detail so as to leave no doubt in the mind of the station that he heard its transmission. Thus, the station should be informed whether the DX'er heard a newscast, an interview, music, or other type of program. Also, the contents or details of these programs should be described. Who was being interviewed, or what subject matter was generally discussed? What type of music was played? Commercials and public service announcements are vital items of information, and an accurate description of these program details gives the station solid evidence on which to issue a QSL. Exact wording of the station identifications is also important.

When writing to a station, a listener should use very polite language in requesting a QSL, since he is asking the station for a favor. When writing to a smaller station, the term "QSL" might not be familiar, so that a DX'er might want to ask for a card or letter verifying, confirming, or certifying that he heard the station. Of course, in such a case, the term "QSL" can be put in parentheses, as many of the smaller stations do recognize and sometimes use that term.

It is customary to include return postage with a reception report. When writing to the large international broadcasters, International Reply Coupons are generally used. However, when reporting to smaller stations, particularly in Latin America and Indonesia, it is much better to send mint stamps of the station's country. Mint stamps may be obtained from the DX Stamp Service, 7661 Roder Parkway, Ontario, New York 14519. Some DX'ers affix the mint stamps to self-addressed envelopes so as to make it more convenient for the stations to reply.

If the DX'er knows the name of the person who generally answers reception reports on behalf of a particular station (this person is commonly called a "verification signer"), it may be advisable to address the reception report directly to him. However, one should make sure that the information concerning a verification signer is reasonably recent (perhaps no more than two or three months old), because a person who was a verification signer in the past might no longer work at the station. This is especially true if the report is sent by registered mail, because the station might not accept a registered letter addressed to a person who no longer works there.

Of course, it is also sound practice to send the report to the correct address. Most of the addresses may be obtained from the *World Radio TV Handbook* and from the *World Broadcast Station Address Book*, but the DX'er also should carefully examine the QSL columns in the various radio club bulletins in order to make sure that he has the most recent address of a radio station and, also, to obtain addresses of new stations which are not yet listed in the above-mentioned publications.

Finally, a DX'er should try to find some way of making his report unique or special—some means of making it stand out from the rest. This is especially true when writing to the smaller stations. For example, it may be helpful to include some personal data, such as the listener's age, occupation, hobbies, marital status, number of children, and perhaps some information about his town, city, or state. This gives the reception report a personal touch which often is appreciated by persons at smaller stations. Various enclosures, such as postcards, tourist brochures, and photographs are sometimes included to add more originality and uniqueness to a report. This subject will be further discussed in the section on *follow-ups*.

The Smaller Stations

Thus far, we have discussed QSL collecting in general terms as related to all shortwave stations, including the large, powerful stations with international services (many of which beam English-language transmissions to North America).

Generally speaking, these stations will readily verify reports and it is usually easy to obtain QSLs from them (although there are a few exceptions). For this reason, a DX'er's first logs and QSLs are usually from these stations. However, after a year or two, he will have acquired QSLs from most of the stations. At this time, he will either be compelled to drastically reduce his efforts to collect QSLs, or he can expand his efforts to include the various smaller stations around the world—stations that broadcast to local audiences and do not target their transmissions to points outside the boundaries of their respective countries.

Acquiring QSLs from the smaller stations is one of the most challenging and fascinating aspects of the SWL hobby. However, the DX'er who has been accustomed to collecting QSLs from the large international broadcasters will find striking differences and a multitude of problems. Reception of any of the smaller stations is difficult and is highly dependent upon propagation conditions. Furthermore, a high proportion of these stations are audible only at the most inconvenient times, forcing a person to DX late into the night or very early in the morning. Still another difficulty is the language barrier: most of the smaller stations either broadcast in English only occasionally or not at all. But probably the most difficult obstacle for most DX'ers to hurdle is the fact that many of the smaller stations, particularly those in Latin America, do not readily respond to reception reports.

The DX'er who considers embarking upon the task of collecting QSLs from the smaller stations must make an initial decision: how intense is his desire to succeed in this undertaking? This is a vital question, because the collecting of QSLs from the smaller stations can be demanding, difficult, time-consuming, expensive, and frustrating. This decision really is not that different from those faced by persons taking up certain other hobbies. For example, a person who plays chess as a hobby can either play purely for pleasure, not caring if he wins or he loses, or he can be extremely serious about playing the game well and serious about winning, in which case he may read numerous books on chess; learn the various openings, strategies and tactics; and even record all the games he plays so that he can continually improve his ability. Similarly, a person who seriously takes up tennis or any other competitive sport will train for many hours each week so that he can excel at playing and winning the game. The same type of dedication is required of the DX'er who desires to attain a high level of success in collecting QSLs from the smaller stations. It will require a substantial amount of effort—frankly, more effort than many persons are willing to sacrifice. Whether the objective is worth the time, effort, and expense is a decision which each individual DX'er must make.

For this reason, the purpose of the remainder of this article is to offer suggestions which hopefully will assist DX'ers in successfully collecting QSLs from the smaller stations. Since I have gained most of my experience while obtaining QSLs from Latin American stations, these suggestions will tend to be most applicable to acquiring QSLs from stations in this region. However, many of the suggestions apply to other stations and countries as well, particularly the numerous broadcasters in Indonesia.

Learning the Languages

A DX'er planning to collect QSLs from the smaller stations in various areas will find it beneficial to learn at least some of the basic vocabulary and grammar of certain languages. Otherwise, he will need to use the foreign-language reception report forms supplied by the various radio clubs, and his reception reports will tend to become rather "mechanical" letters in which it is not possible to express anything outside the rigid confines of the report form. Such letters would probably be less effective when received by the manager of a small station in a country where English is neither generally spoken nor understood. Furthermore, without any knowledge of the vocabulary or basic grammar of the foreign language contained in the reception report form, a DX'er will have difficulty in adequately describing the programs he hears or in giving the station details about himself and his city. Still another reason for learning the language is to better understand the replies received from the station.

Of course, a very important benefit obtained from learning the basics of a language is that the DX'er will be able to understand parts, at least, of the radio transmissions he hears. This is especially true with respect to the many Spanish-speaking Latin American stations, because Spanish is a very phonetic language that many people find is easier to understand than various other languages.

The four languages that are perhaps the most helpful in collecting QSLs from the smaller stations are Spanish, Portuguese, French, and Indonesian. If a DX'er makes an effort to learn some of the basic vocabulary and grammar of any of these languages, he will often find that his knowledge and understanding of such vocabulary and grammar improves as he continues to monitor and write to stations in that language. This is particularly true in the use of certain words and phrases (such as time, frequency, and the style of language used in station identifications) which are frequently encountered when monitoring radio stations. Also, by carefully translating the letters received from such stations, the DX'er soon gains an additional insight into the language and picks up many useful phrases that he can incorporate into future reception reports. As time goes by, the DX'er will be pleasantly surprised to find how much easier it becomes to write reports in a language that is frequently used. While many stations broadcast in languages other than Spanish, Portuguese, French, and Indonesian, such stations are more likely to respond to reception reports written in English.

Sending a Tape Recording

One weapon that is at the disposal of the DX'er who is willing to spend the extra time, effort, and expense is the ability to send a cassette or open-reel tape as part of his reception report, on which is recorded part or all of the radio transmission received.

There will be times when a DX'er logs a small station but the reception is simply not good enough to enable him to gather sufficient details for a satisfactory reception report. Or, the DX'er may monitor a transmission in a language or dialect which is so unfamiliar as to be virtually incomprehensible. (Surprisingly, this sometimes occurs in transmissions from stations in certain Latin American nations, such as Peru and Guatemala, where Indian dialects are widely spoken.) In such cases, it is extremely difficult for the DX'er to include in his reception report sufficient program details so that the station can realistically certify that he has heard its broadcast. In such cases, the use of a recording can be invaluable. Sometimes, stations are able to issue a QSL on the basis of a recording. Even when reception of a station is not good and the announcements are difficult to understand, the station personnel often can recognize the voice of the announcer (or perhaps certain theme music which is sometimes played immediately before each announcement) and, thus, can accurately certify that it was their broadcast that the listener heard and recorded.

A similar situation arises when a DX'er makes a "tentative" logging—that is, he hears a broadcast which, because of a number of factors and clues, he strongly believes was transmitted by a certain station. However, despite these factors and clues, the listener is not completely certain of the identity of the station because he failed to hear or clearly understand the station identification. In such cases, it is the accepted practice to send a reception report in which the DX'er states that he believes he heard the station but acknowledges that he is not certain. Such a "tentative" report can be greatly strengthened by a recording, and this often enables the station personnel to more accurately determine whether it was really their station that was heard by the DX'er.

A listener might want to send a cassette tape to strengthen his reception report, even if a clear station identification or other announcement has been heard. Many stations genuinely appreciate receiving recordings because they can actually hear just how their station sounds at the listener's location. No written description of the quality of a signal can possibly compare with the dramatic effect of hearing exactly how one's own broadcast was received in a distant country. In some instances, stations have even played such recordings over the air to show their audiences how well their transmissions were heard in faraway lands.

In making a recording to send to a station, the DX'er might consider putting the best station identifications (if any were heard) and the clearest announcements first on the tape. In this way, when a station manager listens to the tape, he will be able to determine within a few seconds whether or not it is a recording of his station's transmission. After the initial segment of one or more station identifications and/or clear announcements, the rest of the tape can be filled with several minutes of the transmission, played without interruption. In this way, the station personnel can evaluate how their station was received over a fairly significant period of time. The DX'er might consider describing the contents of the recording in his letter so that the person listening to it at the station can more easily follow it.

A cassette tape should not take the place of program detail. A DX'er should still give as many program details as possible in his letter. That way, the recording is an additional proof of his reception and gives him two items of evidence upon which to base his claim that he has heard the station.

After making a tape for the station, it is a good idea for the DX'er to make an identical recording for himself, perhaps putting it onto a master tape along with recordings of other stations. As will be noted later in this chapter, keeping a recording of one's loggings can help a DX'er in his follow-up program.

Of course, the making of a recording involves considerable time and effort. It also can be quite expensive, when one considers the cost of the cassette and the weight it adds to the reception report. Furthermore, since it is much more preferable to send such a package by registered mail, this results in an even greater expense. Frequently, a DX'er who has sent a recording will learn from the QSL columns in various club bulletins that others have obtained QSLs from the same station without sending recordings, so that it appears that in many cases the sending of a tape is unnecessary. All of these negative factors should be weighed by the DX'er when considering whether or not to send a tape recording. However, the sending of a recording has sometimes resulted in excellent rare QSLs, which otherwise might not have been received.

The Prepared Card

A prepared card is a card on which the DX'er, himself, types or prints a verification statement in the language spoken at the station. This statement will include the date, time, and frequency of reception (Fig. 8-4); the card is then sent to the station for its signature and seal. For example, a prepared card sent to Radio Valera in Venezuela for its seal and signature might say (in Spanish): "We verify your reception of Radio Valera on 4840 kHz on June 1, 1984, from 9:00 P.M. to 10:00 P.M., Venezuelan time." (Of course, this is just a hypothetical example.) There is no special verification language which must be used on a prepared card. Rather, the only criterion is that the language satisfy the DX'er who prepares the card, since this will constitute his QSL from the station—if it is returned to him duly signed and/or sealed. Indeed, some DX'ers prefer to leave blank spaces for the date, time, and frequency and will ask the station to fill in these spaces (Fig. 8-5).

There are several reasons that a DX'er might want to send a prepared card to a small station. He might want to ensure that he receives a true unquestionable certification of his reception of the radio station, and, thus, he guards against the possibility of receiving nothing more than a personal letter which does not contain a verification statement or any verification language. Or, he might want to send a prepared card in order to make it easier and more convenient for the station to verify his reception report, especially if the station has failed to verify previous reports or is known not to readily respond to reports. Then, another reason for sending a prepared card is that the card itself may serve as a further means of communicating to the station exactly what a DX'er wants when he asks that a QSL be sent to him to verify, confirm, or certify his reception of its transmission.

The first reason stated above (the desire to obtain an indisputable QSL and avoid receiving only a letter that does not contain a satisfactory verification statement) is significant, because it brings into focus one of the more common problems confronted by the DX'er who attempts to obtain QSLs from the smaller stations: the "thank you for your report" letter. Almost all of the stations with international services have their own QSL cards, and many of these cards are quite attractive. Most of the African stations also have their own QSL cards, as do the majority of the government-operated stations in other parts of the world. However, many of the smaller stations in Latin America and Indonesia do not have their own QSL cards and, instead, answer reception reports by writing personal letters. Many DX'ers find these letters more desirable than the printed QSL cards because of the personal nature, friendliness, and uniqueness of such letters.

However, a personal letter can present a problem if the station manager (or other person writing the letter on behalf of the station) fails to include language which qualifies as a verification statement. These letters are almost always quite friendly and, usually, they sincerely thank the listener for his reception report; but the verification language is sometimes omitted, either through oversight or because the writer does not fully understand exactly what the DX'er wants in the way of a reply.

In the beginning of this chapter, a few of the commonly used verification statements were set forth; and it was also stated that sometimes a DX'er is compelled to relax his standards to a certain extent and accept as a QSL a letter which, in his judgment, demonstrates an intent on the part of the station to confirm that he has heard its transmission. However, while most DX'ers recognize the practical necessity for making such a compromise, many QSL collectors are not completely satisfied with the so-called "thank you for your report" letters and desire a more concrete verification. A prepared card duly signed and stamped by a station provides such a precise and unequivocal verification. It must be emphasized that this is a very subjective matter, and the *essence* of the hobby of QSL collecting is self-satisfaction. The letter which satisfies one DX'er may not satisfy another. It is not the purpose of this chapter to pass judgment on this issue, but only to point it out and suggest that prepared cards are an excellent method of securing a QSL which contains a properly worded and unquestionable verification statement.

Some DX'ers hesitate to use prepared cards because of a concern that they might lose the opportunity of receiving a personal response which is unique to a particular station and characteristic of its country. A prepared card is rather cold and artificial, especially if the DX'er himself fills out the date, time, and frequency. The DX'er who receives a prepared card, but nothing more, certainly has received a bona fide definite QSL, but, in so doing, he sacrifices much of that enjoyment of the hobby which is based on the uniqueness of collecting exotic and distinctive correspondence from all over the world. Furthermore, some stations do not have stamps or seals, so that all that the DX'er may receive is a signature, without any insignia that identifies the station.

Thus, the decision of whether or not to use prepared cards is not always an easy one, and it might be advisable for a DX'er to consider the use of prepared cards only in certain circumstances. One method is to send a prepared card only after receiving a reply from a station which does not, in the opinion of the DX'er, contain any satisfactory verification language. However, this method

Fig. 8-4. An example of a prepared card written in Spanish. This card was received from La Voz de Apure in San Fernando de Apure, Venezuela.

VENEZUELA

Verificación de Recepción

La Voz de Apure

Esta tarjeta sirve para comprobar que

escuchó: YVRC

Frecuencia: 4.820 MHz

Fecha: 25 diciembre, 1973

Hora: 0407 - 0444 GMT

Firma y Sellar

results in an increase in the time, effort, and cost spent by the DX'er, and there is no guarantee that a station will answer a DX'er once, much less twice. Another practice might be to send a prepared card only after a station has failed to respond after a certain number of follow-ups. A DX'er choosing this course of action takes the position that it is more desirable to obtain a prepared card in the near future than to receive a personal reply only after many years (or possibly not at all).

Still another possibility for a DX'er is to send a prepared card with his reception report, but indicate to the station that he would like to receive a letter or card from the station in addition to the prepared card. This method often has led to the fortunate result of receiving both a personal reply and a definite QSL (the prepared card) from the station. A DX'er could also ask the station to send him the prepared card only if it does not have a QSL card of its own.

If a DX'er does decide to send a prepared card to a station, he should ask the station to sign and stamp the card with the station's seal or stamp. Most Latin American and Indonesian stations have such stamps or seals, and this gives authenticity to the prepared card when it is returned. The station's stamp or seal on a prepared card is excellent proof that the card is the official reply of the station (Fig. 8-5), whereas a signature without a stamp or seal may leave some doubt in this regard.

Another question which arises is whether a DX'er should put his return address on the reverse of the prepared card and affix a stamp thereto. If this procedure is followed, the DX'er may increase the chances of it being returned to him, because it is so easy and convenient for a station manager to answer the reception report or follow-up simply by stamping, signing, and mailing the prepared card. However, this practice may well preclude a DX'er from receiving anything else from the station (such as a personal letter, pennant, or post-

Fig. 8-5. An example of a prepared card written in Indonesian.

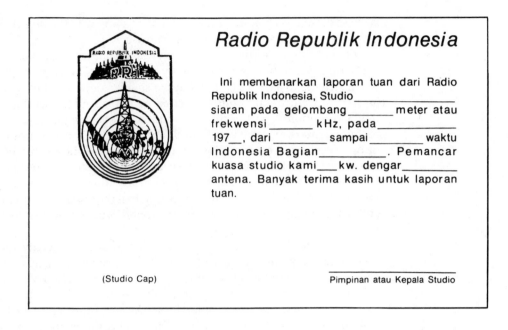

Radio Republik Indonesia

Ini membenarkan laporan tuan dari Radio Republik Indonesia, Studio_____
siaran pada gelombang_____ meter atau frekwensi_____ kHz, pada_____
197__, dari_____ sampai_____ waktu Indonesia Bagian_____. Pemancar kuasa studio kami___ kw. dengar_____ antena. Banyak terima kasih untuk laporan tuan.

(Studio Cap)

Pimpinan atau Kepala Studio

card), because the practice seems to carry the implied message that the DX'er wants the prepared card returned and nothing else.

It is good practice, however, to put one's name and address somewhere on the prepared card (even if a mint stamp is not affixed thereto) as an extra precaution to guard against loss of the card. Sometimes a station will accidentally send one listener's prepared card to another DX'er. By putting one's name and address on the prepared card, the DX'er lessens the chance that the station will make this mistake. Also, in the event that such an error should occur, the person to whom the prepared card has inadvertently been sent will know where to forward it. Another reason for putting one's name and address on the prepared card is due to the possibility that the card might fall out of the envelope while en route from the station to the DX'er. While this possibility may seem remote, it has been known to happen.

The Follow-up

A *follow-up* is a letter which a DX'er writes to a station because the station did not answer his original reception report. The letter is properly accompanied by a copy of the original report and contains a polite request that the station verify the report. It is no exaggeration to say that one of the most important keys to success in obtaining QSLs from the smaller stations is the ability of the DX'er to skillfully prepare and properly use follow-ups.

One of the more sobering experiences encountered by many QSL collectors is the final realization of the difficulty there is in achieving a high rate of return when striving to acquire QSLs from the smaller radio stations, particularly those located in Central and South America. A person might send an extremely detailed reception report, skillfully written in the language of the station, and including interesting data on himself and his city. He also sends along a cassette or open-reel tape recording, which unquestionably shows that he has heard the station, and includes other enclosures, such as stamps, postcards, or tourist brochures. Although he might have faithfully followed the best possible reporting techniques and procedures in putting together this reception report, there is still a very significant chance that the report will not be answered until he has sent one or more follow-ups.

It is quite difficult to adjust to this realization, especially for those DX'ers who previously have directed their efforts to collecting QSLs from the various government-operated stations throughout the world. Whereas one may expect a relatively high rate of return when writing to such stations, the reverse is true in the case of many of the smaller broadcasters, especially the Latin American stations. While it is difficult to predict a successful ratio that will be true in all cases for everyone, it is quite likely that the rate of return for a DX'er in obtaining QSLs from the Latin American stations (in response to original reception reports) will be no more than 35% to 40%. This rate of return is so low as to discourage many DX'ers, particularly in light of the time, effort, and expense involved in the careful preparation and submission of a reception report. However, by utilizing an organized, skillful, and efficient follow-up program, a DX'er can reasonably expect to greatly improve his rate of return to a much higher percentage over a period of several years, perhaps even exceeding 85%. This process is not easy, and it entails considerable ingenuity, patience, and persistence, but the DX'er who sticks with an effective follow-up program generally gets that much-desired QSL.

The Mental Attitude

The ultimate success of a DX'er in using follow-ups is closely related to his ability to develop a certain mental and emotional attitude which is essential to the successful collection of QSLs from the smaller stations. This mental and emotional attitude is primarily based on the understanding that, as a general rule, many of the smaller stations, particularly the Latin American broadcasters, do not as a rule verify reception reports without the necessity of follow-ups. Most of these stations are commercial, religious, or educational operations which transmit on shortwave only to reach the outlying areas of their own countries. Their advertisements, messages, and programs are directed toward their own countrymen, and it is irrelevant to their purpose to be concerned with how their signals are received in other parts of the world. Sometimes a new station will at first readily verify reception reports, perhaps because of the novelty of receiving letters from other countries and knowing that its signal reaches such distant locations. However, as too many DX'ers have learned, this original interest usually wanes rather quickly.

This does not mean that it is impossible to obtain a QSL from a small station. On the contrary, by the persistent use of follow-ups, most stations will eventually verify. However, in order to obtain QSLs from these stations, it is sometimes necessary that a DX'er persist in his efforts, even if it takes a number of years. This is where the importance of having the proper mental and emotional attitude is the greatest. Therefore, when a DX'er hears a small station, he must caution himself that the station probably will not verify on the first try. If he expects a 90% or 100% return on his original reception reports, he will likely become disappointed and discouraged and will, perhaps, give up the quest altogether. This is the attitude which *must* be resisted. Instead, the DX'er should expect the worst even as he prepares his original reception report. He should fully expect that his report will probably not be answered at first, but, at the same time, he should realize that the chances are excellent that the station will eventually respond. It is perhaps best to look upon the original reception report as the important first step in the process of obtaining a QSL from a small station. Consequently, it is unwise to send a reception report to a small station without making at least one copy of the letter for follow-up purposes. It is also a good idea to keep a recording of the broadcast (or selected parts thereof) for possible future use. Everything should be done with the expectation that follow-ups will be needed. Above all, the DX'er should resist the natural tendency to count the days in which a response could conceivably arrive from a small station. In most cases, this will result in disappointment.

It is also important to develop the attitude and understanding that the failure of a station to reply does not necessarily signify a rejection of the DX report; that is, a judgment by the station that the report was incorrect or insufficient to prove reception. Often this is not the reason for a station's failure to reply. A surprising amount of reception reports and QSLs really do get lost in the mail. Sometimes stations do not have the personnel to handle large amounts of correspondence and reports remain unanswered for long periods of time, regardless of the quality of such reports. Also, as noted earlier, there are some small stations that are so locally oriented that they are not especially interested in receiving or answering reception reports, regardless of how well written or detailed they may be; follow-ups are almost always necessary to obtain QSLs from such stations.

For all of the previous reasons, a DX'er who does not receive an immediate reply should not take the negative attitude that it is simply impossible to obtain a QSL from such and such station. Also, he should resist the tendency to discontinue or significantly reduce his efforts to obtain QSLs from small stations during those discouraging periods when no QSLs are received. When it seems as if everything that he has tried has failed, and all of his efforts have come to nothing, that is the time when the mental and emotional attitude of the DX'er is put to the extreme test.

In conclusion, it is probably best for a DX'er to take the stoic attitude that his original reception report, and even his first few follow-ups, probably will not be answered. However, he must couple this stoicism with a strong determination not to give up until he obtains that much-wanted QSL.

The Letter

When writing a follow-up letter, there are a few essential elements which should be included. First, a copy of the original reception report should be enclosed. Second, a letter should be written to the station stating that an original report was sent earlier, but the DX'er has not as yet received a response. The letter should also include a polite request for the station to verify the report. Often DX'ers include in the letter a statement to the effect that perhaps the original report or the station's answer was lost in the mail, although this may seem a bit far-fetched when writing a station for the tenth or twentieth time.

In the section on reception reports, the importance of making a report unique or special was stressed. The same principle is even more applicable to follow-ups, because the stations to which follow-ups are sent are often the broadcasters who respond only rarely to reports. Consequently, when a DX'er prepares a follow-up to a smaller station, he should think about the person who will probably read his letter and try to put himself in the place of such a person. In the case of a small station in a country such as Ecuador or Peru, there is often a limited staff that is primarily or solely interested in directing its programs only to audiences within the boundaries of such country. The task of answering foreign correspondence is not essential to such an operation and, therefore, is probably not part of any staff member's job. Furthermore, the writing of a letter is burdensome to many people. How many times have you put off answering a letter? When a small station receives reception reports from other countries, the station personnel do not know the people who sent the reports and, consequently, do not even have the usual social obligation to respond. Furthermore, some of these stations receive many reception reports, so answering all of them is almost impossible for a small staff.

With all of this in mind, a DX'er has to think of ways in which to make his follow-up stand out from the rest in such a way that the manager (or another person at the station) will want to answer it. There are many ways that a person can accomplish this objective. As noted earlier in this chapter, the ability to write a good letter, using the language of the personnel at the station, is helpful. The inclusion, in the letter, of some information about the DX'er and his family gives it a personal touch, which might make a favorable impression on the station personnel. Similarly, a paragraph describing the writer's city, town, and state could be of interest. Likewise, a few sentences to the effect that he (the DX'er) likes to listen to the type of music generally played by the station (or stations in that country) may be well received. The enclosure of postcards and photographs is also a friendly gesture. Often the tourism commission or Chamber of Commerce of a city may provide free literature which the DX'er can send, and some of this literature might possibly be in the language spoken at the station (particularly in the case of the Spanish-speaking stations). Many DX'ers send large diploma-sized certificates, which are in the nature of awards. Others have various cards and certificates printed with designs, their photographs, and some personal information. A list of ideas of this type can be virtually endless and will depend upon just how ingenious a particular QSL collector can be.

One method that can provide results in certain isolated instances is the sending of a cassette tape recording of the original reception along with the follow-up. While this method may not be particularly successful with the Latin American stations, it sometimes has good results in prying QSLs out of some of the smaller government stations in other parts of the world, particularly when the transmission on which the original report was based is in a language which was incomprehensible to the DX'er. In such a case, the enclosure of the original report along with the follow-up does not give the station much evidence on which to base its decision to send a QSL, because of the lack of details in the original report that results from the language barrier. However, if the reception of such a station was quite excellent, the station staff can recognize a recording of its own transmission, especially if it includes something truly distinctive, such as an interval signal. Sometimes a station, which has failed to reply to numerous follow-ups, will send a QSL in response to a follow-up that is accompanied by a recording.

A DX'er could also consider the possibility of writing a completely new reception report to send to a station which has not responded either to his original report nor to his follow-ups. Often such a task is futile and time-consuming but, sometimes, it is quite effective. This is particularly true in two instances: the original reception report is quite old (more than three years is a good yardstick), or the original report was not based on a good reception, while the more recent logging is far superior and gives the DX'er the means of writing a much better reception report. Also, by sending a new and superior report, the DX'er increases his chances of obtaining a QSL somewhere down the line, even if the station does not immediately answer the new reception report, making more follow-ups necessary.

Certain of the preceding methods may result in an answer from certain stations, but none of them are magic. The DX'er has to continue to send a station follow-ups (written in different ways and/or with different enclosures) until, hopefully, he "presses the right button" and prompts a reply.

Recordkeeping

Another vital aspect for achieving a successful follow-up campaign is to have a well-organized program of sending follow-ups on a regular periodic basis. The importance of this practice cannot be overemphasized. *Follow-ups should be sent to each station on a periodic basis until the station finally answers.* Different DX'ers observe different time intervals for sending new follow-ups. Perhaps a good guideline is three months. Thus, if a station does not answer an original reception report within three months after it was sent, a follow-up should be mailed to such station. If the follow-up is not answered within another three months, still another follow-up should be sent. This procedure should be regularly repeated until the QSL is finally received.

Careful records should be kept, and each station to which a reception report has been sent should be entered in such records, along with the date on which the most recent follow-up or original report was sent to such station. Once the selected time interval (such as three months) for a particular station has elapsed without the receipt of a QSL, a new follow-up should be sent. When and if a station on the list responds with a QSL, it should be taken off the list. In the event that while keeping such a list, you learn from a fellow DX'er or from a radio club bulletin that a reply has been received from a station on your list, it is a good idea to send a follow-up immediately even though the time interval has not yet elapsed. This is particularly true if such information includes the name of the verification signer, in which case, the follow-up should be addressed directly to such verification signer.

A program of sending out regular periodic follow-ups is quite important because of the way in which a small station tends to operate. For years, such a station may have no interest whatsoever in verifying reports. Then, a new person may be employed by the station, the station might be sold to a new owner, or a member of the owner's family may begin working at the station for the first time. In the event that this new person is interested in answering reception reports, QSLs will suddenly be received by a number of DX'ers, and reports of such QSLs will begin to reach the QSL columns of the various radio-club periodicals. However, the new person's interest usually wanes after a relatively short of period of time, and the station then reverts to its usual practice of not answering reports. By sending follow-ups on a regular basis— every three months (or whatever time interval suits the individual DX'er)—the DX'er stands an excellent chance of having his follow-up near the top of the pile of unanswered reports (which may be quite considerable) at that very special and unusual time when such reports are being answered. Also, a follow-up will sometimes produce a favorable impression, but either the station personnel do not get around to answering the letter or the response is lost in the mail. When the station receives another follow-up letter and realizes that either a response to the previous letter was not sent as intended or that its response was lost in the mail, the chances are very good that the new follow-up letter will be answered.

Realistically speaking, once a DX'er has built up a substantial number of stations that have not verified his reports (the number ranging between 40 and 55), it simply is not possible, within the time limitations imposed on most people, to make every follow-up "special." However, this should not stop a DX'er from sending out at least a short simple follow-up immediately after the time interval for a particular station has elapsed. Regularity in sending follow-ups is absolutely vital, and it often happens that the most difficult stations will reply to a short, simple, no-frills follow-up letter after virtually every other device imaginable has been tried—without success.

When a DX'er sends out a group of follow-up letters at one time, there is a tendency to begin looking for replies after about 17 or 18 days. Again, the stoic mental attitude mentioned earlier is important, and one should try not to be too discouraged if nothing comes at that time, or even one or two weeks later. It should be remembered that when sending follow-ups, a DX'er is writing stations that have already evidenced an unwillingness to answer reports, so a low rate of return must be expected. If only a 35% to 40% rate of return can be expected from the original reception reports sent to Latin American stations, logically, a significantly lower rate of return should be anticipated for follow-ups to such stations. This is particularly true if the stations in question have already failed to answer at least one follow-up (in addition to not responding to the original report). While statistics vary with each DX'er, a person should not realistically expect more than a 20% to 25% rate of return on any large group of follow-ups; a 12% to 15% return should be quite satisfactory. While these percentage figures are admittedly low, the overall rate of return will rise dramatically after sending follow-ups on a regular basis over a period of several years. Follow-up success cannot be measured on the basis of one batch of follow-ups but must be judged over a much longer period of time. A 15% rate of return for each of six or seven sets of follow-up letters yields a surprisingly high number of rare QSLs.

Occasionally, it happens that a DX'er may be completely shut out and will not receive even one reply in response to 15 or 20 follow-up letters. This is something which unfortunately happens every once in a while. One of the rather curious aspects of QSL collecting is that people sometimes have extended periods in which they receive very few QSLs, or none at all, and, then, these periods might be followed immediately by a veritable flood of verifications. These "streaks" are completely unpredictable, but they do happen. While a DX'er cannot help being disappointed and discouraged when he has a bad period or cannot give up. One of the true "secrets" of success in collecting QSLs is *perseverance* and *a refusal to quit*. This point has been mentioned numerous times earlier in this chapter, but it is extremely important and cannot be overemphasized or stressed too often.

There is no question that an intensive type of follow-up program is exceedingly time-consuming and requires considerable effort. It can also be quite costly, especially if the DX'er includes return postage with each follow-up letter. But it is necessary to continuously adhere to such a program in order to build up a truly high rate of return with the smaller stations. The unremitting pursuit of a persistent, effective, follow-up campaign, even in the face of adversity, is of the utmost importance, and, usually, the persevering tenacious DX'er is ultimately rewarded with a surprisingly large number of excellent and rare QSLs.

Conclusion

The purpose of this chapter has been to offer suggestions on how to improve the rate of return that a DX'er has in acquiring QSLs, particularly from the smaller stations. Of course, none of these ideas can guarantee success, and no one can say with any degree of certainty what is the best method of approaching a particular station for its QSL. One of the more fascinating aspects of the hobby of QSL collecting is its sheer unpredictability, and, occasionally, people seem to obtain the rarest and most difficult QSLs when they disregard even the most basic reporting techniques. However, the use of sound methods in reporting and in the preparation of follow-ups over an extended period of time should result in a considerably higher rate of return and a much greater degree of success in obtaining QSLs from the numerous shortwave stations around the world.

9

THE ASSOCIATION OF NORTH AMERICAN RADIO CLUBS

Richard T. Colgan

Like so many of us who began our DX'ing careers in the 1950s, it was a Hallicrafters S-38-series receiver that got **Terry Colgan** started in shortwave listening. He has been an avid shortwave listener and DX'er ever since. In addition, his monitoring interests have carried him into mediumwave DX'ing and listening to the VHF/UHF Public Service frequencies.

Terry has held editing positions with the Worldwide TV-FM DX Association and the North American Shortwave Association. He is currently the Executive Secretary of the Association of North American Radio Clubs (ANARC), a position which requires a great deal of his time and energy.

Terry and his wife, Andrea, make their home in Austin, Texas, where he is employed in government service.

Created over two decades ago to unify the radio listening hobby in North America, the Association of American Radio Clubs (ANARC) is one of three similar federations located worldwide. ANARC is an umbrella organization—a club *of clubs*—promoting all aspects of the hobby through its members and its own actions. As the hobby has changed and advanced, so has ANARC, to the point that, in some ways, it does not much resemble the early organization.

Early History

The concept of an alliance of radio DX and listening clubs was not really new. One of the oldest such organizations was the DX-Alliansen in Stockholm, Sweden. But the idea had not been seriously proposed in America until 1963, when Hank Bennett, writing in the *Newark News Radio Club Bulletin*, urged the formation of such an organization to end dissension among the United States clubs. Bennett's comments started Don Jensen thinking, and this led him to draft the constitution for a strictly voluntary organization that would have an appealing set of objectives:

- To promote closer ties between radio clubs.
- To promote the interchange of ideas and information among the clubs.
- To work for the common good of the DX'ing hobby.
- To provide a forum to work out the differences and problems involving radio clubs.
- To provide a medium that could speak for radio clubs and listeners in North America.

Writing the document was the easy part. Jensen then had to "sell" his concept to the clubs. Fortunately, he was highly regarded in the hobby community, both as a DX'er and as an organizer, and this made it possible for him to talk to club officers, board members, and publishers on a one-to-one personal basis. But, a number of stumbling blocks lay in his way.

First and foremost, the clubs had to be convinced that the new organization would not be a "super club" which would compete with them for members. For this reason, Jensen limited membership to clubs *only*. Individual DX'ers could not join and it was over seven years before individuals could even subscribe to the *ANARC Newsletter*.

Second, it was important that the clubs understood that the alliance would not attempt to run them or to dictate to them in any way. In fact, just the opposite would be true. The clubs, through their representatives, would approve the new organization's budgets, vote on all important matters, and elect an Executive Secretary to take care of the day-to-day business. Jensen chose the title Executive Secretary—rather than Executive Director, for example—very carefully, so as not to convey the sense of a position with a great deal of power. The governmental structure was such that a club's participation was to be totally voluntary. Once a member of the alliance, and so long as it met membership requirements, a club could participate in association business as much or as little as it liked.

Third, the idea had to be presented in such a way that each club, regardless of what its radio interests were, could easily see how it could benefit by joining. This may actually have been the most formidable obstacle to adoption of the idea since it existed only on paper!

Despite the problems, Jensen proved adept at persuasion and, on April 20, 1964, eleven clubs approved the constitution of an organization to be called the Association of North American Radio Clubs. The North American Shortwave Association, the Canadian DX Club, the Newark News Radio Club, the American Shortwave Listeners Club, the Kentucky DX'ers Association, the National Radio Club, Inc., the Canadian International DX Radio Club, the Shortwave Listeners-Certificate Hunters Club, the Folcroft Radio Club, and the Great Lakes Shortwave DX Club were the charter members. Don agreed to become Acting Executive Secretary for 6 months to start up the organization; he then was to turn over control to a more permanent Executive Secretary. When the 6 months were up (October, 1964), the clubs found that his talents were still needed and he was elected to a full 2-year term, thereby becoming ANARC's first Executive Secretary. At that time, each club had one, two, or three representatives to ANARC, depending on the number of members it had. These representatives were the organization's decision makers.

Jensen's term of office was a flurry of activity as ANARC generated a great deal of interest and many hobby-related problems were studied. Three special committees were formed to meet hobby needs. The Broadcaster's Liaison Committee was created to open a dialogue with the international shortwave broadcast stations on programming and frequencies. A link between listeners and the manufacturers and dealers of radio equipment was made possible by the Manufacturer's Liaison Committee. The Ethics Study Committee looked into the possibility of a code of ethics for DX'ers.

In 1966, the first ANARC convention was held July 27–29 in Kansas City, MO (Fig. 9-1). The event attracted 16 hobbyists—representing six member clubs —including Don Jensen, Glenn Hauser, and Richard Wood. The program was similar to that of present-day conventions, with lectures, equipment displays, and discussions.

Later that year, with his term drawing to an end, Don Jensen felt that he had taken the Association as far as he could. Despite his original intention to serve only 6 months, he had spent more than 30 months getting ANARC off the ground. Now that it was aloft, he could see that it needed new leadership. Gerry Dexter volunteered for the position and was elected by the clubs, assuring ANARC's continuing operation.

Dexter continued the work begun by Jensen, ANARC's endeavors were expanded, and its status in the hobby world grew. Other Executive Secretaries— Gray Scrimongeur, Wendel Craighead, Al Reynolds, and Dave Browne—put in many thousands of hours on ANARC work. Dave Browne, for instance, served for 8 years—from 1974 to 1982—the longest that anyone has held the position!

Fig. 9-1.
ANARC's first
convention was
held in 1966 in
Kansas City.
This photo
shows the
convention
group, with Don
Jensen, the first
Executive
Secretary,
shown in the
suit and tie at
the left.

ANARC Today

ANARC is essentially the same organization today that it was in 1964. It still carries out most of its operations behind the scenes, letting its member clubs have their well-deserved publicity. Occasionally, ANARC itself surfaces in its publications, radio broadcasts, or its annual conventions.

The Association has enlarged the scope of its work to keep up with the needs of the hobby, and has made some changes in its organizational structure (clubs now have one representative, regardless of their membership size), but there have been no changes in ANARC's basic objectives.

ANARC Member Clubs

With over 10,000 members worldwide, the twenty or so ANARC member clubs are the basic units and the most visible parts of the Association. For a club to become part of ANARC, it must have a certain number of members, publish a newsletter or bulletin on a regular basis, have been in operation for at least one year, and not be a "for profit" venture. Once admitted to ANARC, the club must serve a probationary period before it can become a Full Member. If a club does not continue to meet ANARC's standards, it is dropped from membership. While ANARC membership does not guarantee a club's performance or its long life, it does mean that the club had to meet and maintain a prescribed level of service to its members.

The ANARC clubs cover the spectrum from the very low frequencies to the superhigh frequencies. Whether a listener's interest is tuning the public-service bands, ferreting out a seldom-heard aeronautical beacon, or listening to stereo/AM stations, there will be at least one club that has the kind of information he wants. Some clubs cover only one area of the spectrum—for example, long wave or TV/FM. Others combine several specialties in their bulletins. Medium-wave, shortwave, and high-frequency utility stations make up one such combination. Hobbyists join ANARC clubs for one or more reasons—station program and frequency information, technical data, feature articles, tips for listening and DX'ing, or contact with other listeners. A modest investment in club dues will put all this and more at the listener's fingertips.

Most clubs publish a monthly bulletin, which usually ranges in size from 12 to 90 pages, and a few publish mid-month "updates" that contain up-to-the-minute information. Many clubs put out other publications. Often these are just station listings for one of the "specialty" bands—long wave, medium wave, TV, or FM. Others are technical books or pamphlets on receivers and receiver modifications, antennas and listening accessories, or DX'ing techniques. Several of the clubs also offer reprints of articles from their bulletins. All of these publications are reasonably priced and make an excellent reference library for the listener and DX'er.

A list of the ANARC clubs, with complete information on frequency coverages, membership dues, and sample bulletin prices, is available (in the United States) for 25 cents in coin and a business-size self-addressed stamped envelope (SASE); in Canada, for fifty cents in mint Canadian stamps; and elsewhere, for three International Reply Coupons (IRCs). Requests should be sent to ANARC Club List, 1500 Bunbury Drive, Whittier, CA 90601, USA.

Annual ANARC Conventions

Beginning in 1966 in Kansas City, Missouri (Fig. 9-1), a yearly ANARC convention (now called ANARCON) has been held. In 1967, the site was Chicago; then in the successive years, the locations were:

1968	Omaha, Nebraska	1976	Los Angeles, California
1969	Toronto, Ontario, Canada	1977	Palatine, Illinois
1970	Chicago, Illinois	1978	Montreal, Quebec, Canada
1971	Indianapolis, Indiana	1979	Minneapolis, Minnesota
1972	Boston, Massachusetts	1980	Irvine, California
1973	San Diego, California	1981	Thunder Bay, Ont., Canada
1974	Bronxville, New York	1982	Montreal, Quebec, Canada
1975	Montreal, Quebec, Canada	1983	Washington, DC

ANARCON '84 in Toronto, Canada, and ANARCON '85 in Milwaukee, Wisconsin, round out the list of ANARC's first 20 conventions. Fig. 9-2 shows Ian Mc Farland of Radio Canada International speaking at a seminar publicizing the radio hobby during the 1983 convention at Washington, DC.

Fig. 9-2. The 1983 ANARC convention in Washington, DC drew over 300 attendees. Here, Radio Canada International's Ian Mc Farland speaks at a convention seminar. (*Photo by David Klein*.)

These conventions have long been gathering places for shortwave DX'ers and international broadcast personalities. A major reason for this shortwave orientation is that—unlike the medium-wave or TV/FM clubs—few of the North American shortwave clubs hold annual meetings. Thus, the ANARC conventions have filled this need. But that does *not* mean that other DX'ing interests are forgotten or ignored at ANARCONs. Each year's program includes something for everyone.

A typical ANARC convention extends over a weekend (Friday-Saturday-Sunday), usually in mid- to late July. The schedule on Friday begins at noon, and includes equipment exhibits and displays, the ANARC annual business meeting, the opening ceremonies, and an evening reception. Saturday is a full day of presentations and discussions on shortwave, medium wave, utilities, and much more. Included in this schedule the always-popular Broadcaster's Forum, with a chance to meet and ask questions of such shortwave personalities as Ian Mc Farland of Radio Canada International, Jeff White of Radio Earth, and Jonathan Marks of Radio Netherlands. The day ends with the ANARC banquet, awards, and the ANARC raffle. Sunday features a seminar or two, followed by what many consider to be the highlight of the conventions—the HAP (Handicapped Aid Program) auction. Collectors of bumper stickers, QSLs, pennants, or anything else relating to radio and TV stations have a chance to bid on these items at the auction.

ANARCON sites are selected two years in advance, with the events hosted by ANARC member clubs or international broadcasters. The best way to keep current with these conventions is to listen to the major DX programs or to read the ANARC member-club bulletins.

ANARC Newsletter

What began as a means of sharing information among the ANARC clubs was taken "public" in 1971, when subscriptions were offered to individuals. The

ANARC Newsletter is the official publication of the Association and is the only comprehensive source of news about ANARC and its clubs.

Regular features include "Clubscan," which highlights items from member-club bulletins and lists selected articles of interest; "Marketplace Report," which describes the latest receivers and listening accessories; and "Club News," which includes a calendar of regional and national club conventions and meetings.

The *ANARC Newsletter* is available for an annual fee of U.S. $7.50 to subscribers in North America (U.S. $10.00 by airmail elsewhere). A sample copy is available for 60 cents in coin and a business-size SASE (in the U.S.); 75 cents in mint stamps (in Canada); or 4 IRCs (elsewhere). The address is ANARC Newletter, 1500 Bunbury Drive, Whittier, CA 90601 USA.

ANARC Broadcasts

Each month, ANARC has two features as part of Ian Mc Farland's *SWL Digest* program over Radio Canada International. The "ANARC Report" is broadcast the first weekend of the month and the "Marketplace Report" is heard on the third weekend. These broadcasts contain information and news from the *ANARC Newsletter*, club bulletins, and the ANARC committees.

SWL Digest can be heard every weekend in the African, European, North American, and Caribbean services of RCI. A schedule of times and frequencies can be obtained by writing to Radio Canada International, P.O. Box 6000, Montreal, PQ H3C 3A8, Canada.

ANARC Committees

Much of the work done by ANARC is carried out by its volunteer committees. Even though they are not seen, they are an integral part of ANARC's functioning.

The *DX Equipment Information Committee* keeps the hobby informed of the latest developments in receivers, antennas, and accessories. Its work is reported in the "Marketplace Report" in the *ANARC Newsletter* and over RCI.

Contacting shortwave broadcasters to determine what their verification policies are, and what is necessary to get QSLs from them, is the job of ANARC's *QSL Committee*.

The *Frequency Recommendation Committee* works closely with international broadcasters to monitor and report on their signals, and to help them select useful frequencies to avoid interference and adverse propagation.

The *Over-the-Horizon Radar Committee* researches and reports on the U.S. and Soviet over-the-horizon (OTH) radar systems. Information is furnished to interested individuals in *Backscatter*, an occasional publication of the Committee.

ANARC's *Computer Information Committee* disseminates noncopyrighted computer programs relating to the DX'ing and listening hobby. Most programs are for the Commodore C-64 and VIC-20, but there are several excellent ones for the Timex-Sinclair, the Apple, the TRS-80, and other major brands. All software is available in hard copy; some is also available on cassette or disk.

Regular reports on the committees are broadcast in the ANARC Report and printed in the *ANARC Newsletter*. The September issue of the *Newsletter* usually contains the annual activity reports of all the committees.

Summary

Much more could be said about the ANARC, its history and its operations, but to do justice to the organization would require a separate book. The best way to learn more is to read the Association's *Newsletter*, listen to its broadcasts, attend the ANARCONs, and—most importantly—join its clubs. These things will also add to knowledge and enjoyment of the listening hobby. And that is really the bottom line to what ANARC is all about!

10

COMPUTERS AND DX'ING

Bill Krause, Ph.D.

Like many of us, **Bill Krause** started his DX'ing career at an early age—13, in his case. His special areas of interest include the Tropical Band stations of Indonesia, South Asia, and sub-Saharian Africa.

For several years, Bill has been deeply involved in computers and their ever-growing links to radio monitoring and DX'ing. He chairs the ANARC Computer Information Committee which explores this field and makes information available to those desiring to add computers to their gear to use with their hobby.

Bill feels very strongly that computers are on the verge of revolutionizing our hobby. He stresses the need to become comfortable in using computers so that we may continue to enjoy listening and DX'ing as the hobby changes in response to new technologies.

Bill lives with his wife and four children in a suburb of Minneapolis, Minnesota.

Five years ago, when home computers were just beginning to become popular, a very good friend of mine told me what wondrous machines they were, and how I ought to have one. I politely informed him that I couldn't possibly think of enough uses to justify buying one. Now, five years later, our family owns four computers and a room full of peripherals, accessories, and software. I have had to eat my words, because I have thought of more than enough uses, just relating to my radio hobby, to justify the cost of my computer equipment, and that's without considering all the nonradio uses our family makes of the computers.

I am a very serious DX'er, and have been for a number of years. A little more than three years ago, I invested $1500 in a quality receiver. Its performance has not disappointed me, and has enabled me to get some very excellent DX. Then, about a year ago, after having first experimented with a more modestly priced computer, I spent about $200 on a larger-capacity computer. Its performance has been equal to that of my receiver. First with the smaller-capacity computer, and now with the somewhat larger model, I have learned to use the computer to make my DX'ing simpler and more effective.

The Computer as an Accessory

Most of you reading this will probably not have spent quite as much on your radio equipment as I have, but you will still have a very sizeable investment in equipment. Also, most of you will have invested in equipment and accessories designed to improve the performance and operation of the receiver. And, like me, most of you will be on the lookout for that one special piece of equipment that will give you that extra edge in performance that all DX'ers always seem to be looking for. The point I am trying to make is that a relatively modest-priced home computer is a very cost-effective radio accessory that will significantly improve both your equipment's and your performance, and will add measurably to your enjoyment of the radio hobby.

Using a Computer

First, I will introduce you to the many and varied ways that you can use a computer in the radio hobby. Then, I will go over specific applications for specific special areas of the radio hobby, such as SWL'ing, DX'ing, and the utilities. We'll look at some simple programs, and we'll view the output of some longer, more complex programs. If this whets your appetite and you're still interested, I'll review some of the computer equipment you'll want to use with your radio, and mention some of the best places to get it. Then, we'll have a closer look at some of the more unusual and specialized applications of computer technology to the radio hobby. Finally, we'll conclude with more sources of information, a look into the future, and a few parting words.

We'll assume that you don't have any use for a computer, other than a need to use it for your radio hobby. Most people will eventually find a considerable number of applications for a home computer beyond just using it as a radio accessory but, for the purposes of this discussion, we will assume that you are using the computer only as an accessory to your radio and *that* is your only reason for getting one.

Why spend the money, even the relatively modest amount required, to get a computer to use with your radio? How can this bundle of IC chips attached to a typewriter-like keyboard help with your radio? In answer, there are a number of ways you can use a computer with your radio, and which of them you use depends to a large degree on just how you use your radio.

Cataloging Your Schedules

None of us, of course, are just DX'ers, or just SWL'ers, or just into utilities. Each of us in the radio hobby has his or her own unique blend of emphasis. I am primarily a Tropical Band DX'er, but I listen to the BBC all the time. Thus, how the computer is used as part of the radio hobby will, to a very large extent, depend on just how we listen. The SWL'er, who is primarily interested in program listening, will find different uses for a computer than will a DX'er. But there are uses for each particular emphasis, and there are some uses that all radio enthusiasts will share. Thus, there is something for everyone.

I have a desk drawer full of program schedules, some of which are current, but many of which are long outdated. Most SWL'ers go to rather elaborate efforts to make their favorite program schedules as easily accessible as possible. It is inherent in their nature to want to know just where to tune to, with as little hassle as possible in finding that information. Some, like me, stuff a drawer full of the schedules that friendly broadcasters are constantly sending out. Then, they spend an indeterminable amount of time searching through the drawer looking for just the right schedule. Others clip charts out of some of the hobby magazines that publish this information. Some make charts that they tape to the walls of their shack, or even to their radios. Some use index cards. Finally, quite a number of SWL'ers subscribe to various services that will do this for them. What it all boils down to is that most SWL'ers go to a great deal of time, effort, and, sometimes, expense to make program and scheduling information as easily accessible as possible, but usually with only limited degrees of success.

Organizing Your Schedules

One thing that every computer can do well is organize data. Thus, home computers are well suited to fulfilling this need of the SWL'er. An advantage that computers have over the traditional means of carrying out this task is that they can usually do it much faster, and in a more satisfying manner than other methods. Speed is particularly significant since looking through a drawer for just the right schedule or checking a stack of index cards can take a fair amount of time. Even checking a chart taped to a radio can be a hassle, particularly if you're looking for the schedule of more than one country. Of course, it takes a bit of time to load a schedule-type program into the computer, but once it's loaded, it's there as long as the computer is on, and it provides almost instant access to the desired information. Many computers have built-in clocks that can be used as timers.

SWL Shortwave Programs

Public domain software, which will automatically display the schedules of the stations broadcasting to North America and the times when they are actually broadcasting, is currently available for several computers. Also, several programs have been written for the SWL'er by fellow SWL'ers to do the sort of things I just mentioned. The pioneer in this field is Bill Cole of North Cape May, New Jersey, who first worked with a VIC-20 and later with a Commodore 64. His first program was "English Language Broadcasts to North America." He wrote this program incorporating the data appearing in *English Language Broadcasts to North America*, a regular feature by Roger Legge that appears in *FRENDX*. Bill's program, by the same name, was originally published in *FRENDX*. It is now widely circulated, and is available from the ANARC Computer Information Committee, from Bill Cole, himself, and from other sources as well.

Bill Cole's program was originally written for the VIC-20, but has now been translated for use on several other types of computers. Bill's program will enable the user to access the information by typing in either the time period of interest, the name of the station, or the frequency. The program will also use the VIC's built-in clock to automatically display the stations broadcasting in English at a given time, assuming the user has selected that option and has set the computer's clock. Bill has revised this program several times to reflect the latest data, and to incorporate faster and more sophisticated data sorting techniques. Samples of the output of this program can be seen in Fig. 10-1. George Wood, of Radio Sweden International, also using a C64, has developed a similar program for English broadcasts to Europe. His program is formatted somewhat differently, but it also provides a very useful output.

Other Programs

Now that Bill is working with a Commodore 64, he has utilized the computer and his programming skills to develop a new series of programs even more useful to the SWL'er. In this series of programs, he incorporates a World Map Clock as a menu. This menu displays a world map, local time, GMT, and the local time for selected cities around the world. The user can select from the program options found in the ELBNA programs. An additional feature of the "SWL Guide" program allows the user to instantly access propagation data such as sunrise/sunset, distance, bearing, and MUF for any of the cities whose schedules are included as part of the program. Also, with the input of proper data, independent propagation calculations may be made for sites other than those included in the program. The program is a little long, and may take tape users some time to load, but it has been found to be tremendously useful.

Another question the SWL'er often faces is the time when a certain program or type of program is on. When, for instance, is your favorite DX program on, or when are music programs of different specific stations being broadcast? Some

Fig. 10-1.
Computer
printout of
English
language
shortwave
broadcasts.

```
30 - 115 GMT
5910 KHZ    9880 KHZ    0 KHZ
BRT - BRUSSELS BELGIUM
BEAMED TO NORTH AMERICA
DAYS ON: MONDAY THRU SUNDAY


100 - 157 GMT
5930 KHZ    7345 KHZ    9540 KHZ
R PRAGUE - PRAGUE CZECHOSLOVAKIA
BEAMED TO EAST NORTH AMERICA
DAYS ON: MONDAY THRU SUNDAY


100 - 145 GMT
9730 KHZ    11975 KHZ   0 KHZ
RBI - BERLIN GDR
BEAMED TO EAST NORTH AMERICA
DAYS ON: MONDAY THRU SUNDAY
```

magazines, such as *RIB* and *DX ONTARIO* attempt to address this, but, generally, this information is not readily available as mere scheduling information. There are no programs, that I am aware of, which address this problem with quite the same degree of polish as does Bill Cole's program for mere frequency scheduling, although George Wood does approach it to a very limited basis in his program. However, several programs are available that do deal with the time-of-day problem to a larger extent. Generally, these programs are set up to organize and sort data in this matter of broadcasting times. Generally, what they lack is the specific data itself, so that while necessary data is already included in the ELBNA family of programs, the individual users of these programs must themselves fill in the appropriate data. This involves a bit of work on the user's part, but it is really necessary, because due to the nature of the issue they address, these programs generally must be personalized to suit the individual requirements of the user. Generally speaking, instructions for incorporating data into a program is usually included with the program. For an illustration of the wording of these programs, see Fig. 10-2. I have found that the work involved in customizing the program is minimal and the results very satisfactory.

World Radio TV Handbook

Most SWL'ers (indeed, almost everyone who is serious about shortwave radio) are familiar with the *World Radio TV Handbook*. This almanac-like book, which is published yearly, contains a wealth of information about radio and TV stations throughout the world. It also contains a great deal of material about the

Fig. 10-2. A computer program designed to handle program information and types.

```
100 PRINT"(CLR)":REM ***ARRAY SORT***
105 GOSUB2000: REM TITLE
110 PRINT"(C/DN)(C/DN)(C/DN)(C/DN)(C/DN)(C/DN)(C/DN)(C/DN)(C/DN)(C/R
T)(C/RT)(C/RT)(C/RT)NEED INFORMATION (Y/N)?"
111 INPUTA$
112 IFA$<>"Y"THEN115
113 IFA$="Y"THEN GOSUB 3000
115 PRINT"(CLR)(C/DN)(C/DN)(C/DN)(C/DN)(C/DN)INPUT A NUMBER EQUAL TO OR GR
EATER THAN";
117 PRINT"THE # OF LOGGINGS IN YOUR DATA LISTINGS"
119 INPUT DL
120 F=10:DIMA$(DL,F)
130 I=1
135 PRINT"(CLR)(C/DN)(C/DN)(C/DN)(C/DN)(C/DN)(C/DN)(C/DN)(C/DN)(C/RT
)(C/RT)(C/RT)(C/RT)(C/RT)READING DATA"
140 FOR J=1 TO F
150 READ A$(I,J)
160 IFA$(I,J)="END"THEN 300
170 NEXT J
180 I=I+1
190 GOTO 140
200 LE=0:REM<LE=LOG ENTRY PRINT COUNTER:REF LINES 340,535>
300 PRINT"(CLR)(C/DN)(C/DN)SELECT ONE OF THE FOLLOWING NUMBERS (1 TO 8)"
302 PRINT"(C/DN)(1)OPENING TIME(GMT)":PRINT"(2)CLOSING TIME(GMT)"
303 PRINT"(3)STATION":PRINT"(4)COUNTRY":PRINT"(5)BEST FQ"
304 PRINT"(6)OTHER FQS":PRINT"(7)DX PROGRAMS"
306 PRINT"(8)DAY OF THE WEEK"
320 INPUT X
321 ON X GOTO 322,323,324,325,326,327,328,329
322 PRINT"OPENS AT (GMT)":GOTO330
323 PRINT"CLOSES AT(GMT)":GOTO 330
324 PRINT"WHICH STATION":GOTO 330
325 PRINT"WHAT COUNTRY":GOTO 330
326 PRINT"WHAT FREQUENCY":GOTO 330
327 PRINT"ALTERNATE FREQUENCY":GOTO 330
328 PRINT"TYPE THE LETTERS DX TO LIST DX PROGRAMS":GOTO 330
329 PRINT"WHAT DAY OF THE WEEK(SUN,MON,TU,WED,THU,FRI,SAT)":GOTO 330
330 INPUT K$
332 PRINT"(CLR)":FORL=1 TO I
340 IFA$(L,X)=K$ THEN Q=1:LE=LE+1:GOSUB500
350 NEXT L
360 IFQ<>1THENPRINTK$"NOT FOUND. PL. CHECK YOUR INPUT"
370 Q=0
375 PRINT"(C/DN)(C/DN)ANOTHER SELECTION(Y/N)":INPUT B$
380 IF B$="" THEN 380
390 IF B$="Y" THEN 300
400 IF B$="N" THEN PRINT"(CLR)HOPE YOU FOUND THE PROGRAM USEFUL"
410 PRINT"(C/DN)(C/DN)YOUR SUGGESTIONS ARE MOST WELCOME."
415 PRINT"PLEASE CONTACT ME AT THIS ADDRESS:"
420 PRINT"(C/DN)(C/DN)(C/DN)(C/RT)(C/RT)(C/RT)NEVILLE DENETTO"
425 PRINT"(C/RT)(C/RT)(C/RT)(C/RT)KAKABEKA FALLS"
430 PRINT"(C/RT)(C/RT)(C/RT)(C/RT)ONTARIO, CANADA"
435 PRINT"(C/RT)(C/RT)(C/RT)(C/RT)POT 1W0"
450 END
500 Q=1
510 FORJ=1TOF
515 IFLE=(6) THEN LE=0 :GOSUB 1000
520 IFJ=XTHENPRINTCHR$(32);
530 PRINTA$(L,J)CHR$(144)CHR$(32);
540 NEXT:PRINT:RETURN
550 :
1000 PRINT"(CLR)(C/DN)(C/DN)(C/RT)(C/RT)(C/RT)(C/RT)PRESS ANY KEY"
1010 GETA$:IFA$=""THEN1010
1020 RETURN
1030 :
2000 PRINT"(CLR)":A$="(C/DN)*"
2020 FORN=5TO15:PRINTTAB(N)A$;:FOR1=1TO100:NEXT:NEXT
2030 FORM=1TO8:PRINTTAB(5)A$;:FOR1TO100:NEXT:NEXT
2040 FORP=5TO17:PRINTTAB(P)A$;:FORJ=1TO100:NEXT:NEXT:PRINT
2050 PRINT"(C/DN)(C/DN)(C/DN)(C/DN)(C/DN)(C/DN)(C/DN)(C/RT)(C/RT)(C/RT)(C/RT)(C/
RT)(C/RT)(C/RT)(C/DN)(C/RT)SUPER":PRINT
2055 FORI=1TO500:NEXT
2060 PRINT"(C/RT)(C/RT)(C/RT)(C/RT)(C/RT)(C/RT)(C/RT)(C/RT)(C/RT)(C/RT)(C/DN)SEL
ECT"
2070 FORI=1TO3000:NEXT
2100 RETURN
3000 PRINT"(CLR)(C/RT)(C/RT)(C/RT)THIS PROGRAM ALLOWS  YOU TO SELECT LOGOINGS BY
:"
```

Fig. 10-2 (cont.)

```
3010 PRINT"  1. STARTING GMT"
3020 PRINT"  2. CLOSING TIME"
3030 PRINT"  3. STATION NAME"
3040 PRINT    "  4. COUNTRY"
3050 PRINT    "  5. BEST FQ"
3052 PRINT    "  7. DX PROGRAMS"
3053 PRINT    "  8. DAY OF THE WEEK"
3055 GOSUB1000
3060 PRINT"(CLR)(C/DN)(C/DN)(C/DN) INITIALLY, YOU WILL  ASKED TO INPUT A NUMBER
BASED ON ";
3062 PRINT"THE NUMBER OF LOGGINGS IN YOUR PROGRAM"
3064 PRINT"(C/DN) PLEASE PAY STRICT    ATTENTION TO THIS    INPUT"
3066 GOSUB1000
3068 PRINT"(CLR)(C/DN)(C/DN) MAKE SURE THAT ALL   DATA ENTRIES ARE   CORRECT; IN
 PARTICULAR"
3070 PRINT" 1.THAT YOU HAVE NO     MORE THAN (C/DN)10(C/RT)       FIELDS"
3072 PRINT" 2.THERE SHOULD NOT     BE MORE THAN      (C/DN)9 COMMAS(C/UP) IN ANY
LOG ";
3074 PRINT"ENTRY"
3076 PRINT"(C/DN)(C/DN)STUDY THE ENTRIES   MADE IN THIS PROGRAM, TO SERVE AS A G
UIDE"
3078 GOSUB 1000
3079 RETURN
5000 REM<<< DATA FORMAT FOR 'ARRAY SORT'>>>
5002 REM<GMT(START),GMT(END),STN, COUNTRY,BEST FQ, OTHER FQS,DX(=DX
PGM),DAY,OTHER INFO>
5004 REM<ALL ITEMS MUST BE SEPARATED BY COMMAS EXCEPT
UNDERCLOSECLOSECLOSE'OTHER INF>
5005 REM <ITEMS IN 'OTHER INFO' MAY BE SEPARATED BY A SPACE...NEVER BY A
COMMA>>>
5006 :
5007 REM<TYPICAL ENTRY:0100,1200,INDIA,AIR,11620,9912,XX,DLY,PR NORMALLY,5007>
5008 :
5010 DATA0915,1100,FR3,NEW CAL,11710,7170,XX,DLY,XXX,(5010)
5015 DATA1000,1100,RK,KOREA,9570,0000,XX,DLY,XXXXX,(5015)
5020 DATA1000,1100,AIR,INDIA,17875,0000,XX,DLY,PR NORMALLY,(5020)
5025 DATA1000,1620,PERTH,AUSTRALIA,9610,6140,XX,DLY,XX,(5025)
5030 DATA1000,1700,VON,NIGERIA,15120,0000,XX,DLY,XX,(5030)
5035 DATA1030,1130,SLBC,SRI LANKA,17850,15120 11835,XX,DLY,XX,(5035)
5040 DATA1100,1130,R.MOGADISHU,SOMALIA,9585,0000,XX,DLY,XX,(5040)
5045 DATA1100,1200,V.O.ASIA,TAIWAN,5980,0000,XX,DLY,XXX,(5045)
5055 DATA1100,1400,4VEH,HAITI,11835,9770,XX,DLY,XXX,(5055)
5060 DATA1130,1155,R.NAC.,ANGOLA,11955,9535,XX,MON-FRI,XX,(5060)
5065 DATA1130,1210,R.POLONIA,POLAND,17865,11840 9675 9525,XXDLY,XXX,(5065)
5070 DATA1130,1230,R.THAILAND,THAILAND,11905,9535,XX,DLY,XX,(5070)
5075 DATA1300,1600,R.MALAYSIA,SABAH,5980,4970,XX,DLY,XXX.(5075)
5085 DATA1400,1425,R.FINLAND,FINLAND,25950,15400,XX,MON-SAT,XX,(5085)
5090 DATA1230,1257,ORF,AUSTRIA,21615,0000,XX,DLY,XXX,(5090)
5100 DATA1230,1430,SLBO,SRI LANKA,15425,9720,XX,DLY,XX,(5100)
5110 DATA1300,1325,R.FINLAND,FINLAND,2145,15400,XX,MON-SAT,XXX,(5110)
5115 DATA1300,1330,R.BUCHAREST,ROMANIA,17850,15250 11940,XX,DLY,XXX,(5115)
5120 DATA1330,1415,RBI,ODR,214565,0000,XX,DLY,XXX,(5120)
5125 DATA1330,1420,RN,NETHERLANDS,17605,11930,DX,THU,XXX,(5125)
5130 DATA1330,1500,AIR,INDIA,15335,11810,XX,DLY,XXX,(5130)
5135 DATA1330 .1525.RF.FOM:AMO.21475.15400,XX,SUN,XXX,(5135)
5140 DATA1330,1000,BBC,UK,21710,21600,21470 15070,XX,DLY,XXX,(5140)
5150 DATA1400,1430,KTWR,GUAM,21730,21700,26030,25730,17830,XX,SUN,XX,(5150)
5155 DATA1400,1430,RS,SWEDEN,2165,0000,DX,TUE,XX,(5155)
5160 DATA1400,1430,RTASHKENT,CSSR,11765,9600,9540,6025,5945,XX,DLY,XX,(5160)
5165 DATA1400,1430,RULAN BATOR,OUTER MONGOLIA,12070,9575,XX,NOT SUN,XX,(5155)
6000 DATA1400,XX,XX,XX,XX,XX,X,X,X,X,
6010 DATA1400,Q,Q,Q,Q,Q,Q,Q,Q,Q
6020 DATA1400,W,W,W,W,W,W,W,W,W
6030 DATA1400,R,R,R,R,R,R,R,R,R
6040 DATA1400,T,T,T,T,T,T,T,T,T
6050 DATA1400,Y,Y,Y,Y,Y,Y,Y,Y,Y
6060 DATA1400,U,U,U,U,U,U,U,U,U
6070 DATA1400,I,I,I,I,I,I,I,I,I
10000 DATA END
20550 FORI-12100;NEXT
63000 REM***VIC BI-DIRECIONAL SCROLL***ANY MEMORY***'COMPUTE',FEB 1983,P.189***
63001 SA-PEK(44)*256+PEEK(43)-1
63002 LN-PEK(SA+3)+PEEK(SA4)*256
63003 PRINT"(CLR)(C/RT)30TO63010":PRINT"LIST";LN;
63010 IF PEEK(197)-5 THEN 63100:REM TEST FOR '-' KEY
63020 IF PEEK(197)-61 THEN 63200:REM TEST FOR '+' KEY
63030 GOTO 63010
63100 SA-SA+5:GOTO 63002
63200 SA-SA-L:IF PEEK(SA)-0 AND PEEK(SA-4)<>0 AND PEEK(SA-3)<>0 THEN 63002
63210 GOTO 63200
```

radio hobby in general. A great many of us are currently shelling out nearly $20.00 annually to purchase a copy. It has been suggested that information about the various countries and their radio operations could be easily stored in files on a computer disk. Also, the information would probably be easier to access, more attractively displayed, and would take up less space than the *WRTVH* currently does. Since much of the information in the *WRTVH* does not change from year to year, updating the information would also be fairly convenient and would be much less expensive than buying a new *WRTVH* every year. Some SWL'ers are already working on projects of this nature. Perhaps in the future, a complete file will be readily available to the SWL community.

Is a Computer a Good Buy?

A computer isn't the answer to every need of the SWL'er. It won't make static go away, it won't eliminate interference, and it won't even make the broadcasters choose their frequency more wisely. It can, though, as you will see, be a very useful tool. Does every SWL'er need one? That depends. Generally speaking, if you're only going to listen to a few major international broadcasters and have absolutely no other use for a computer or any interest in radio, beyond listening to a small handful of broadcasters, then it might not be worth it. However, if you have any possible use for a computer, beyond the minor problem of keeping track of the schedule of a few broadcasters, or if you can possibly use it for other than radio, then a computer can be a very useful tool. However, how "much" computer you get will depend on a number of different factors that we will look at a bit later on in the chapter.

I have just gone to some length to tell you how useful a computer is to the SWL'er. Now I am going to let you in on a little secret. A computer is an even more useful tool for the DX'er. This judgment is based not only on the fact that I am primarily a DX'er, myself, but also due to the fact that the computer's ability to organize and effectively present data is a characteristic that is just as important to the DX'er as it is to the SWL'er. Add to this the computer's incomparable ability to make complex calculations far more rapidly and accurately than is possible with any other tool that is available to the DX'er and you will see its potential. Since calculation and data organization are different utilizations of the computer for the DX'er, let's take a moment to look at them separately.

Data Storage

First, let's see how the DX'er can use the computer's ability to sort and present data to his or her advantage. In some respects, this usage is similar to the way that an SWL'er uses the computer, and, in some ways, it is different. First, of course, data may be sorted in ways similar to the methods used by the SWL'er— data may be sorted by time, frequency, station, etc. Often, both the SWL'er and the DX'er will use the computer to replace their index cards. However, while the SWL'er may seek programs with the data already built into them, such as

Bill Cole's ELBNA programs, the DX'er must always use programs that lack data (the data must be entered on prompting) because no two DX'ers will have the same data. Every DX'er must personalize his or her programs. This, of course, means some familiarity and work with the computer for the DX'er entering the data. (Of course, the DX'er will have to record that data somewhere, whether the data are on index cards or on a computer disk. My experience has been that it is just as easy and as quick to enter the data on the computer and then save the data to diskette as it is to record data on index cards. The computer's big advantage is the speed and manner in which it presents the data upon recall.)

Programs

One result of a program which lacks data is that the programs themselves tend to be reasonably short. The simplest programs may not be much more than a dozen or so lines long; even the most complex and sophisticated of these programs are hardly ever more than several tens of lines long. The difference in the programs is the speed with which they operate and the variety of ways they present the data. Fig. 10-3 shows examples of both a simple and a more complex program. Even the slowest and simplest of programs operates very fast compared to the manual means that have been the only alternative in the past.

Using one of the programs, I can obtain in several seconds the following information on any of the countries I am interested in: times and frequencies, number of loggings, and dates of logging. The ability to do this quickly enables me to focus my attention more precisely than was previously the case, and it

Fig. 10-3. Data management programs.

```
10 REM***************** *        PROGRAM BY
20 REM PAUL BLANCHARD 148 CAROLINE ST. ROCHESTER N.Y.14620
30 REM FEEL FREE TO CONTACT ME FOR MORE INFO OR TO MODIFY
35 REM THIS PROGRAM.
40 REM PLEASE DON'T CONTACT ANARC. BEST OF 73'S.
50 POKE36875,24:PRINT"(CLR)     "S.W.RADIO LOG."
60 FORQ=1TO750:NEXT
70 PRINT"(C/DN)(C/DN) USE FREQ,GMT,STATION,OR TYPE OF PROGRAM.":K$=" KHZ."
80 PRINT"(C/DN) EXAMPLE...FORMAT"
90 PRINT"(C/DN) STATION...R.CANADA  GMT...0200  FREQ...5960 (KHZ)"
100 PRINT"      TYPE...DX"
110 PRINT"     '  '...MAILBAG"
120 PRINT"(C/UP)(C/UP)";X$;:INPUT"SELECTION";S$:PRINT"(CLR)";S$;"(C/UP)"
130 READA$;B$;C$;D$;E$;F$
140 IFA$="-1"THEN RESTORE:X$="   ANOTHER ":GOTO120
150 IFA$=S$THENPRINTD$;" GMT"TAB(9)B$TAB(15)C$
160 IFB$=S$THENPRINTD$;" GMT"TAB(10)A$
170 IFC$=S$THENPRINTD$;" GMT"TAB(10)A$
180 IFE$=S$THENPRINTA$TAB(14)D$TAB(13)F$TAB(22)B$TAB(28)C$TAB(33)K$TAB(44)"--"
190 IFD$=S$THENPRINTC$TAB(6)B$TAB(12)A$
200 GOTO130
210 DATAR.MOSCOW,15100,15150,1600,,
220 DATAR.MOSCOW,9640,9600,0200,,
230 DATAR.CANADA,5960,9755,0300,MAILBAG,SUN
240 DATAR.CANADA,5960,9755,0200,DX,SUN
250 DATAR.CANADA,5960,9755,0000,MAILBAG,SUN
260 DATANEDERLAND,6165,9590,0300,DX,THU
270 DATAW.GERMANY,6040,6145,0100,,
60000 DATA-1,,,,,DON'T REMOVE THIS DATA LINE
```

(A) Simple program.

Fig. 10-3 (cont.)

```
5 PRINT"{CLR}{C/RT}{C/RT}{C/RT}{C/RT}"
10 PRINT"   *************"
12 PRINT"   {C/RT}*           {C/RT}{C/RT}*"
15 PRINT"   {C/RT}*{C/RT}{C/RT}DUAL  SORT{C/RT}{C/RT}*"
16 PRINT"   {C/RT}*{C/RT}{C/RT}         {C/RT}{C/RT}*"
20 PRINT"   {C/RT}*************"
21 FORI=1TO4000:NEXT:PRINT"{CLR}"
25 PRINT"THIS PROGRAM LETS YOU SORT LOGGINGS BY ----":PRINT"DAY OF THE
WEEK---";
26 PRINT"   GMT---      STATION---OR  FREQUENCY"
30 GOSUB 3030
75 CLR:RESTORE
76 DIMC$(100):REM<<<<<DIMC$(Z)..'Z'MUST HAVE A VALUE GREATER THAN
77 REM...THE NUMBER OF DATA LINES
80 PRINT"{CLR}{C/DN}{C/DN}{C/DN}{C/DN}{C/DN}{C/DN}{C/DN}{C/RT}{C/RT}{C/RT}{C/RT}
DATA BEING READ"
90 N=1
91 READ C$(N):REM C$(J)=C$(N)
92 IFC$(N)="END"THEN 100
93 N=N+1
94 GOTO 91
100 N=N-1
240 PRINT"{CLR}{C/RT}{C/RT}{C/RT}"
250 PRINT"SELECT ONE OF THE FOLLOWING LETTERS DEPENDING ON HOW YOU
WANT THE LOG"
251 PRINT"SORTED"
255 PRINT:PRINT
260 PRINT"D=DAY OF THE WEEK"
261 PRINT" G=GMT"
262 PRINT" S=STATION"
263 PRINT" F=FREQUENCY"
264 PRINT
265 PRINT"SELECT EITHER  D, G, F, OR S ":PRINT
267 PRINT"YOUR SELECTION IS...?"
290 INPUT D$
293 IFD$="D" THEN A=1:GOTO 320
294 IFD$="G" THEN A=6:GOTO 320
295 IFD$="S" THEN A=12:GOTO 320
296 IFD$="F"THEN A=17:GOTO 320
300 REM: SET THE VALUE OF EACH FIELD TO THE VALUE OF (5)
320 B=5
321 PRINT"{CLR}{C/DN} S=SHELL SORT, OR B=BUBBLE SORT..SELECT {C/RT}S OR B ":INPU
T A$
322 IFA$="B"THEN 330
323 IFA$="S"THEN 1200
330 PRINT"{C/DN}BUBBLE SORT":REM<<<BUBBLE SORT BEGINS>>>
331 TI$="000000"
340 FOR J=1TON-1
350 IF(MID$(C$(J),A,B)(MID$(C$(J+1)),A,B,))  THEN 420
360 T$=C$(J+1)
370 FOR K=J TO 1 STEP-1
380 IF(MID$(C$(K),A,B,)(MID$(T$,A,B,)) THEN C$(K+1)=T$: GOTO 420
390 C$(K+1)=C$(K)
400 NEXTL
410 C$(1)=T$
420 NEXT J
422 AA=TI
425 T1=INT ((((AA)/60))*100)/100: PRINT"BUBBLE SORT TIME=";T1;"SEC"
426 PRINT"SHELL SORT TIME =";T2
430 REM****SORT ENDS****
435 IF=17 THEN B=30
436 PRINT"{C/DN}DAY GMT STN FQ"
437 PRINT"{C/RT}——— ——— ——— ——"
438 Q=1
440 FORJ=1TON
445 PRINT CHR$(144)C$(J)
450 IF LEN(C$(J))(23THENPRINT"{CLR}"CHR$(28)TAB(A-1)MID$(C$(J),A,B):GOTO457
457 IF J=Q*10 THEN Q=Q+1:GOSUB 3030
460 NEXT J
466 PRINT"***ASTERISKS INDICATE DX-PROGRAMS***"
480 PRINT "{C/DN}DO YOU WANT THE SORT CHANGED(Y/N)" :INPUTA$
481 IFA$="Y"THEN 240
482 IF A$="N"THEN 1000
1000 PRINT"{CLR}{C/DN}{C/DN}{C/DN}{C/DN}{C/DN}{C/DN}{C/DN}{C/DN}HOPE YOU FOUND T
```

(B) Complex program

Fig. 10-3 (cont.)

```
HE PROGRAM USEFUL":END
1010 :
1200 PRINT"(C/DN)SHELL SORT":FORI=1TO300:NEXT
1201 CC=TI
1205 G=N/2
1210 IF G=0 THEN 1350
1220 FOR J=1 TO N-G
1230 IF MID$(C$(J),A,) <= MID$(C$(J+G),A,B,>THEN 1290
1250 T$=C$(J)
1260 C$(J)=C$(J+G)
1270 C$(J+G)=T$
1280 E=J
1290 NEXT J
1300 IF E=0 THEN 1330
1310 E=0
1320 GOTO 1210
1330 G=INT(G/2)
1340 GOTO 1210
1350 DD=TI:T2=INT (((DD-CC)/60)*100)/100
1352 PRINT"SHELL SORT TIME=";T2;"SEC"
1353 PRINT"BUBBLE SORT TIME=";T1
1355 IFA=17 THENB=30
1359 R=1
1360 FORX=1 TO N
1370 PRINTCHR$(144)C$(X)
1372 IF LEN(C$(X))<23 THENPRINT"(CLR)"TAB(A-1)CHR$(28)MID$(C$(X),A,B):GOTO 1375
1374 PRINT"(CLR)(CLR)"TAB(A-1)CHR$(28)MID$(C$(X),A,B)
1375 IF X=R*10 THEN R=R+1:GOSUB 3030
1380 NEXT X
1390 PRINT"WANT TO CHANGE THE SORT(Y/N)":INPUT A$
1400 IFA$="Y" THEN 240
1500 IFA$="N"THEN PRINT"(CLR)":END
3030 PRINT"(C/DN) PRESS ANY KEY"
3040 GET A$:IFA$=""THEN 3040
3050 RETURN
5000 REM(((START OF DATA...ENTRIES MUST START AT COLUM  'S 1,5,7, & 12 ONLY)))
5040 DATA "SUN 0230* HCJB 09745,9580,15300,17675"
5050 DATA "SUN 0249* RN 09590,5995,15325,17715"
5060 DATA "SUN 0112 RA 21740"
5070 DATA "SUN 1231* ORF 21535"
5080 DATA "MON 0010* KOLIS11640"
5090 DATA "MON 0045* SFR  11880"
5100 DATA "DLY 0900 BBC 15070,11955"
5110 DATA "DLY 0900 RJ  15195"
5120 DATA "DLY 1000 RA  5995"
5130 DATA "DLY 1010 VON 15250"
5140 DATA "SAT 1100 TWR 11815"
5150 DATA "WED 1205 VAT 21485"
5160 DATA "FRI 1230 SLBC 15425"
5170 DATA "TU  1235 VOG 11645"
5180 DATA "SAT 1300 RSA 21535"
5190 DATA "SUN 1430 HCJB 17885"
5200 DATA "SAT 1430 UNR 2160,21265"
5210 DATA "SUN 1700 WYFR 21615"
5220 DATA "DLY 0230 RA 17795"
5230 DATA "FRI 2130 BUD 9533"
5240 DATA "SAT 2030 RN  21685,17695"
5260 DATA "SUN 0230 BBC 11750,9915"
9999 DATA "END"
```

(B cont.) Complex program

enables me to use my DX time much more wisely. I can spend all of my time looking for that station which is most likely to be hearable at just one particular time. It, of course, doesn't guarantee my hearing a tough DX, but it does increase my chances of hearing what I am looking for. There is no question in mind that it has produced results for me. A sample of the output of these programs can be seen in Fig. 10-4.

Fig. 10-4.
Personal scores
and recent
loggings by
others can be
called on
demand.

```
BURMA-HEARD NOT VER.
BBS 4725*,5040,5985

REPORTED
FRE 12/1 4725 1315-1330
FI  11/20 4725 1210
FRE 3/5 4725 1125
FRE 2/2 4725 1240
FRE 2/1 4725 1150-1225

FRE 12/22 5985-1445
FRE 12/3 4725 1250
FRE 11/7 5040 1430
FRE 11/7 5985 1445

MORE INFORMATION?
YES OR NO

KAMPUCEA-HEARD NOT VERIFIED
VOPK 11938KZ*,9695KHZ

REPORTED
FRE 11/14 11938 1245-1300
 (GONDERSON)
FRE 8/10 9695 1300+
FRE 3/18 11938 1200-1220
FRE 2/7 11938 1200-1215
FRE 11/23 11938 1200-1215

MORE INFORMATION?
YES OR NO

  1.   ANTARCTICA
  2.   BURMA
  3.   CAPE VERDE
  4.   CONGO
  5.   DOMINICAN REPUBLIC
  6.   EQUATORIAL GUINEA
  7.   ESTONIA
  8.   FALKLAND ISLANDS
  9.   GREENLAND
 10.   GUINEA
 11.   KAMPUCHIA
 12.   KIRIBAT
 13.   LAOS
 14.   MALAWI
 15.   MALI
 16.   SABATH
 17.   SUMATRA
 18.   UPPER VOLTA
 19.   URUGUAY
 20.   YUGOSLAVIA
MORE INFORMATION?
YES OR NO
```

Station Logs

A data-sorting feature which is unique to the DX'er is *log keeping*. SWL'ers are more interested in hearing the same station over and over again, any time they want to, but the DX'er is always on the lookout for new and never-before-heard stations. Most DX'ers, whether they be shortwave, medium wave, FM, TV, utility, or whatever, are interested in keeping track of their "catches," and, in some cases, the verifications or QSLs.

A computer, of course, is quite good at this, but, realistically, it's easy enough to do this manually. The advantage of the computer here is its ability to store quite a lot of this data in a relatively small area and then recall it to either a video display or printed output at the convenience of the user. I have found that it makes my shack a whole lot neater. Gone are the loose sheets of loggings that I never could find when I wanted them. Gone, too, are the overstuffed spiral notebooks. Also, I have been able to easily make duplicate copies of all the information and store them in a secure place.

The programs to do these wondrous deeds range from quite simple to the somewhat complex, depending on how many different kinds of data you want to store, and how many different ways you want to be able to recall the data. A variety of programs are available from several sources. They will all require the entering of data, but logging has always been a chore for me and I think the computer is helping me keep more complete and accurate logs. A sample of such programs can be found in Fig. 10-5.

Calculating Programs

Because DX'ing by its nature involves listening for stations whose ability to be heard is at best variable, DX'ers are forever trying to calculate the various factors which might affect their ability to hear a given station at a given time. I speak of such things as Sunrise/Sunset times, MUF, distance to the station, and solar activity.

Coming up with accurate measurements of these things usually involves some rather complex calculations. DX'ers have always been looking for cheaper, easier ways to make these calculations. To date, they have not really been too successful, although such items as *The DX Edge* represent an attempt at this. One of the great strengths of the computer is its ability to make complex calculations rapidly and accurately. Very simple and relatively short programs, such as the Sunrise/Sunset program shown in Fig. 10-6, can provide the user with some of the more basic (no pun intended) calculations. Longer programs will produce a complete set of information for the DX'er, as can be seen by the sample output from such a program, shown in Fig. 10-7. These programs are readily available in printed form for most types of computers, and, in some cases, are available ready-to-run from a cassette tape or a floppy disk. Use of these calculating programs will unquestionably make DX'ing easier and more successful to those who use them.

Fig. 10-5.
"Logbook"
computer
program of the
National Radio
Club.

```
0 REM"LOGBOOK"
1 PRINT"[CLR/HOME] ":PRINT"DO YOU WISH TO QUERIE ITEMS BY
  'FREQ' OR BY 'STATE'":INPUT X$
2 IF X$="STATE" THEN 9
3 PRINT"[CLR/HOME]" :PRINT"WHAT IS THE FREQ"
4 INPUT Q$:PRINT"[CLR/HOME]":FOR I=1 TO 2500: READ A$
5 IF A$="END" THEN 15
6 READ B$,C$,D$,E$,F$,G$,:IFA$<>Q$ THEN 8
7 PRINT A$;TAB(5);B$;TAB(14);C$;TAB(20);D$;TAB(25);E$:PRINT
  F$;TAB(9);G$
8 NEXT I
9 PRINT"[CLR/HOME]":PRINT"ENTER STATE(--)OR COUNTRY NAME":
  INPUT X$: PRINT "[CLR/HOME]"
10 FOR I=1 TO 2500: PRINT "[CLR/HOME]"
11 IF A$="END" THEN 15
12 READ B$,C$,D$,E$,F$,G$: IF F$<>X$ THEN 14
13 PRINT A$;TAB(5);B$;TAB(14);C$;TAB(20);D$;TAB(25);E$:
   PRINT F$;TAB(9);G$:PRINT
14 NEXT I
15 PRINT:PRINT "THANK YOU....END OF PROGRAM....TO RESTART
   TYPE 'RUN'":STOP
```

The next lines are your DATA statements, as many as you
want. Don't use commas to separate words, unless you enclose
those words in " ", okay?

```
20 DATA 525, 11/21/78, 01412, R. RUMBOS,,COSTA RICA, LISTEN
   TO CASSETTE
21 DATA 540, 12/18/79, 04452, KWMT, FT. DODGE, IA, ID FULL
    WX & ID
22 DATA END
```

FOOTNOTES:

A$= Frequency

B$= Date

C$= Time

D$= Call letters

E$= Location (city)

F$= State or Country

G$= Remarks section of your log

Notice the two ,, after R.
RUMBOS in line 20,
this is cause I didn't
enter the city or E$
information. A comma
without anything entered
to the machine is a
blank.
Any problems with this let
me know.

Mike

THIS PROGRAM WAS WRITTEN ON A VIC-20 AND COMMODORE 64. Works
with others as well.

Word Processing

At this time, I want to close my discussion on using a computer in DX'ing by looking at the subject of "Word Processing for the DX'er." Most DX'ers aren't secretaries or typists, and, thus, are probably not well acquainted with word processing. However, they should become more acquainted with the subject, because word processing on a home computer is not quite the complex business that secretaries must deal with. As many DX'ers are not particularly competent typists, word processing, using a home computer, will make the whole business of typing a much simpler and easier task. I would have been hard pressed to do a chapter like this without the use of the free word processor I received from a computer magazine.

While I took typing in Junior High School, I never felt I had developed enough skill to really use it on any kind of a regular basis. I was too slow and I

Fig. 10-6. The
National Radio
Club's
Sunset/Sunrise
times computer
program.

SUNTIMES--A COMPUTER PROGRAM

Ronald F. Schatz

The following BASIC computer program is especially
designed for use with the Sharp PC-1211 or the TRS-80 pocket
computers. It is a simple but reasonably accurate program
that will calculate the times of sunrise and sunset for any
point on earth. While the version of BASIC employed by these
units is surprisingly sophisticated for a computer that is
only twice the size of most calculators, their limited
memory and display area demand a relatively conservative,
economical program philosophy as compared to larger systems;
this is reflected in the foregoing program.

Setting Mode to PROGRAM:

```
  2: "A"                            Optional step for "DEF"
                                    program.
  5: PRINT "SUNRISE/SUNSET PROGRAM"
 10: INPUT "MONTH (#): ";M          Asks for the month as a
                                    number.
 15: INPUT "DAY OF MONTH: ";D
 20: INPUT "LATITUDE: ";T           E.g., input 39°15' as
                                    "39.15"
 30: INPUT "LONGITUDE: ";L          Same format as above.
 35: T=DEG T:L=DEG L                Converts 0-'-" to
                                    decimal form.
 40: INPUT "SEASON (S=1, W=0):";Z   Enter "1" if spring or
                                    summer, anything else
                                    otherwise.
 45: IF M=1LET Y=100                Month-value assignments.
 50: IF M=2LET Y=131
 55: IF M=3GOTO 110
 60: IF M=4LET Y=11
 65: IF M=5LET Y=41
 70: IF M=6LET Y=72
 75: IF M=7LET Y=102
 80: IF M=8LET Y=133
 85: IF M=9GOTO 125
 90: IF M=10LET Y=8
 95: IF M=11LET Y=39
100: IF M=12LET Y=69
105: GOTO 125
110: IF D<21LET Y=160               Decision for March.
115: IF D>20LET Y= -20
120: GOTO 135
125: IF D<22LET Y=164               Decision for September
130: IF D>21LET Y= -21
135: Y=Y+D
140: IF Z=1LET Y=Y*30/31            Summer perihelion
                                    correction.
145: A=90-SIN Y*23.45              Solar declination
                                    formula.
150: G=TAN T/TAN A
155: IF G>1GOTO 225                 "Midnight-sun" detector.
160: G=(ASN G)/15                   Time-differential
                                    conversion.
165: H=L/15:J=6-H:E=18-H
170: IF E>24LET E=E-24
175: IF J<0LET J=J+24
180: IF Z=1GOTO 195
185: R=J+G:S=E-G
190: GOTO 200
195: R=J-G:S=E+G
200: R=100*DMS R:S=100*DMS S        Converts times to 4-digit
                                    integrals.
205: BEEP 1                         Wake-up call.
210: PRINT "SUNRISE AT ";R
215: PRINT "SUNSET AT ";S
220: GOTO 10                        Returns to beginning for
                                    more data.
225: BEEP 3                         "Midnight-sun" warning.
230: PRINT "POLAR REGION - NO
     SR NOR SS"
235: GOTO 10
240: END
```

Times are automatically calculated in UTC (GMT);
local-time calculations can be performed by adjusting the
longitude to your prime meridian; e.g., add "75" to the
longitude for EST. Enter all "S" latitudes and "W"
longitudes as negatives. The "Season" entry refers to the
Northern hemisphere only.

Comments on SUNTIMES are invited, preferably via
Musings.

- 30 -

Fig. 10-7.
Output from a more extensive program for providing beam direction, propagation data, sunrise/sunset times tailored to the user's location.

```
                    PROPAGATION DATA

          DATE -  3 / 22

          HOME QTH - ANOKA MN
          DX QTH - ZANZIBAR
           ( 6.13 S,  39.12 E )

          BEAM FROM HOME QTH - 61 DEGREES
          PATH LENGTH - 13697 KM OR 8510 MILES

          SOLAR FLUX -   128 SUNSPOT # - 79
          NUMBER OF SKIPS - 4

          ANOKA MN SUNRISE - 1211 GMT
          ZANZIBAR SUNRISE -  322 GMT

          ANOKA MN SUNSET - 0 15 GMT
          ZANZIBAR SUNSET - 1525 GMT
```

HOUR(GMT)	MUF(MHZ)
0	17.471
1	15.028
2	14.197
3	13.514
4	12.957
5	12.506
6	12.144
7	11.853
8	11.623
9	12.582
10	22.466
11	25.89
12	28.057
13	29.569
14	30.627
15	31.321
16	31.695
17	31.77
18	27.099
19	24.742
20	22.738
21	21.045
22	19.626
23	18.445

```
               END DATA
```

Fig. 10-8.
Spanish
reception report
written with a
word processor.

6700 153rd Lane NW

Anoka, Minnesota 55303

April 11, 1983

Estimados Senores:

Me complace transmitir un informe sobre su difusion

recibida aqui en Anoka, Minnesota (U.S.A.). Esta es la

primera vez que escucho su estacion de radio y fue una

agradable experiencia.

La recepcion de sue emisora fue el 12 de setiembre 1982

de 0002 de 0030 hora del Meridiano de Greenwich en una

frecuencia de 15476 kilohertz.

Mi equipo consiste de lo siguiente: un receptor de

comunicacoes modelo R7/DR7 Drake y mi antena dipolo de 33

metros.

CALIDAD DE SENAL

SINPO 25343

Fuerza de la Senal: Mala"

Interferencia: Ninguna

Ruido Atomospheric: Moderado

Propagaciones: Pequena

Merito General: bastante buena

Espero me perdonen las faltas que se deben a que no

conozco la lengua espanola muy bien Seguidamente los

detalles lo que he eschuchado sus transmisiones.

made too many errors. So, while I have the English-speaking world's worst handwriting, I just made do with it. As a DX'er, problems arose when I sent reception reports to various stations around the world. Certainly those stations whose native tongue was English would have difficulty with my handwriting, so it doesn't take much imagination to realize that someone whose native tongue wasn't English would find my handwriting very difficult to understand, indeed. There is no way of knowing for certain how many QSLs my handwriting cost me, but I am afraid the number might be considerable.

Typing on a computer keyboard seems to me to go much faster than typing on a conventional typewriter. The most significant factor about the word-processing process is the ease with which errors can be corrected. It's almost unbelievable. Another factor to be considered is the ability to save finished

Fig. 10-8 (cont.)

```
                    0002 G.M.T. Locutora

                    0003 G.M.T. Seleccion vocal por un cantante

                    0011 G.M.T. Seleccion instrumental

                    0008 G.M.T. Seleccion vocal por un cantante

                    0011 G.M.T. Seleccion vocal por un cantante ELECCION

                    0008 G.M.T. SelEccioneccion vocal por un cantante

                    0011 G.M.T. Seleccion vocal por un cantante

                    0002-0015 G.M.T. Programa musical sin anuncios

                    0015 G.M.T. Locutora con Identificacion de la estacion.

        frecuencia anunciada 6030khz & 15476khz

                    0016 G.M.T. Locutora con Identicacion de la estacion

                    0018 G.M.T. Seleccion por un cantante

                    0021 G.M.T. Seleccion instrumental

                    0023 G.M.T. Locutora

                    0024 G.M.T. Marcha militar

                    0025 G.M.T. Locutora breve

                    0025 G.M.T. Marcha militar

                    0028 G.M.T. Seleccion "Taps"

                    0029 G.M.T. Himno nacional

                    0030 G.M.T. Dejo el aire

              Les ruego me perdonen si mis informes no son mas

        detallados, pero la causa de ello es que no entiendo muy

        bien el espanol hablado.

              Espero que este informe sea de interes para Uds. Si mi

        informe es correcto y util para el Ingeniero de su Estacion

        de Radio, le agradecere me haga llegar la tarjeta o carta

        QSL de su Estacion. Una tarjeta o carta QSL, para ser de
```

letters on either tape or disk. It enables you to send follow-ups that have the same quality as the original, and much better looking than the photocopies commonly used for this purpose.

The most significant use of word processing in DX'ing, for me, is the storing of foreign language form letters on tape or disk. I used to spend hours copying form letters from various sources by hand, or filling out impersonal forms. Now I can recall almost instantly a reception letter in any one of several languages. Perhaps I want to edit it a bit or change it around and add some appropriate details, and then I quickly have a personalized reception report in any language I might desire, from Spanish to Indonesian (and several others as well). Also, I always save the original letter on tape or disk, so I can make another original-in-appearance follow-up, if necessary. When I get my QSL, I just scratch the particular file for that letter. A sample of such a letter can be seen in Fig. 10-8.

Another use that the DX'er has for a word processor is in the production of

Fig. 10-8 (cont.)

valor para el coleccionista, debe contener los detalles
indicados mencionado mas arriba. Tambien debera contener la
siguiente declaracion: "Los detalles de su informe han sido
revisados y se ha comprobado que son coorectos. Por la
presente, verificamos su recepcion de su emisora."

Le agradecere quiera anotar la tarjeta/carta con la
frecuencia de estacion de radio y ubicacion del transmisor.
Tambien soy coleccionista de insignias y banderines de
estaciones. Si dispone de alguno, me encantaria receibirlo.
Las tarjetas y cartas QSL, que yo colrcciono, se usan con
propositos, y se reguieren detalles completos sobre la
verificacion.

Mejores augurios a usted y al personal de Estacion de
Radio. Saludo a usted muy atentamente.

N.B. Patece que mi reporte anterior no llego a manos de
Uts., y por ello, les vuelvo a escribir de nuevo, con el fin
de obtener su verificacion, que tanto me gustaria recibir.

report forms. I, like many DX'ers, belong to, and report to, several hobby organizations. By using my computer to generate the forms, and then using it to fill in the blanks with the appropriate data, I can report to these various clubs without any of them feeling that they got a copy of a report that somebody else received first. Also, I can easily print and keep a copy in my files. It's neat, it's quick, and it's inexpensive. It's hard to ask for more than that.

A great many word-processing programs are available commercially for all brands of home computers. Some are quite expensive, but most, particularly those with the features that a typical DX'er might require, are quite modestly priced. Good printers for most brands of computers can be purchased for from $200 to $400. I wouldn't recommend buying word-processing capability just for its DX applications, though. But, if you have any other conceivable use for a word-processing capability, the DX applications make it a particularly attractive and valuable use of the home computer.

For the SWL'er, the home computer is a useful tool. For the DX'er, it opens a whole new world and greatly expands his capabilities. I think every serious DX'er, who can financially afford one, ought to consider adding some form of home computer to their shack.

Utilities and the Computer

Now we'll see how a person interested in "utilities" can use a home computer. After that we'll touch on "medium-wave" applications.

The "utility" fan shares many uses of the computer in common with other types of radio buffs, particularly DX'ers. I don't want to go over the same ground too hard, but let's look at how a person interested in utilities would use a computer in the same way as a DX'er. Then we'll look at the special uses of the computer for the utility fan.

Logs and Data Storage

I am not, as yet, a true dyed-in-the-wool utility fan, so my experience in the field is somewhat limited. However, because utilities are often where the action is, I find myself becoming more attracted to them. Often, but not always, utility work is a lot like "normal" DX'ing; that is, scanning the bands and looking for action or new stations, or for countries never heard. For these types of situations, the logging type of uses discussed earlier in the section on "DX uses of the computer" are operative. As I mentioned, the computer is terrific for storing lots of data, and then arranging and presenting the data back to you in the most meaningful and useful of ways. Many programs, both those commercially produced and those in the public domain, are available for this purpose.

On the other hand, there are times when you might want to listen to a specific event and want to be able to recall the specific frequencies that might be used for this event, such as, perhaps, a space shuttle mission. In this case, a different type of logging program or data-storage program would probably serve the purpose and several programs are available that would work; they, of course, would need to be tailored to your particular situation, but I don't think this would be too difficult.

The "utility" DX'er, like any other type of DX'er, would also benefit from the calculating programs that I have already mentioned.

From the standpoint of programming and "software," it is the computer's ability to speedily deal with and organize large amounts of data that is the greatest benefit to the utility buff, who possibly has to deal with a greater volume of raw data than does any other specialized field of the radio hobby. However, it is probably in the area of "hardware" that the computer holds the greatest significance for the utility fan. With the addition of some very modestly priced hardware, the computer can be used to decode CW (code) and/or RTTY (radio teletype).

CW/RTTY Decoders

Normally, the so-called "dedicated" or stand-alone CW/RTTY decoder ranges in price from around $400 to over $700. Hardware/software packages for most popular models of computers begin at not much more that $100. Now, if you use your "package" with one of the low-cost computers in, say, the $100 range, you can see that you've got quite a savings. In addition to the savings gained over the dedicated converters, you end up with a computer that you can use for any of a myriad of other wonderful things. And, of course, if you already have a computer, your additional investment is really quite modest.

The devices are quite easy to hook up and use. You buy a converter which will operate with all computers. These range in price from $50.00 to $200. These converters are usually hooked to the audio output jack of your receiver. Then, you buy some software for your particular model of computer (probably in the form of a module or game cartridge). The software module (program module) usually plugs into one of the "ports" or openings on your computer. The software ranges in price from $20.00 to $100. The converter and software/hardware package are connected by cables. All necessary cables are supplied by the manufacturers. For the uninitiated, it should be mentioned here that *hardware* refers to the machine (mechanical device), *software* refers to the written program (written instructions and/or procedures), and *firmware* refers to software installed in a device (i.e., a software program coded onto an IC chip which is physically installed in the computer).

A printer will be needed for a "hard copy" printout, as would be the case with the use of a dedicated CW/RTTY decoder unit. The converter software package I use cost a little over $100 and seems to produce very satisfactory results. It should be understood that copying CW, and, particularly, RTTY, is perhaps not quite the simple matter that listening to the BBC is. It usually requires a bit of practice to develop a "feel" for code, but most people quickly gain skill and proficiency in it with a little practice. Still, there will always be a certain element of DX'ing about it.

Fred Osterman, of Universal Amateur Radio, who is an acknowledged expert in the field, feels that the more expensive CW/RTTY dedicated units offer a significantly better performance than do the computer add-on units. That might be so, but I feel strongly that the computer units offer more than adequate performance for the needs of the overwhelming majority of users. The cost factor of the computer units makes them affordable to a large segment of the radio community that cannot afford the dedicated units. Thus, the computer enables many to taste the fascinating world of CW/RTTY that would otherwise not experience it.

I hope that these last few paragraphs have shown how the computer may be revolutionizing utility listening and DX'ing. Now, let's look at what the computer can do for all you medium-wave FM DX'ers.

Medium Wave and FM

The medium-wave and FM DX'er makes use of a home computer in much the same manner as some of the other types of specialized radio users. But medium-wave and FM DX'ing are an extremely important part of the radio hobby, and are a part of the hobby that I don't wish to ignore. Thus, I am going to spend a bit of time looking at the ways this important part of our hobby community can make use of a computer. Some of these ideas will have been discussed previously, but I will show how they relate to this particular aspect of the hobby.

FM and medium-wave DX'ing is DX'ing in perhaps the purest sense of the word. Conditions fluctuate widely from day-to-day, and, indeed, often from hour-to-hour. This means that the medium-wave or FM DX'er needs persistence, skill, and a lot of luck. My experience is that these folks are as dedicated and serious about their specialty as any group in our hobby. The computer can do nothing for their persistence or luck, and can help only a little in the area of skill.

Logs

FM and medium-wave DX'ers are almost fanatical in their log keeping, and, really, they have many more stations to keep track of than do those of us who are merely concerned with shortwave stations. Generally, however, they're not concerned with program or frequency schedules. Most of those that I have met have all of that relevant information stored in that great computer between their ears. However, it seems particularly important to this group to keep track of their achievements. Thus, log-keeping programs which speed up, or simplify, or, in some other way, make this an easier task are important to them. And, I am pleased to report that there are a great many of such programs available for all types of computers. There are many generalized log-keeping programs available that can easily be adapted to the specialized requirements of medium-wave or FM DX'ing. There are also a fair number of such programs that are written specifically for the medium-wave and FM DX'er. There is even one program (at least) written for the requirements of a particular club's contest. There is no question in my mind that a computer can make the keeping of the logs of a medium-wave or FM DX'er a whole lot simpler.

Calculating Programs

The calculating programs that can be so important to a shortwave DX'er are of some value, but not as much, to an FM or medium-wave DX'er. The reason is that the factors which affect medium-wave DX and, particularly, FM and TV DX are much more variable and complex, and are much less understood than the factors which affect shortwave DX.

Whether or not a computer is of sufficient usefulness to justify its purchase by a medium-wave DX'er is a hard call for me to make. But one of the most prominent medium-wave clubs, the National Radio Club, has published many programs and articles on the subject. They were one of the very earliest clubs to recognize the usefulness of the computer and its applications to the radio hobby. This, as much as anything, suggests to me that the computer might indeed be an effective accessory for the medium-wave DX'er.

Amateur Radio

It isn't the function of this book to deal with amateur radio on a large-scale basis and, thus, I am not going to go into amateur radio applications for computer. However, I do want to mention for the casual reader that there are a great many computer applications used in amateur radio—probably more than any other aspect of radio. Amateurs as a group tend to be more technically oriented and sophisticated than SWL'ers or even DX'ers, and, possibly for this reason, they have latched on to the computer as a means of dealing with a variety of factors that they must take into account. This means that they are often very knowledgeable on the subject, and can be quite helpful. Also, amateur publications seem to have been dealing with the subject of computers for a longer period of time and with more technical support behind them, and thus should be considered a particularly valuable source of information.

Telecommunications

Before I delve into what to do if you already own a computer (or, if I have convinced you to rush right out and buy one), I want to deal with an aspect of computers and radios that really transcends all of the various specialties within the radio hobby, and which might be for some a significant use for the computer, providing perhaps a reason for deciding to buy a particular setup. We'll call this section "Telecommunications and Radios" or "My Favorite BBS in Columbus, Ohio, isn't CompuServe."

Telecomputing is an area of computing which is ignored by some, but which many computer users find to be quite useful and a lot of fun. Personally, my original thought was that hooking a computer up to the telephone was some kind of a gimmick. Once again, I spoke too soon. After I broke down and bought a *modem* (*m*odulator/*dem*odulator), I found the whole business of telecomputing surprisingly useful and a lot of fun. Telecomputing has a great many applications for the radio hobbyist, particularly for the DX'er. But, it is useful for virtually all elements of the radio hobby.

When the radio hobbyist hooks his computer into the phone lines, he can access other computers which may be able to store and sort much larger amounts of data than his home unit can. This is possible if he connects to the mini- or mainframe computer of a college or a nonline service source, or to the computer of a BBS. The advantages of this are that the hobbyist can then control when he receives his radio data. He can receive the latest information monthly, weekly, or daily, if he chooses. He decides what's best for his situation. He doesn't have to wait until a certain time of month for his favorite magazine to show up in the mail box. Because he can control the times when he receives his data, it is much easier for the hobbyist to stay current and receive the latest information fresh, as it's happening, rather than waiting a month or longer to read it in a favorite magazine.

The central computer that stores and sorts all this data is usually referred to as a BBS, or Bulletin Board System. Across the country there are hundreds, if not thousands, of BBSs. Many of these BBSs are generalized social operations. Others specialize in a particular brand or model of computer; still others specialize in a specific topic.

Telecommunications can be quite beneficial to the radio hobbyist. As modems now sell for as little as $50.00 it is clear to me that telecommunications will play an increasingly significant role in the radio hobby.

At this point, if you already own a computer, maybe you're saying, "Gee, I didn't know I could do all those things with my computer." There will be more for you in just a "bit" or "byte" as the case might be, but for now, I want to talk with those who have just decided to rush out and get a computer (those who, heaven forbid, are not sure they can really handle a computer).

For those who might think that a computer would be useful to them in the radio hobby, but are not sure they're technically capable of operating one, I offer reassurance. One would think that anyone who can operate a radio ought not be afraid of a computer, but my experience with the ANARC Computer Information Committee suggests that such people do exist. After all, modern radio receivers are pretty simple to operate—but so are modern personal computers.

I have a four-year-old son who won't enter kindergarten for a few months yet, and who can neither read nor write, yet he shares his own VIC 20 computer with his big sister, who is all of eight. My young son is quite able to operate his computer with only occasional help from his sister and none at all from me. If a four-year-old child, and a not-all-that exceptional one at that, can operate his own computer, then I would think that no adult ought to fear the machine. *Operating the computer* refers to turning on the machine, and then loading and running the appropriate programs. This is a task simple enough for a four-year-old.

Programming a computer, or entering instructions into the computer to accomplish whatever task you want it to (whether that task is remembering a program schedule, sorting logs, writing Indonesian reception reports, or calculating sunrise-sunset times), does require a certain amount of skill, which must be acquired through some study. My eight-year-old daughter has already begun to learn the rudiments of programming, and I would think that anyone capable of spinning a radio dial could pick up programming fairly fast, if they have the time to devote to its study. However, the crucial thing to remember is that there is already so much ready-to-use software available, which is related to the radio hobby, that there is no need for the radio hobbyist to become involved with computer programming unless, of course, he or she wants to. The material is already out there waiting to be used, computers are easy to operate (if my beloved four-year-old can operate his own computer, then anyone reading this can), and a computer is a wondrous tool.

Which Computer ?

Hopefully, now, any of you who may have had questions about using a computer for the radio hobby have had those questions resolved. So what computer should you use? That, of course, depends on a lot of things. There really are no hard and fast rules, but I will try and give you a little general guidance to go on.

First, if you already have a computer of some sort, whether it be a small Timex Sinclair, or a large IBM, don't buy another computer just for use with your radio, as tempting as that might be. Generally speaking, all the popular home computer systems, and some of the larger business systems, can easily be adapted for use with the radio hobby. There is an ample supply of hardware and software for all makes and models of computers now available, and more is coming out every day. Use what you already have.

If you plan to buy a computer with some other purpose in mind, such as business or as an educational tool for your children or even as a game machine, let this purpose dictate your choice of computer. Select your computer on the basis of how well-suited it is for performing the other tasks you have in mind for it. You can then adapt it to radio hobby use as I mentioned in the previous paragraph.

If, however, your exclusive or, possibly, primary use of the computer will be as a radio accessory, then cost ought to be a primary consideration. The least-expensive computers, such as some of the Timex models, sell for $40.00, or even less, new. These machines, even with their limited memory, will still enable you to do quite a bit. Mike Witkowski, of Stevens Point, Wisconsin, has been largely responsible for developing and adapting quite a large amount of very useful software for the Timex family of computers. However, their limited ability to use hardware accessories, such as RTTY/CW decoders, and their limited ability to be adapted to a wide range of nonradio uses is a consideration. If you can afford $40.00, but not much more, this is the way to go. A Timex Sinclair 1000 or Sinclair ZX81 will serve you well.

If you can afford to spend a bit more, you probably should consider the Commodore family of computers, the VIC 20 and the C64. The Vic 20 also comes with a limited memory, but it can easily be expanded. A wide range of hardware and software is available for it. The VIC 20 is currently priced at about $90.00 and is a very good buy. My personal favorite is the Commodore 64, which sells for less than $200. It has a wide range of radio-related software and hardware available for it, and comes with both a very large memory and a useful 40-character display. ANARC Computer Committee research shows that these two Commodore computers are by far and away the most popular computers among radio hobbyists.

The Atari group of computers is reasonably priced, but they don't have quite as much radio-related software or hardware available. They don't seem to be as popular with radio hobbyists as I would have assumed, but they are good computers.

Radio Shack's TRS-80, the Apples, IBMs, and Kaypros are all good computers with varying amounts of software and hardware available. But, they are all more expensive than can be reasonably justified just for radio use alone. If, however, they are intended for other purposes besides the radio hobby, then they will probably work just as well for radio.

Accessories

Now before we discuss where to get the computer, let's have a quick look at what accessories you might want to consider, and how they will relate to computer usage for the radio hobby.

First, you'll need some way of permanently storing your programs and data. Essentially you have two choices, cassette tape or floppy disk. Tape units will run from to $40.00–$100. They are simple to use and very reliable. They are, however, incredibly slow. This can be a significant problem when using long programs, storing large amounts of data, or storing a number of programs on one tape. The seriousness of the slow operation should not be underestimated. However, if you're going to use your computer exclusively, or almost exclusively, for radio use, then, because of the cost factor, I think you ought to limit your choice to tape storage.

If you're going to use your system for a variety of uses that are particularly, but not exclusively, business purposes, you might want to consider purchasing a disk drive. Disk drives are incredibly fast, and allow you to store a large number of programs or a large amount of data in a very small space, with quick and easy access to the programs and/or data. Disk drives are more complex machines, though, and some have a reputation for not being as mechanically reliable as a tape system. They range in price from a little more than $200 to more than $400. Still, I recommend them for the serious computer user or for those individuals who will also use their machines for business purposes.

You'll need word-processing software and a printer if you're going to use your computer to write reception reports or other correspondence. The software should be no problem, however, with adequate software costing about $40.00, or less, maybe much less. The least-expensive printers cost about $200, with the better ones selling for $300–$400, and even more if you want to spend it. It's not worth the money just for reception reports, but if you're planning on doing a significant amount of typing, a more expensive printer might be worth it to you. Also, consider the fact that word processing on a home computer will make typing easier than you have ever dreamed.

Monitors

A monitor simply provides a video display of the computer output. Generally speaking, any TV set you have around the home will work with most of the less expensive home computers. But remember that TV sets, particularly the older TVs, tend to give off a lot of RF interference. Also, if you're going to be doing a fair amount of word processing, you might want to consider using some sort of dedicated black-and-white monitor rather that a TV set. You should be able to get a good one for less than $100. Color is nice if you can afford it.

Where to Get Your Computer

If you are absolutely convinced by now that you must have a computer to use with your radio, you may be saying, "Where is the best place to get one of these gadgets?" After all, computers are sold everywhere these days, from the corner drugstore to specialized computer stores. Let's look now at where to get the hardware, and then where to get the software.

In my opinion, the best place to buy the basic computer itself, and only the basic computer, is the cheapest place you can find it. This is usually, but not always, a discount department store, such as K-Mart or some other mass merchandiser, that buys huge quantities of items and can afford a smaller mark-up. This is particularly true with the less expensive machines, such as the Commodore line, but less true of the more expensive machines, such as the Apple Computer, which are generally not available at mass merchandisers. In large metropolitan areas, such as where I live, the computer market is presently very competitive and the price range for a given model of computer may vary by 10% to 20%. In the past, the variance has been much greater, sometimes more than 50%. In rural areas, you may be forced to buy from a catalog, or travel to a nearby metropolitan area. Even considering a small margin of difference in price on a relatively modestly priced computer, the difference in terms of dollars makes shopping around a worthwhile endeavor.

Help

I know that many of you shudder at the prospect of buying a relatively technically sophisticated piece of equipment from a store where the clerks don't have the foggiest idea about what they're selling (and they don't), but believe me it's all right. First, as I have alluded earlier, the operation of the basic computer unit is just not that complicated. While the manuals that come with the computers are not always the examples of literary clarity that they ought to be, any person who operates a radio ought not have any trouble operating a computer. In those rare instances where this generalization doesn't apply, you should have no trouble finding a neighbor, friend, relative, or radio buddy who can help you. There are a lot of computers and computer users out there.

Servicing

Don't worry about servicing, either. Most computers that go bad, go bad right away, and you simply take the machine back where you got it and they give you a new one. That's pretty much true no matter where you buy it. Computers have few moving parts and will hold up as well as, if not better than, most modern radios. In any case, few computer stores have in-house servicing, so if something does go wrong with the machine, chances are they're going to send it to the same place for servicing that the department store does.

Peripherals

Buying peripherals, such as disk drives or printers, is not quite the same thing as buying the basic computer. There are different considerations. Generally speaking, these peripherals have many more mechanical parts and, as a result, are much more likely to break down. Perhaps more important, they generally are much more complicated. It is likely that you'll need more help in operating a disk drive than you will for the basic computer itself, and, in this case, you'll definitely want a store where the clerks know what they're doing so they can show you how to connect and operate the drives. Also, in some cases, such as with a printer, you'll have a choice of several models that will work with your computer, and you'll want intelligent advice as to which model is best for your particular needs.

In the case of those peripherals directly related to radio usage, such as RTTY/CW convertors, I think you'll find that both the discount store and the computer store are going to leave you out in the cold. There just isn't enough demand for converters to warrant these stores carrying these specialized products. Therefore, you'll have to find them where you would expect to find radio accessories, at a good radio shop. Not all radio stores carry computer accessories, but the better ones do. You should shop around. Also look in hobby publications, particularly the amateur magazines, for reviews and information about specific pieces of equipment. This will help you choose what's best for your needs.

When you buy computer gear to use with your radio, approach the purchase in the same manner that you approached buying your radio. Use your intelligence and good judgment and you'll do well.

Software

Once you get everything hooked up, where do you get all the programs you need to get your computer system to do all the wonderful things I have been describing? The number of sources is surprisingly limited but, fortunately, the available sources do offer a great quantity and variety of software. Without doubt the best source of software for the radio hobbyists is the ANARC Computer Information Committee. They publish an extensive offering of public domain (noncopyrighted) software. At present, they offer over 50 programs for all models of computers. These programs cover the whole range of hobby applications and are modestly priced. Details are available from the committee for return postage. Their address is 4347 29th Street S.E., Rochester, Minnesota 55904. A limited amount of commercial software is also available. Much of this commercial software is Ham-operator orientated, but may be adapted to nonamateur uses. An example of this type of software is the *Ham Data* line available from EGE Inc. in Woodbridge, VA. Other radio stores should also be able to help you. Another source of software is the various hobby publications. Amateur publications, in particular, have published extensive listings of software. Much of this software can also be adapted for nonamateur radio applications. Radio hobby clubs have been a little slow in picking up on computer applications in the hobby. However, a couple of clubs, the Ontario DX Association and the National Radio Club, are doing a very good job, and publishing extensive amounts of software. It is perhaps not coincidental that these two clubs are the next two hosts of the ANARC convention. There may be others that I am not aware of.

A source of programs that I mustn't forget is the *Media Network* program of Radio Netherlands. *Media Network*, under the guidance of its producer, Jonathan Marks, has done a great deal of work in this field. They have even conducted experiments in transmitting computer programs over the radio. In cooperation with a Dutch hobby group, they have developed a program that allows programs to be transmitted in a sort of universal form of BASIC. The programs can then be translated for use with various types of computers. The station has published a booklet containing a modest selection of programs written in a number of different popular versions of BASIC. Anyone interested should write to Radio Netherlands, P.O. Box 222, Hilversum 1200JG, Holland.

If you'll just pick up a copy of ANARC's listing, you'll find that while the sources of software are limited, the quantity and quality of programs available are not.

Very briefly, I want to summarize for you sources of information and help on the subject of using computers and radio. These sources are largely the same as those just mentioned as sources or software, but they are worth mentioning again. First, in my biased opinion, the best source of information is the ANARC Computer Information Committee. Another excellent source of information is the various DX programs of the major broadcasters, particularly those of Radio Netherlands and Radio Sweden International, and, to a slightly less extent, Radio Canada International. Next, the amateur radio community, particularly through, but not limited by any means to, its publications. If you have a question, ask a Ham operator and he can probably help you, or will refer you to someone who can. Also, Ham operators operate several nets which deal with computer subjects. Next, we have the hobby clubs. Some are better than others. If yours hasn't done anything yet, tell them to get on the stick and get going. Then, there are the radio shops and all the other quality radio-equipment distributors. Finally, if all else fails, you can always get in touch with me, care of the ANARC Computer Information Committee, at the address mentioned earlier.

Future of Computers in Radio

Before I conclude this chapter, I want to take a quick peek at the future and see where computers and radio are going in the future.

Since I have become so heavily involved in the use of home computers with the radio hobby, I have been asked on a number of occasions to look into my crystal ball and see what the future holds for this relationship. That might be all well and good for someone like Merlin the Magician, but everytime I attempt to predict the future I discover that my prediction has come to pass the next day. Predicting the future in a fast changing world is very risky business for mere mortals like me.

In the general sense though, it is clear that the relationship between the radio and the computer will develop and mature. The use of computers with radios will become much more prevalent than is now the case. I would think that the computer, within just a few years, will become as common an accessory as an antenna tuner or preamp. And, by the end of this decade or the beginning of the next, it will be as common an accessory as a long-wire antenna. I believe that use of the computer, and its relationship to the radio hobby, will largely continue to develop along the lines that I have outlined in this chapter. However, I believe there will be one new dramatic element added to that relationship—an element that is just beginning to surface now, but an element which will have a profound effect on the hobby within just a few years.

Over the past ten years, the most significant trends in radio equipment have been the digitalizing of receivers and the adding of microprocessors to all manner of radio equipment. The big change coming is that, instead of adding microprocessors to radios, radios will be added to microprocessors. This will give the user the greatest possible control over the radio, perhaps to the point where radio may control the user, rather than the other way around. Using a computer, the operator will be able to preprogram all aspects of a receiver's performance. He'll be able to set these parameters and then go on to other things. It will enable him to fine-tune the receiver as never before possible. It will also, for instance, enable the ''DX'er'' to set his ''radio'' to check all frequencies of all the DX targets he might be interested in. If any of the frequencies have the specified programming at specified quality levels, the receiver will then record the program and even write a reception report in the proper language—all without the ''DX'er'' having heard a single syllable or crackle of static, if he so chooses. Is this DX'ing? Well, the hobby will, before long, have to answer that question.

Is this really coming? As they say in the ''valley,'' ''fer sherr.'' ICOM has already marketed receivers designed to have the capability to be interfaced with home computers. These are the very popular IC-R70 and the new IC-R71A. I have also seen information from Japan Radio that refers to the computer control of their new NRD-525 receiver.

Summary

It's time to conclude this chapter, and rather than rehashing all I've gone over, let me pass you on to the next chapter with just a brief thought. The computer and the radio are now irrevocably connected. The radio hobbyist, whatever his special interest within the hobby might be, can use this relationship to further his or her enjoyment in the hobby, or they can ignore the relationship and be passed by. I hope that this chapter will give you the stimulus and information to further your enjoyment of our special hobby.

11

OVER-THE-HORIZON RADAR

Robert Horvitz

The arrival of the Russian "Woodpecker" over-the-horizon radar system on the shortwave bands some years ago caught the interest of **Robert Horvitz**, and he has been involved in researching and studying such systems ever since. Born in Massachusetts in 1947, his monitoring activities have included regular contributions to the *Review of International Broadcasting*. He was, at one time, the "Below 30 MHz" editor for the Radio Communications Monitoring Association.

Bob Horvitz is Chairman of the Association of North American Radio Clubs' OTH (Over-the-Horizon) Radar Committee and publishes the committee's newsletter, *Backscatter*. A resident of Washington, DC, he is employed as Art Editor of the *Whole Earth Review*.

You can't tune around the shortwave band for very long without soon encountering an annoying air-hammer-type sound. It's present 24 hours a day, but doesn't seem to be targeted against any particular station or stations, as jammers are. Nor does it avoid occupied channels, as it probably would if it were engaged in some form of communication. The sound seems totally indifferent to the havoc it wreaks as it jumps around the spectrum, causing severe interference to international broadcasters and their audiences, to amateur radio operators, and to aeronautical, maritime, and fixed stations. Because of its distinctive sound, this signal has been dubbed the "Woodpecker."

Fig.11-1 illustrates an impression of the Woodpecker's moment-to-moment behavior. It often seems to jump around within a band, and then disappears for a while, turning up in a nearby band, with perhaps a louder and wider signal that may be from the same site or a different site.

Fig. 11-1. An impression of the Woodpecker's moment-to-moment behavior.

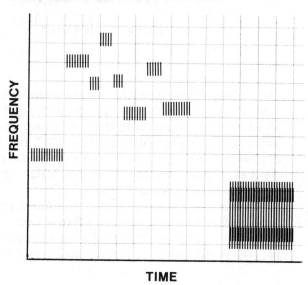

The first instance of Woodpecker interference that was reported to the FCC occurred around 0300 UTC on 12 July 1976. The complainant's name has been lost to history, but he probably was a Ham operator, as the complaint concerned the band at 14,000–14,200 kHz. More complaints soon followed, concerning other bands, and the FCC began a full-scale investigation. By November 1976, they had learned enough about the signals to send a report to the National Security Agency. A summary of this report was obtained through a Freedom of Information Act request and it describes the Woodpecker in terms that are still largely true today:

Virtually every user of the HF spectrum has been affected. The signal has a repetition rate of ten pulses per second. Its bandwidth varies greatly. At times, it may have a bandwidth of only 30 kHz. Other times, it may be as wide as 300 to 500 kHz. Its activity does not appear to have any definite pattern. It may stay on a given band for only a few seconds or it may remain on constant for several hours. At one instant, it may be heard on 12 kHz, then just a few seconds later may appear on 18 MHz. We have observed it throughout the HF spectrum from 6 MHz through 28 MHz. It appears to operate around the clock....The strength of the signal, as received in this country, varies with propagation (time of day and band in use), but under optimum conditions it is extremely strong and totally disrupts other communications.[1]

A letter from the FCC to the State Department a week later gave additional details:

After thirty-seven separate direction-finding alerts, the station's fix has averaged out to be 51° North by 31° East, the vicinity of Kiev. Information received from the British Broadcasting Corporation indicates the station is located at Gomel [between Kiev and Minsk]. Communication with Radio-Suisse Ltd. indicates the Soviet station is located in the Kiev–Gomel area. The location the Commission has been supplying to the public has simply been the general area of Minsk.[2]

When the source was determined to be in the USSR, the U.S. government sent a complaint—and another, and another, and another. Other nations complained, too—among them, Denmark, Sweden, Norway, West Germany, and Great Britain. (Although the Woodpecker is heard worldwide, even in Antarctica, its interference is most severe in northern Europe.)

The Soviet Union responded to the complaints on 3 December 1976, sending this terse note to the ITU's International Frequency Registration Board:

In the Soviet Union tests are being carried out with radio installations operating in the HF bands. These tests may cause interference to radio installations for short periods. The necessary measures are being taken to reduce any such interference. The reports which you have sent us will be carefully studied.

But the interference did not diminish, and the number of complaints continued to grow. When no further response came from the USSR, the IFRB issued a report declaring:

The station or stations in question should cease operation until such steps have been taken to ensure that any interference that may result from the resumption of such tests shall be below the level that would be considered as harmful interference.[3]

Unfortunately, although the IFRB coordinates international use of the radio spectrum, it has no enforcement power. It depends on the cooperativeness and good will of the ITU member nations to resolve conflicts. In this case, the Soviet Union has proven totally intransigent. They brushed off the IFRB report with a

---

1. "Summary of broadband pulse emission emanating from the USSR," (3 November 1976), FCC Woodpecker file.
2. Letter from Robert L. Cutts, Chief, International & Operations Division, FCC, to Gordon Huffcutt, US State Department (12 November 1976), FCC Woodpecker file.
3. "Report on Harmful Interference in the High Frequency Bands Caused by Emissions Originating in the USSR," International Frequency Registration Board (10 November 1977), includes the Soviet acknowledgment of responsibility quoted here.

note saying their monitoring showed that the measures they'd already taken to reduce interference had been effective.

Signals from a second site were detected by the FCC starting in mid-April, 1979. After 33 bearing measurements, this new source proved to be near Khabarovsk or Komsomolsk on the east coast of Siberia (average fix: 48° 33′ North by 135° 13′ East). A third transmitter site was reported by the Associated Press in an article dated 15 May 1980—though it probably had been detected somewhat earlier.[4] The location of the third site is still less certain than the first two, but if one assumes that the Woodpeckers are actually over-the-horizon radars—and this is the most reasonable explanation—then a map in the 1983 edition of the U.S. Department of Defense booklet, *Soviet Military Power*, shows the third site as just north of the Black Sea with a beam aimed toward China.

The three known Woodpecker sites, based on data from the FCC and Department of Defense are illustrated in Fig.11-2. Beam headings are as shown on a map on page 28 of *Soviet Military Power* (Government Printing Office, 1983 edition), they may be variable. Beam paths do not necessarily represent target detection zones. If one assumes they have the same range as the USAF OTH-B radars (1800 miles), their detection limits would be about as indicated by the dotted lines, even though their emissions are audible much farther away. The Pentagon claims that the Woodpeckers are designed to detect ICBMs, but there are technical reasons to suspect they are better able to detect bombers and battleships.

Shortly after the first Woodpecker started transmitting, Dafydd Williams, BBC External Broadcasting's chief engineer, estimated its power to be in the 20–40 megawatt range. That estimate seems to have stuck through time–although it isn't clear exactly what it means. Since the Woodpeckers use pulse modulation, Williams' estimate is often interpreted as peak pulse power. If the emissions were continuous, instead of being crammed into very brief (3 to 10 millisecond) bursts every tenth of a second, this would be equal to a mean continuous power between 600 and 4000 kW—this is greater than the most powerful transmitters used by international broadcasters. But complicating any estimate of power is our lack of knowledge about the antennas used. If the Woodpeckers are indeed radars, their beams must be highly directional. That is, their antennas must have very high forward gain. Not knowing the gain, we cannot accurately correlate transmitter power with effective radiated power (it's never been clear whether Williams' estimate referred to the former or the latter).

Experiments in the United States have shown that engineering a system to handle millions of watts of power is actually quite difficult. Arc-overs, burnouts, and grass fires are hard to avoid. A much better approach is to feed the signal to several transmitters operating in parallel, each driving a separate antenna. If the antennas are properly configured and phased, their emissions will combine in space to form a single powerful radiation pattern, without any element having to bear the entire load. Multiple antennas have an additional

---

4. Norman Black, "Russian Woodpecker," Associated Press (15 May 1980).

Fig. 11-2. Three known Woodpecker sites.

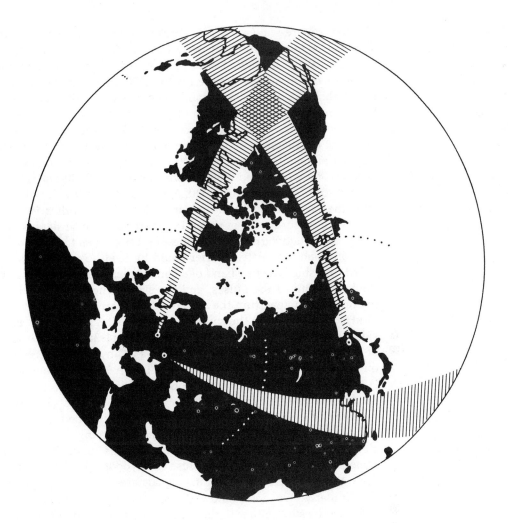

benefit for radar systems: by varying the phase of the signals fed to each antenna, it is possible to steer the composite beam in different directions to increase azimuthal coverage without physically moving the antennas. This is especially important for shortwave radars, where the antenna dimensions are huge.

There is some evidence that at least the first Woodpecker site has several antennas, transmitters, pulse- and carrier-generators that can be switched and combined in different ways. In 1977, S. A. Cook of Intruder Watch (an interference-monitoring group that watches over the ham bands) said that "the 10 p/s signal comprises a pulse train of up to 20 different square-wave pulses, some less than 2 ms in length," and he was "convinced that there are as many as four sources."[5] While the signals monitored by Cook may have contained "up to 20

---

5. "Mystery Soviet Over-the-Horizon Tests," *Wireless World* (February 1977), pp. 53 & 68.

different square-wave pulses," others have found the pulses within a train to be identical or more limited in variety. The Woodpecker signals vary more than most shortwave listeners realize, and in ways not always obvious to the ear.

A spectrum analysis made for the FCC by Spectrum Control, Inc., in 1978 found that some very wideband Woodpecker emissions had several equally large amplitude peaks (loud spots) within the overall pulse bandwidth; that there were lesser peaks symmetrically flanking the main peaks at 10 kHz intervals; that the distance between the main peaks was freely variable; and that the number and center frequency of the main peaks could change abruptly. In one instance, they observed a signal with peaks at 10920 and 11220 kHz suddenly change to three peaks (at 10920, 11040, and 11170 kHz), then about a minute later the peak at 10920 disappeared, leaving the other two unchanged. This suggests a pulse signal—or several synchronized pulse signals—modulating separately controlled carrier frequencies, grouped so as to cover a wider bandwidth than any one modulated carrier could by itself. Such signals are still regularly heard. But since the passband of a communication receiver is narrow, relative to the bandwidth of the pulses, and the signals move around, it's always hard to tell how much spectrum they're covering at any moment, or if peaks hundreds of kHz apart are from the same or different sites. It appears as though there are sidebands something like the diagram in Fig.11-3; it is not to scale. This diagram of a Woodpecker signal was sent to the FCC by James C. Shaw, W6JQX, based on observations made at 0636–0658 UTC on 13 November 1979.

A number of monitors have observed double pulse signals. In some cases, the second pulse might actually be a strong echo, or some sort of ionospheric reverberation, of the first pulse. The FCC has described this as a "ringing effect," and the oscillogram by F. C. Judd accompanying this article shows it quite clearly. In other cases, separate transmissions might be superimposed on the same part of the spectrum. Because the Woodpecker pulse rate is so stable, we can't generally tell if overlapping signals are deliberately synchronized or not.

Similarly, we don't know if there is any coordination of frequency use between the sites, either in the sense of staying out of each other's way or in the sense of joint target coverage. *Flight International* has pointed out that

Fig. 11-3. Diagram of the Woodpecker signal sent to FCC by James C. Shaw in 1979. (From FCC Woodpecker file.)

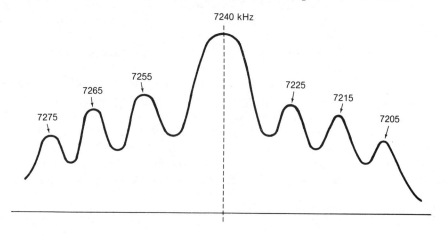

A possible role for the multiple transmitters is suggested by recent U.S. research into the use of radar to determine the geometric shape of targets. Tests have shown that—provided the frequencies used are such that the wavelength is longer than the largest dimension of the target, and that two or three different frequencies are used—examination of the amplitude, phase and polarization of the reflected signal gives each target a recognizable 'signature.' During trials with models it has been possible to differentiate the MiG-19, MiG-21, F-104, and F-4....

The wavelength of HF transmissions (10–150 m) is too short to permit the use of this technique against ships, but they are ideal against such targets as the B-1 and the Cruise missiles.

An early-warning radar which could discriminate between these weapons and other air targets would be a valuable addition to the Russian air defenses. If the development of such a system is the goal of the current Russian test transmissions, it is understandable that Western protests over the interference have been ignored.[6]

Since there appeared to be several transmissions coming from the Kiev/Gomel/Minsk site, this was offered as an explanation. The argument is more tenuous as an explanation for there being several different sites. The beams at the three known sites seem to point in different directions, and, as already noted, we lack evidence for frequency coordination between them.

All the variations—and theories about their significance—aside, there does seem to be a basic emission that typifies the Woodpecker. This is a train of square-wave pulses in which most of the energy is concentrated in a bandwidth of about 15–18 kHz, with a series of weaker peaks at 10 kHz intervals on either side of the primary emission. In periods of mild interference, the sidebands may be inaudible. When the basic modules are ganged together to cover a wider bandwidth, the lesser peaks may overlap, blending into a single pulse-scape.

Many people have examined the Woodpecker signals with oscilloscopes. Systems Control found that the pulses "have surprisingly sharp leading and trailing edges. Some evidence of fading and multipath structure can be seen in the somewhat erratic shape of the pulse at maximum signal level."[7] Others have interpreted the "erratic shape" as an intentional coding—perhaps this is to identify different pulse sources or to enhance some aspect of radar performance. These interpretations are highly speculative, of course, but not totally implausible.

J. P. Martinez has an ingenious theory based on his discovery that when the pulses are greatly dilated with a spectrum analyzer, they seem to have brief amplitude "glitches" at multiples of 100 microseconds. His plot of amplitude (log scale) versus time appeared in *Wireless World*, April 1982 (Fig. 11-4). By arbitrarily identifying the initial modulation phase as "0" and its reverse as "1," and, assuming each "glitch" represents a phase reversal, the pulse could be binary-coded. He attributes the glitches to phase reversals in the transmitted

6. "USSR Develops Anti-B-1 Radar?", *Flight International* (8 January 1977), p. 50.
7. W. R. Vincent, "Spectral Characteristics of the Woodpecker Signal," Systems Control, Inc. (23 February 1978), FCC Woodpecker file.

signals. If this is the case (his equipment apparently couldn't register phase directly), and if the signal were sampled every 100 microseconds, the phase modulation could function as a binary code. He further found that the code sequence was always 31 units long (implying a pulse duration of 3.1 milliseconds), that the sequence was exactly repeated from one pulse to another, and that there were four different sequences—implying that there were four separate transmitters. Analyzing the sequences mathematically, Martinez concluded that they were each "maximum-length, pseudo-random binary sequences" of a sort that could be generated by a bit shift register:

> The interesting point about this use of p.r.b. codes arises from the shape of their autocorrelation function. If such a sequence is compared bit-for-bit, with a shifted version of itself, at all possible shifts, then, apart from the position where all 31 bits match, at all other shifts no more than 1 bit matches between the two sequences....
>
> The conclusion from all this, it seems to me, is that the woodpecker must be simply a pulse compression radar system, with a resolution of 100 μs (10 miles), but the sensitivity 31 times that of a 100 μs radar of the same power. Not only does the p.r.b. sequence cancel out shifted versions of itself in order to achieve its performance, but it has a high immunity to other codes in the same family,

Fig. 11-4. Plot of the amplitude (log scale) versus time by J. P. Martinez. (From J.P. Martinez, *Wireless World*, April 1982.)

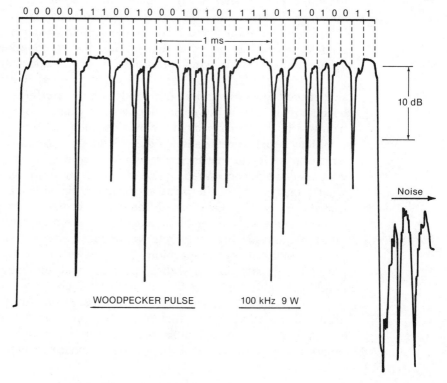

thus reducing cross-interference between separately sited radars on the same frequency....[8]

It may also help cancel the effect of multipath propagation, which in the polar region can cause time spreads of 100–400 μs in the return echo and degrade range resolution.*

Martinez's elegant theory rests on two claims: first, that the glitches represent an intentional phase coding, and second, that their sequence is repeated from pulse to pulse. A more recent study by F. C. Judd seems to corroborate the second claim, although it disagrees with Martinez's duration measurements. Judd also sheds new light on how the Woodpecker signals' dwells begin:

> Prior to most transmissions of modulated pulses there are often a few seconds of unmodulated multiple pulse transmission....These may appear on any frequency and could simply be ionospheric sounding transmissions....
>
> After the initial "sounding"...the "Woodpecker" signals change to a four-pulse format, each pulse being modulated as is shown in [the illustration]. These pulses (1–4) vary in amplitude even during quite short periods...which suggests a "search mode" for best signal return from a target....The modulation on each pulse is quite different but as shown by the expanded oscillogram...it varies continuously....[9]

"These pulses last 3–4 milliseconds each," Judd says. Within each group, the interval between pulses is also 3–4 ms, with the groups coming every 100 ms. Judd adds that "when a 'target' has been located (echo received), the pulse transmission usually reverts to a single pulse of about 4 ms duration but still with the p.r. time of 100 ms."

Apparently Judd hadn't seen Martinez's letter to *Wireless World*, so it's even more fortuitous that the pulse he chose to expand to illustrate his article shows exactly the same pattern of glitches as the one Martinez presented. The glitches aren't quite as narrow, and some of the depths are different, but the sequence is definitely the same. But Judd measured its duration as 4 ms. For Martinez's theory to work, the rhythm of phase change and the sampling rate must be consistent. The discrepancy in their measurements of the pulse duration has yet to be resolved.

Judd's oscillogram of a single pulse signal (Fig. 11-5) shows a strong "pseudo-pulse" (I_s) immediately after the primary pulse (P_1). This is almost certainly what the FCC has noted as the "ringing effect." Judd attributes it to ionospheric scattering near the transmitter site. Also visible are two echoes (E_c), one strong and one weak, at different delay-times before the next pulse (P_2). Fig. 11-6 shows two more oscillograms from F. C. Judd's *Practical Wireless* article. In

---

8. J. P. Martinez (letter), *Wireless World* (April 1982), p. 59.
* Author's supposition, not Martinez's.
9. F. C. Judd, "Over-the-Horizon Radar Systems—Beyond the Blue Horizon (Part 2)," *Practical Wireless* (September 1983), pp. 44–47.

**Fig. 11-5.
Unrectified
Woodpecker
signals.** (From
F.C. Judd,
*Practical
Wireless,*
September
1983.)

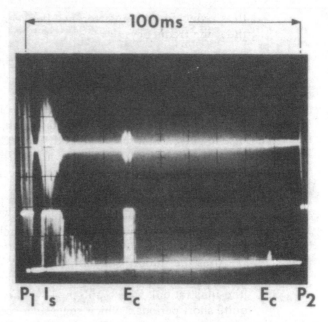

Fig. 11-6A an example of a four pulse modulated Woodpecker signal is given. The four pulses have different waveforms, but similar durations (3–4 ms). An expanded view of the initial pulse in Fig. 11-6B reveals it has a series of amplitude "glitches" with exactly the same time-pattern as observed by Martinez.

The discovery of what may be sounding signals adds weight to the argument that the Woodpeckers are OTH radars. Current propagation data is essential to OTH radar operation, and having a sounding signal different from the main emission implies that the purpose of the main emission isn't sounding as it might be if the Woodpeckers were for ionospheric research.

Sounding brings up the question of what determines which frequencies are used by the Woodpeckers. Although their moment-to-moment behavior is unpredictable, it is far from random. Early on, the FCC identified three modes of operation:

1. Operation which covers a 100–200 kHz band for extended periods of time.

2. An On-Off mode where transmissions are on for one minute and off for five minutes covering a given band.

3. A Sweep mode where emission may be 30 kHz wide and sweeps through a one megahertz band.[10]

The FCC gathered new data about the Woodpeckers' band use during the winter of 1982–3. A plot of 62 encounters with Woodpecker emissions during routine

10. Letter from Robert L. Cutts, FCC, to Gordon Huffcutt, State Department (24 February 1977), FCC Woodpecker file.

Fig. 11-6. The four-pulse modulated Woodpecker signal. (From F.C. Judd, *Practical Wireless*, September 1983.)

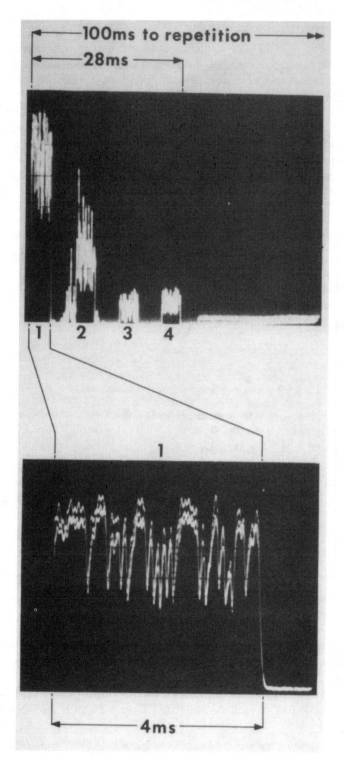

(A) The signal showing the four different pulse waveforms.

(B) Expanded view of the initial pulse.

cruises of the shortwave band by FCC monitors during the winter of 1982–3 is given in Fig. 11-7. There is an unexplained gap in the data between 1:30 and 4 P.M. Whenever one was heard during routine cruises of the spectrum, the time of day, strongest frequency, overall bandwidth, etc., were recorded. When strongest frequency was plotted against time of day for a sample of 62 encounters, a clear pattern emerged. Although this sample may be biased, one can see that the Woodpeckers' band use conforms to the daily cycle of ionospheric propagation. (There is an unexplained gap in the data between 1:30 and 4 P.M.; perhaps no spectrum-cruising occurred then.) The FCC engineer in charge of the study concluded that "the seemingly random frequency changes could be the result of computer control and propagation data being fed from a sounder."[11]

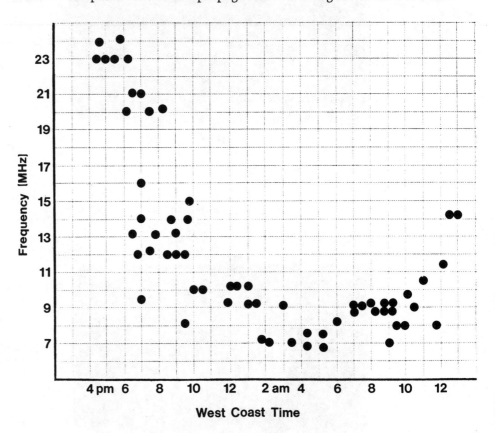

Fig. 11-7. Plot of the 62 encounters with Woodpecker emissions during routine cruises of the shortwave band by FCC monitors during the winter of 1982–83. (From FCC Woodpecker file.)

The Soviet Union has never explained the purpose of the Woodpeckers beyond saying they are for "tests." This explanation is vague to the point of being meaningless, but Soviet scientists do regularly publish papers about the effect of powerful pulse emissions on the ionosphere, and Radio Moscow's broadcasting has obviously benefited from research and development in the

11. David J. Smith, "Study of the Russian Woodpecker Signal" (8 June 1983), FCC Woodpecker file.

field of shortwave transmission. However, it is hard to imagine any research program that would justify running these complicated, electricity-guzzling, multimegawatt noisemakers continuously, for years on end, in defiance of the international community. Given the cost of building and running them, the signal characteristics and bullish band-use, it is all but certain that the Woodpeckers are operational OTH (over-the-horizon) radars.

The British built the world's first air defense radar system in 1938. It also operated in the shortwave band, but only because the technology for using even shorter wavelengths hadn't been developed. Nor had they figured out how to exploit ionospheric refraction for over-the-horizon surveillance. The "Chain Home" system was limited to line of sight. Even so, it proved so effective against the Luftwaffe that radar development became a top priority for many national military forces during and after World War II.

As it developed, radar tended to move up-band: from HF to VHF to UHF and eventually to microwave. Shortening the wavelength dramatically reduces the size of the antenna needed to form a narrow beam. A narrow beam concentrates RF energy into a "searchlight," thereby reducing "clutter" (echo unrelated to the target of interest) and increasing directional resolution. Shorter wavelengths also mean smaller targets can be detected with less transmitter power.

Inverting these arguments shows some of the reasons why the shortwave band is ill-suited for radar. But it has one special property that has always tantalized radar designers: skywave propagation. Radar signals of the proper frequency, launched to the ionosphere at the proper angle, will come to Earth hundreds (even thousands) of miles beyond the range of line-of-sight radars. And if there is an ionospheric path from the radar to a reflective object in the far-field, then there will also be an ionospheric path for an echo from the object back to the radar a split-second later. Fig. 11-8 illustrates the general concept of an over-the-horizon radar. A small portion of the emission is reflected back towards the transmitter, including Doppler-shifted echoes from aircraft flying through the illuminated far-field. However, most of the emission continues travelling away from the transmitter, leaving the detection zone as "forward-scatter," which can come back to Earth and cause interference far beyond the radar's surveillance range.

Very little of the radar signal will be "backscattered" toward its source. That's why extremely powerful transmitters, very sensitive receivers, and highly directional antennas are required. Even then, the echo of an airplane or battleship is minute compared with the backscatter from the land and sea—it's 40 to 80 dB weaker. Add in natural and man-made noise, multipath propagation, ionospheric storms, aurorae, everything that degrades the quality of shortwave channels, and you have some idea of how difficult it is to find the echo of a target of interest. Nevertheless, military planners in various countries felt that being able to detect a fleet of enemy aircraft a thousand miles away would be so valuable that it would be worthwhile to try to overcome the obstacles.

We don't know when the USSR began experimenting with OTH radar, but work in the U.S. began in the late 1940s, sponsored mainly by the Army Air

Fig. 11-8. General concept of an over-the-horizon radar.

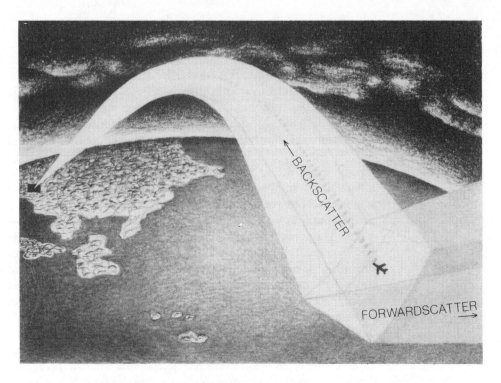

Force (the Air Force had not yet become a separate service). In the 1950s, the U.S. Naval Research Laboratory established that none of the technical obstacles were insurmountable. The echo of an aircraft at shortwave frequencies, though weak, is coherent, and the Doppler shift caused by the target's motion is distinctive (not many radar-reflective objects move at 400 miles/hour). The key to successful development of OTH radar was—and still is—the processing of the backscattered echoes. Computer technology was in its infancy then, so the Navy devised "a cross-correlation signal processor that utilized a magnetic drum as the storage medium. Under Air Force and Navy sponsorship, a high-power transmitter and antenna suitable for testing aircraft detection feasibility were added, and in the fall of 1961 aircraft were detected and range tracked over the major portion of their flights across the Atlantic."[12]

Further development of OTH radar depended largely on breakthroughs in automatic data-processing and ionospheric research. The Air Force began experimenting with a somewhat simpler technique, forwardscatter OTH radar, in the mid-1960s. As the name suggests, this technique takes advantage of the fact that most OTH radar energy is not scattered back toward the transmitter; rather, it caroms forward into another ionospheric hop. The much higher power of the forwardscatter means the receiver doesn't have to be nearly as sensitive as it

---

12. James M. Headrick and Merrill I. Skolnick, "Over-the-Horizon Radar in the HF Band," *Proceedings of the IEEE*, Vol. 62, No. 6 (June 1974), pp. 664–672.

would have to be to pick up backscatter. But that second hop makes extracting Doppler information much more difficult, so the technique lends itself to a somewhat different application—instead of looking for an echo from a distant object, forwardscatter can be used to detect the appearance of a hole in the ionosphere, as might be caused by a nuclear explosion or a rocket launch.

The Naval Research Laboratories' MADRE OTH-B radar, located at Chesapeake Bay, Maryland, is pictured in Fig. 11-9. The large array measures about 140 feet tall by about 320 feet wide and consists of 20 corner reflector elements arranged in two rows of ten. The beam can be steered $\pm 30°$ with "mechanically actuated line stretchers." Above and behind the large array is a rotatable corner reflector antenna for experiments in directions not within the main antenna's coverage. Built in 1961, MADRE's present status is unknown.

Fig. 11-9. The Naval Research Laboratories MADRE OTH-B radar in Chesapeake Bay, Maryland. (From *Aviation Week & Space Technology*, 6 December 1971.)

With four transmitter sites in the Far East and five receiver sites and a correlation center in Western Europe from 1968 to 1975, the Air Force used Project 440-L, a forwardscatter OTH system, for this very purpose. In other words, the continuous-wave emissions skipped across the Soviet Union.

Thus, the USSR was exposed to American OTH signals for years before the appearance of the Woodpecker. In 1971, the USAF's Director of Command, Control, and Communications was asked if the Russians knew about this system. "Yes sir, I'm sure they do," he replied.[13]

The Air Force's OTH station in Japan was subsequently dismantled and shipped to Australia, where it became the core of Project Jindalee, the most

---

13. Philip J. Klass, "HF Radar Detects Soviet ICBMs," *Aviation Week & Space Technology* (6 December 1971), pp. 38–40.

ambitious OTH backscatter radar project to date, outside of the U.S. and USSR. Desmond Ball described it in 1978:

> The transmitter site, consisting of a phased array antenna 185 meters long, several other masts to support communication and radar calibration antennas, power house, etc., is located in the Hart's Range about 160 km northeast of Alice Springs. The main receiving site, in the Mt. Everard area about 15 km north of Alice Springs, features a 600 m long receiving array, calibration and communication aerials, and large earth mats providing stable ground planes; some 300 km of wire was used to make these mats. About six or seven technicians from WRE [the Weapons Research Establishment] reside in Alice Springs, and these provide a 24 hour a day manning of both the transmitter and receiver sites.
>
> Under Stage One of the project, the transmitting station at Hart's Range operates on a power output of some 50 kW, provided by five 10 kW subtransmitters—40 kW is utilized in the actual radar transmission, and 10 kW feeds into the "ionospheric sounder" dedicated to monitoring the behavior of the ionosphere. Sixteen further subtransmitters are being installed under Stage Two....The Hart's Range station should eventually operate on at least 1 MW. The basic operating frequency spectrum of the installation is expected to be 5–29.5 MHz....It could at times go up to 60 MHz. On overseas experience, the operating frequency will centre around 14 MHz. The bandwidth will be quite wide, probably varying from about 30 kHz to more than 300 kHz at some times. The basic pulse repetition rate will depend to some extent on how the system is eventually optimized for the detection and tracking of high-Mach aircraft and relatively slow-moving ships. It should be somewhere from about 3 pulses per second to about 10/sec.
>
> For early warning, an actual operating system would require two or three such "Jindalee" installations....An actual working system could be operational within five years from today....[14]

One ham radio operator who wishes to remain anonymous said, "the Australian OTHR has a quieter attack and usually stays on a channel longer than any of the other Woodpeckers." This remark was made early in 1983, but we don't know anything more about signal characteristics or the system's present status. Signals like the woodpeckers, but with different pulse rates and bandwidths, are heard from time to time. Their source remains a mystery, but some may be from Australia.

According to Dr. Ball, Jindalee, like the Woodpeckers, was designed to transmit variable but potentially very wide bandwidths throughout most of the shortwave band. This suggests that spectrum-hogging isn't just due to a bad attitude toward other users of the band on the part of the Soviet Union but is in

---

14. Desmond J. Ball, *Electronics Today International* (February 1978), pp. 35–40.

the nature of OTH radar technology. Many different combinations of wavelength and beam elevation angle are needed to illuminate different air spaces—the best combination depends on ionospheric conditions at a given moment—and wide bandwidths are necessary for range resolution. Propagation data is obtained with sounders that sweep through all the frequencies available to the radar. A sounder receiver at the radar receiver site, with a passband that sweeps in tandem with the sounder transmitter, picks up the echoes so radar operators can check the time-delay (ionospheric height and distance to reflection point), path loss, distortion, etc., of each frequency. Shortwave listeners are thus more likely to encounter signals from a sounder operating in support of an OTH radar than from the radar itself. Fortunately, sounders don't have to be very powerful, and in order to test every frequency between 5 and 28 MHz every 20 minutes, it must sweep at least 19 kHz per second. That means its signal will cross a 10 kHz-wide channel in about half a second. Many of the brief zips one hears crossing a shortwave channel are ionospheric sounders—they are not necessarily connected with OTH radars, however. The data they yield is valuable for any radio service that relies on skywave propagation, as well as for scientific research. Their use is rapidly spreading.

One system definitely not active today is Polar Cap III. The Canadian Defense Research Board and the U.S. Air Force collaborated on this project "to determine the cost and effectiveness of OTH radar to detect airborne targets in the polar region."[15] From a military perspective, a long frozen border between Canada and the USSR is where OTH capability would be most valuable—as a super DEW Line able to "look" right into Soviet air space. The DEW Line stations, based on 1950s technology, are limited in coverage and expensive to maintain. The idea of replacing them all with one or two OTH radars had obvious appeal.

The Polar Cap III transmitter was located at Hall Beach on Melville Peninsula in Canada's Northwest Territories. There were two receiver sites, to increase detection possibilities (this configuration is called "bistatic" radar). One was near the transmitter, and the other was at Cambridge Bay on Victoria Island, 550 miles to the west. Equipment and supplies were air-lifted in during the summer of 1972. Although very large, the antennas went up quickly and tests began in October. The results proved disappointing. When presenting its Continental Air Defense Plan to the House Armed Services Committee in 1981, the Air Force explained, "We have found that when OTH is looking north, the auroral interference degrades its performance so the OTH may not give us the quality warning we need."[16]

This must raise questions about the Woodpeckers' effectiveness as radars, since some of their beams skirt the polar region. However, ionospheric turbulence is believed to have a worse impact on Doppler-based systems like the Air Force's than on time delay systems like the Woodpeckers. That may be why the Soviets use the latter.

15. Scotty Yool, "Polar Cap III," *Sentinel* (January 1973), pp. 24–25.
16. Lt. Gen. James V. Hartinger, *Full Committee Hearing on Continental Air Defense*, Committee on Armed Services, US House of Representatives (22 July 1981), p. 31.

Given our unhappy familiarity with the Woodpeckers, it's not surprising that a shock wave swept through the shortwave community in 1980 when the Air Force announced it was starting to test a powerful OTH backscatter radar in Maine, and if the tests were successful, more and bigger radars would be built near the Atlantic, Pacific, and Gulf coasts. According to O. G. Villard:

> The U.S. counterpart of the woodpecker is still experimental....The mission...is to give early warning of bomber attack via the northeast oceanic approaches to the U.S. and Canada....Since all aircraft flying over the water must routinely file flight plans in advance, those seen by the radar...are then compared with their reported plan; if a target or targets cannot be associated with such a plan, appropriate action is taken....
>
> Although the Soviet radar transmits on-off pulses (a conventional approach that permits the transmitting and receiving functions to share one antenna), the U.S. radar employs separate transmitting and receiving sites so that the transmitter can radiate a continuous signal which is FM-modulated or 'chirped' for range resolution.
>
> ...the broadside receiving array...located at Columbia Falls, Maine, is about 4000 feet in length. Since mechanical rotation would hardly be practical, scanning is accomplished by digital beamforming and slewing. The beam is made unidirectional by means of a reflecting screen located about a quarter-wave behind the radiating elements.
>
> Essentially the same design is used at the transmitting location, except that the transmitting beam is made wider so as to floodlight the area within which multiple receiving beams conduct a fine-scale search. The average transmitter power is roughly 1 megawatt....
>
> Economic considerations have dictated the present operating frequency range—6.7 to 22 MHz. It would be nice to be able to go both lower and higher, and perhaps this will be possible in the future....
>
> The radar is, of course, computer-controlled and extraordinarily flexible. It is provided with a low-power auxiliary sounder which repeatedly checks all the available frequencies to determine what propagation paths are present....A separate facility automatically checks all the available channels for interference, measuring not only the level at any given instant, but also the time history of that channel's usage.
>
> All these input pieces of information plus many, many more (such as worldwide geophysical data) are taken into account by the radar's main computer....
>
> So long as it is performing its mission satisfactorily, the radar's operating guideline is to cause as little interference as possible. Thus, when propagation is good, it can reduce power....
>
> There is little problem in recognizing the Soviet OTHR signals, so long as they continue the present format....The American radar signals, by way of contrast, will sound on an AM receiver more like power-line hum, but at any one of several modulation frequencies

from 20 Hz to 60 Hz. They should be much less irritating than on-off pulsing. Because they will be coming from a source physically closer to some U.S. amateurs [and shortwave listeners], however, they may be troublesome when propagation is exceptionally good....[17]

A typical USAF OTH modulation waveform is given in Fig. 11-10A and the power spectrum profile is given in Fig. 11-10B.

Fig. 11-10. The USAF OTH-B radar signal.

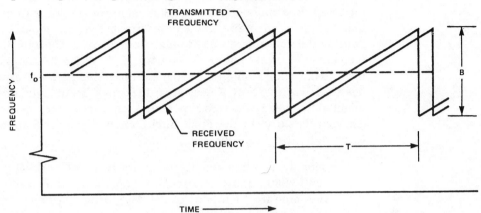

(A) Typical modulation waveform. B = nominal bandwidth, f_o= carrier frequency, and 1/T = waveform repetition rate. (From Draft, Environmental Impact Statement, West Coast, pg. A-6.)

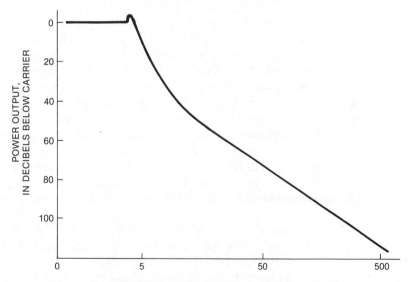

(B) Power spectrum profile of a (nominally) 10-kHz wide emission from the experimental OTH in Maine. (From O.G. Villard, Jr., QST, April 1980.)

17. O. G. Villard, Jr., "Over-the-Horizon or Ionospheric Radar," *QST* (April 1980), pp. 39–43 *(Courtesy QST).*

The initial furor caused by this article faded as the radar tests, much to everyone's surprise, ran more or less unnoticed. The Air Force says it got over a dozen interference complaints, but only one was verified: a Military Affiliate Radio System station in Florida, using 20937.5 kHz on 30 January 1981. All the other complaints were for bands or times where the radar wasn't operating.

The significance of this excellent noninterference record is hard to gauge because the Air Force has released contradictory information about the tests. In the Draft Environmental Impact Statement for the West Coast radar, they said that the radar in Maine transmitted for a total of about 900 hours, usually "during the ionosphere's day-to-night and night-to-day transition periods," over the course of a year ending in January 1981. When it was pointed out that the night-to-day transition on the radar's propagation path corresponds to the hours before dawn in North America, so the small number of interference reports might only mean that most potential reporters were asleep, the Air Force changed its story. In the Final Environmental Impact Statement for the West Coast radar, they said the Maine radar

> actually operated about 2300 hours over 330 working days beginning June 1980...and the cumulative operating time was approximately the same for each hour of the day, instead of being concentrated in the day-night and night-day transition periods....Thus, there was much more opportunity than previously stated for ham radio operators, shortwave listeners, and Fixed Service providers to notice interference.[18]

The last sentence is the crux of the matter. Aside from the fact that there weren't 330 working days between June 1980 and January 1981, if the second text is true, we may not have anything to worry about, so far as interference is concerned. If the first text is true, the tests were not representative of radar operation at times of peak band use and we cannot assume that an operational radar, working around the clock, would be similarly benign.

Of course, the primary purpose of the tests wasn't to see how many interference complaints came in but to see how well the system worked—how well the radar could distinguish between unidentified and identified aircraft in the heavily trafficked North Atlantic, how well it coped with auroral disturbances, sea-surface clutter, varying band conditions, etc. According to the General Accounting Office,

> Initial testing of the experimental system...showed that the radar would not meet performance requirements for an operational system....*ACQUISITION MAY NOT BE JUSTIFIED*...The threat of a precursor Soviet bomber attack against the United States is the scenario used to justify the need for the OTH-B radar system. Considering the threat described in intelligence reports, along with alternatives to

---

18. *Construction and Operation of the West Coast OTH-B Radar System—Final Environmental Impact Statement*, Air Force Systems Command, Electronic Systems Division (February 1984), p. 330.

OTH-B, GAO questions the need to acquire the OTH-B radar system as now planned....[19]

Since the bulk of the GAO report was classified, we don't know what shortcomings were revealed by the tests, but *Aviation Week & Space Technology* said that when the radar was demonstrated for them, "severe ionospheric disturbances crossing Newfoundland and extending about 200 mi. south of Greenland adversely affected system performance. Not long after, however, tracks of aircraft flying the organized route system could be observed roughly 1,000 mi. farther to the east...."[20] Apparently, the radar was blinded for periods of time by ionospheric disturbances. Because the experimental radar's coverage extends deeper into the auroral zone than the rest of the proposed continental system, the problem may be less acute in other sectors.

The Air Force claimed that the tests were successful and that moving from the experimental radar to an operational system would require only refinements of the design. The GAO disagreed, saying that the tests were only successful because the Air Force tailored its criteria to the capabilities of the system, not to its operational requirements, and "a substantial redesign effort for both hardware and software is involved" in moving toward an operational OTH radar.[21]

The U.S. Air Force's plan for continental air defense is based on OTH-B radar sectors that cover 60° each as shown in Fig. 11-11. Three of them are based in Maine, three in the Pacific northwest, two are slated for Alaska, and four for the northern midwest. Each sector is subdivided into eight subsectors (each 7½° wide). The radars will step from one subsector to another about every ten seconds. A different frequency may be used in each step.

OTH radar emissions don't stop at the edge of the surveillance area, they carry forward into a second (even a third) ionospheric "hop." Comparing Fig. 11-12 with the map of the U.S. Air Force OTH-B radar deployment plan, one can see Western Europe would be regularly "illuminated" by the second hop of the northern-most sector of the East Coast system.

The controversy surrounding these projects thus includes much more fundamental issues than interference to other services: Are they reliable? Does the benefit to national security outweigh the cost? Are there better alternatives? The way these questions are answered by the U.S. Congress will determine the extent and pace of implementation. This should be kept in mind as we review the Air Force's plans because a majority of members of the House of Representatives has shown it agrees with the GAO, so the Air Force may not get all it wants—at least not right away.

Because the law requires public disclosure of plans for projects likely to have a significant impact on the "human environment," the Air Force has published detailed descriptions of their proposed continental OTH-B systems. The experimental radar is now being upgraded into the first of three adjacent

19. "Acquisition of the Over-the-Horizon Backscatter Radar System Should Be Reevaluated," GAO/C-MASAD-83-14, General Accounting Office (15 March 1983), pp. i–ii.
20. Kenneth J. Stein, "Backscatter Radar Unit Enters Production Phase," *Aviation Week & Space Technology* (16 August 1982), p. 77.
21. GAO/C-MASAD-83-14, p. iii.

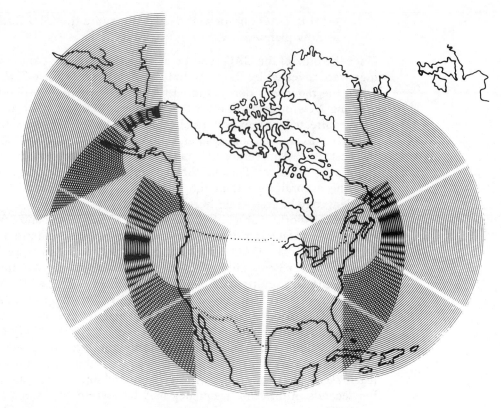

Fig. 11-11. The USAF OTH-B radar deployment plan.

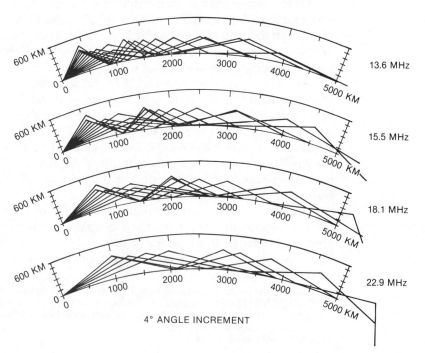

Fig. 11-12. Chart of the OTH radar emissions. (Courtesy *Proceedings of IEEE,* 62 (1964), pg. 668.)

600 KM
0
0   1000   2000   3000   4000
5000 KM
13.6 MHz

600 KM
0
0   1000   2000   3000   4000
5000 KM
15.5 MHz

600 KM
0
0   1000   2000   3000   4000
5000 KM
18.1 MHz

600 KM
0
0   1000   2000   3000   4000
5000 KM
22.9 MHz

4° ANGLE INCREMENT

identical sectors to be based in Maine, each covering 60° of azimuth out to a distance of 1800 miles. (There is an inner limit to the detection range of 500 miles, due to the length of the ionospheric "skip" zone.) Each sector will have twelve 100 kW transmitters for a maximum power output of 1.2 megawatts; six separate phased-array antennas, only one of which would be driven at a time; a 10 kW sounder; and a receiving station with a mile-long linear array of active antennas. The transmitters will be located at 45° 10′ 38″ North by 69° 52′ 23″ West, in Somerset County—oddly enough, near a small town named Moscow—and the receivers will be about 100 miles east, in Washington County, near the Maine coast. An operations center at Bangor International Airport will house the data processing and control the transmitters and receivers of all three sectors.

Essentially the same design has been proposed for the West Coast. The plan for the transmitting arrays of the three sectors in the West Coast OTH system is given in Fig. 11-13. The receiving arrays would be similarly arranged. The three sectors of the East Coast system are more dispersed, presumably to fit the landscape better. (Diagram from the Draft EIS, West Coast, p. A-8) There the operations center would be at Mountain Home Air Force Base in Idaho; the transmitters at Buffalo Flat, Oregon (43° 16′ North by 120° 30′ West) and the receivers at Rimrock Lake, Oregon as shown in Fig. 11-14.

Unlike the East Coast system, which is aimed out over the ocean, the West Coast system's surveillance coverage includes huge tracts of U.S. territory—the southern half of California and virtually the entire state of Alaska. These areas could be "illuminated" by the radar beams about every 80 seconds. Shortwave listeners in Alaska and southern California thus would be more likely to hear the radars—and hear them louder—than listeners elsewhere.

The Air Force has also expressed the desire to build a 120° south-facing OTH system somewhere in northern Texas or Oklahoma by the late 1980s. If built, the two sectors of this system would illuminate much of Central America.

Each radar sector would be on the air continuously and simultaneously. The detection beams would not sweep across their sectors but rather would step in 7.5° increments from azimuth to azimuth, "sometimes in a seemingly random manner."[22] A total of eight steps are required to cover each 60° sector. The frequency and elevation angle used at each step are chosen independently so that up to 24 frequencies could be used during each 180° work-cycle, three at once. If the East, West, and Gulf Coast systems are all built, up to 64 frequencies could be involved in the total work-cycle, eight of them on the air at once.

"Frequency" here means frequency *band*, of course, as the emissions are variable in bandwidth from a minimum of 5 kHz to a maximum of 40 kHz, with some weak overspill into adjacent bands. The typical waveform would be a continuous-repeat frequency-modulated (FM) sawtooth. When the bandwidth is 5 or 10 kHz, the waveform repetition rate could vary from 10–20 per second in 2.5 Hz increments; when the bandwidth is 20 or 40 kHz, the rate could vary from 20–60 per second in 5 Hz increments. On an audio receiver, these would all sound like low frequency buzzes, the pitch set by the repetition rate.

---

22. *Construction and Operation of the West Coast OTH-B Radar System—Draft Environmental Impact Statement*, Air Force Systems Command, Electronic Systems Division (March 1983), p. C-1.

Fig. 11-13. Plan for the transmitting arrays of the three sectors in the West Coast OTH system. (From Draft, Environmental Impact Statement, West Coast, pg. A-8.)

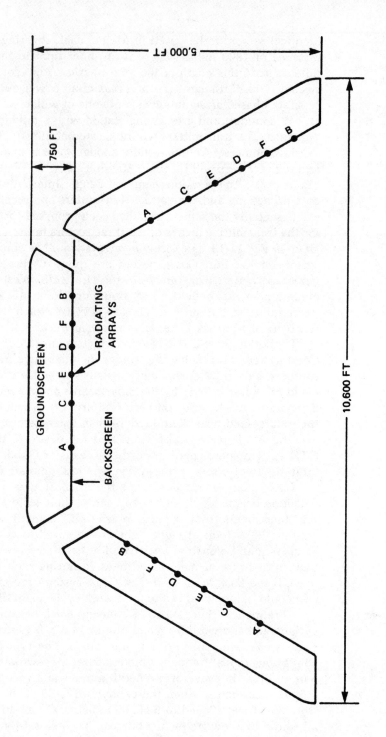

**Fig. 11-14.
Transmitter and
receiver sites
for the Air
Force's
proposed West
Coast OTH-B
radar system.**
(From Draft,
Environmental
Impact
Statement,
West Coast, pg.
2–8.)

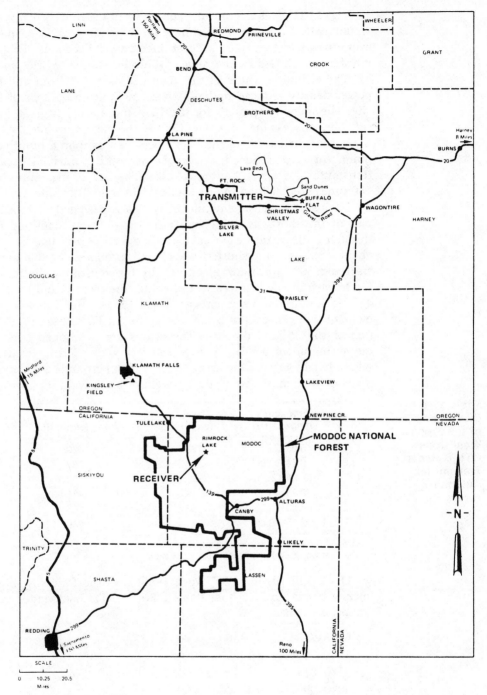

Fig. 11-15 gives the top and side views of the radar's main beam. The beams would dwell on a particular azimuth for up to 10 seconds, so in each sector, each azimuth would be visited at least once very 80 seconds. That means that any interference caused by the radars would be relatively brief and episodic.

The azimuthal width of the main beams is 19° from null to null, with the power density reduced by half, ±4.2° away from the axis of maximum intensity. The four-field pattern for the 12-element array is given in Fig. 11-16. The sidelobes are less than one-twentieth of the maximum intensity, and because of the antenna backscreen, the backlobes are less than a hundredth of the maximum. Although reduced, these lobes are still nontrivial because of the high transmitter powers and antenna gains. The axis of maximum intensity is some 160 times greater than the theoretical isotropic distribution.

Each sector has six antenna arrays, each designed to transmit in a different part of the shortwave spectrum (see Table 11-1). Identified by the letters A through F, the arrays each have 12 dipole radiating elements spaced evenly at about 0.58 times the shortest wavelength produced by that array. The beam is steered ±30° in azimuth electronically by controlling the timing of the signal fed to each dipole. Viewed from behind, the band C and D elements are tilted 45° clockwise, the B elements are tilted 45° counterclockwise, and the E and F elements are vertical (for some reason the Air Force hasn't revealed the orientation of the A band dipoles). The arrays are positioned over a groundscreen measuring about a mile long by 750 feet wide, and in front of a backscreen whose height varies from array to array. Both groundscreen and backscreen are made of wire mesh, either copperweld or alumiweld for corrosion resistance.

Table 11-1. Characteristics of the Proposed West Coast OTH-B Radar Transmitter Antenna Arrays*

Band	Frequency range (MHz)	Mid-band wavelength (ft.)	Array length (ft.)	Backscreen height (ft.)
A	5.00–6.74	170	1020	135
B	6.74–9.09	126	756	100
C	9.09–12.25	93.3	560	75
D	12.25–16.50	69.3	416	55
E	16.50–22.25	51.4	308	45
F	22.25–28.00	39.4	236	35

* From the Draft Environmental Impact Statement, West Coast OTH system (p. B-4). Figures for the East Coast system are essentially the same.

O. G. Villard, Jr. (*QST*, April 1980) says the following bands are the frequency bands (in kHz) that the Air Force OTH backscatter radars have been authorized to use:

6765–7000	13360–14000
7300–7508.4	14350–14990
7530–8195	15100–16460
9040–9995	17360–17900
10100–11175	18030–19990
11400–11975	20010–21000
12000–12330	21450–21850

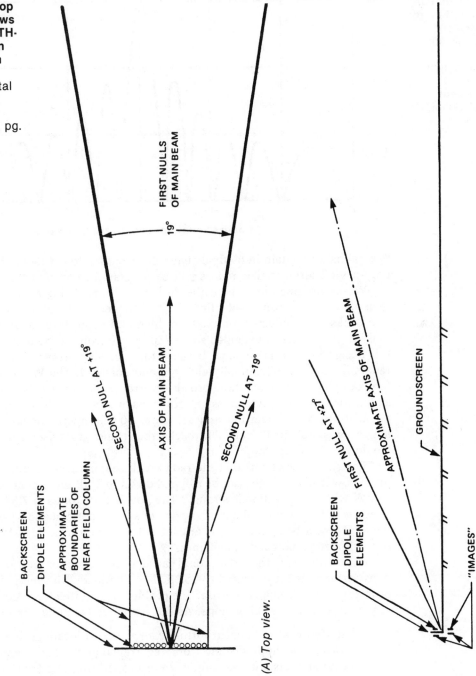

Fig. 11-15. Top and side views of the US OTH-B radar main beam. (From Draft, Environmental Impact Statement, West Coast, pg. A-11.)

FIRST NULLS OF MAIN BEAM

19°

SECOND NULL AT +19°

AXIS OF MAIN BEAM

SECOND NULL AT –19°

BACKSCREEN

DIPOLE ELEMENTS

APPROXIMATE BOUNDARIES OF NEAR FIELD COLUMN

(A) Top view.

FIRST NULL AT +2°

APPROXIMATE AXIS OF MAIN BEAM

GROUNDSCREEN

BACKSCREEN

DIPOLE ELEMENTS

"IMAGES"

(B) Side view.

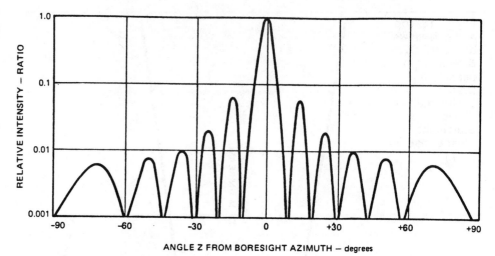

Fig. 11-16. Far-field pattern for 12-element array. (From Draft, Environmental Impact Statement, West Coast, pg. B-15.)

ANGLE Z FROM BORESIGHT AZIMUTH — degrees

Several bands contain individual channels forbidden to the radars because they are for vital services. The radars will be programmed to omit them.

We know much less about the design of the receiving arrays. The experimental system in Maine had a line of 137 V-shaped monopoles mounted in front of a backscreen reflector some 50 feet high and about three-quarters of a mile long. A groundscreen extended some 250 feet in front of the array. Apparently this design proved inadequate, because the receiving arrays of the operational radars have been substantially changed and enlarged. The West Coast system, for example, would have three broadside arrays, each facing the same direction as the corresponding transmission arrays. Each sector would resemble a row of 84 steel towers, 65 feet tall and set about 60 feet apart, for an overall array length of about 5000 feet. The backscreens and groundscreens presumably would be scaled up, too.

The Air Force claims that the radars will seldom need to transmit at maximum power. Each sector can be stepped down in 1 dB increments to as little as 37 kW when propagation is strong. "The power used for each 7.5° surveillance sector is independently chosen to optimize system performance. Consequently, the total power being radiated at any instant may vary from 111 kW to 3.6 MW, but will rarely approach either limit."[23]

It is obvious that the radars could cause massive interference to communications services, if they wanted to—or if they were run with indifference to the problem, as the Woodpeckers are. The problem is that their emissions don't stop at the edge of the detection zone. As the Air Force candidly admits:

> Interference could occur virtually any place at locations one or more ionospheric hops distant from the transmitter. Interference could occur if a distant transmitter were propagating at signal...that, because of ionospheric propagation conditions, did not include the radar site....In that situation, although the radar system monitors the

23. Draft EIS, West Coast, p. A-12.

spectrum before transmitting, the radar operators would believe that the in-use channel was clear and would feel free to use it. They could not realize that the radar signal could be producing interference at some far distant location....Because the radar signal involved could be emitted by the main lobe, backlobes, or sidelobes and because the signal could propagate by more than one hop, predicting when or where this may occur is impossible.[24]

As for interference caused to the radars by other transmitters, Villard has pointed out that

OTHRs are designed to work through some level of interference, no matter what its origin. For example, there are circuits which convert interference carriers (from Teletype or shortwave broadcasting stations, for example) into what the radar receiver perceives as relatively harmless broadband noise.[25]

From a technical perspective, then, the radars don't need to avoid already-occupied channels in order to function. The level of interference caused by them depends decisively on how sensitive the radar operators are to the spectrum rights of others, and how effective their methods for avoiding interference are in practice.

The National Telecommunications and Information Administration has authorized the Air Forces's OTH radars to use any of about 30 bands between 5 and 28 MHz on a noninterference basis. All these bands are allocated to the Fixed and Broadcasting Services. Among them, 9500–9900, 11650–12050, 13600–13800, 15100–15600, 17550–17900, and 21450–21850 kHz are (or soon will be) allocated to the Broadcasting Service. Unlike the Woodpeckers, the Air Force radars will not use any band where the safety of life or property is at risk: the maritime and aeronautical bands, distress calling, hurricane warning, search and rescue channels, etc.

Operation of the radar is intended to produce minimal interference with other users of these same portions of the spectrum, and operation of the ERS [experimental radar system] in Maine indicates that this can be achieved....Air Force operating procedure will be to attempt to 'find an unused frequency in the fixed (point-to-point) section of the HF spectrum...' [and] 'to conscientiously avoid the use of the SWBC [shortwave broadcast bands] wherever possible'....When operating in the Fixed band, the radar will stay far enough away from the band edges to minimize the possibility of interference to other users, such as broadcast receivers....

The Air Force wishes to operate the radar without interference to others. Those who believe that they are experiencing interference from it are urged to keep an accurate log of the times and the fre-

24. Draft EIS, West Coast, p. C-21.
25. O. G. Villard, Jr., "Over-the-Horizon or Ionospheric Radar," *QST* (April 1980), p. 42 *(Courtesy QST)*.

quencies and to provide that information to the Air Force. The address is: HQ ESD/SCU-4, Hanscomb Air Force Base, MA 01731.[26]

These are certainly commendable statements, so far as shortwave listeners are concerned. Unfortunately, other more recent statements are less reassuring. When they were asked for clarification of the phrase "wherever possible" in the above quote—which implies that under certain conditions the radars could transmit in the broadcasting bands—the Air Force replied:

> Whenever possible, the radar would operate in the Fixed part of the spectrum. However, if at some time the Air Force considers it to be in the national interest to use a frequency in the International Broadcast band, this would be done—without prior notification and probably without producing interference. Military systems of all nations have that right; U.S. systems do not abuse it.[27]

In support of their claim that all military forces "have that right," they cited Article 38 of the International Telecommunications Convention, which says that nations signing the Convention "retain their entire freedom with regard to military radio installations of their Army, Naval, and Air Forces." According to the Air Force, "This means that a nation's military forces are not bound by the 'no harmful interference' rule in any case where it is determined, unilaterally by the military, that a given transmission is important to the nation's defense."[28]

It is discouraging that they base such a bold interpretation on just one phrase from Article 38—especially since the very next phrase reads, "Nevertheless, these installations must, so far as possible, observe statutory provisions relative to...the measures to be taken to prevent harmful interference...."[29]

The Air Force's claim that they can unilaterally ignore NTIA's noninterference stipulation "in any case ... important to the nation's defense," gives them a loophole large enough to negate the stipulation entirely. It's the kind of statement one might expect from the Russians—if they were willing to admit that the Woodpeckers are military systems (they've never done that). We can only hope that the Air Force will rarely invoke it.

The risk of interference obviously depends on how the radars' operating frequencies are chosen. Villard described the process this way:

> First, the useful band of frequencies is determined from soundings and other data. Then the open [i.e., not already occupied] allocated channels within that band are identified and ranked in order of desirability. Low path loss, low multipath and low noise are the chief criteria. The radar then comes up on the most desirable frequency and continues there until conditions change or interference develops, at which point [it] instantaneously changes to the frequency which at

---

26. Draft Environmental Impact Statement (EIS), West Coast, pp. C-19 through C-22.
27. Final EIS, West Coast, p. 332.
28. Final EIS, West Coast, p. 328.
29. International Telecommunication Convention, *US Treaties and Other International Agreements*, 1976–77, Vol. 28 Part 3, p. 2531.

that moment is at the top of the continuously updated 'most desirable allowable frequency' list.

Often the radar will shift from an otherwise excellent frequency in order to resolve range or Doppler (velocity) ambiguities.[30]

The radar may also alternate quickly between two or three frequencies as it switches between search mode and tracking mode (this might occur if a suspicious target has been spotted).

The process described by Villard occurs separately for each azimuth, so when the radars change azimuth—every ten seconds or so—they'll probably change frequencies, too. The Air Force adds that the radars will generally operate near the highest frequency that will propagate to the desired range at a given time.

The keys to avoiding interference are not using frequencies already being used by others—this requires effective monitoring of the spectrum prior to radar transmission—and responding quickly to valid complaints of interference. The Association of North American Radio Clubs' OTH Radar Committee sent several suggestions along these lines to the Air Force during the environmental impact assessment process. The first suggestion was to eliminate the problem of radar operators not being able to detect distant stations using a channel because their signals didn't propagate to the radar sites: give the operators real-time access to monitoring assets located elsewhere in the world—particularly in areas that would be regularly reached by the radar beams in one hop (southern California and Alaska) or two hops (western Europe and Japan). The Air Force rejected this, saying, in essence, that the radar receiver is so much more sensitive than a remote monitoring station that the latter is unnecessary. Aside from the fact that this contradicts the Air Force's own description of how they could fail to identify a channel as occupied, and ignore the splotchiness of propagation, it hints at a capability discussed openly during the 1970s but rarely since. As Desmond Ball put it:

> Because of the great sensitivity, wide bandwidths and ability to cover such bandwidths over a wide frequency range, OTH-B receivers are admirably suited for gathering electronic intelligence....This is especially significant because of the amount of military signals and communications which are transmitted on the HF band....It may even be possible to modulate the frequencies so that the radar detection and the elint intercept capabilities can be operated simultaneously....[31]

The Air Force has never mentioned the gathering of electronic intelligence as part of the mission of their OTH radars, and considering some of the comments made by critics of the program in Congress and the GAO, the classified statements of mission may not contain it either. But it's plausible.

Our second suggestion had to do with the prompt processing of complaints. We suggested that there be an "800" telephone number and/or a Telex address

---

30. O. G. Villard, Jr., "Over-the-Horizon or Ionospheric Radar," *QST* (April 1980), p. 42 *(Courtesy QST)*.
31. Desmond J. Ball, *Electronics Today International* (February 1978), p. 40.

so that reports of interference could reach the radar operators with the least possible delay. They replied:

> The Air Force is currently developing procedures by which authorized users of the HF bands who believe that the radar is creating interference would have the opportunity to report their complaint in real time. Their complaint would be received and evaluated by a person who would have the responsibility and authority to immediately eliminate interference caused by the radar....
>
> At this time, neither the communication link from the complainant to the interference-mitigating authority nor the operational procedures for establishing the validity of a complaint have been determined because testing and operation of the radar would not begin until 1985–1986. However, toll-free (i.e., 800-area code) phone numbers are under consideration for the communication links. When the procedures have been developed for receiving and evaluating reports of interference to authorized users, and for acting on complaints that are valid, the Air Force will make existence of these procedures widely known. ANARC will be able to help in this notification process.[32]

However, they rejected our suggestion that something like an "800" number be provided for potential complainants in northern and western Europe and Japan—areas likely to be affected by the second-hop signals.

Our final suggestion was that the Air Force express its commitment to this policy in the Final Environmental Impact Statement for the West Coast system:

> If a complaint of interference is received, it will be evaluated promptly. If found to be valid, and if the interference condition still exists at the time of validation, the radar will be taken off the offending frequency immediately.

In the Final EIS they wrote: "The paragraph suggested by ANARC...describes precisely the Air Force's philosophy and obligations in the matter."[33]

So far, then, the Air Force's attitude toward the concerns of shortwave listeners has been quite accommodating—though they pointedly assert the right to suspend their commitment to noninterference whenever they think "a given transmission is important to the nation's defense," and they haven't actually developed procedures for responding to complaints in real-time, though they've promised to do so by the time the first radar sector goes on the air in 1985–6. We won't know how well their good attitude will translate into behavior for a while yet.

The first sector is the experimental radar, upgraded. Work on the second 60° sector in Maine began in 1984 and will continue to 1987–8. In granting funds for the second sector, Congress made clear that "The purpose of this

---

32. Final EIS, West Coast, p. 333.
33. Final EIS, West Coast, p. 333.

action is to expand the test database available to ensure that subsequent procurement decisions can be made on the basis of the most complete information possible."[34] In other words, they want to see how well the first two sectors perform before they authorize money to start additional ones. This is only prudent, as the results of the experimental radar's tests appear to have been mixed, and the cost of building both the East and West Coast systems is expected to total over a billion dollars.

As this book goes to press, the Air Force announced it was dropping its request for authorization to begin building the West Coast system. The question now facing the Congress is whether to appropriate money to start the third sector of the East Coast system before the first two have been tested. Their decision will have been made by the time you read this.

Meanwhile, the U.S. Navy, which played a pivotal role in the development of OTH backscatter technology, isn't sitting by idly. They're developing transportable OTH systems "to cover several critical ocean gaps and choke points around the world."[35] Gerald Green explains that the

> primary mission would be in support of maritime air defense of the Navy's carrier battle groups and selected sea lanes. The [relocatable radars] also could be assigned to support Rapid Deployment Force (now US Central Command) operations in remote areas of the world. ROTHRs are expected to be the primary wide-area search systems for regional air defense operations in critical ocean areas not covered by existing or planned surveillance systems. The initial ROTHRs are scheduled to become operational in the mid- to late 1980s.[36]

We don't know anything about the design or emission characteristics of the Navy systems yet, but since they say they're building upon the WARF and Air Force experience, one can assume that they will probably use continuous-wave FM modulation and some sort of frequency-hopping. Because transportability imposes limits on antenna size, we suppose these radars will mainly operate in the upper two-thirds of the shortwave band. *Aviation Week & Space Technology* has said that the first area covered by a Navy ROTHR will probably be the sea between Iceland and Norway, with construction of an automated command center to begin in Iceland in Fiscal 1985.[37]

Several OTH-type systems have been around for years. The WARF (Wide Aperture Research Facility) has been operated by SRI International since the 1970s. With transmitters and log periodic antenna arrays near Lost Hills, in south-central California, the repeater/transponders at several sites in the western part of North America, WARF was the test-bed for many of the OTH concepts now being applied by its sponsor, the U.S. military. Because of the

34. *Conference Report, Fiscal 1984 Authorization, Department of Defense* (1 August 1983), Senate Report 98-213, p. 170.
35. Melvin R. Paisley, *Appropriations Hearings*, Armed Services Committee, US House of Representatives (15 March 1983), p. 497.
36. Gerald Green, "C³I: The Invisible Hardware," *Seapower* (April 1983), pp. 121-2.
37. Clarence A. Robinson, Jr., "Defense Decision Hikes Strategic Funds," *Aviation Week & Space Technology* (23 August 1982), pp. 17-19.

type of research done there, WARF's band use, emission characteristics, and schedule all vary, but at times it probably sounds something like the Air Force radars.

The Rome Air Development Center at Ada, New York has a number of transmitters and antennas for radar experiments and ionospheric research. They have been known to transmit in a series of 200 kHz bands at approximately 2 MHz intervals between 6.5 and 30 MHz, at powers up to 100 kW. Though RADC has been used to track aircraft approaching Canada's east coast, it is more often used to generate backscatter from the ocean surface and aurorae.

OTH radars designed for aircraft detection treat environmental backscatter as clutter, to be suppressed and removed by filters, but in fact, the clutter contains information useful for other purposes. Ocean waves happen to reflect the frequencies used by OTH radars quite well. By charting their echoes, it is possible to deduce the surface wind pattern over hundreds of square miles of ocean. Scientists have long known that the interacton between the atmosphere and the oceans is the pump that churns the weather, but they couldn't monitor this interaction synoptically and in real time until techniques like OTH radar evolved. As the air/ocean relationship becomes better understood, and radar signal-processing techniques improve, the meteorological use of OTH radar is likely to increase. Right now there are perhaps a dozen experimental sea-state monitoring programs, dispersed worldwide.

The use of shortwave radar for ionospheric research is also growing. OTH radar relies on the ionosphere simply to carry signals to and from a distant air space. But if the beam is aimed higher, or straight up, scatter from the ionosphere itself can be processed to map its structure in various ways—to reveal the interaction between the solar wind and the Earth's magnetic field, to help us understand how spacecraft in low orbit become electrically charged, etc. One of the largest ionospheric research programs now underway is sponsored by the Defense Nuclear Agency, which wants to know how nuclear war would affect ionospheric support of long-distance radio communications. Several relocatable ionosonde transmitters, operated by SRI International, are involved.[38]

Since 1970, scientists have been using high-powered shortwave transmitters to heat up parts of the ionosphere. This might seem to be a long way from OTH radar, but listeners could easily mistake an RF heating signal for a radar signal. Tests have shown that a few hundred kilowatts fed into simple dipoles tuned to 0.5 to 1.0 times the ionosphere's critical frequency (the highest frequency that will be reflected when the incident beam is vertical) can raise the electron temperature in the overhead region enough to refract obliquely aimed signals over the horizon, all the way up to the UHF band. "Voice, teletype, and facsimile transmissions have been sent, by means of the scattering region above the modifier, between ground terminals separated by several thousands of km and using frequencies which would not otherwise have been useful for those

---

38. C. L. Rino, J. F. Vickrey, R. C. Livingston and R. T. Tsunoda, "High-Latitude and Equatorial Research Support for Predicting High-Altitude Nuclear Effects," Report DNA-TR-81-176, Defense Nuclear Agency (1982).

paths."[39] NTIA has a 2 megawatt facility at Platteville, Colorado (40° 10' 48" North by 104°43' 48" West) for ionospheric modification experiments, with two ring-array antennas that cover 4.5–10 MHz and 2.7–3.5 MHz, respectively. Since the ionosphere's local critical frequency varies, the modifier's emitted frequency must vary, too. The process is regulated by monitoring backscatter from the modified region, and Platteville's frequency coverage is broad enough to permit 24-hour-a-day operation. Pulse, continuous-wave, and chirp modulation have all been used, so the facility doesn't have a readily identifiable "sound," but in general it must choose lower frequencies than an OTH radar would.

As you can see, noncommunication uses of the shortwave band are increasing and are likely to increase substantially in coming years. A striking aspect of this trend is that it is occurring without there being any allocation for radar or radio determination anywhere in the shortwave band. Shortwave radars and radar-related systems use channels assigned to communications services. And they use many of them, consuming far more spectrum than the stations to which they are assigned. The Air Force has laid claim to some 30 bands between 5 and 28 MHz because their OTH-B radars need different frequencies at different times to cover different ranges. Of course, the same could be said for international broadcasters wishing to beam programs to audiences in different parts of the world. Need alone doesn't establish a spectrum right. Radars using other parts of the spectrum do so in bands specifically authorized by the International Table of Frequency Allocations—but not OTH radars. And regarding this particular band, Article 9 of the ITU Radio Regulations says:

> Members recognize that among frequencies which have long-distance propagation characteristics, those in the bands between 5 MHz and 30 MHz are particularly useful for long-distance communications; they agree to make every possible effort to reserve these bands for such communications.[40]

What right, you may ask, does any nation have to operate OTH radars in channels reserved for long-distance communication?

When the Soviet Ministry of Posts and Telecommunications told the IFRB that the Woodpeckers were for tests, the IFRB replied that this doesn't entitle them to disrupt authorized services. Article 19 of the Regulations says, "Any harmful interference resulting from tests and experiments shall be eliminated with the least possible delay."

The U.S. Air Force cited the International Telecommunication Convention about nations' retaining their "entire freedom with regard to military radio installations." This would certainly allow a government to authorize its military to use any part of the spectrum. But contrary to the Air Force's assertion that

39. W. F. Utlaut, forward to "Special Issue: Ionospheric Modification by High Power Transmitters," *Radio Science*, Vol. 9, No. 11 (1974), p. 881.
40. *Final Acts of the World Administrative Radio Conference, Geneva (1979)*, International Telecommunication Union (1980), p. 156.

"entire freedom" includes the right to cause interference, the Convention specifically denies this.

It would seem, then, that OTH radars are only allowed to use the shortwave band on a noninterference basis, whether they are for tests, experiments, or national defense. Unfortunately, for some reason, "not allowed" means very little in the shortwave band.

We aren't sure how much interference will be caused by the American systems, but the Woodpeckers have been doing it on an unprecedented scale for nearly a decade. The Russians have set a horrendous example, and their refusal even to admit there is a problem is infuriating. None of the planned or existing OTH systems that we know about are likely to cause anywhere near as much harmful interference as the Woodpeckers. Which is good news, in that we've probably already seen the worst of the problem. But can't something be done about those wretched pulsers?

There have in fact been several types of protest. In 1977–78, communications workers in ten European countries organized a boycott of Soviet shipping. "After that threat of boycott the transmissions were reduced for a prolonged period," says the *Oslo Aftenposten*[41], which recalled the event in connection with renewed threats of a boycott by labor unions in northern Europe, as a result of what they perceive as an increase in Woodpecker interference in 1983. The Norwegian Federation of Trade Unions and the Norwegian government have been leading voices in protesting the Woodpeckers—no doubt because Norway is subjected to the worst interference.

Norway, Sweden, Denmark, Switzerland, and West Germany were among the nations endorsing a statement in the Final Protocol of the 1978 World Administrative Radio Conference for Aeronautical Mobile Services that condemned the Woodpeckers' blocking of enroute civil aviation channels—a particularly dangerous practice. Once again, the USSR brushed this aside with the claim that the measures they'd already taken had been effective.

The ANARC OTH Radar Committee submitted a proposal to the State Department's Office of International Communications Policy arguing that the U.S. Delegation to the 1984 WARC for HF Broadcasting should offer the Conference this resolution:

> High-powered pulse transmissions within the HF Broadcasting Service bands are incompatible with the rational utilization of those bands by stations in the Broadcasting Service. Elimination of these emissions is essential to the development of effective plans for the future use of the HF broadcasting bands.[42]

We pointed out that since the U.S. Air Force radars won't use pulse modulation, they would not be encumbered by this resolution in any way. Although inter-

---

41. Rolf L. Larsen, "Protest Action Against Kiev Transmitter Considered—Boycott of Soviet Ships?," *Oslo Aftenposten* (6 July 1983), p. 40; translation in *Worldwide Report: Telecommunications Policy, Research and Development* #283.
42. Memorandum to Earl S. Barbely, Office of International Communications Policy, US State Department, from Robert Horvitz, ANARC OTH Radar Committee (July, 1983), p. 4.

ested in the idea, one State Department official responded that such a resolution would probably get more support if it were to be introduced by a delegation other than the one from the U.S. Too little time was left to find another sponsor after this polite rejection, but we have a second chance with the follow-up WARC in 1986.

One tactic that has proved to be a waste of effort—although it's probably the most common response among U.S. citizens—is writing to the FCC. They didn't issue the Woodpecker's license. Government-to-government complaints are handled primarily by the State Department, so if you want to write someone in the U.S. Government, try:

> Office of the Coordinator
> International Communication and Information Policy
> U.S. Department of State
> Washington, DC 20520

Better yet, send your comments to Radio Moscow, or to:

> Embassy of the Soviet Union
> 1825 Phelps Place NW
> Washington, DC 20008

There has been speculation in the amateur radio community for many years about the possibility of forcing the Woodpecker to change frequency by sending counter-transmissions to confuse or spook the operators. Most often the idea has been to send Morse Code dots at 10 per second on a frequency where a Woodpecker is audible. Some have even claimed success with this method. The published literature on OTH radar, however, suggests that these systems have little trouble rejecting any waveform that isn't part of their own backscatter. Martinez's analysis suggests how fine their discrimination could be—so it is especially significant that he thinks that it would be possible to devise a counter-transmission capable of "jamming these signals, or at least puzzling the distant radar operator."[43] This is a strategy that certainly deserves more study. Its main drawback is that the counter-transmissions themselves would cause some interference.

At the moment, there's no evidence that any form of protest short of an international shipping boycott is likely to subdue the Woodpeckers. On the other hand, not protesting guarantees that the situation won't improve.

Over the longer term, all systems eventually wear out, and with the rapid development of electrotechnology, systems are often superceded before they fail. If we knew more about Soviet progress in radar, what the true purpose of the Woodpeckers is and how well they perform, we might be able to foresee when they'll be taken off the air. The limitations of the USAF OTH-B radars—and incidents like the Korean airliner tragedy in 1983—suggest that their performance is less than spectacular. If this is the case, the Russians could be working on improvements and alternatives right now.

---

43. J. P. Martinez (letter), *Wireless World* (April 1982), p. 59.

The U.S. Air Force is. Even as it asks for a billion-plus dollars for the bicoastal system, it is pushing development of microwave radar satellites for tactical surveillance. Satellites could provide early warning capability comparable to the OTH-B systems but on a global scale.[44] The alternative that the General Accounting Office seems to favor is "using existing airborne warning assets"—that is, AWACS planes patrolling the approaches to U.S. territory—"until a more endurable system than OTH-B can be deployed." GAO doesn't say what "more durable" system it has in mind, but it points out that "Both the Air Force and Navy plan to develop tactical warning systems for use during the 1990s that will withstand a greater threat environment than the OTH-B."[45]

OTH-B radars for early warning air defense thus are likely to be superceded in the next decade, and even if development of follow-on systems runs into trouble, they may well be rendered useless by progress in "stealth" techniques, which can reduce the radar cross-section of a bomber or Cruise missile a thousandfold. According to the *Wall Street Journal*, "The Air Force admits it can't even be positive that the new over-the-horizon radar it is building on America's coasts will be powerful enough to pick up small Cruise missiles now being developed...."[46] It is ironic—though hardly unusual—that the same arms race that stimulated the development of OTH radars may soon make their air defense application obsolete.

There is something truly marvelous about being able to "see" a thousand miles over the curve of the Earth. Shortwave listeners, who hear over the same distances, are uniquely able to appreciate the difficulties that have been overcome, the sophisticated frequency-selection process required for continuous operation, and the incredible sensitivity of the receivers. And international broadcasting could benefit from the adoption of sounding and backscatter-monitoring techniques derived from radar experience. Stanley Leinwoll has suggested that Radio Moscow monitors its tune-up signals' backscatter to determine the best beam elevation angle to reach its target area.[47] If this is true, they probably learned how to do it from their Woodpeckers.

The problem, of course, is interference. There would seem to be enough "holes" in the Fixed Service bands to handle all the OTH radars likely to be built in the 1980s, but doing so without interference would require more international cooperation and skillful spectrum management than we've seen so far. The proliferation of shortwave radars and radar-like systems is going to continue, no doubt about that. It is possible to operate them in ways that minimize interference, though not all radar operators wish to do so. We must not lose sight of the fact that they have no right to cause harmful interference to other authorized users of the band. Since the ITU has no enforcement power, some responsibility for enforcement devolves to individual nations and spectrum

---

44. James A. Calder, "Space-Based Radar Comes Over the Horizon," *Aerospace America* (January 1984), pp. 112–114.
45. GAO/C-MASAD-83-14, p. ii.
46. Gerald F. Seib, "Unfriendly Skies—Worried by Russians, US Plans to Sharpen Its Weak Air Defenses," *Wall Street Journal* (28 March 1984), pp. 1 & 29.
47. Stanley Leinwoll, "Why Radio Moscow is Winning the dB War," *Radio-Electronics* (December 1981), pp. 55–57.

users. Shortwave listeners should learn to identify the various man-made noises that litter the spectrum. Remember: a complaint to an interference source takes no more time to write than a reception report to a station offering QSLs.

There have been some important developments since this chapter was written. The U.S. Air Force has expanded its plans for deployment of OTH B radars. Two 60° sectors would be built near the coast in Alaska, with surveillance coverage extending into eastern Siberia, and four sectors would be put in the northern midwest. The midwestern sectors are intended to cover the southern approaches to U.S. territory and fill in the "skip-zones" of the east and west coast OTH installations. However, they would also cover about 80% of the continental U.S. with millions of watts of shortwave power, 24 hours a day. The midwestern sectors have the greatest potential for causing harmful interference to shortwave listeners living in North America.

Testing of the east coast OTH radar in Maine began late in 1985, without causing any complaints of interference in the first few months.

In February, 1984, the Australian government awarded study contracts for the conversion of their Jindalee system from an experimental to an operational configuration. Contrary to Dr. Desmond Ball's article describing the system in the late 1970s (quoted previously), more recent articles say that Jindalee uses continuous-wave FM modulation similar to the US OTH-B radars. We don't yet know when it will go into regular operation.

Japan has decided to deploy OTH radars on its territory, in cooperation with the US Navy, and has committed about $173 million for this purpose. The islands around Okinawa were mentioned as a likely site. The *Nichibai Times* reported that the U.S. Navy also plans to install moveable OTH radars on Amchitka Island in the Aleutian archipelago, and on Guam.

A cooperative research program involving the U.S. and England may result in OTH radar being deployed in the United Kingdom. Several possible sites have been mentioned: Orford Ness (for coverage of the Norwegian and Arctic Seas), Cricklade (for coverage of the North Atlantic), and northern Scotland. British scientists are also working on development of a *ground-wave* OTH radar, which would take advantage of radio waves adhering to the sea's surface and traveling along the curve of the Earth. A test system with a detection range of about 200 km has been built at Angle in Wales.

A worldwide, coordinated effort by shortwave listeners to assess the extent of the interference caused by the Soviet OTH radars took place in October, 1985. One hundred and eleven volunteers from 20 countries took part in the "Woodpecker Project," organized by ANARC. In addition to over 2500 specific observations of the Woodpeckers' band-use, we got 115 reports of harmful interference caused to 39 shortwave broadcasting stations. The evidence suggests that the original Woodpecker site, between Minsk and Kiev, was not transmitting during the monitored periods, but the other two sites were. Since these two sites are not well-heard in Europe and the eastern part of the U.S., the overall level of interference seemed less than in earlier years.

12

DX'ING THE U.S. PIRATES

Darren S. Leno

The usual progression in the radio hobby is from shortwave listening (SWL) to Ham operator. For **Darren Leno**, it went the other way. Although he still has his Amateur Radio license (WD0EWJ), Darren says he finds shortwave listening more interesting, informative, and rewarding.

Darren was shanghied by pirate broadcasting in 1978 when he happened across his first such station. In the years since he has heard and verified dozens of them. Active in the now-defunct Free Radio Campaign (FRC-USA) for a few years, Darren later founded the Association of Clandestine Enthusiasts (ACE) which, although Darren is no longer involved in day-to-day operations, continues to publish a monthly bulletin covering pirates, clandestines, and number stations.

Daren is a recent graduate of the University of Minnesota.

One of the more fascinating aspects of the radio listening hobby is underground or "pirate" radio DX'ing. Searching for, hearing, and verifying these unlicensed broadcasters presents an exciting challenge that involves a knowledge of when and where these stations will appear. In this chapter, I will introduce you to the world of underground broadcasting and will present tactics you may wish to use to make your "pirate DX'ing" time more worthwhile.

Since the advent of radio, there has been, and probably always will be, pirate radio stations. The question that has been presented to me several times since I began listening to these stations many years ago is, "Why do these people risk going on the air when the maximum legal penalty is so severe; a $10,000 fine and one-year imprisonment for EACH violation?"

I don't suppose there is a uniform answer for this question or any other where radio pirates are concerned. Their reasons for going on the air are as varied and unique as the stations themselves. I think it would be realistic to say that many radio pirates are not aware of the severity of the crime they are committing. Others just don't think about it, or feel that there is no way that they can be tracked down by the Federal Communications Commission (FCC). Some pirate stations are operated for the sole amusement of the operators, while others try to provide a true alternative to commercial broadcasting. There are as many reasons for going on the air as there are pirates.

Who Are These Pirates?

Who are the people behind the transmitters? This is another complex question with no uniform answer, since pirate radio involves people from many different walks of life. After talking and meeting with many radio pirates, I would have to say, if I were to stereotype them, that most are college students, although quite a large number still attend high school. I would also go out on a limb and say that the average age of a pirate operator would be somewhere in the early twenties.

Again, I must emphasize that this stereotyped view is not completely accurate. Although most pirates are in their early twenties, I personally know others who are in their thirties, and one who is over forty years of age. A few actually work within the broadcasting industry, while others are unemployed. Some have a Ph.D., while others have not yet finished the tenth grade.

No matter who is behind the transmitter, one fact becomes evident after a person has tuned into many of these stations. The people who operate them usually have strong ideological convictions and definite musical tastes.

Free Radio Movement

Unlike the European "Free Radio" pirate stations, which were basically professional operations trying to break a government stranglehold on the broadcasting media, a growing group of U.S. pirate stations claim to be part of an American Free Radio movement. These pirates see the U. S. radio industry made stagnant not only by excessive government regulation, but also by strong vested commercial interests to which the radio stations cater their programming for the purpose of attracting advertising. The result of this is a strictly limited access to the radio media by individuals and groups who may not conform completely to the norms of society. In short, the Free Radio pirates feel that the Fathers of our Country could not have foreseen the radio broadcasting media when the Constitution of the United States was written. Had radio existed back then, they feel that the Freedom of the Press amendment would apply to radio, since a radio transmitter (Fig. 12-1) is the equivalent of the 1776 printing press.

Fig. 12-1. The transmitter installation for KFAT (The Fat One), showing a modified Heath DX-100 and a Macintosh 60-watt audio driver amp.

Of course, there are some definite problems that would be created were everyone legally able to run out, buy a transmitter, and, with little or no technical experience, put it on the air on any frequency they wished to use to broadcast whatever programming they wished, whenever they wanted to. The result would be complete chaos and disorganization on the radio bands.

However, after talking with many of these Free Radio pirates, I've concluded that few, if any, pirates are for complete deregulation. Instead, they would like to see an "Amateur Broadcasting Service" organized to provide a medium through which a low-powered, inexpensive, radio station could be legally organized and operated by any group or individual who has something to say to all those who will listen. The stations would be licensed, but obtaining the license would not be an impossible task. Few programming restrictions would be imposed upon the stations. Of course, one of these restrictions would have to be no advertising.

The Crystal Ship (Fig. 12-2), a pirate station that enthusiastically supports the creation of such a broadcasting band, says, "We would be happy to do what we're doing now in a legal manner if only the Fed's would give us the means. Unfortunately, they show no sign of doing it soon."

Fig. 12-2. Verification card from the Crystal Ship, a pirate radio station.

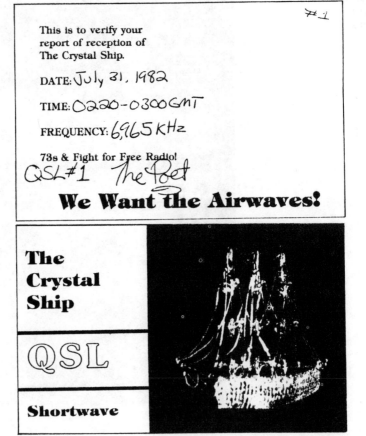

This is to verify your report of reception of The Crystal Ship.

DATE: July 31, 1982

TIME: 0220-0300 GMT

FREQUENCY: 6,965 kHz

73s & Fight for Free Radio!

QSL #1   The Poet

We Want the Airwaves!

The Crystal Ship

QSL

Shortwave

Opponents of the Free Radio Movement point out that the radio spectrum is a limited resource, which should not be squandered away just to fulfill the broadcasting ambitions of people not serious enough to pursue them legally. They also point out that many pirates enjoy being pirates and would probably not bother to become licensed even if they could, and that the many poor past performances of the pirate stations show them unworthy of legal status. Unfortunately, there was one pirate in recent times who distinctly reinforced the claims of the Free Radio opponents. Pirate station WRNR, in an article which appeared in *THE ACE*, was quoted as saying, "I personally ... don't fight for anything but a good time. We don't scream and holler, 'Free Speech, Free Radio.' If we get caught, we get busted and pay the fine and then go back on the air."

It certainly seems unlikely that we will see the implementation of an Amateur Broadcasting Service in the near future. However, the FCC has shown a fairly liberal attitude toward the broadcasting industry lately, which has brought forth program deregulation, low-powered television stations, and, as of this writing, consideration of low-powered FM broadcast stations. Meanwhile, some pirates continue to broadcast week after week, hoping that some day they will be able to offer their programming to listeners without the fear of Federal prosecution.

Cooperation

Many people may be surprised to learn that there is cooperation among many of the pirate radio broadcasters. This cooperation often includes passing along sensitive information among themselves, such as who can't be trusted, what other pirates are operating tonight, "telco loop" telephone numbers, and technical advice. Many stations even exchange taped programs or relay transmissions of other pirates. This often happens when a transmitter breaks down, or when it is felt that the FCC may be closing in on a particular station's location (in an effort to thwart FCC radio-location efforts).

Never has pirate cooperation been so obvious as it was on the evening of July 4, 1982. On that date, the first and most successful "piratefest" to date took place. On that memorable Fourth of July, at least nine different pirates appeared, one after the other. They often gave hints to listeners at the end of their program as to where the listeners should tune their receivers to hear another pirate. The schedule for that night went something like this:

GMT	Frequency	Station
0230	6978	R. Free San Francisco
0330	7425	KQSB
0400	7373	Voice of Syncom
0430	7390	Radio Indiana
0500	6280	Voice of the Pyramids
0530	6975	WCRS
0600	7436	Radio Free Radio
0630	7413	WOIS
0700	6857	KCFR

The lengths of the broadcasts during this historic piratefest ranged anywhere from 15 minutes to a full hour. July 4, 1982 was an exciting night for the SWL community, and SW club bulletins were filled with pirate loggings in August.

When and Where to Hear Pirate Broadcasters

If you were to go back over several months of pirate loggings, you would find distinct patterns in the frequencies, times, and days that the pirate stations operate. As I stated earlier, we're going to give you some hints that will help you tune into these pirate broadcasts.

Pirates have been known to frequent the following ranges of frequencies, almost exclusively, on local Friday and Saturday nights:

*1600–1630 kHz,

3400–3500 kHz

6200–6300 kHz

6900–7000 kHz

*7325–7500 kHz

* equals the most active frequencies.

Operation on these frequencies has been as early as 2100 GMT and as late as 1000 GMT. However, you'll have your best chances of hearing a pirate in these frequency ranges if you tune in between 0100 and 0400.

Recently, two new frequency ranges have been frequented by the pirates and it is worth keeping a casual watch over them. The band used by legal international broadcasters, 9550–10000 kHz, has recently been invaded by pirates. Pay close attention and watch for weak stations sandwiched between the "big gun" broadcasters, especially those playing rock n' roll music. 15000–15100 kHz has also been the site of pirate activity lately, and a station or two may turn up if watched closely. The best time to watch these two new pirate bands is in the early afternoons, from 1800–2200 GMT (local weekends).

With a lot of patience and a little luck, you will be able to hear many different pirate broadcasters. Keep an eye on the current pirate loggings that SWL'ers report to SW club bulletins. An excellent source of pirate loggings is the Association of Clandestine Radio Enthusiasts (A.C.E.) monthly publication, *THE ACE*. For current membership information, send a large SASE to A.C.E., P. O. Box 46199, Baton Rouge, LA 70895-46199. A.C.E. is a member of ANARC.

The pirate radio scene is subject to constant and frequent change. Some stations will only be around for a year or two—until the operator loses interest or is closed by the FCC. New pirates are always waiting in the wings, however, ready to pick up where another leaves off.

There have been hundreds and hundreds of pirate stations on the air since the early 1970s, although a complete history of these stations could take us back well over half a century and could cover thousands of these illicit broadcasters. Having neither the resources, energy, nor space in this book to undertake such a project, I'll only mention a few of the more interesting pirate stations (from modern times) in this chapter.

The Voice of the Voyager

Although the Voice of the Voyager (VOV) was closed by FCC enforcement engineers in 1979, this energetic and popular pirate (Fig. 12-3), who entertained listeners like no other pirate before, yielded to the cries of its fans across the nation and returned to the airwaves in 1982.

Fig. 12-3. The famous "Voice of the Voyager" pirates.

"Once pirate radio is in your blood, it's hard to keep off the air," Scott Blixt told me shortly after the Voyager was "busted" for the second time on May 9, 1982. Scott Blixt, along with his partner, Mike Martin, had been on the air an hour and a half when FCC engineers from St. Paul decided that they had heard enough and closed the station down. The two men received fines of $3000, and were fortunate not to have had their equipment confiscated. The VOV used a 500-watt Hallicrafters HT-20 transmitter, and an expensive vertical antenna.

Both Scott and Mike were, at the time, unable to pay their large fines and attend college classes at the same time. After arguing this point with the FCC for over a year, an agreement was reached whereby the men would pay off their fines on a generous time scale; $25.00 a month until the balance was reached.

The Voice of the Voyager was a special pirate in several ways. It was widely heard because of the fairly consistent and predictable schedule it followed. People knew when and where to tune their receivers to hear the VOV. Several different on-the-air personalities incorporated a unique blend of comical and musical programming. The VOV promptly verified all correct reception reports. Their signal was usually clean and intelligible.

All these ingredients added together made the VOV one of the more widely heard and popular shortwave pirates.

Jolly Roger Radio

Jolly Roger Radio (JRR) gained national attention after they were closed by the FCC on November 10, 1980, after serving the Bloomington, Indiana community for at least a decade. The station operators were fined $750 each; a figure which was appealed and finally reduced to $250.

Shortly after JRR was "busted," the operators tried to put the station back on the air legally. Jolly Roger Radio Group Inc. was formed, and lawyers and consultants were hired to help with the complicated red tape to which every application is subjected. After all the work was completed and the application submitted, JRR Group Inc. was denied a license because they couldn't raise the money to finance the equipment the FCC would require them to purchase.

Bruce Quinn, JRR's famous and outspoken personality, was more than a little upset about this unfortunate turn of events. In an A.C.E. interview, Mr. Quinn stated "The cost (of putting a station on the air) need not be so high! The FCC required me to buy so-called 'type approved' equipment from their buddy manufacturers. For example, I could make an antenna for $2.00 that would (perform) just as well on FM as the one costing $1000, which the FCC would require me buy from an equipment company."

Mr. Quinn further stated that "only millionaires and those who have owned stations for a long time have access to the airwaves in our land. Radio belongs to the people, not the government."

Fed up with trying to work within the system, Mr. Quinn, with the help of 25 or 30 students from a Bloomington university, decided that with or without federal permission, JRR was going to make a comeback. By early 1982, JRR was being heard almost daily on the FM band on 94.1 MHz. FCC enforcement agent George Skloam, after hearing that the station he had helped locate and bust two years earlier was again on the air, vowed that if he had to go back to Bloomington, he would make sure that "somebody does some time."

Shortly after Mr. Skloam publicized his threat, JRR announced it was moving its base of operation to London, England, where Free Radio stations are usually tolerated by the government. After nearly six months of broadcasting, JRR again left the air. It is reported (quite vaguely) that various problems and disappointments were encountered with JRR's attempted move to England and, ultimately, Mr. Quinn returned to Bloomington, Indiana.

Throughout JRR's history, which spans a decade, it was heard on the FM and AM broadcast bands, and even on shortwave. JRR was famous for the fun, quality programs that Mr. Quinn and his help produced, and also for the variety of music they played—everything from punk rock to Irish folk ballads to Iranian banshir. One could never tell what was going to happen on JRR, and this constant variety and uniqueness made them a popular station among the listeners.

Purple Pumpkin

The Voice of the Purple Pumpkin is a very interesting shortwave pirate station with a history nearly two decades long. First heard in the 1960s, this station will still crop up occasionally, delight DX'ers for a brief time, and then vanish as mysteriously as it appeared, sometimes for several years, before it is heard again.

In July, 1974, Terry Provance—an active pirate DX'er—wrote to the Federal Communications Commission requesting information on the Purple Pumpkin after not hearing the station for a couple of years. Shortly thereafter, he received a letter from the FCC stating that the station had been closed down on March 26, 1972, and that it had been situated in Maryland.

But one decade later, on Halloween night, 1982, this mysterious station (or another with the same name) was again broadcasting from "Downtown you-know-where" to the delight of SWL'ers. However, only five broadcasts were logged by members of A.C.E. before the station vanished, exactly one month later.

Will the Purple Pumpkin again return to haunt the airwaves, or have, this time, we heard the last of it? No one knows for sure, but it is very likely that in the future, a station calling itself the Purple Pumpkin will again take to the air to keep the name and the tradition alive.

Voice of To-morrow

The Voice of To-morrow (VOT) has been called America's most controversial pirate station. In spite of a professional-sounding announcer, strong and clean signals, well-produced slick programming, and an attractive QSL card (Fig. 12-4), VOT was not greeted with open arms by the SWL or pirate radio community when they first took to the air on June 18, 1983, at 2000 GMT on 7410 kHz. *Why?*

VOT is operated by advocates of white supremacy, and broadcast messages offensive to non-whites, and those of the Jewish faith. VOT is a far cry from a high-school kid "playing radio" on his Dad's Ham radio setup, and that's probably what worried many SWL'ers the most—the realization that the VOT is a very professional station, and not "just a gag."

Fig. 12-4. A QSL
from The Voice
of To-morrow.

VOT claims to have studios located in Providence, RI, and a 2000-watt transmitter in Baltimore, MD. Wherever they are located and whatever power they run, listeners from around the country often report extremely strong signals. This pirate station has been heard in many of the pirate bands, and it frequently changes operating times and frequencies. As of this writing, the VOT is still active.

QSL'ing Pirates

Collecting QSL cards from pirate stations can be as challenging as hearing them. Most stations try to QSL as soon as possible in order to build good feelings among their listeners (Fig. 12-5). Other pirates (like the VO Venus) prefer to make DX'ers sweat a year or two before letting a QSL magically appear in the mail when you least expect it.

Most pirates will give their QSL policy on the air, along with an address that reception reports can be sent to. It is a good practice and common courtesy to enclose at least 3 first-class stamps with your reception reports. All stations will appreciate this, and your chances of getting a QSL card, in return, will be improved.

Fig. 12-5. Pirate
QSLs are often
handmade.

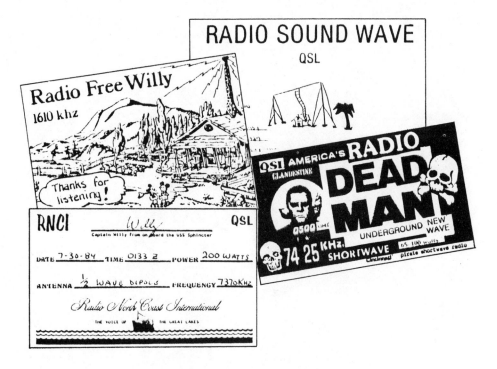

Conclusion

In this chapter, we have taken a glimpse at the exciting and challenging world of pirate radio stations. If, for some reason, you feel you would like to be a pirate yourself, think again. Pirates face one year in prison and a $10,000 fine for each time they go on the air. I surely wouldn't encourage you to operate a transmitter illegally. Besides, you'll have just as much fun DX'ing and QSL'ing these illicit and interesting broadcasters.

13

AN OVERVIEW OF THE NUMBERS STATIONS MYSTERY

Harry L. Helms

The mysterious "numbers" stations and other unidentified radio communications have held the interest of **Harry Helms** for a number of years. He has spent tremendous amounts of time in exploring, analyzing, and classifying such stations in an attempt to solve what must be the ultimate listening puzzle.

Harry Helms is a consulting editor for McGraw-Hill Book Company and has written a number of books including the *Computer Language Reference Guide*, *The BASIC Book*, and *Electronics Applications Sourcebook*. Thousands of radio hobbyists have read his *How to Tune the Secret Shortwave Spectrum* and *The SWL's Manual of Non-Broadcast Stations*. He also holds Extra Class Amateur License K2RH.

Harry's other interests include scuba diving, astronomy, and computers. He notes that, although living on the 36th floor of a high-rise along the Hudson River is nice for many things, DX'ing is not one of them.

If you do any tuning whatsoever outside the standard shortwave broadcasting or amateur bands, you'll eventually come across a woman reading groups of numbers in a stilted, mechanical voice. Most often the numbers will be in Spanish; you may also hear German, English, or other languages used. The numbers may be in groups of three, four, or five digits. Sometimes you'll hear musical effects with these stations. You can even find stations transmitting groups of numbers in CW (code), or using a man's voice. Regardless of the exact type of "numbers" station you hear, you will have eavesdropped on the longest running and most persistent mystery on shortwave.

What is the purpose of such transmissions? That question has tantalized listeners since the early 1960s, when they started being reported in North America in large numbers. The oldest and most common theory holds such transmissions to be coded messages directed toward espionage agents in the field. Other theories have attributed such transmissions to broadcasts of lottery numbers, navigation information, bank account and business data, etc. There was a period in the latter half of the 1970s when listeners favored such arcane explanations for numbers stations over the espionage theory. Now the pendulum has swung back the other way, bolstered by impressive new evidence, and it is now apparent that most numbers stations are indeed transmissions directed toward spies!

If such stations transmit to spies, where do they transmit from? For many years, Cuba was a prime suspect, particularly for Spanish-language messages. East Germany has also been targeted as the source for many of the German and English numbers stations. However, there is now abundance of solid, reliable evidence that many numbers stations transmit from the United States and its territories!

Types of Numbers Stations

The term "numbers station" is a misnomer in some cases. Certainly the bulk of activity you can hear will involve groupings of various numbers. But you may also hear groupings that are composed of letters from the international phonetic alphabet. You may also hear tones or music. Sometimes, numbers may be read in a continuous stream without any discernable grouping.

However, most activity you'll hear will fall into several broad categories. Among these are the following.

Five-Digit Stations

This will be the most commonly heard type of numbers station that listeners in North America will hear. Most five-digit stations will be heard speaking Spanish, with a sprinkling of others heard in German and English. As the name implies, transmissions from these stations consist of groups of five numbers, usually preceded by an identifier that is repeated for several minutes before the actual message. For Spanish stations, the opening identifier will be the word "atencion" followed by two number groups. The second group will always be the number of five-digit groups in the message. The purpose of the first number block after "atencion" is unclear, but it is likely either the addressee the message is intended for or the number key to be used to decode the message. Each Spanish message closes with the words "final, final." German and English transmissions usually open with two words from the international phonetic alphabet, such as "Kilo Whiskey" or "Papa November," repeated. Tones and musical intervals may also be used.

Four-Digit Stations

Spanish is again the most-common language for these, followed by German and English. For Spanish stations, the transmission opens with a three-digit number repeated three times followed by a count from one through zero in Spanish. The German and English stations open with beeps and tones followed by "achtung" or two words from the international phonetic alphabet. Four-digit numbers stations are much less commonly heard in North America than five-digit stations.

"3/2" Stations

A new type of numbers station was first reported in 1983. This station transmitted five-digit number groups with a distinct and apparently deliberate pause between the third and fourth digits of each group. Thus, these stations became known as "3/2" stations. This pattern was first reported on German stations, followed soon by some Spanish and English stations. A rare variation—with a pause between the second and third digits—has been reported and is known as the "2/3" type.

CW Stations

Number groups have been reported on CW for several years. These number groups may consist of four or five digits, although the digits may be "cut." This means that letters of the alphabet are substituted for numbers (this is because the Morse characters for letters are shorter than those for numbers). The opening message of CW numbers stations consists of a three-digit identifier repeated three times followed by the number of groups in the message—which is also repeated three times. The message closes with "TTT" sent three times. One distinctive element of many CW numbers stations is that the keying is actually an amplitude-modulated tone rather than an unmodulated carrier. This is very distinctive on the shortwave bands.

Phonetics Stations

These are very similar to numbers stations, but letters from the international phonetic alphabet are substituted for numbers in various groups and combinations. Sometimes all that can be heard is a phrase such as "Sierra November

Yankee" repeated continuously, although such transmissions may be interrupted for numbers messages.

Variations and Patterns

There are numerous variations on these basic types. Most stations transmit in AM (amplitude modulation), although some SSB (single sideband) transmissions can also be heard (especially on the German stations). Most stations also use a female announcer, although the use of male announcers has become more common in recent years. Further variations and the emergence of more types can be expected in the future.

An interesting pattern to numbers station activity was discovered in the early 1970s by listener Mike Chabak, then based in Colorado. He found that some numbers-stations transmissions are repeated, digit for digit, days or even several weeks apart. He also found close correlations between the opening identification of a numbers station message and the time and frequency of the message transmission. These patterns still hold to some degree, and messages are repeated days and weeks apart. However, the encloding scheme believed to be widely used by numbers stations (the one-time pad system) would permit several entirely different messages to be decoded from the same number groups merely by the use of different pads. This coding method will be discussed later.

A Typical Numbers Station Transmission

What will you hear if you run across a numbers station transmission? The following is the transcript of an actual Spanish-language numbers station message heard on 11533 kHz from 0238 to 0321 GMT. As you will see, the transmission was one of the four-digit variety.

0238: "Mensaje, mensaje, mensaje 9999999 mensaje, mensaje 9" (announced by a woman, followed by rapid CW or RTTY).

0239: A string of pulses lasting approximately five-eighths of a second each.

0300: "457 457 457 1234567890" (read in Spanish by a woman; repeated until 0310).

0310: Ten long pulses spaced approximately three seconds apart. Then a woman began reading the following: "Grupo 63, grupo 63. 4614 2580 6627 3535 6906 6712 5552 2152 1041 4197 1622 5590 3340 8848 4190 2333 7098 3910 8217 3137 7323 1041 9805 5795 7555 9811 1122 8241 4265 0910 1704 9912 1357 6240 4357 1245 6595 4434 5063 7961 4837 6333 5577 3516 7143 4297 3746 5569 2447 8822 7238 4594 7553 2962 9195 9368 7488 5143 6822 8164 3453 8485 4178."

0316: Woman announces "repetir grupo 63, grupo 63" and the entire set of four-digit groups were repeated.

0321: "La fin" announced by woman, followed by series of short pulses; an open carrier for a few minutes thereafter.

Note that there are exactly 63 four-digit groups in this message. "457" may well be either the key to decoding the message or the identifier for the intended recipient.

How Are Numbers Messages Encoded?

Many listeners to numbers stations have often wondered if it is possible to "crack" one of these coded messages. However, it is highly unlikely that casual listeners (or even expert cryptographers) could decode one of these messages, since they are all apparently transmitted using one of the variants of the "one-time pad" coding scheme.

In the one-time pad system, both the sender and receiver of a message have copies of the correct pads. As a general rule, each number group will represent a complete word or phrase rather than each number representing a letter. To confuse unintended recipients of the message, meaningless groups of numbers are also inserted into the message.

With time, a professional cryptographer will break such a code. To avoid this, the recipient of the message is instructed to add or subtract another digit group to each group transmitted. A one-time pad will contain several pages of random numbers, and each page is used for addition/subtraction only once, after which it is destroyed. As long as one of the pads does not fall into the hands of the opposition, such a system is almost unbreakable. There are also rumors that microprocessor-based decoding devices are in use, replacing one-time pads. These devices can overcome the main drawbacks of the one-time pad system—slow speed and the chance for an arithmetic error causing a garbling of the message.

It can be reasonably assumed that messages transmitted in Spanish, German, etc., will be in that language when decoded. However, a properly encoded one-time pad message will give no clue as to whether the original was in Spanish, English, or any other language. The number groups will be statistically random. Indeed, there is no way to tell what language a decoded CW numbers transmission would be in just by examining the number groups composing the message.

Frequencies for Numbers Stations

Numbers stations may pop up on any frequency and at any time. However, certain times and frequencies are more productive than others.

Spanish numbers stations are most often heard during the hours of darkness in North America (0001–1200 GMT). Some of the more reliable frequencies for five-digit stations are 3060, 3090, 4030, 6892, 7845, 9462, and 10020 kHz. Four-digit stations are often found on 4670, 5812, 6800, 9074, and 11532 kHz. One fascinating aspect of Spanish four-digit stations is that some transmit simultaneously on two different frequencies.

German numbers stations are most frequently reported from 1400 to 0500 GMT. German five-digit stations far outnumber the four-digit ones. Common frequencies for the five-digit stations include 3260, 5015, 6508, 6860, 7410, 8120, 11545, and 12952 kHz. The four-digit stations have been reported on 4990 and 5780 kHz.

English five-digit stations can be heard on 6875, 7632, 9265, and 9435 kHz, while the four-digit stations can be heard on 7375, 8078, and 9465 kHz. Transmissions may sometimes be made simultaneously on two different frequencies.

CW numbers transmissions have been reported on 6840, 8900, and 12163 kHz. Phonetics have been observed on 6840, 7918, 9325, and 13291 kHz.

Recent Trends in Numbers Activity

Numbers station activity is an evolving phenomenon, and new wrinkles appear each year. 1982 and 1983 were no exceptions.

One widely heard mystery appeared that was known simply as "The Babbler." The Babbler was a male who rapidly read a string of Spanish numbers, much like an auctioneer. The numbers were not single digits, as is the case with "conventional" numbers stations, but sometimes ran into the hundreds! The most common frequency for The Babbler was 3646 kHz, followed by 5895 and 7397 kHz. Reception was generally in the 0400–0600 time period.

One surprising discovery, first made by Thad Adamaszek of Ohio, was the mixing of English and German numbers in the same transmission! First noted on 6785 kHz at 0430 hours, these transmissions sounded like conventional five-digit Spanish numbers transmissions except that the German "funef" (five) was substituted for the Spanish "cinco"! Later Thad and other DX'ers monitored a similar transmission on 7845 kHz. A similar discovery was a five-digit English numbers station that used "grupen" instead of "group" and "ende" in place of "end." This station was heard on 9050 kHz at 0105 hours.

An increasing number of two-way numbers transmissions have also been reported of late. Most of these are on SSB and operate in frequencies adjacent to the Amateur Radio bands; it is likely that these stations are using amateur transceivers. Frequencies for these stations include 10157 and 10466 kHz.

A few numbers stations were noted alternating with other modes of transmission, usually RTTY (radioteletype). An active frequency for this was 4640 kHz round 0100 hours, where a Spanish-speaking male could be heard alternating with RTTY. Sometimes the alternating periods of numbers and RTTY lasted as little as 20 seconds each!

Several clear links between the numbers stations activity and unusual voice markers were reported beginning in late 1983. The most common frequency for this was 4670 kHz, where a four-digit Spanish numbers station was noted mixing with a station transmitting a female voice that spoke "Victor Lima Bravo

Two'' repeatedly. Often the "Victor Lima Bravo Two" station continued transmitting after the four-digit station left the air.

Early 1984 saw the first positive reception of Chinese-language numbers stations in North America. David Crawford of Florida reported four-digit Chinese groups (read by a woman) on 8300 and 10721 kHz around 1000 hours. Transmissions opened with flute music.

Another intriguing mystery from early 1984 was the "545" station. This consisted of nothing more than the number group "545" continuously repeated by a woman! George Osier of New York noted this station (speaking in Spanish) on 4670 kHz at 0200 GMT. Dennis Rutowski of Connecticut heard a similar station on 5089 kHz at 0110 GMT speaking in English. No other numbers or words are ever heard from this station; "545" is all!

Of course, it is quite likely that in the interval between when this was written and the time you are reading this that several bizarre new types of numbers stations have sprung up.

Where Do Numbers Stations Transmit From?

This has been a question that has remained unanswered for over two decades. Since Spanish numbers activity began around the time that Castro came to power in Cuba, Cuba has long been assumed to be the location for the Spanish stations so often heard in North America. Evidence recently developed through direction-finding procedures indeed points to Cuba as the source of some numbers transmissions. But, there is an equally weighty body of evidence that many numbers transmissions operate from within the United States!

The evidence for Cuba is lengthy. In 1975, for example, listeners reported hearing audio Radio Havana Cuba in the background on a five-digit numbers transmission. The audio was parallel to a Radio Havana Cuba shortwave broadcast, and the clear implication was that the numbers station and Radio Havana Cuba shared a common transmitter site and cross-modulation was taking place there. Listeners reported similar incidents of joint numbers/Radio Havana Cuba reception in early 1984.

The Federal Communications Commission (FCC) has also indicated that numbers stations transmit from Cuba. In early 1978, David Crawford of Florida wrote the FCC asking about the location of some five-digit Spanish numbers stations that he heard on 3060 and 3090 kHz. The FCC's reply read in part, "...as a result of our routine monitoring of the radio spectrum, we have on previous occasion encountered the signals you describe. Through means of long-range direction-finding bearings, we determined the signals were emanating from Cuba. Consequently, they were of no further interest...." (However, the FCC isn't consistent in making this claim. Your author wrote the FCC shortly after David received the reply to his letter asking for confirmation of the location of the stations on 3060 and 3090 kHz, and specifically mentioned the reply received by David. The FCC blandly replied that 3060 and 3090 kHz were allocated to the aeronautical mobile-radio service and the signals could be coming from anywhere in Central or South America.)

Several clues pointed to a location outside of Cuba for some Spanish numbers stations, however. Many numbers stations are quite loud; listeners in such locations as Ohio and Washington, DC noted numbers stations with loud fade-free signals which indicated ground-wave or space-wave propagation. Such reception would be impossible if all the numbers stations were in Cuba (except for listeners located in extreme southern Florida).

In 1978, your author wrote a monthly column on SWL/DX topics for a now-defunct electronics magazine. After one column on numbers stations, a letter was received from a listener in southern Florida. He claimed that he had driven around with a portable shortwave receiver when various numbers stations were on the air and had discovered that his receiver overloaded to the point where no other signal could be received anywhere else on the shortwave bands! This indicated that he was almost certainly within a few hundreds yards (at most) of the numbers station transmitter. This overloading happened adjacent to government and military areas which would be ideal for hiding radio facilities.

Since most local-quality Spanish numbers stations were of the four-digit variety, the consensus which evolved in the DX'ing community held that five-digit numbers stations transmitted from Cuba while four-digit stations originated somewhere in American territory. This remained the case until the debut of *Popular Communications* magazine in late 1982. *Popular Communications* was the first national publication to devote extensive coverage to the numbers station mystery and it helped generate renewed interest and discoveries.

One important revelation came from the shortwave broadcaster, *La Voz de Nicaragua. Popular Communications* reader Joe Erwin of Virginia sent a reception report to them and received a verification and a copy of a book titled "CIA Conspiracy in Nicaragua." In the book, the Nicaraguans claimed that the CIA sent coded messages to its agents in Central America via shortwave radio. A frequency mentioned in the book, 9074 kHz, is an active one for four-digit numbers station; an illustration of the code pads used to decode messages showed pads filled with columns of four-digit numbers! If the Nicaraguan report was fabricated, the Nicaraguans must have done a great deal of research into SWL reports on numbers stations.

By 1983, many listeners had equipped themselves with directional-loop antennas and were capable of direction finding. An appeal in *Popular Communications* for listeners equipped with loops to take direction bearings on various numbers stations brought forth some fascinating results. One such listener was Ron Weiss, W9OFF, of Indiana. He took bearings on a four-digit station transmitting on 4670 kHz, using WWV* on 5000 kHz and various tropical band broadcasters as his reference points. Roy reported his results in *Popular Communications* as follows:

> We tuned in 4670 kHz for the numbers station to begin searching for the direction and angle of the sky wave. Of course, we assumed that it was from some southerly direction. We searched and searched, but

---

* WWV is the call letters of the National Bureau of Standards radio station at Ft. Collins, Colorado.

obtained no nulls from the south, the east, or the west. We really had not considered north, but before quitting altogether decided to try north. Voila! We obtained consistent and deep nulls, sometimes reducing a S9+30 signal to S0! Needless to say, we were (and are) quite perplexed. We ruled out a transmission from over the North Pole (no polar flutter). Further exploration indicated the sky wave nulled best when the RDF antenna was facing north and tilted about 70 to 80 degrees from the horizontal. It appears that the signal was coming from almost overhead (straight up). No, I'm not suggesting a source from outer space, but rather a sky wave from a transmitter nearby Indianapolis but somewhere north.

Ron later took a bearing on the four-digit station on 5812 kHz and found a clear bearing toward the east.

Similar efforts were made by Ron Ricketts from his listening post near Forth Worth, Texas. One important discovery by Ron was that bearings for a number station on a certain frequency changed with time. The different bearings were far beyond those that could be reasonably expected due to measurement errors (the four-digit Spanish numbers station on 11533 kHz varied from 84 to 126 degrees in less than a month). Such results strongly point to the use of multiple transmitter sites or mobile/portable transmitters, such as shipboard facilities.

The common assumption that all five-digit Spanish stations originate from Cuba was brought into question by receptions by Ron Rickets and Jim Thornton of Tennessee. They both got bearings on five-digit stations toward the northeast —in one case, very clearly toward northern Virginia, southern Maryland, and Washington, DC.

Other direction-finding efforts placed the bulk of five-digit Spanish activity in Cuba with the four-digit Spanish stations apparently scattered between sites near Washington, DC, southern Florida, and Puerto Rico.

Some enterprising listeners, in response to a suggestion in *Popular Communications*, called various FCC monitoring facilities about the numbers stations. One such listener, Ken Navarre, Jr. of California, called the FCC in Livermore, California about a four-digit Spanish station he was hearing on 6802 kHz. The engineer on duty told Ken he was also hearing the station and discussed various possibilities as to what it could be (the engineer told Ken he also read *Popular Communications*). Ken then called the FCC's Fort Lauderdale, Florida office and was told that they also had the signal and that it was very strong there. Note that the FCC didn't offer an opinion as to where the signal was coming from, unlike their efforts in connection with the five-digit stations.

The most successful effort along such lines was recorded by David Batcho of New Mexico. He managed to get some FCC personnel to take bearings on both four-digit and five-digit stations using the FCC's full complement of direction-finding equipment. The five-digit stations were located at a transmitting site near Havana, Cuba while the four-digit stations were located near Washington, DC.

One transmitting site for four-digit numbers transmissions has definitely been traced to a U.S. Army facility near Warrenton, Virginia. This same site is also used for transmissions by KKN50, the U.S. Department of State station supposedly used for communications with U.S. embassies and consulates abroad. The location was discovered through the efforts of a mathemathics professor at a small Connecticut college, who actually visited the site armed with a portable receiver. Efforts by your author to obtain more information about this site, and the entire subject of numbers stations, through Freedom of Information Act queries to various government agencies have been fruitless so far.

Thus, for Spanish numbers stations, it now seems safe to say that most five-digit stations transmit from Cuba while four-digit stations operate from the United States and Puerto Rico. This is what has long been assumed, but it now appears that the pattern may sometimes be broken. If a five-digit transmission can apparently originate occasionally from the United States, it may well be that a four-digit transmission may originate from Cuba or other non-American sites. And it now appears that portable/mobile transmitting facilities are used to some extent.

The situation with the use of German, English, and other languages is considerably less clear. For many years, East Germany has been suspected of being the source of many German and English numbers stations. Confirmation of this has come from several East German agents who have been apprehended in the West. For example, in 1976, the West German magazine "Der Spiegel" ran a series on the confessions of an East German spy who told how he received coded messages, which consisted of number groups, from a station in East Germany. More recently, the 1982 spy trial of England's Geoffrey Prime brought out how he received instructions from an East German station which transmitted five-digit English number groups. (At the time of his arrest, Prime had KGB-issued one-time pads in his wallet.)

Unfortunately, European listeners have yet to do direction-finding work with the various numbers stations audible in Europe on par with the work done by North American listeners. Thus, the probable location of some of these stations (some of which must surely be in western Europe) is still open to question. There is also evidence that some English and German numbers stations transmit from somewhere in North or South America due to the signal strength and reception times.

One possibility that is beginning to get serious attention is that some numbers stations could well transmit from embassies. Under international law, embassies have the right to establish and maintain their own radio communications links without the approval, sanction, or even knowledge of their host countries. An example of how seriously some countries take this right can be found by glancing at the incredible antenna system atop the Soviet embassy in Washington, DC. The United States also maintains embassy radio facilities around the world, including facilities at the embassies in London and Tokyo—friendly nations with no shortage of the most modern communications systems. Having radio stations in such areas, just to communicate with Washington, seem a bit anachronistic, unless they have other purposes.

Hopefully, such questions regarding the locations of numbers stations will be resolved in the future as listeners equip themselves with increasingly sophisticated listening gear and participate in coordinated monitoring efforts. The purpose or purposes of numbers stations are still largely the subject of speculation, and will likely remain so until a major "leak" is made by a participant in the activity. The one safe statement to make is that listeners will still continue to be intrigued and perplexed by these strange transmissions for years to come!

14

CHALLENGES AND CHANGES IN MEDIUM-WAVE LISTENING

Karl D. Forth

Karl D. Forth has spent the last 15 years as an active medium-wave DX'er. During that time, he has amassed the enviable total of about 2500 stations heard, in 49 states and 38 countries. Karl's QSL count runs to some 1300 stations verified.

He has edited columns and written articles for several clubs and is a consistent contributor to the major medium-wave DX clubs. His interests extend to most other forms of electronic communication as well.

Karl lives in Chicago and is an editor for a trade magazine.

The familiar AM broadcast band is changing. That may come as surprising news to those who have focused their attention on the newer technologies, such as FM radio, cable, and satellites, but it's true.

It's also true that one of the medium's strengths, the capacity for distant nighttime reception, has become one of its weaknesses. This has come about as the limits to the band's capacity for accommodating new radio stations have been reached. The Federal Communications Commission, the government's regulatory agency, has tried to accommodate more stations while protecting current nighttime service.

The most visible change is the breaking up of the clear channels, those frequencies on which there are few radio stations operating at night that can be heard at great distances, sometimes all the way across the country. This process began slowly, in the early 1960s, when the FCC licensed a number of stations for nighttime operation in the west on the same frequencies as the major clear-channel stations in the northeast and midwest. (This was done to eliminate what the FCC believed were "white areas" without nighttime radio service.) This process has accelerated, especially since 1980, with a number of nighttime authorizations being granted for clear channel operation in locations all over the country.

Another change has come very suddenly, and may have even more of an impact on the medium. This change affects the 2400-odd stations that are licensed for daytime-only operation in the United States. They have been authorized to expand broadcast hours beyond what has been considered "daytime" in the past. (Daytime-only stations leave the air at sunset to lessen nighttime sky-wave interference.)

These developments will also change the hobby of BCB DX'ing, the pursuit of distant radio signals on the standard AM band. DX means distance, and a DX'er is one who tries to hear distant signals. It is quite likely that DX'ing this band, referred to here and elsewhere as the broadcast band (BCB), medium wave (MW), or simply AM, will become much more of a challenge.

The prospect of further challenges should come as no surprise to the experienced medium-wave DX'er—he or she is already accustomed to the vagaries of the AM band and the difficulties in chasing those elusive, low-powered stations. And he is aware of the crowded condition of today's band (4700 U.S. stations occupy the same 107 channels that accommodated some 800 broadcasters when the band was expanded to its present limits). But he is also aware of the enjoyment and rewards that come from listening to those familiar AM signals.

Why the interest in medium wave? The sports fan can hear a multitude of broadcasts covering every athletic event from an NHL hockey game broadcast in French from Québec, to a high-school basketball game over a small Arkansas radio station. The listener can choose from every conceivable music format and talk show, from all-news stations from all over the country, and even from broadcasts in Filipino or Croatian. The casual listener may even become interested in DX'ing on its own.

Most American DX'ers are interested in tuning domestic (United States and Canada) stations, but it is possible to hear broadcasts from Europe, South

America, Oceania, and beyond. Many DX'ers collect verification cards or letters from the stations they hear, or keep comprehensive tape libraries recording their listening efforts, but the main attraction for most listeners is the reception itself, trying to hear and identify that weak signal.

This fascination with distant radio signals is nothing new. In the 1920s, when broadcasting as we know it began, everyone with a radio was an experimenter and a DX'er. In fact, the hobby enjoyed its greatest popularity in the 1920s and 1930s. There are still many listeners interested in medium-wave reception, but they pursue a hobby that is very different from the one enjoyed by their grandfathers 60 years earlier. It's a hobby of sophisticated equipment, precise scientific studies, and advanced communications techniques.

DX'ing's Start, and Its Mass Appeal

AM stations operate within the relatively low medium-wave frequencies (300 to 3000 kHz). There is some debate as to when the first voice broadcasts were made, with claims ranging from 1892 to 1906. The most recognized date for the start of regular licensed AM broadcasting is November 7, 1921, when KDKA in Pittsburgh was granted a license and it broadcast returns from the Harding-Cox presidential election. Other stations with claims to being the first regular AM broadcaster include WWJ in Detroit, WHA in Madison, Wisconsin, and WBZ in Springfield, Massachusetts. The first experiments in network broadcasting came about in 1922, when WGY in Schenectady, New York, and WJZ in New York City were joined by telephone lines to broadcast the World Series.

Early government regulation of the medium dates back to the Wireless Ship Act of 1910. After the Radio Act of 1912, the Secretary of Commerce and Labor was responsible for licensing broadcasters. The first frequency assigned was 360 meters (approximately 833 kHz) in 1922. Another frequency (750 kHz) was later added and, in 1924, the Department of Commerce allocated 550 to 1500 kHz for standard broadcasting use, with a maximum power of 5000 watts. These early broadcasters changed frequencies, power, and schedules almost at will, creating chaos on the then-new broadcast band.

The government's answer to this disorder was the creation of the Federal Radio Commission (FRC) in 1927. The FRC had the authority to allocate frequencies, issue licenses, and assign power levels, and was primarily concerned with bringing order out of the chaos.

It was during this early period that DX'ing enjoyed its greatest popularity. Everyone who had a radio was a DX'er, and everyone tried to hear the new distant stations coming on the air. The stations cooperated by leaving the air at certain times so that local DX'ers could try for other stations normally blocked by their hometown broadcasters. (In Chicago, "Silent Night" was Monday evening.) The hobby was popular enough to support several national magazines, numerous local and regional clubs, and columns in newspapers.

The successor to the FRC, the Federal Communications Commission, was a Roosevelt New Deal agency created to regulate electronic communications. The

FCC, brought into being by the 1934 Communications Act, took over the FRC's functions and also regulated telephone, cable, and wire communications.

The popularity and growth of the hobby continued during the 1930s, but many of the clubs folded during the Second World War because so many of their members were serving overseas. After the war, many took up the hobby again, but were faced with a band that was much more crowded as hundreds of new stations took to the airwaves. (The number of AM stations on the air actually doubled in only five years, from 1945 to 1950.) There were also other activities, most notably television, and later FM radio, that competed for leisure time.

Structure of AM Broadcasting

Today's AM broadcast band extends from 535 to 1605 kHz, with stations occupying frequencies every 10 kHz from 540 to 1600 kHz, a total of 107 broadcast channels. AM stations in the United States use power levels ranging from 100 to 50,000 watts, with 250, 500, 1000, 5000, and 50,000 watts of power being the most common.

There are four basic types of AM radio stations and channels now in use. Class I broadcasters are the clear channel stations transmitting on what have become the not-so-clear channels. These stations use high power levels (most have 50,000 watts) and are heard over large areas. These stations usually possess the "dominant" signals that other stations on the frequency must "protect" by using directional antenna patterns which radiate lower power levels in certain directions. The clear-channel Class I stations are mostly big-city operations and are among the oldest and most recognizable AM broadcasters. Up until relatively recently, many of these stations could be heard over very large parts of the North American continent at night, but recent efforts by the FCC are aimed at cutting the coverage area of these stations back to 750 miles by licensing other nighttime operations on the same frequency.

This breaking up of the clear channels, the latest round of which began in 1980, is sure to have some far-reaching consequences for DX'ers. The most immediate is the short-term gain in the number of stations to hear.

In the long run, though, these policies can only hurt the hobby. It is the reception of these clear-channel stations that brings new listeners into the hobby, and while some distant reception will still be possible for the non-DX'er casually tuning the radio dial, it won't be the same.

For the experienced DX'er, this could be a good time to catch many of the new stations as they move to the clear channels from other frequencies or come on the air as brand new stations. Eventually, interference will be too great to allow easy reception of many of these stations. The 60 clear channels are: 540, 640–780, 800–900, 940, 990–1140, 1160–1220, and 1500–1580 kHz.

Class II stations are secondary broadcasters that share the clear channels. These stations must protect the coverage areas of the major clear-channel stations. There are also many daytime-only stations (Class II-D) licensed to some of the clear channels.

The regional channels have larger numbers of lower-powered (up to 5000 watts in the U.S.) Class III stations that are meant to cover their city of license and the surrounding area. Nearly half of the AM stations currently on the air broadcast on the 41 regional frequencies (550–630, 790, 910–980, 1150, 1250–1330, 1350–1390, 1410–1440, 1460–1480, 1590, and 1600 kHz).

The major change on the regional channels in recent years has been the elevation of many daytime-only (Class III-D) stations to fulltime status, typically with low nighttime power (500 to 1000 watts), and a very restrictive antenna pattern to protect the signals of other stations on the frequency.

Broadcasters in the final category of stations in the U.S. are classified as Class IV, or local stations, and transmit on the six local channels (1230, 1240, 1340, 1400, 1450, and 1490 kHz). Interference can be severe on these crowded frequencies, with close to 200 stations licensed on each channel in the U.S. alone. Local stations, called "graveyarders" by DX'ers because there are so many of them, are limited to 1000 watts of power.

Almost all U.S. stations are licensed for either daytime-only or fulltime (unlimited) operation, but there are a few exceptions. "Limited-time" stations are basically daytime stations operating on clear channels with the sign-on or sign-off time determined by the local sunrise or sunset time at the dominant Class I station. There are very few of the stations, and most have been granted unlimited licenses over the past few years.

Another rare type of station is known as a "share time" broadcaster. One or more stations in the same or nearby cities will share the same frequency, splitting up the broadcast day or week. This arrangement was more common in some large cities many years ago.

A "specified hours" station is usually a fulltime outlet that does not meet the FCC's requirement for a minimum amount of broadcasting time per day. A few of these stations receive this type of license because of some special situation.

Some stations issue coverage maps as sales tools. The four representative samples in Fig. 14-1 indicate a variety of coverage patterns.

Canadian stations are similar to their counterparts in the United States, but there are a few differences. Many Canadian stations on regional frequencies broadcast with up to 10,000 watts of power, but employ directional antenna patterns that usually beam the majority of this power to the north, away from the U.S. There are also very few daytime-only broadcasters in Canada (CHYR-710 Leamington, Ont.; CHRS-1090 St. Jean, Que.; CHBR-1110 Hawkesbury, Ont.; CFML-1170 Cornwall, Ont.; CJMR-1190 Mississauga, Ont.; CKOT-1510 Tillsonburg, Ont.; and CHIN-1540 Toronto, which has been granted a fulltime license). Hearing this small group is a major challenge for most stateside listeners.

But the ultimate challenge for North American listeners comes from the Low-Powered Relay Transmitters (LPRTs) that operate in remote parts of Canada. These "stations" use only 40 watts of power and are truly local because they are designed to serve only a single small town or settlement. LPRTs receive programming from regular Canadian broadcasters by telephone line and rebroadcast the programs on different frequencies than those used by the original

Fig. 14-1. Coverage maps used as sales tools by radio stations.

station. These unattended transmitters occupy many different frequencies, including many clear channels, which makes them prime targets for DX'ers. If a listener hears an identification for CBU-690 Vancouver on 1070 kHz, chances are the station being received is 40-watt CBUV from Fort St. James, British Columbia!

Canada also has a government-run network, the Canadian Broadcasting Corp. (CBC), which operates many of the most widely heard stations, in both

French and English languages. Canada's equivalent to the FCC is the Canadian Radio-Television Commission (CRTC).

Both U.S. and Canadian stations are also under the rules and regulations of the North American Regional Broadcasting Agreement (NARBA) that was created to reduce interference between nearby nations and provide protected channels for several countries. Most of the frequencies are dominated by U.S. stations, with Mexican, Canadian, and Cuban stations filling the role of the "dominant" on the rest. (Fidel Castro has chosen to largely ignore this agreement.) The Bahamas have a single clear channel (1540 kHz) on which other North American stations must not interfere.

Call Letters and Slogans

Broadcasting stations in the United States are assigned call letters beginning with K or W (K west of the Mississippi and W east of it). A few stations (KDKA Pittsburgh, KYW Philadelphia, WDAF Kansas City, and WBAP Fort Worth, among others) are exceptions to this rule. Canada has most of the C series, and AM station call letters in Mexico begin with an XE prefix. In Cuba (which has the CM prefix), and in most of the rest of the world, call letters are rarely used by broadcasting stations.

This is another area affected by deregulation. In the past, the FCC would not assign similar call letters for the same area, or call letters that were the same as the initials of federal agencies. The FCC will now assign any call letter combination, as long as it is not already being used by another station.

Another change in the call letter rules will definitely have a negative effect on DX'ers. There used to be a clear period between the time a station applied for and received new call letters, enough time for the DX clubs to report both the call letter application and the grant. Under the new call letter nonrules, a station may apply for, and receive, new call letters in a matter of days. Under these rules, a DX'er will never be quite sure what call letters a distant station may be using and will have to be careful when identifying stations.

A trend that has grown appreciably over the past ten years involves stations using their call letters very infrequently, opting to identify with a slogan instead. These slogans, Newsradio 95, Musicradio, 13-Q, Magic 97, The Mighty 1090, 15-X, etc., can also make positive identification of some stations difficult.

Networks and Programming

The networks are back on AM radio, but they're not the networks that entertained the entire nation before the rise of television. Instead, they broadcast news, talk, and music to fit specialized station formats. The DX'er should become familiar with the sound, length, and timing of the major network newscasts (CBS, NBC, Mutual, RKO, the various ABC Network feeds, AP and UPI audio services, and the CNN news service), and the times that the other programs break for local identification announcements.

There are also several regional networks and about 20 state networks. Most of the state networks broadcast state news with a heavy emphasis on agricultural programming. In fact, almost all of the state networks are located in states with large agricultural economies. Most of these network programs are only broadcast during the day.

Only about 40 AM stations are noncommercial operations that carry National Public Radio programming. (Most of these stations are on FM.) The best known broadcasters in this category are located in large college towns.

Sports continue to be an AM programming staple. Major league baseball is carried by radio networks for each team. A "flagship" station, usually in the same city as the franchise, may feed the game to dozens of far-flung stations, hundreds of miles from the home city. Stations on the network may air local announcements and usually identify on the hour. Other sports networks cover professional hockey, basketball, football, and college football and basketball.

The most significant trend in AM programming over the past five years has been the proliferation of talk programs, both on the network and local level. The best-known talk show is hosted by Larry King, and can be heard over the Mutual network. ABC has jumped into the field with its Talkradio program, and NBC has launched its own program called Talknet.

Religious programs continue to be popular on AM radio, with the oldest and best known being the "Back to the Bible" series. One trend in this area has been the adoption of a religious format as a for-profit venture by stations that may not have been successful using more traditional programming. Some stations act only as time brokers, selling blocks of a half hour, hour, or more, time to religious broadcasters who, in turn, ask for funds from listeners to continue their efforts.

Radio stations have an ever-widening choice of music programming available by satellite, including the Satellite Music Network and RKO's Nighttime America show. Other taped offerings, such as the American Top 40 program or the American Country Countdown, continue to be popular with AM broadcasters. The use of satellite programs means fewer identifications (as few as one an hour) for DX'ers to listen for.

One of the hottest music trends to hit AM radio in the past five years is not disco, modern country, or hot hits, but a compilation of old music from the 1930s, 40s, and 50s, packaged into a "Music of Your Life" format. But the overall trend in AM programming, even with the "MYL" stations, is to more news and talk.

Beacons, Traveler's Stations, and Pirates

Most of the stations that concern the medium-wave DX'er are standard broadcasting stations with programs intended for the general public. There are a few additional stations, mostly on the fringes of the band, that provide some interesting signals for DX'ers to try and receive.

Radio beacons, which fall into the "utility" category of radio stations, operate mostly just below the standard AM band. These stations are used for direction-finding by ships and aircraft, and may broadcast continuous Morse code identification, or tone and code—often on a 24-hour schedule. The most popular beacon frequencies in North America are 512, 515, 521, 524, 526, and 530 kHz, although stations can be heard anywhere in this range.

Another type of specialty station that has come into prominence in the last few years is known as the "Traveler's Information Station" (TIS). These stations broadcast traffic, parking, and tourist information at airports, highway interchanges, and national parks. Most of the TIS outlets operate on 1610 kHz, with 530 kHz being used by some others. Operating with very low power, these stations are designed to cover only a few square miles. TIS stations may be found in places such as the Grand Canyon National Park, the Tampa International Airport, and at the intersection of Interstates 75 and 285 in Forest Park, Georgia.

A final type of station is not licensed by the FCC and is, in fact, illegal. These stations, known as pirates, are operated by electronics hobbyists or thrill-seekers. Pirates can be heard anywhere within the AM broadcast band, but usually transmit in the upper part of the band or at 10 to 20 kHz above the band. Recently, there have been a number of these stations broadcasting in the New York City area, just above 1600 kHz. These stations are operated by the young and usually feature punk rock and other irreverent programming.

Developments in AM Broadcasting

Despite a number of changes in station power and operating schedules, there appears to be no immediate move toward any structural change in the Western Hemisphere broadcast band. After several conferences, the nations of the Western hemisphere elected not to change the present 10 kHz spacing standard on the AM band. In 1978, the rest of the world changed to a 9 kHz spacing standard.

Another action which may affect the band was agreed upon by nations of the International Telecommunications Union (ITU) and the World Administrative Radio Conference (WARC) in 1979. This meeting resulted in new frequencies being added to the upper end of the band (in the 1610–1700-kHz range). Use of some of these channels may take place before the end of this decade.

One of the most publicized changes affecting AM broadcasting, the introduction of AM stereo, has had little impact on BCB DX'ing so far. In the spirit of deregulation, the FCC decided to "let the marketplace decide" which system was the best, and approved several different systems and companies for AM operation. The most popular system, perfected by Leonard Kahn, was also the first to come on the air during the summer of 1982 on KDKA in Pittsburgh. Most of the other stations now on the air with AM stereo are using this system, with others opting for the Harris, Magnavox, or Motorola systems.

Equipment and Clubs

Some potential BCB DX'ers (especially those who have had experience DX'ing other media) want to start out with the most expensive receiver and the most sophisticated antenna system. This is a mistake. The best way to learn the band, and ultimately the most rewarding, is to start out with modest equipment, progressing slowly to more advanced gear. The receiver itself is not as important in snaring the rare DX as experience, patience, and luck.

Most veteran BCB DX'ers prefer a combination of a good communications receiver and a loop antenna. For a number of years, the most popular receivers were models made by Hammarlund, Collins, or National. These tube-type receivers provide good selectivity and sensitivity, and are easily modified to a number of applications. They are still available (Hammerlund and National are out of business and Collins manufactures military and amateur gear) and a quality used receiver can usually be had for $200–$400.

Recently, receivers manufactured by such Japanese companies as Panasonic, Yaesu, Icom, and Kenwood have found their way into the serious BCB DX'ers radio shack. These receivers vary as far as BCB performance is concerned, but generally have proven adequate in most situations. One problem that sometimes surfaces on these and other modern receivers is overloading in the presence of very strong signals, or difficulty with image rejection and poor selectivity. These problems can be especially annoying if the listener lives in a large metropolitan area near many high-powered stations.

For the beginner, though, almost any radio will do. Automobile radios often have superior selectivity when compared to the average table model or clock radio. Two portables, the General Electric Superadio and the Sony ICF-S5W, have received good reviews from DX'ers, and are recommended for the BCB DX'er who does not want to spend a lot of money on equipment. The Radio Shack TRF, for years a favorite of portable users, is now out of production.

A long-wire antenna, similar to any shortwave antenna, is usually sufficient to begin BCB DX'ing. Be sure to follow all the rules as far as grounding and insulating are concerned. Another type of long wire, a Beverage (named after its inventor, Harold Beverage), is similar to a long-wire antenna but is several thousand feet in length. These antennas have very high gain and unique directional characteristics, allowing stations from very long distances to be received. Since it is impractical for most DX'ers to set up a Beverage, several listeners will get together and set up this long aerial during a weekend on a cooperative farmer's field or other open space.

The loop antenna is much smaller than a Beverage, but has special qualities of its own. Loops exist in a number of different styles, some built around a ferrite bar, others employing a variety of wire arrangements wound around a wooden frame. The main advantage of a loop antenna is that it is directional. Turning a loop in a certain direction will cause stations from that direction to be received at greater strength. For example, if a ferrite bar loop is set with the ends of the bar pointed towards the east and west, stations to the north and south of the bar will be received, and stations from the east and west will be nulled out. A loop is especially important when DX'ing today's crowded band, and it is usually possible to completely null out or loop out all but the strongest stations with a good one. By carefully positioning a loop on a crowded frequency, the listener can receive a number of different stations from different directions. Portables also use a ferrite rod antenna, so turning the entire radio will have the same effect. Long wires should generally not be attached to portables. The major BCB DX clubs have information on receivers and plans for building a loop or other type of antenna.

Speaking of clubs, membership in one or both of the national medium-wave-only organizations is a must for the serious BCB DX'er. The National Radio Club (Membership Center, P.O. Box 118, Poquonock, CT 06064) and the International Radio Club of America (P.O. Box 21074, Seattle, WA, 98111) both offer a wealth of information for the BCB DX'er.

Each club has regular columns covering what is being heard in different parts of the country from the U.S. and Canada, and sections detailing reception from other parts of the world. Members are also encouraged to write in and share their hobby exploits with the rest of the club members. Other features of the club bulletins (issued weekly during the winter DX season) include technical reports, station profiles, network lists, and data on the latest station changes. The NRC Domestic Log is the best compilation of U.S. and Canadian station information for the BCB DX'er. The log contains information on station call letters, location, day and night power, directional/nondirectional antenna, network, address, and operating hours—all arranged by frequency. The NRC also publishes receiver and antenna reference books and a Night Antenna Pattern Book, with maps showing the directional-antenna patterns for stations on each frequency.

The IRCA Almanac has a lot of information relevant to the DX'er, including basic explanations of different facets of BCB DX'ing, network information, and sports and music programs. The IRCA also publishes the IRCA Foreign Log, a compilation of all foreign loggings reported to the club's bulletin during the previous DX season.

A final advantage to club membership is the annual convention held by each club. At these gatherings, the listener has the chance to compare notes and talk with others interested in BCB DX'ing.

Other reference books available to the BCB DX'er include the *Broadcasting Yearbook* (published by *Broadcasting* magazine), the *World Radio-TV Handbook* (useful for foreign station information), and *White's Radio Log*.

Medium-Wave Propagation

There are several different ways that an AM radio signal can be received. The first involves simple line-of-sight reception between transmitter and receiver. Another is called "groundwave" and can produce reception of a signal up to 150 miles or so as it follows the ground. Various areas of the country have different ground conductivity levels, with bodies of water, particularly seawater, having far greater conductivity. Groundwave reception is possible during both day and night, and may extend for long distances on low frequencies during the winter.

Ionospheric propagation, or skip, is the most interesting form of reception as far as the DX'er is concerned. During the day, the D layer of the ionosphere (30–50 miles above the earth's surface) absorbs all medium-wave radio signals. Fortunately, the D layer is a daytime phenomenon only and disappears as the sun sets. Another daytime layer, the E layer (60–80 miles above the earth's surface) may exist in a weakened state during darkness, but does not greatly affect medium-wave signals. The most important ionospheric region is known as the F layer, which lies in the 150–250-mile range at night. The F layer is split into two layers during the day, into F1 and F2, but combines during darkness and is responsible for most distant reception of medium-wave signals.

The quality of signals reflected by the F layer is dependent on the electron density in the region, which comes from the sun. A high electron density level attenuates, or dampens, radio signals. The electron density is much higher in the summer and lower in the winter, one of the reasons why winter is considered to be the "DX season." In the winter, there is also less solar radiation to disturb the D layer. Another factor affecting seasonal reception is the amount of thunderstorm static, which can often make summertime medium-wave DX'ing impossible.

In addition to the seasonal variations caused in part by the sun, radio signals are affected by the 11-year sunspot cycle. In theory, medium-wave DX should be best during the sunspot-cycle minimums when absorption is at a low level. The amount of solar activity, which enters the ionosphere through the earth's magnetic pole regions, can affect reception by blocking northern signal paths.

The most spectacular solar condition is known as the *aurora*. This is the same phenomenon that produces the *Aurora Borealis*, or *Northern Lights*. Emissions of highly charged particles from the sun can spur these storms, which may last for days, disrupting radio communications over some parts of the globe. For BCB DX'ers, the principal effect of these auroral storms is the blockage of signal paths from the north. If an aurora is severe (and depending on the listener's location), there may be no sky-wave reception noted at all. The only thing that will be audible will be signals from the nearby stations within groundwave range, sounding much stronger than usual because all the nighttime sky-wave interference is absent.

This can be good or bad for the DX'er. Sometimes, low-powered stations from the south and from Latin America will be blasting in on clear regional, and even local, channels, with all the usual interference from the north eliminated. Other times, nothing of interest will be received from any direction.

If an aurora lasts for two or three nights, it has been observed that the first night is usually the best for southern reception. Often, after the first night, skip from all directions tends to be dampened. Listeners have also noticed that unusually good reception from some northern latitude areas can occur just before an aurora begins. This type of northern reception may only affect a small geographic area. Extended periods of low geomagnetic activity are usually required for reception of European signals in North America.

It's possible to predict, with a certain degree of accuracy, what conditions will be like by tuning to WWV (5000, 10,000, and 15,000 kHz shortwave), which offers current A and K indices (measures of solar activity) and solar activity forecast information at 18 minutes past each hour.

What is more difficult to predict, putting auroras and unusual propagation aside, is when conditions will be good and when they will be poor. Often, an opening will exist to a certain area that may last for a few minutes, or for several hours or even a few days, causing stations in that region to be received with much stronger signals than usual. Other times, the entire band will seem to be "live," with signals being received with better-than-usual strength from all directions; at other times, it will seem "dead."

As mentioned earlier, the fall and winter months will be "good" more often than the summer months. October, November, and December seem to always be good for domestic DX'ing, with January, February, and early March usually good. Late September and early April can also produce good DX. The months of May, June, July, and August are usually the worst months as far as propagation is concerned, and many BCB DX'ers forego the hobby entirely during this period. Graveyard DX and sunrise DX up to 500 miles may actually be improved during the summer because of the lower levels of interference from more distant stations.

Daytime and Evening DX

One of the first things a listener should do is to become familiar with the local band during the day. Medium-wave daytime conditions can best be observed in the 10 A.M. to 2 P.M. period, well after any post-sunrise propagation and before the sky wave begins to fade up toward local sunset. At this time, the listener will note many stations that will be heard at night also, but can get a good picture of which stations are within groundwave range of his listening post.

During the day, some daytime-only stations will be heard from locations up to 150 miles away or more. Although the upper part of the AM band is the best for nighttime sky-wave reception, the low end of the band (particularly below 600 kHz) provides the best groundwave reception. Obviously, many nearby stations that are licensed for daytime-only operation will be heard only during this period. During the "critical hours" period, two hours after sunrise and two hours before sunset, some weaker stations may be covered by stronger, but more distant stations.

It may also be possible to hear fulltime stations during the day that are very near the listener's location, and which are not audible at all during darkness. Such stations may have to drastically reduce power after sunset, or have an antenna pattern that does not favor the listener's location.

At some coastal locations and in mid-winter, when the days are shortest, it is possible to receive some distant signals up to 1000 miles away and beyond during the day, especially from powerful clear channel stations on open frequencies.

A DX'er out in the country, away from a large number of stations, may find midday DX'ing worthwhile in some situations. For most listeners, however, the daytime period is more important as a time to establish an inventory of local and semi-local stations that can be received rather than chasing rare DX.

The other time period that a beginning listener should check is the evening. Many of the local stations noted during the day will still be audible, of course, but there will also be some type of signal on every frequency. The first stations to seek out will be the distant clear channel broadcasters. It is also important to learn the band during the evening, because this will be the basis for all other periods. Listen carefully to each frequency. On many frequencies, especially the clear channels, there may be only one station audible. On others, two or three stations may be regularly heard. The listener will become familiar with each frequency by slowly and deliberately identifying, logging, and noting the signal quality and format of each station received.

Evening conditions do not vary greatly from night to night, partially because all fulltime stations are on the air, masking true reception patterns. Occasionally, there is an opening after sunset to a particular area that may last an hour or two, or there may be unusually good conditions noted throughout the evening to an area or all over the country. It is sometimes worthwhile to check the graveyard channels during the evening because they will occasionally yield a rare station that is off the air later at night. If good conditions are spotted during the evening, it is often an indication that things will be really interesting later on, after midnight. One exception to this observation is when an aurora is underway, and stations from Latin America are being heard. The evening may be the best time to hear these stations because, unlike most North American broadcasters, stations in Latin America do not maintain all-night schedules and most are off the air by 10 or 11 P.M. Eastern Time. (There are a few exceptions, but very few Latin American stations are normally heard during the early morning hours.)

If you are unfamiliar with the band or with the stations heard during the evening, take your time and log each station deliberately. Memorize the call letters, city, format, and network. These are stations that will be heard hundreds of times in the future, and it will be beneficial to be able to tune right to the frequency and know which station is being received, or know if the dominant station is off the air.

By having a working knowledge of the band at these times, the listener will be able to quickly recognize unusual conditions as they are developing and take advantage of them.

Silent Periods, Graveyarders, and Equipment Tests

After midnight, several things happen that make BCB DX'ing much more interesting than in the evening hours. One thing that happens is that some stations leave the air, allowing others to be heard on the same frequency which are normally blocked by the dominant stations. Stations that leave the air every night or once a week have a "silent period" (SP), stations with no scheduled silent period are said to be "no silent period" (NSP) stations. Stations which go off the air every night may sign off the air at midnight, 1 A.M., or later. A few fulltime stations may be off the air as early as 10 P.M.. If the listener is in the Central Time Zone, especially if he is near the Eastern Time Zone, he may notice that some of the stations that are normally dominant during the evening hours have signed off the air at 11 P.M. CT. The same is true in other time zones, especially if the listener is located near stations that are one hour ahead of him.

Don't expect many stations to be off the air every night. Most stations are licensed for unlimited operation broadcast on a 24-hour basis. Nearly all the clear-channel and regional-channel stations in large cities are on the air 24 hours a day, but many broadcasters located in smaller towns and many of those operating on the graveyard channels still go off the air nightly.

The post-midnight period is also the best time for involved graveyard channel DX'ing, since this is the time that fewest stations are on the air. There are still many 24-hour operations, and some stations on the local channels will be heard regularly because of this schedule even though they may be up to 400 or 500 miles away from the listener. This type of DX'ing requires a great deal of patience, but it can also be very rewarding. A tape recorder is very helpful, and special attention should be paid to the top-of-the-hour period, and a few minutes after, when an identification or sign-off announcement is most likely. When first tuning to a graveyard frequency, try to identify the strongest signal; then, use that as a reference point. For example, if station WCCC is playing country music and loops to the northeast from your location, try to identify the station that has a talk program that's looping to the southeast.

After many graveyard sessions, the listener should be able to identify the usual dominant stations by format and loop angle. Being able to tell when "something is there that shouldn't be there" is especially important for graveyard DX'ing. Special attention should also be paid to the operating schedule of any local station that may be blocking graveyard frequencies. When one of these stations goes off the air, be sure to put that time to good use, as there are dozens of potential stations to be logged.

An example of the value of a single graveyard channel was demonstrated when WVON in Cicero, IL, left 1450 kHz to assume the 1390-kHz frequency of WNUS in Chicago. (WVON had purchased WNUS and didn't have a buyer for the 1450-kHz facility.) When WVON, a 24-hour-a-day NSP station, left 1450 kHz one afternoon in February, 1975, area DX'ers immediately concentrated their efforts on this channel.

Stations were logged from the eastern two-thirds of the continent by DX'ers in the Chicago area on this frequency, from Beaumont, Texas, to Rugby, North Dakota to Hartsville, South Carolina and Front Royal, Virginia. In all, some 40 new stations were logged by various area listeners on the frequency before WFMT, a Chicago FM station, was granted an interim license to operate on 1450 kHz in the fall of 1976.

Part of this excitement was due to the fact that two of the Chicago graveyard channels (1240 and 1490 kHz) were blocked by local stations and two others

(1230 and 1340 kHz) were occupied by semi-local broadcasters, so DX'ers recognized an unusual opportunity.

Although the best shot at rare DX on the local channels is during the night, distant stations can sometimes be heard signing on in the morning (these frequencies are very crowded at this time) or during the sunset DX period. Stations logged at sunset will generally be within a few hundred miles, although unusual conditions noted on regional frequencies will sometimes result in stations from the same area being heard on the local channels.

One day is special on the after-midnight calendar, and that is Monday morning. Only a few years ago, many fulltime stations left the air once a week, and that was usually on Monday morning between approximately midnight and 5 A.M. local time. Sunday morning was also popular with some stations. Although the number of stations with regular Monday morning silent periods has declined significantly since the late 1960s and early 1970s, this day is still important. When a station does go off the air (perhaps only once or twice a year), it is often on a Monday morning; it is necessary for stations to leave the air periodically for routine maintenance and testing.

A beginning DX'er's first Monday morning session should produce some DX that cannot be heard at other times during the week, and he should be able to note that at least a half dozen stations are off the air. Sunday morning is also worth checking, but serious all-night DX'ing can be inconvenient for those on a regular work or school schedule. The listener should scan the entire band during these periods, noting anything unusual and then concentrating on what appears to be the most promising frequency. The rest of the band should then be checked at regular intervals, especially just after the hour, when stations are most likely to leave the air.

In addition to being a good time to hear dominant stations off the air, the midnight-to-sunrise period is the only time a listener can hear radio stations testing their equipment. These tests offer a unique opportunity to hear many stations, and often are the only way to log particularly hard-to-hear broadcasters.

All post-midnight tests fall into the broad "equipment test" category. Basically, an equipment test is just that, a station testing its equipment—over the air —for purposes of adjustment and maintenance. During these tests, daytime-only stations are allowed to use their facilities and fulltime broadcasters may use their daytime power and antenna pattern. Since the daytime power is almost always greater and the antenna pattern less restrictive, this means that the station stands a better chance of being heard at a distance.

Another factor that increases the chance of a testing station being received is what is being broadcast. Test tone (often 1000 Hz) music that is not part of the station's format, and periods of open carrier may all be part of the rest. Testing stations are required to identify at the beginning and end of the test (not all do), which may last for 10 minutes, an hour, or all morning. Identifications may also be heard at irregular intervals during the test, sometimes at or near the hour.

One special type of equipment is known as the Proof of Performance (PoP) test. This is an annual test of equipment which is required by the FCC. During a "proof," a station will broadcast at a variety of modulation levels and with a number of different frequencies (tones) to determine if the equipment is capable of broadcasting the different frequency ranges clearly. A listener encountering a test of this type will usually hear different pitched tones, with periods of silence while adjustments are made.

A proof of performance test may last for hours, but another type of equipment test is a regularly scheduled measurement that lasts for no more than 15 minutes and may last only a single minute! These tests, known as frequency checks, are run for outside monitors that measure the frequency on which the station is transmitting to ensure that it is within the ±20-Hz tolerance allowed by the FCC. Most stations that run frequency checks are monitored once a month, on a scheduled day, such as the second Tuesday or the fourth Friday. A test tone, usually 1000 Hz, is preferred by most stations running frequency checks, although some stations use different pitches of tone, and a few use music or an open carrier. In parts of the South, especially Alabama and Mississippi, a telephone dial tone is used for frequency checks.

Many frequency checks are scheduled to last for 15 minutes, with identification announcements coming at the beginning and end, every five minutes or every minute. A recent trend has been for the station to run tone for only a minute or so, and wait for the monitor to call and confirm that the signal was received and measured. Another tendency is toward fewer frequency checks being run in the post-midnight experimental period. Many stations test their own frequency during regular broadcast hours with their own equipment.

Another type of equipment test which is of special interest to the BCB DX'er is called a DX Test. A DX Test is an equipment test that is scheduled and publicized well ahead of the test date, allowing DX'ers to try and hear that specific station. Frequent identification announcements, test tone, easily recognizable music (Souza marches or Dixieland jazz), and even Morse code identifications may all be a part of the DX Test and may help the listener identify the signal. The major clubs each maintain a committee that contacts prospective stations and asks them to test for this purpose. Stations that do broadcast special tests for listeners are often staffed by an amateur radio operator or even a member of the NRC or IRCA.

The FCC frowns on a station running a test specifically for the purpose of being heard at distant points, so many DX Tests are now run as part of the station's regular testing schedule, with more identifications than usual. There are still some of these tests, however, that are particularly interesting for DX'ers. Some have even taken phone calls during the test from DX'ers all over the country and put them on the air.

Although they are a good way to hear many new and interesting stations, there are definitely fewer tests of all types, particularly frequency checks and DX Tests. One reason is that modern broadcasting equipment requires less maintenance. Another is that many of the old timers in the business were hobbyists and enjoyed experimenting with the equipment. Today's station is much more likely to have a contract engineer not based at the transmitter.

A final type of special broadcast is authorized only during an emergency, and may be heard at any time. During floods, hurricanes or blizzards, daytime-only stations may stay on the air and fulltime stations may use their daytime facilities to broadcast important messages and emergency announcements. Some of the most widely heard emergency broadcasts have been during hurricanes, when small stations in Florida, Louisiana and Texas have been heard on the air providing emergency information and warning the local populace to seek shelter.

A recent, rather curious application of this emergency rule occurred during the Grenada invasion in 1983. WIAF, a daytime-only station on 1500 kHz in Clarkesville, Georgia, was heard throughout the east and central portions of the country broadcasting conversations with an amateur who had been listening to another amateur broadcasting from Grenada during the invasions.

Changes in Twilight DX'ing

The two twilight DX'ing modes, sunrise skip (SRS) and sunset skip (SSS) are two of the most popular times for medium-wave DX'ers to listen, and are the most productive in terms of logging new stations. The existence of so many daytime-only stations and the change from daylight to darkness and vice versa are essential components in the BCB DX equation. The structure of American broadcasting was such that these two DX'ing modes remained unchanged for a number of years. That is changing, however, with great speed.

In the fall of 1983, the FCC decided to allow daytime-only stations to begin broadcasting after local sunset. The decision went into effect after an agreement with the Canadian government was signed, and nearly all of the 2400 daytime-only stations in the U.S. have been granted postsunset authorizations (PSSAs) to broadcast with reduced power up until 6 P.M. local time. Many of these same stations have held presunrise authorizations (PSRAs) since the 1960s, which allow them to broadcast with up to 500 watts of power beginning at 6 A.M. local time. (The actual name of this sanction is the Presunrise Service Authority, which has been abbreviated PSA prior to the new postsunset authorization. It will be abbreviated PSRA here to avoid confusion with the sunset rule.)

Most Class III-D stations have presunrise power levels at or close to the 500 watt maximum, but many of the Class II-D clear channel broadcasters must use very small amounts of power in the presunrise period.

To deal with the goals of protecting fulltime radio broadcasters from interference and extending the hours of the daytime-only outlets, the FCC used computerized data based on "diurnal curves," which were intended to determine levels of acceptable interference to the fulltime stations. The result has been a complicated, and not yet completed, plan in which daytime-only stations may have to use different power levels for each month of the year or have none at all for certain months. Under the current agreement, daytime stations will be able to use their PSSA power levels for only a few months out of the year, as local sunset in most areas is past 6 P.M. for most of the spring, summer, and fall. Once an agreement is reached with Mexico, daytime stations will be able to use their PSSA powers up to two hours after local sunset, no matter when that occurs. This means that some daytime stations may not have to leave the air until 10 or 11 P.M. in the summer. (Sunrise times are very early in the summer, and many daytime stations are able to sign on before 5 A.M. local time.) Daytime stations on 1540 kHz will have to wait for a similar agreement to be concluded with the Bahamian government before PSSAs can be granted for use on that frequency.

Initially, only about 90 stations were granted PSSA powers in the 400 to 500 watt range. About 50 stations received 300 to 400 watt PSSA powers, with quite a few broadcasters (about 500) getting PSSA powers in the 100 to 300 watt range. A total of 600 stations will be using PSSA powers between 50 and 100 watts, with another 1000 authorized to use less than 50 watts after sunset. Of this last group, approximately half must get by with less than 25 watts. Stations with very low PSSA powers (a few watts) will probably not find the investment or the coverage worthwhile, and are expected to sign off at the normal time.

At sunrise, some additional clear channel daytime stations will be granted PSRA powers, but will be restricted depending on their location in relation to the dominant protected broadcaster on the frequency. Some 200 daytime stations operating on Canadian clear channels will be allowed to come on the air with PSRA power at 6 A.M. local time for the first time. Some stations may have to reduce current PSRA levels due to the new diurnal curve measurements.

The whole transition of regulation regarding North American BCB stations during the twilight periods is not yet complete, but BCB DX'ers will soon be able to judge how these legislative changes have helped, or hurt, the medium-wave DX hobby.

How to DX Sunset Skip

Because of the difficulty many have in getting up or staying up for late-night or early-morning DX sessions, one of the most attractive types of BCB DX'ing is known as sunset skip (SSS). It occurs, as the name implies, before, during, and after the listener's local sunset. As of this writing, there are several factors that will change the DX possibilities of this mode. The greatest change will be the granting of the postsunset authorizations for daytime stations, but many of the principles of sunset DX'ing will remain the same.

A good road atlas and a set of sunrise/sunset maps (available from the major clubs) are a great help for both sunset and sunrise DX'ing. These maps show when each area will sign on and sign off for each of the 12 months of the year.

The primary targets for the sunset DX'er are the 2400-daytime only stations in the United States. These stations have, up until recently, signed off the air at approximately local sunset. Sunset times change every day, but to avoid confusion the FCC assigns a single sunrise and sunset time for each month of the year for every city with a radio station. This time is rounded off to the nearest quarter-hour, and is based on the actual sunset or sunrise time for the fifteenth of the month. If the sun sets in Smithville at 5:11 P.M. on October 15, then all daytime stations in that city will have to leave the air at 5:15 P.M. every day in

October, and fulltime stations will have to change from daytime to nighttime facilities at that same time.

The amount of daylight is changing almost constantly from day to day. The days are either getting shorter in the fall or longer in the spring. This constant change allows a darkness path to exist between station and listener, often for only a few minutes. In the fall months, up to December, this darkness path exists after the fifteenth of the month. In the spring months after January, the days before the fifteenth offer the greatest darkness path. The actual sunset time of the station on the fifteenth will be a factor, but the last day in November should be the best in that month while the best day in March should be the first. Of course, conditions vary from day to day and there will be "good" days in the "bad" half of the month and vice versa.

The fall months are generally better than the spring months for sunset DX'ing, with the October–January period usually producing the most interesting openings. Although sunrise can be good during the summer, sunset DX'ing is usually a waste of time for all but the most inexperienced DX'er, who may need relatively easy-to-hear nearby stations.

The sunset DX period can cover a time span extending from up to an hour before local sunset to two hours after sunset, depending on the listener's location. Under normal conditions, the best time to try is just before local sunset and for about 45 minutes to an hour after the local stations have signed off or changed to nighttime power.

During the fall and winter months, the angle of the sun at sunset helps determine which stations will be received. Since the angle of the sunset line in December, for example, runs in a northwest/southeast line rather than a north/south line, listeners in the northern states will be able to receive stations in the south and southeast after their local stations are off the air or operating with reduced power.

The listener will also discover that certain areas of the country and certain stations are received often at sunset. A listener in New York may often notice openings to Virginia and North Carolina; an Indiana DX'er may often experience good reception from Texas and Oklahoma; and a DX'er in Minnesota may enjoy unusually good signals from stations in Colorado and Wyoming. The sunset DX'er will be able to tell when a good opening is going to occur by noting certain "indicator" stations that are received with much better signals than normal. It is especially important to have a good working knowledge of what can be received under different conditions at sunset because the listener may have only a few minutes to take advantage of an opening before stations in that area sign off or reduce power.

The easiest stations to identify at sunset are the daytime stations, since they end their broadcast day with a sign off announcement. The recent rule changes regarding daytime operation will give the DX'er two chances to hear a station, just before the PSSA power reduction and at actual sign off. The listener will have to be alert to catch an identification before many of these stations reduce power. Fulltime stations reducing power and/or changing antenna patterns typically "pull the switch" without warning—often in the middle of a song.

Frequencies at the upper end of the band, in the 1500–1600 kHz range, are a good place to start sunset DX'ing. Regional frequencies in the 1200–1500 kHz range are good in most locations. Other frequencies may prove to be particularly good, depending on the listener's location and his local interference.

How to DX Sunrise Skip

Even if the regulatory changes described earlier have much more far-reaching and negative consequences than anyone expects, it's still a safe bet that sunrise skip (SRS) DX'ing will continue to be of great interest to the medium-wave listener.

Stations are assigned sunrise times in quarter-hour gradations based on the sunrise time on the fifteenth of the month, like the sunset time assignments. If sunrise in Smithville is at 6:30 A.M. on February 15, then stations in Smithville with PSRA powers can switch to full power at 6:45 A.M., daytime stations without PSRAs can sign on at this time, and fulltime stations may change from nighttime to daytime facilities at 6:45 A.M.

The first frequencies a listener should check at sunrise are the regional channels. DX'ers in the Central and Eastern Time Zones will find the 6 A.M. ET period a very good time to try for stations with PSRAs in the Eastern Time Zone. DX'ers should start listening a few minutes before the hour (about 5:55 A.M. in the East, 4:55 A.M. in the Central Time Zone) and check for what sounds like a good frequency, one without a lot of interference from fulltime stations already on the air. Once a good frequency is found, the DX'er should simply wait for stations to sign on with their PSRAs. Some come on a few minutes before the hour, some right on the hour, and a few sign on a minute or so later. If the signal is being received particularly well, an open carrier may be audible before the sign on announcement is made. Many stations sign on with the national anthem, America the Beautiful, Dixie, or even a rooster crow. If conditions are good and the listener has selected a good frequency, he may hear three or four sign ons all at once. When conditions are poor no sign ons may be audible. Often the national anthem is heard, but the signal will be too weak to identify. Most regional stations use PSRA powers at or close to the 500-watt limit and this is usually sufficient to be heard at some distance.

There are a few clear channels with large numbers of daytime stations with PSRAs that are worth checking (900, 1050, 1220, and 1570 are usually the best to try). Most of these stations use PSRA powers much lower than the regional stations, and they are somewhat more difficult to hear. On some of the other clear channels, daytime stations may not sign on the air until sunrise occurs at the dominant station. On 810 kHz, for example, daytime stations must wait until sunrise (the nearest quarter hour) in Schenectady, New York (WGY) before they sign on. (Some stations have PSRA s on this frequency.) Other frequencies on which some stations must wait for the dominant station's sunrise include: 710 (WOR, New York); 1060 (KYW, Philadelphia); 1080 (WTIC, Hartford, Connecticut); 1130 (WNEW, New York); 1140 (WRVA, Richmond, Virginia); 1210 (WCAU, Philadelphia); and 1560 (WQXR, New York).

Many daytime stations on the clear channels do not have PSRAs and are unlikely to get big power grants under the new rules. These stations sign on at actual local sunrise. DX'ing this group requires a great deal of patience and is generally not as rewarding in terms of numbers of stations heard.

A tape recorder may be helpful in deciphering sign on announcements on the crowded regional channels, and is almost a necessity on the clear channel frequencies because the sign on announcements may be received with very weak signals under the dominant fulltime station. These clear channel daytime stations, especially the ones with no PSRAs, will not be logged by casual tuning. To log this group it is necessary to determine targets and diligently pursue them. One trick is to find a month in which two or three stations sign on at the same time on the same frequency. Since the angle of sunrise is always changing, there are different groups of stations within a sign on period each month. The more there are, the greater the odds that the listener will be able to catch at least one, and over time, receive all of the ''possible'' stations and a few that were initially judged ''impossible.''

The most productive time periods for the sunrise DX'er will be during the sign on times of the PSRA time zone stations one hour ahead of his own time, and in his own time zone. Non-PSRA sign ons in approximately the same area will also be the most productive.

DX'ers in the Eastern Time Zone will find their own 6 A.M. PSRA sign on period the most productive, while Central Time Zone listeners will find the ET and CT PSRA sign ons the best. (Obviously, a DX'er in eastern Wisconsin or western Kentucky, both Central Time Zone areas, will have better luck with ET PSRAs and non-PSRAs than listeners in North Dakota or southern Texas, also in the Central Time Zone.)

Mountain Time Zone listeners will obtain the best results by DX'ing CT and MT stations at sunrise, and DX'ers in the Pacific Time Zone will note most sunrise catches from MT and PT broadcasters. Some really good catches have been made on the west coast of stations signing on in the east and midwest, but this requires unusually good conditions and clear frequencies. (And dedication on the part of the western DX'er. Eastern Time Zone PSRAs sign on at 3 A.M., Pacific Time.)

Sunrise reception in the opposite direction is also possible. DX'ers in the east or central areas of the country may note an occasional sign on from the Mountain Time Zone, but this is almost as rare as western reception of eastern sign ons.

One of the main attractions of sunrise skip is that there are not usually regular stations which can be heard signing on day after day. Conditions on the same frequency change every day, and listening to a certain frequency with a lot of potential sign on possibilities will produce many different stations. An experienced DX'er will be able to tell moments after turning on the receiver whether conditions are good, normal, or below average to a certain area or overall. The experienced listener will also be able to select which channels will produce the best DX under a variety of different reception conditions. Even when sunrise conditions are not particularly good, sunrise skip may produce a new station for the logbook. Sunrise may also be good during the summer, providing reception of stations that are fairly close (within 500 miles) and from different areas since the angle of the sun at sunrise is different in the summer than it is in the winter.

An aurora can produce very good conditions at sunrise. A good aurora will bring in stations from Georgia, Florida, South Carolina, and other Eastern Time Zone areas in the southern states. Signals from these areas are often heard with outstanding strength in the midwest and northeast during an aurora. Central Time Zone areas such as Alabama, Mississippi, Louisiana, and Texas have also been heard with very good signals at sunrise by listeners in the northern states.

Hearing the Difficult States

One of the main goals of the advanced BCB DX'er is to hear as many stations as possible, but one of the first goals is to see how many states and provinces can be heard. The accompanying table shows the best bets for each of the 50 states and 10 provinces. Here are more detailed profiles of the ten toughest states, and the stations most often reported by DX'ers at some distance from these locations.

Alaska is the toughest state to hear east of the Rockies, not only because of sheer distance (Anchorage is 3400 miles from New York, 4000 from Miami and 2800 from Denver), but its northerly location requires unusually quiet geomagnetic conditions and very good propagation. Actually, a window may be developing in the span of time between the granting of power boosts to some of the larger Alaskan stations and the clear channels being filled up with stations in the "lower 48." The best bet seems to be KFQD-750 in Anchorage. This station has been heard several times in the midwest and plains states with WSB in Atlanta on the air, but the most likely time would be with WSB off. Additional targets include KYAK-650 and KTNX-1080 (both in Anchorage), KFAR-660 in Fairbanks and KJNP-1170 in North Pole.

One of the toughest states to hear from listeners west of the Mississippi is *Delaware*. It is difficult for many farther east also because the state has only 10 AM stations. The most likely station to be heard at a distance is WDEL-1150 from Wilmington, which has been heard at sunrise and on special tests. WDOV-1410 in Dover has been heard in the midwest in the evening, WSUX-1280 in Seaford has been heard in several areas at sunrise, and WKEN-1600 from Dover has an irregular frequency check which is occasionally reported.

Hawaii was possible for eastern listeners when WSM-650 in Nashville left the air every Monday morning, allowing a shot at KORL-650 in Honolulu. KORL is still the best bet for midwestern and eastern DX'ers, but it will require a great opening and a good loop angle to null WSM to hear it. On the west coast, most of the Hawaiian stations are relatively easy. Other possibilities for those east of the Rockies include KGU-760, KIKI-830, and KIFH-1040, all from Honolulu.

Difficult in the east, but definitely possible in *Idaho*. KGEM-1140 from Boise can be heard when WRVA Richmond, Virginia is off the air and conditions to the west are well above average. Another to try for is KFXD-580, Nampa, which has been reported by some midwestern and eastern DX'ers. What would seem to be the most logical candidate for distant reception, KBOI-670 in Boise, is very rare in the midwest and east, even with WMAQ in Chicago off the air.

Maine is the most difficult state to hear from New England for DX'ers outside that region. The best bet now is WYNZ-970 in Portland, during the early morning hours. WMER-1440, also from Portland, is another possibility. Several of the other 5000 watt regional channel stations have been heard well at sunrise.

The best-heard broadcaster from *Nevada* in the east and midwest is KMJJ-1140 in North Las Vegas, which can sometimes be heard testing with WRVA off the air. This one was widely heard during the early 1970s on a series of tests under the call letters KLUC. Longshots from Nevada include KDWN-720 in Las Vegas and KROW-780 in Reno.

New Hampshire has been noted by some DX'ers at sunset via WHEB-750, a limited-time station in Portsmouth. WHEB stays on the air until sunset in Atlanta, the location of this frequency's dominant station, WSB. Another one to try for is WFEA-1370 from Manchester, which is occasionally reported in the midwest during the night or at sunrise. This one even made it all the way to the west coast one morning.

Another difficult state for easterners is *Oregon*. The two most likely stations here are KPNW-1120 from Eugene (suffering interference from KMOX in St. Louis in the midwest and east) and KEX-1190 from Portland (covered by WOWO in Fort Wayne, Indiana and KLIF in Dallas). If KMOX and WOWO are off and conditions are good to the northwest, these two should be heard.

There are also two stations from *Rhode Island* that are often heard east of the Mississippi, WPRO-630 and WHJJ-920, both in Providence. These two stations represent the best bets for this state for DX'ers farther west also.

Vermont is another tough New England state, difficult to hear at any distance. The best bets are all at sunrise: WDEV-550 in Waterbury, WVMT-620 in Burlington and WHWB-1000 in Rutland. All have been heard in the midwest.

50 States on Medium Wave

Here are the most frequently reported stations from each of the 50 states:

Alabama	WERC-960	Birmingham
Alaska	KFQD-750	Anchorage
Arizona	KVVA-860	Phoenix
Arkansas	KAAY-1090	Little Rock
California	KFI-640	Los Angeles
Colorado	KOA-850	Denver
Connecticut	WTIC-1080	Hartford
Delaware	WDEL-1150	Wilmington
Florida	WQIK-1320	Jacksonville
Georgia	WSB-750	Atlanta
Hawaii	KORL-650	Honolulu
Idaho	KGEM-1140	Boise

Illinois	WLS-890	Chicago
Indiana	WOWO-1190	Fort Wayne
Iowa	WHO-1040	Des Moines
Kansas	KFH-1330	Wichita
Kentucky	WHAS-840	Louisville
Louisiana	WWL-870	New Orleans
Maine	WYNZ-970	Portland
Maryland	WBAL-1090	Baltimore
Massachusetts	WBZ-1030	Boston
Michigan	WJR-760	Detroit
Minnesota	WCCO-830	Minneapolis
Mississippi	WJDX-620	Jackson
Missouri	KMOX-1120	St. Louis
Montana	KGHL-790	Billings
Nebraska	KFAB-1110	Omaha
Nevada	KMJJ-1140	North Las Vegas
New Hampshire	WFEA-1370	Manchester
New Jersey	WVNJ-620	Newark (now WSKQ)
New Mexico	KOB-770	Albuquerque
New York	WABC-770	New York
North Carolina	WBT-1110	Charlotte
North Dakota	KFYR-550	Bismarck
Ohio	WLW-700	Cincinnati
Oregon	KPNW-1220	Eugene
Oklahoma	KOMA-1520	Oklahoma City
Pennsylvania	KDKA-1020	Pittsburgh
Rhode Island	WHJJ-920	Providence
South Carolina	WFBC-1330	Greenville
South Dakota	KSCO-1140	Sioux Falls
Tennessee	WSM-650	Nashville
Texas	WOAI-1200	San Antonio
Utah	KSL-1160	Salt Lake City
Vermont	WVMT-620	Burlington
Virginia	WRVA-1140	Richmond
Washington	KOMO-1000	Seattle
West Virginia	WWVA-1170	Wheeling
Wisconsin	WTMJ-620	Milwaukee
Wyoming	KTWO-1030	Casper

...and 10 provinces:

Alberta	CBX-740	Edmonton
British Columbia	CFUN-1410	Vancouver
Manitoba	CBW-990	Winnipeg
New Brunswick	CBA-1070	Monoton
Newfoundland	CTB-540	Grand Falls
Nova Scotia	CBI-1140	Sydney
Ontario	CBL-740	Toronto

Prince Edward Island	CHTN-1190	Charlottetown
Quebec	CBM-940	Montreal
Saskatchewan	CBK-540	Regina

Foreign BCB DX'ing

So far this chapter has primarily dealt with domestic (U.S. and Canada) BCB DX'ing, but it's possible to hear almost any country in the world on medium-wave! Foreign BCB DX'ing is divided into three broad areas, classified by signal path. The most common form of foreign BCB DX'ing for North American listeners is Latin American (LA). This category includes all other nondomestic stations in the western hemisphere, even those not in a Spanish or Portuguese-speaking country.

Another type of foreign BCB DX'ing is called Trans-Atlantic (TA) and this includes all European and African signals heard across the Atlantic Ocean. The final classification includes the Trans-Pacific (TP) stations that are located in Asia and the Pacific and are heard over the Pacific Ocean.

Location is very important in foreign DX'ing. Obviously, those in the east will have a better shot at Europeans and west coast listeners will have a great advantage in Pacific reception, but a listener's proximity to the coastline also plays a major role in reception. Listeners in such places as Cape Cod, Massachusetts, and Long Island, New York regularly log much more foreign DX of interest from Europe, Africa, and Latin America than someone in Albany, New York or Springfield, Massachusetts. A DX'er on the west coast (right on the coast) will have a great advantage in hearing stations from Asia and the Pacific over someone only 100 miles inland.

One of the main factors aiding foreign DX, no matter what the listener's location, is the difference in the way frequencies are arranged in the rest of the world. Stations in Europe, Africa, Asia, and Oceania operate on a broadcast band that extends from 531 to 1602 kHz, with stations operating every 9 kHz. This spacing allows many stations to be heard in North America on these so-called "split" frequencies. Between the North American "even" frequencies of 670, 680, 690, and 700 kHz there are foreign broadcasters operating on 675, 684, and 693 kHz. When two signals are received that are close in frequency (680 and 684 kHz, for example) a heterodyne will be produced. The tone from the heterodyne is often the first indication that a station is being received on a split frequency.

Although the Western Hemisphere is on a 10 kHz frequency spacing plan, there are a few splits here also. One of the most common is Radio Cayman, from the Cayman Islands, south of Cuba. This station, located at Gun Bluff in the British colony, can also be heard with good signals during the evening on 1555 kHz.

Another well-heard split is Radio Paradise, broadcasting from the Caribbean island of St. Kitts, with religious programming on 825 kHz. One of the best heard splits in the past, Radio Belize-834 from the Central American nation of

Belize, has moved to 830 kHz. Other countries that had a lot of splits, most notably Costa Rica and the Netherlands Antilles, have moved their stations to even frequencies.

The two countries that will probably provide the first foreign reception for beginning DX'ers are Cuba and Mexico. Cuban stations are well-heard on such frequencies as 590, 600, 640, 670, 710, 740, 860, and 1160 kHz from Panar del Rio, Urano Noris, Camaguey, Santa Clara, La Julia and Havana. Their programming (political programs and no commercials) is easily distinguished from that of all other Latin American stations.

Mexico is heard on a number of frequencies, primarily in the western two-thirds of the continent. The first Mexican stations heard may not be broadcasting in Spanish, but may be one of the large border stations with English programs for American listeners. The most widely heard of this group includes XERF-1570, Ciudad Acuna (across the river from Del Rio, Tex.), XEG-1050 Monterrey, and XETRA-690 Tijuana. Other well-heard Mexicans are XEX-730, XEQ-940, and XEB-1220, all in Mexico City, XEWA-540 from San Luis Potosi, XEROK-800 in Ciudad Juarez, and XEDM-1580 from Hermosillo.

Central American stations are somewhat more difficult to hear than some of the Mexican and Cuban stations, but are possible under the right conditions. Some stations to try for include La Voz de Guatemala-640 and Radio Mundial-700 from Guatemala; YSS/Radio El Salvador-655 in El Salvador; Radio-Sandino-750 in Nicaragua; Radio Reloj-730 from Costa Rica; and Radio Nacional-770 and Radio Reforma-860 in Panama.

Turning back to the Caribbean, even though there are no longer any split frequency broadcasters from the Netherlands Antilles, one of the most widely heard stations in the region is located in these islands. PJB-800 from the island of Bonaire broadcasts with 500,000 watts of power and is heard almost everywhere in North America with good signals. Religious programming from Trans World Radio is featured in English, Spanish and Portuguese.

Jamaica's Radio Jamaica is reported on 580, 720, and 770 kHz, and the Jamaican Broadcasting Corp. is heard on 700 and 750 kHz. The nation of Bahamas, just east of Florida, can be heard by tuning ZNS1-1540 from Nassau and ZNS3-810 in Freeport. Grenada's Radio Grenada-535 is often heard in North America.

From the island of Anguilla, the Caribbean Beacon, another religious station, operates on 690 and 1610 kHz and is often heard in the east and midwest. Another common radio country is Puerto Rico, with many medium-wave stations on the air. Two that are reported often by DX'ers are WKAQ-580 and WQBS-630, both in San Juan. Other countries that are audible from this area include Antigua (ZDK-1100 and Caribbean Radio Lighthouse-1165); Barbados (Caribbean Broadcasting Corp-900); Dominican Republic (HIJB-830, R. Clarin-860 and Radiolandia-1160); and Haiti (4VEH-1030, RadNationale-1080, R. Soleil-1170).

Moving to South America, there are many stations that are audible in North America from the countries of Colombia and Venezuela, but almost every other nation on this continent is very difficult to hear. Colombian and Venezuelan stations will dominate many clear and regional channels during good auroral conditions. Several high-powered Brazilian stations are logged in some east coast locations, and Ecuador and Peru are occasionally reported.

The split frequencies are much more important when DX'ing the European stations than they are with the Latins. Some of the European stations operate on even frequencies, but most of the widely heard stations are well between the Western hemisphere stations. A receiver with good selectivity is almost as important as a good location for TA DX'ing.

European stations that are heard in the central and western portions of the continent tend to be toward the top end of the band, are logged late in the evening, just before sunrise in Europe (around 1 A.M. ET), and are noted during the mid-winter (December–January) period. On the east coast, TA reception may be possible from early fall well into the spring, and may occur as early as local sunset.

The most common European stations heard in North America are Westdeutscher Rundfunk (WDR) on 1593 kHz from Langenburg, West Germany; Radio France (RF), Nice, France on 1557 kHz, and Trans World Radio, Monte Carlo, Monaco on 1467 kHz. Others to try for include Radiodifusao Portuguesa (RDP) in Lisbon, Portugal on 666 kHz; Radio Nacional from Madrid, Spain on 585 and Sevilla on 684 kHz; Radio France from Toulouse on 945, Bordeaux on 1206, and Lille on 1377 kHz, VOA relay, Munich on 1197 and Deutschlandfunk from Mainflingen on 1539 kHz from West Germany; Radiotelevisione Shqiptar, Lushjne, Albania on 1395 kHz and Osterreichischer Rundfunk (ORF) in Vienna, Austria on 1476 kHz. Listeners who have never heard a European medium-wave station should concentrate on hearing the Langenburg and Nice transmitters.

Receiving medium-wave signals from Africa is often easier than European reception, because the signal paths are farther south, away from auroral conditions which may eliminate the European signals. The more common Africans are Radiodiffusion Television Tunisienne-1566 from Sfax, Tunisia; Radiodiffusion Television Algerienne-891 from Algiers; Radiodiffusion-Television du Senegal-765 in Dakar, Senegal; and La Voix de la Revolution-1404, Conakry, Guinea.

Consistent reception of Asian signals is generally limited to areas west of the Rockies. Listeners with a good coastal location may be able to receive dozens of Australian and New Zealand broadcasters, along with more exotic stations from such Pacific islands as Fiji, New Caledonia, and Kiribati. Japan, Korea, China, and the Asiatic portion of the USSR are also heard on the west coast, but rarely make it farther east.

A listener in the midwest or plains states will probably not receive a fraction of the TAs someone on the east coast logs, or many of the Asian and Pacific stations heard on the west coast. The interior locations are the best for domestic DX'ing, however, and a person living on either coast will hear significantly fewer U.S. and Canadian stations.

Reception Reports, Verifications, and Tapes

Many BCB DX'ers collect verifications from medium-wave stations. Most stations reply by letter, but a number of broadcasters, including most of the larger clear channel stations, issue QSL cards in response to correct reports. Examples are shown in Fig. 14-2. Since this subject is covered in another chapter, it will not be dealt with at length here.

Fig. 14-2. An assortment of QSL cards sent in response to reception reports sent by DX'ers.

Most of the rules regarding reception reports are similar for both shortwave and medium-wave stations. A report to a BCB station should include the date and time, expressed in the station's local time, a brief description of reception conditions with an emphasis on any interfering stations, and verifiable program information. Commercials (called "spots" in the industry), public service announcements, promotional or contest announcements, and names of announcers or newscasters are all good verifiable information and should be included in the report. Song titles, news, or weather items are not. (Information on the music played can be included in reports to Canadian stations, since they must log music played in order to prove compliance with the Canadian content rule.) There is no absolute minimum amount of time a report should cover. A detailed description of a 1-minute local commercial would be preferable to a 45-minute listing of song titles.

One difference between shortwave and medium-wave reports involves such radio terms as DX, S-units, QSLs, SINPO codes, etc. Don't use them in reports to AM stations, because station personnel are unfamiliar with them. Do make an effort to be neat, polite, and always include return postage. Mail the report within a day or two of reception.

At one time, a DX'er would not count a station as being heard unless it was also verified. There are fewer stations that verify reports today, but quite a few are still glad to hear from DX'ers. A fair number of current BCB DX'ers do collect verifications from domestic stations, but relatively few now verify foreign broadcasters.

While on the subject of verifying or proving reception, some DX'ers enjoy taping their DX. Tapes can also be useful in some circumstances when sending a station a reception report, instead of listing program details. Send a cover letter, describing the reception, time, frequency, etc., and use the tape only for the actual programming heard.

Many DX'ers use tape recorders (both reel-to-reel and cassette units are popular) to record stations identifications, and some seek to tape "full identification" announcements which include the station call letters and location or frequency. A few DX'ers consider such a tape to be a "taped verification," which is something of a misnomer, since what is normally considered a verification must come from the station in question. Nonetheless, taping is certainly a worthwhile method to keep rare DX for posterity, and can be used in conjunction with written verifications in chronicling an individual's DX achievements.

In addition to collecting verifications and tapes, some medium-wave hobbyists enjoy amassing collections of a number of radio-related promotional items and related articles. Among the most common radio items collected are coverage maps, record survey charts, and stickers of all types. For most hobbyists this is a casual pursuit, but some have avidly collected thousands of coverage maps or survey charts by writing to and visiting radio stations all over the country and trading with other collectors. Related collectibles include matchbook covers (with the station's call letters or slogan printed on them, of course), balloons, pens and pencils, cups and mugs, T-shirts, hats, cigarette lighters, dial cards, and even station license plates for states that use only one plate.

There is even a group that is dedicated to this part of the hobby, the North American Radio Promotional Bank (NARPB). This organization (Box 540381, Houston, Texas 77254) publishes a twice-monthly bulletin that contains information on recent FCC actions, profiles of frequencies and radio in selected states and cities, and in-depth discussions of formats and programming in use by both AM and FM stations. Some of the members of this club are DX'ers, but the emphasis here is not on reception of distant stations. The bulletin is intended for those who are interested in the radio industry.

Recordkeeping and Logbooks

Most DX'ers maintain a logbook of what they have heard. Such a record of listening usually includes a station's call letters, frequency and location, time and date of reception, quality of signal, a brief description of what was heard, and perhaps the power the station was using at the time of reception. If the listener does send reception reports, a record of that and any verification received should also be made. The listener may also want to keep track of which stations have been taped.

This logbook can be maintained in chronological order, adding each new station as it is received, or lists may be made by state and province, or by frequency. A listing by frequency, or the appropriate notations in such reference books as the NRC Domestic Log, is particularly helpful in DX'ing.

Another method used by some DX'ers is to keep single or multiple index cards for each station received, arranging them in chronological, geographic, or frequency order. Many personal computers now on the market are ideal for tabulating this type of data. Additional records may also be kept, detailing farthest reception, number heard and verified from each state, and target lists of desired stations. Most serious DX'ers have some sort of target list of stations that they have not received. It is surprising how many difficult stations on these "most wanted" lists can be heard with persistence.

Future of Medium-Wave DX'ing

The rewards for most listeners are not as tangible as tapes, stickers, maps, or other items. There is a special thrill to tuning a new station for the first time, whether that station is KCHS-1400 in Truth or Consequences, New Mexico, with 250 watts of nighttime power, or PJB, the half-million watt powerhouse from the tiny Caribbean island of Bonaire.

One of the main appeals of medium-wave DX'ing is this diversity. There are so many different stations to hear from every conceivable location.

During DX'ing's "Golden Age," in the 1920s and 30s, no one was able to hear every station, but a few came very, very close to logging all U.S. broadcasters then on the air. It is unlikely that anyone DX'ing the AM band now or in the future will come as close, but that doesn't mean that DX'ers won't try, in their own way and within their own areas of interest.

BCB DX'ing will continue, despite the possible ominous changes on the horizon, but in a different state than it exists today and certainly different from the DX'ing done in the 1920s.

Even though some of the changes affecting the band may not be good from a DX'ers point of view, there is still plenty to hear, and plenty of ways to enjoy DX'ing those familiar AM frequencies.

FM DX'ING

Bruce F. Elving, Ph.D.

With the passing a few years ago of Dr. Edwin Armstrong, the inventor of FM broadcasting, the unofficial title of "Mr. FM" was set free. That title must certainly now be given to **Bruce Elving**. He was Number One in the FM DX field when I first met him many years ago, and despite what he says, I suspect he is still at or very near the top.

Bruce holds a Ph.D. from Syracuse University in the field of instructional communications, has managed college FM stations, and has worked in commercial radio. He is the publisher of the *FM Atlas and Station Directory*. His firm also offers a number of FM-related electronics items. (For more information, contact him in care of FM Atlas, Adolph, Minnesota 55701.)

Bruce and his wife have three daughters. He has been active in FM Dx'ing since his early teens and has followed the medium through all its growing pains to the strong position it is in today.

FM radio—a medium offering almost unlimited DX potential—developed slowly in the United States. Unlike AM radio, where DX'ing was the chief early means by which stations were received, FM had its start during a period when the public was paying attention to another new medium—television. The popular thought that FM was line-of-sight operation, and not good for over 50 miles, contributed to FM being regarded as strictly a local broadcast medium, thereby stunting what early FM DX'ing activity there could have been. This was at a time when stations were few and far between, and there was the chance to DX "the wide open spaces" of 88 to 108 MHz, which is just not possible on today's crowded FM band (or so it seems).

With the development of better receiving equipment, an understanding of how signals at the FM frequencies were transmitted began to take hold. Listeners installed antennas and snatched signals at distances hitherto thought impossible. Most antenna owners probably had no intention of becoming DX'ers, being content to hear a few predictable stations from a nearby larger city.

These folks, however, could not ignore DX. It intruded into their "woofers" and "tweeters," overtaking the signals of semi-local stations on occasion. Some people regarded it as just that—an intrusion. A few may have been curious, and flipped the dial around in an attempt to discover where the stations, which normally could not be tuned in, were coming from. Still others were DX'ers, or became DX'ers. How a person becomes a DX enthusiast could be the subject of a separate chapter, but suffice it to say that throughout much of FM's history, there have been DX listeners. I like to think that the number is growing, but it is still a very small number proportionately, when you consider that FM is the dominant aural medium, surpassing AM radio in terms of the numbers of listeners and the average audience (nationally).

The same qualities that helped make FM the preferred medium for listening also helped make it appealing for DX'ing. When conditions are right, you can hear DX in stereo, with a crisp static-free quality, and often over longer distances than possible on an AM radio. This is true, especially, when you consider that much of FM's DX activity takes place during daylight hours. You can make DX tapes that scintillate—tapes you can play for your friends to show how well FM sounded at a great distance, as received on your home stereo.

All of the technologies that apply to television are applicable to FM radio. Cable hookups enable you to listen to distant stations with ease, and satellite signals can be picked up from some radio "superstations" (all FM). Cable-originated "stations" exist on FM in some cities, and translators and boosters help spray the signals of hearby FM stations into cities, where the terrain makes direct reception difficult, with some cable systems sending the satellite-derived signals to their customers who have FM radio hookups. FM even has pictures! Yes, some stations, by means of subcarriers, are sending out slow-scan TV and data that can be printed out on computer screens.

The technique of squeezing from a tuner the nonbroadcast component of an FM signal, including talking and music subcarriers, is an exciting challenge to some DX'ers. And, often in college towns, FM boasts pirate stations and translators.

Cable DX'ing

If your FM radio is connected to a cable system, you will find that one of two methods is used to bring the signal to your set. Individual signal processing gives the cable system the greatest control over the signals. Each station is carefully nutured, and placed on a different frequency from where it originated. Usually, only a few of the many available area stations are so accommodated, and new stations find it hard to get included. A far more democratic method of giving the people cable-FM is "allband" transmission. Allband FM involves the use of an antenna and amplifier at the cable "head end" to pick up the signals and pass them on frequencies (where they naturally are) to the customers' radios. Allband FM suffers from a poorer handling of the individual station frequencies, with some fading, but it does [or should] include all of the area FM stations that are strong enough to reach the cable system's antenna. If your cable system has allband, you can probably DX right now, by carefully tuning your receiver and noting what comes in.

Your Own Antenna

Even with access to cable, you probably will find that you can get more stations, often with better quality, using your own antenna. The cable system may be depending on a directional antenna, giving you enhanced reception of stations in a certain favored direction. There could be local interference which would make the signal the cable system sends out not ideal, and losses in the cable could cut down on any DX potential from that source.

Your antenna need not be elaborate. Even a pair of TV-type rabbit ears might work wonders, especially if you can move them about, or face them into a window in the direction from which you might get DX. Eventually, you may want to invest in an outdoor antenna, preferably a directional yagi type having five or more elements and an electronic rotator. Perhaps you would prefer an antenna mounted on a long pole and level with the edge of your porch roof, so it can be rotated by hand, such as shown in Fig. 15-1.

Apartment dwellers may be able to install outdoor antennas, especially in the absence of a lease forbidding this practice. Many a landlord or building manger, in fact, will be willing to lend a hand. The benefits of outdoor FM antenna use do not have to accrue exclusively to home owners! Those persons who own homes or condos in areas having restrictive covenants against outdoor antennae could perform a service to DX'ers and others by getting such rules or laws overturned.

I discussed the antenna before the receiver, because most of us have FM radios, but not too many have outside antennas. Now, let's examine the receiver.

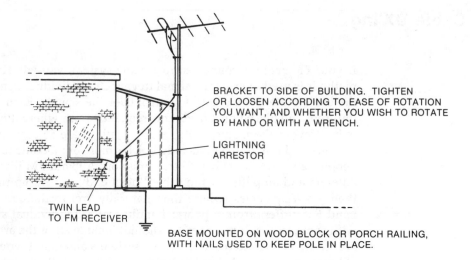

Fig. 15-1. A hand-rotatable patio or porch-mounted antenna.

BRACKET TO SIDE OF BUILDING. TIGHTEN OR LOOSEN ACCORDING TO EASE OF ROTATION YOU WANT, AND WHETHER YOU WISH TO ROTATE BY HAND OR WITH A WRENCH.

LIGHTNING ARRESTOR

TWIN LEAD TO FM RECEIVER

BASE MOUNTED ON WOOD BLOCK OR PORCH RAILING, WITH NAILS USED TO KEEP POLE IN PLACE.

FM Receivers

With several radios capable of tuning in the 88- to 108-MHz segment of the electromagnetic spectrum in the average home, you may want to take an inventory of your radios to see which radio or receiver might be the best candidate for bringing in FM DX. Usually, the choice will be a stereo receiver or an FM tuner, but, in some situations, it may be the FM section of a better-quality portable or table radio. Factors to look for, which may differ from that which makes a receiver work best for stereo music listening, are *sensitivity*, *capture ratio*, and *selectivity*.

Sensitivity

Sensitivity is the ability to tune in weak signals. Usually, for FM tuners or stereo receivers, this is already adequately high, especially for those units sold as separate components. If you have an otherwise good radio or receiver, but it does not seem to compare well to the one your friend has, you can buy an FM booster or RF amplifier to perk up the sensitivity. Be careful of mast-mounted boosters, as they can bring in too much signal, and an overload might cost you some real DX. Boosters in use near the receiver will typically give a lower gain, but they can be added or taken out of the system at will. Some are tunable and others are passive, amplifying the entire band without need for separate tuning of the booster. Tunable boosters, such as the "Magnum" from Canada, can act as traps, cutting down interference from strong stations, but much depends upon your expertise at operating the dials.

Capture Ratio

Another factor is *capture ratio*, or the ability of a tuner to null out interference from another station that is on the same frequency where you are listening to a slightly stronger station. Actually, poor capture ratio is an advantage in DX'ing, because it is good to be able to hear some interference to alert you to the fact that another station may be about to break in. It hasn't been invented yet, but a variable capture-ratio control would be nice to have on a receiver that is used for FM DX'ing.

Selectivity

Selectivity has become the most important quality of a receiver in recent years, what with the burgeoning number of stations virtually everywhere. Selectivity readings are almost always given for the alternate channel that is two frequencies away from where you are dialing, and not for the adjacent channel which is far more stringent. Look for selectivities in the range of 70 dB and higher. Ceramic or crystal filters help to ensure good selectivity, and some of the more expensive receivers have variable selectivity controls. If your radio or receiver has an automatic frequency control (AFC), always DX or tune with that control "off." This will help you reap whatever selectivity the radio has naturally. You can lock in the AFC when tuned to a strong station that you want to listen to over a period of time.

If you are in a metropolitan area, any decent receiver should enable you to tune in stations on the alternate frequencies to your local stations. A person in New York City or Chicago with locals on 98.7 and 99.5 MHz should be able to tune in stations on 99.1; anything less selective is not worth using as a DX instrument. The best receivers will enable you to tune in stations adjacent to your local stations, especially if you can point your antenna in a direction away from the local transmitters and aim for DX. Thus, a New Yorker living south of the Empire State Building, such as on Staten Island, should (with a very selective receiver and with the antenna pointed correctly) be able to get WUSL 98.9 Philadelphia, which is adjacent to WRKS 98.7 New York. If not all the time, he or she should at least be able to tune in Philadelphia when conditions are right, when there is an improvement in the general level of DX activity. If you have a receiver with this type of selectivity, you should be able to DX anywhere, even in the canyons of mid-Manhattan! Not all DX'ers are fortunate enough, however, to live on farms or in small towns where long-distance reception is for the taking. On a whole, the world's population live in big cities containing high levels of electromagnetic pollution.

An enterprising DX'er I know of lives near Washington, DC, which is an area with a heavy concentration of local stations. He DX'es mainly from a vacation cottage that is farther away from the local stations, or in his car after he drives to nearby tall hills or mountain passes. Another DX'er made arrangements with a friendly local farmer, who allows him to visit an empty house in the hills and set up his DX equipment from time to time.

Receiver Selection

If you're looking for a receiver to use for DX'ing, and maybe also a good stereo receiver for music listening, the choices are many and varied. Almost all FM tuners or stereo receivers are good for DX'ing, with some sporting slightly better performances of the previously listed criteria than others.

Be warned, however, that you can spend too much on fancy features—like oscilloscope displays—that do nothing to improve a set's DX abilities. Avoid super-priced tuners or receivers, unless you really know what you are looking for. A mid-priced or top-of-the-line unit from almost any reputable manufacturer should give you all the DX you can use.

Opinions differ on frequency-synthesized receivers, with their digital "presets" for different stations, but no conventional tuning dial. The ability to set up station frequencies is a real boon. You can devote some buttons to your favorite local stations, as well as some buttons to those frequencies that are devoid of local stations but which have good DX potential. Or, during an opening, when DX abounds, you can variously set and reset the buttons to hear as many identifications on the top of the hour as possible and, thus, increase your DX prospects. By tuning electronically, your receiver is always in perfect tune, and there is no dial cord to break.

A drawback of the newer receivers is that the fixed tuning makes it difficult to tune slightly away from a strong local station. However, many can be tuned in increments of 100 kHz, which is half the normal spacing between the U. S. and Canadian channels, or even as low as 50 kHz, sometimes with the help of a behind-the-set switch. I find the digital/light display used for tuning to be less accurate than the tuning meters found on older receivers. This is especially true when rotating an antenna. It is hard to aim an antenna with pinpoint accuracy using a three-light indication, as found on one receiver I tried out at my house, compared to the wide-swinging meter employed by my 1969-vintage receiver. Most of the newer digital-tuned receivers all seemed to perform about the same (having less average selectivity than my old receiver) with the lower-priced units tending to suffer more from strong station overload. But these matters are strictly personal. The vast majority of the receivers on the market will perform admirably for anyone interested in trying FM DX, as well as using the FM for a stereo music source.

It is not my purpose to endorse brands of receivers or antenna. Equip yourself with a receiver you like and try to get an FM antenna (an FM-only antenna, if possible, not a combination TV-FM antenna) which is directional and can be rotated, especially in directions *away* from your strong local stations.

Then, develop a familiarity with all your local and nearby stations. That way, you'll be ready to recognize when DX, or distance, reception is possible. Many FM DX'ers write to tell me that the *FM Atlas and Station Directory* (Fig. 15-2) is indispensible in helping them find the DX that they want. It can be an aid in helping you sort out your nearby stations, and it can be flipped open to the pages showing the areas from where you may be getting DX. I started this publication in 1971, while living in Milwaukee, where the headquarters of the Worldwide TV-FM DX Association was then located. It's been a labor of love since then, and the book has grown considerably in the years since. For information on the current edition, write to "FM Atlas," Box 24, Adolph MN 55701.

Fig. 15-2. The
*FM Atlas and
Station
Directory*,
**written and
compiled by the
author.**

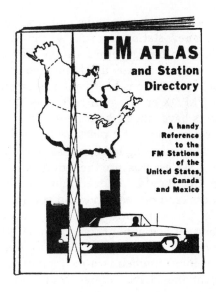

Recognizing FM DX

Once you know the entire FM dial for the area where you live, you can be on the alert for times when conditions are above normal. Often, in the morning—two or three hours after sunrise—you can note tropospheric DX. That's when your semi-local stations, and others that are farther from you, come in with increased signal strength. It happens almost every morning of the year in most of the country, except possibly in the arid mountain regions. Tropo can really be extended at times, and you might hear stations hundreds of kilometers distant. The great growth in stations has meant more targets for DX, but it has also cut down on the opportunities to experience really super tropo, for you might find closer-in stations blocking weaker DX that is trying to come through. Yet, there are times when farther-away stations will still come through; ducts will form and they will help stations override the signals of closer-in stations. While living in central Nebraska, I tuned in Eau Claire, Wisconsin, stations on two Omaha frequencies, with Omaha being considerably closer. By using a directional antenna, you can often null out a closer station to bring in the more distant station at a slightly different angle. Experience will show you how.

You've got to be willing to sample the FM dial, especially those frequencies not occupied by strong local stations, in order to determine when conditions are favorable for DX. Even with the most expensive tuner and the best antenna, unless you search diligently, you will not get the results that a person with modest equipment gets who searches the FM dial daily, or several times a day, in the quest for DX.

FM DX'ing really is a form of collecting. The inveterate DX'er is interested in adding to a list of totals. When I first started DX listening, and not having contact with those in radio clubs, a friend of mine and I called it "grasping." To hear a new station, we said we "grasped" it. I've kept a running log of stations grasped in one general location through the years, and it is interesting to see how the log has grown. At one time, I was the leading FM DX'er, in terms of the number of stations heard at one location, in the country, and I believe I was the first FM DX'er to log (in 1970) the 1000th station in the same geographical area. Others have since taken the lead and, while I enjoy the hobby immensely, it is almost as much of a pleasure to write about the hobby as it is to DX itself.

The collecting instinct asserts itself when it comes to QSLs or verifications. Many DX'ers treasure the letters or cards that come back from stations "verifying" that they did, indeed, tune in a broadcast on such-and-such occasion. The DX'ers send out accurate and courteous letters called "reception reports" (Fig. 15-3). These reports detail the programming heard, listing the commercials, public service announcements, and other details, plus some brief mentions of the receiver and antenna that was used to pick up the station reported. I find it lends interest when writing a report to ask the station if my report was the most distant they have yet received from any listener. Surprisingly, many of the stations, even those quite close to me geographically, will write back and mention that I was the most distant listener who had contacted them.

You don't have to collect "veries" from stations to prove to some skeptic that their signal did get to you on a certain occasion. Yet, the contact with stations is good, and it serves to indicate to the station employees that their signals do get out, possibly more than they, themselves, realize. I know a DX'er who likes to call stations long distance, especially when they have contests, and he has won tee-shirts and other memorabilia from FM stations who flipped out at being received by a listener so far away!

Tropospheric DX

Tropo DX'ing, as alluded to earlier, represents a lengthening of a station's normal service area. It can extend half a continent or more in a favored direction under ducting conditions, and, often during continental fog, can cover much of a section of the continent. A good antenna, a selective receiver, and the utmost in DX skill combine to make tropo DX possible. Seasoned DX'ers often speak of this as their favorite form of FM DX, although skip may be more spectacular.

Tropo has the best chance of happening in coastal areas where temperature inversions are common—in the northern and eastern parts of the United States and the adjoining areas of Canada during the fall, and in the central plains of the U.S. in the spring. It can also happen at other times, such as during the spring and summer, with rare winter tropo a particular treat. The Rocky Mountain area has little tropo, and nobody I know of ever gets trans-Rocky Mountain DX in Colorado. Good tropo conditions can last from just a few hours at a time to several days on end.

Fig. 15-3. A
typical FM
verification
letter.

July 23, 1973

PHONE 362-4433   YORK, NEBRASKA 68467

AM 1370
FM 104.9

Mr. Bruce F. Elving
Post Office Box 24
Adolph, Minnesota 55701

Dear Mr. Elving:

Thank you for your reception report of July 7th noting the program
material contained in periods from 6:58 until 9:07 AM.

I have been on a slight vacation and thus the delay in answering
and verifying your report.

Our transmitter site is located a mile South of York, Nebraska on
land that is just slightly higher than average terrain (about four
or five feet). We use a three kilowatt transmitter running at
2.4 kilowatts. This feeds a three bay circular polarized antenna
system for an E.R.P. of 2.75 kilowatts. All of this has been
assumed from paper and has not been verified by extensive field
strength measurements. Our center of radiation is 310 feet above
average terrain. Our class of license calls for a maximum of
3 kw at 300 feet. The reduction in power to obtain an equal field
has never bothered us much. I'd rather have the height any day.

For the first couple years of operation, we were having external
frequency measurements made on this station from Kansas City and
they never once missed receiving a signal. I have, however, no
idea of what they use in the way of an antenna and receiver.

Aside from that, and a fellow from Omaha who picked us up on rabbit
ears one morning, yours is the only distant report we've received.
Needless to say, it is the _most_ distant.

Once again, thanks for your report, and I hope this letter fulfills
your needs for information and verification.

Best regards,

David L. Johnson
KAWL AM-FM
York, Nebraska.

Nebraska Broadcasters
Association

Skip

With the band filling with stations of all descriptions, spewing variations of rock
and country music on the commercial part of the dial (92–108 MHz), and with
classical and student rock music, public affairs, and religion on the educational

part of the dial (88-92 MHz), opportunities for DX are often confined to getting skip.

Skip can be received in any part of the country, but it is most common during the daylight hours in late spring and early summer. Its distance is usually greater than that of tropo, with stations "skipping" over other stations closer to the listener, whereas tropo often brings in the closer and more distant stations with equally enhanced signal strengths. Therefore, you can often DX skip on a channel where you find a semi-local station, with very little interference. Skip lasts typically an hour or two at a time, but intense openings around the time of its annual peak—June 21—can last all day.

Typical skip-distance examples are when California people tune in Texas or Nebraska, or even British Columbia FM. Kansans may tune in Maryland, Mexico, or Montana. Upstate New Yorkers may get Florida, Alabama, or South Dakota. Skip is two-way, so the distant state you are hearing will be able to tune in your local stations—or, at least, the FM DX'ers in that state will if they have the right equipment.

Skip happens most often during hours of local daylight. I have records of receiving California skip in the Midwest up to 11 P.M. local time, or close to California's summer sundown. Nighttime skip is a rarity, but it can happen. Another time to look for skip is during the annual secondary peak in December–January, centered on Dec. 21. A DX'er friend told me that he never noticed a year without some December skip, but later he admitted that he was counting both TV and FM. Skip is related to frequencies or channels, the lower the better! It affects TV Channel 2 and, as it develops, can work its way higher, until it gets past Channel 6 and into the FM band. Sometimes the MUF (maximum usable frequency) never rises above the educational band, and your skip experience for an opening like that may be confined to educational stations up to 92 MHz. Other openings could go to the top of the band, to 107.9 MHz and beyond, sometimes hitting TV Channels 7 and 8.

I know some FM DX'ers who use a TV set on Channel 2 or 3 to take note of when skip activity is building. Getting skip on FM or TV often does not take a super antenna or receiver—rabbit ears may do—but you will get more and better skip if you have at least a modest outdoor antenna, which is rotatable and, if possible, oriented in the direction from which the skip stations are coming.

Short skip is a particular treat. Some memorable examples are the tuning in of Rapid City, South Dakota from the Duluth, Minnesota, area, and Oklahoma City from Minneapolis. A good DX'er, favorably located, can use tropo and skip, both short and long, to get reception from most of the states in the country, as well as adjacent countries, like Canada, Mexico, Cuba, and the Bahamas. DX'ers in the southern United States have been known to tune in Guatemala and the English-speaking nation of Belize via FM skip, as well as stations far in the interior of Mexico, with their distinctive slogans, musics, and echoey production effects.

FM DX does have its limitations, however. At the beginning of this chapter, I mentioned that it is a band of almost unlimited DX potential. There is such a thing as double-hop skip, caused by ionizations of the E-layer in the ionosphere that are so aligned that one can tune a signal bouncing from one cloud to another at double the normal skip reception distance—such as being in North Carolina and tuning in the state of Washington. Such happenings are rare, however, because after the first hop, the signals are often weak and interfered with by closer stations. There is also tropo-skip, which is the skip signal picking up a weaker tropo signal and extending the range of reception. This is the method by which I believe I received stations from southern Florida in northern Minnesota.

However, even with double-hop and skip-tropo propagations, FM DX is limited. If you live in North America, you will probably never tune in a program from Europe or South America on FM, unless your local station is playing a transcription from a country located on one of those continents! But, within the confines of the time you have for this hobby, and all of the stations now on the air and about to come on the air, you can expect plenty of DX activity on FM for years to come. You will never DX all of the stations there are, or fully exploit all that the best openings can give you. Always, the prized station you want will elude you, lurking there, for the next occasion when conditions may bring it into your DX lair.

Other Forms of FM DX

Skip and tropo are the most common forms of FM DX, but, for the sake of completeness, let me mention that you can also get distant stations in other ways. One of the most common is by *meteor-scatter propagation*. This is the momentary enhancement of signals as a meteor hits the atmosphere and sends out a soon-dissolving ion trail. Distances for meteor-scatter DX vary from tropo through moderate skip. Thus, in Minnesota, I have picked up by meteor-scatter stations in Richmond, VA and Detroit, MI. Since reception may be only for a fraction of a second with up to 30 seconds for a long burst or series of bursts, you have to be tuned in just right, when stations are identifying. Meteor-scatter, or MS, reception is usually quite weak, so it takes good equipment and a quiet frequency, free of nearby stations, to monitor it. This kind of DX happens most often between midnight and local noon, with some DX'ers concentrating on MS reception during the hours before 6 A.M. local time when fewer stations are on the air, and during periods when there are known meteor showers. A few have tried voice-activated tape recorders, listening to the tapes to hear if any station identifications were recorded. *Auroral DX* is another of the minor forms of long-distance reception at FM frequencies. This is largely confined to the "great white north," the northern United States and Canada. Here, you might find DX best if you aim your antenna to the north; you will note stations to the east, west, and south of you coming in with a fizzling or chugging quality. If you listen carefully, you may be able to understand what is said, and will be able to pick out call letters or other program elements sufficient for you to elicit a verification from the station.

When going after auroral, or *northern lights*, DX, it helps to switch off the stereo circuitry to maximize intelligibility, as well as to set the tone controls carefully. Not all DX on FM is static-free and in true stereo, but enough of it is so that you will often thrill at how clear the sound is when hearing stations that have never been received in your city by any other human being before.

I have heard some DX by means of *airplane scatter*, with signals deflecting off the fuselage of planes. This can be a way for a person living in a deep valley to build station totals. There is also said to be *lightning scatter*, with signals reflecting off bolts of lightning, but I prefer disconnecting my antenna and turning off the radio rather than fighting nature and risking a direct hit to my equipment! Some people in mountainous areas experience reception of stations across the mountain ranges by a refraction caused by a sharp mountain peak or range set midway between the receiver and the station. Probably more common in mountain areas, especially in those areas not covered by vegetation, is the reflection of FM signals from one mountainside to another, as happens when driving in the Rockies west of Denver.

FM Translators and Boosters

These are getting to be more numerous but are not well-known, although the *FM Atlas* has listed them ever since they first became legal in the United States in 1971. FM translators operate on frequencies different from the originating station, while boosters operate on-channel. Those units west of the Mississippi can have 10 watts, while those east of the river are allowed only 1 watt, although the FCC has been petitioned to increase that to 10 watts nationwide. Translators can offer limited commercial advertising, with 30-second announcements to help defray expenses, and there are moves afoot to allow more local program origination and advertising—in effect, making FM translators "low-power radio." Expect to see much more translator activity, with translators becoming a DX target on tropo and skip, as well as having the ability to fill up frequencies and make your own DX'ing more difficult.

FM Subcarriers

Hundreds of FM stations in the United States, Canada, and Mexico are sending out special programs of music, speech, and computer data to special groups and places, but not to the general public. This is done by means of a subcarrier—in effect, using one transmitter and station to send out more than one program at a time. In the United States, this is called *SCA* (Subsidiary Communications Authorization), which is a special permit issued by the Federal Communications Commission. Canada has a similar service called "SCMO." SCA has generally meant furnishing background music, like Muzak, to the stores, with the general public not having access to the receivers. Over the last several years, the uses of SCA have broadened to take in such things as talking books for the blind, ethnic and foreign language programming, religion, network relay, voice paging, and data to interface with computers.

Many FM DX'ers find SCA an invaluable part of the FM spectrum and delight in the characteristics of listening on the subcarriers, along with main-channel DX. SCA is sent out with only about 10% of the modulation of the main carrier. This means that if an FM station has 100,000 watts, its SCA component will only be 10,000 watts, effectively. Yet, SCA, if it is properly maintained at the station and installed well in the receiver, is surprisingly uncritical and clear in its reception. You should be able to hear SCA on your home receiver or your car radio for as far a distance as you can get stereo from the same station.

An SCA adapter is usually purchased in kit form and assembled by a DX'er or technician having some knowledge of electronics. It is installed in the detector or discriminator section of the receiver, ahead of the de-emphasis capacitor and resistor. This is before the stereo circuitry takes effect in a stereo set. The audio channel is cut at the radio's volume control, and wires are run to a switch to select normal FM or SCA–FM. In a stereo receiver, it is easiest usually to make a switchless installation, connecting the output of the SCA device to the tape monitor jack, and pushing in the "tape monitor" button to listen to SCA. A source of 10- to 12-volts DC must be found in the receiver to power the adapter, or an external power source must be used.

As DX'ers and electronics hobbyists, you can legally listen to SCA, and even report its contents to the stations you DX. You cannot, however, play it in a business that you may own and, thus, deprive the background-music companies of revenue. The *FM Atlas and Station Directory* lists all stations known to have an SCA subcarrier, and also lists their uses. As more and more DX'ers get SCA adapters, I am anxious to hear from them about the status of the stations in their areas, and what use, specifically, is made of the subcarriers.

Even doing the SCA installations can be fun, with practice, and something that the average person, good with a soldering iron and tools, can do. I get pleasure when confronted with a variety of radios—clock, portable, table, walkperson-type, TV/audio, small-screen TV with AM/FM, stereo receivers, and car radios. The process of installing an SCA can be summarized by this poem I wrote several years ago when I first became interested in FM subcarriers.

> Ah, the joy of probing into a radio
> And discovering a whole new program!
> There all the time, only awaiting
> The magical touch to unlock the
> Elusive, expansive world of SCA!

Automotive DX

You can DX in the car, too. Automobile FM radios are getting better, and certainly more numerous. When taking a trip, you can take note of the many stations that are receivable almost anywhere, and enjoy their programs. Tuning around in the quest for DX can help take the boredom out of a long drive. You won't be able to add stations you hear on a trip to the totals for your home area, of course, but it is still fun to take note of DX when you're on the go!

FM—The DX Way to Go

I find all of the components of an FM signal exciting and relevant to DX: the red or green light that tells you a far-away station is broadcasting in stereo and being received with a strong enough signal to be in true stereo, the defeatable AFC or variable selectivity, the mute defeat switch that enables you to get all the weaker stations, the station presets that help you identify DX, the tuning meters or signal-indication lights that help with the antenna bearing or as a guide to the signal strength of the DX, and, of course, FM–SCA.

I hope you will consider using one of your FM radios or receivers for DX, or will consider getting some new equipment and following the principles presented here. For anyone interested in learning more about the hobby of long-distance FM listening, I suggest contacting the Worldwide TV-FM DX Association. This nonprofit club publishes a monthly bulletin, the *VHF–UHF Digest*, giving reports of people who have heard DX, columns listing the totals of stations heard by DX'ers, news of FM and TV station changes (including new stations and those having changes in program or network affiliations), and technical articles to help you sharpen your DX skills or improve your equipment. Send a self-addressed, stamped envelope to the WTFDA, P. O. Box 514, Buffalo, NY 14205, for information about their publications and membership.

DX'ing is a good leisure-time pursuit, and FM DX'ing is something you will most likely enjoy if you give it a fair trial.

16

EXPLORING THE VHF/UHF ACTION BANDS

Mike Nikolich

Mike Nikolich has been a radio communications hobbyist since 1969. He started in the shortwave-broadcast phase of the hobby and later added an interest in scanning. The two interests have proved quite compatible. Scanning has not interfered with Mike's building a log of over 200 SWBC countries.

As the Executive Editor for *Scan Magazine*, Mike is an expert in the scanning field—an area which, as he notes, has millions of followers throughout the country, making it at least as popular as shortwave, if not more so.

Mike has a considerable lineup of shortwave and scanning equipment in his Chicago shack. He nearly appeared on the ABC "Nightline" program during the time that Owen Garriott was transmitting from the Space Shuttle Columbia, but a late-breaking story bumped the scheduled feature on the space mission.

Mike is a member of several clubs and helped found the Chicago Area DX Club.

Public Service Band monitoring has come a long way from the days when you could listen to the police by simply tuning an AM radio above 1600 kHz.

Indeed, with today's sophisticated scanner radios, VHF and UHF listening has taken on new dimensions. Besides providing a fascinating behind-the-scenes look at public safety agencies in action, the VHF and UHF bands offer a whole world of listening adventures..., with DX possibilities that rival anything on the HF bands.

DX'ing with a scanner? You bet! In late 1983, millions of listeners throughout the world were glued to their scanners as Dr. Owen Garriott (W5LFL) aboard the Space Shuttle Columbia became the first astronaut to establish direct two-way radio contact with Amateur Radio operators on earth. All it took to hear him was a scanner radio capable of monitoring 145.550 MHz..., and a little luck. If you did manage to receive a broadcast from Garriott's tiny 4.5-watt transceiver, your reward was a beautiful commemorative QSL card from the American Radio Relay League.

Although the pastime of Public Service Band monitoring has been enjoyed since the 1930s, it really didn't take off in popularity until the mid-1960s, when inventor Al Lovell conceived the first radio capable of automatically "scanning" frequencies to find active communications channels. Lovell, founder of Electra Company, named his radio "Bearcat," after his love of the classic Stutz-Bearcat automobile. The first Bearcat scanner rolled off Electra's production line in 1968, and a multimillion dollar industry was born. Ten years after its introduction, the scanner has achieved the same degree of popularity that color TV has achieved during the same time frame. Today, scanners are in over 6 million U.S. homes.

Monitoring the Public Service Bands requires different listening techniques than those used by SWL'ers. In most instances, long-distance reception is the exception, not the rule. Most reception will be limited to "line-of-sight" transmissions—anywhere from a few miles, to perhaps 100 miles away. The frequency ranges were selected by Public Service agencies because they are ideally suited for reliable short-range communications. The last thing a policeman at the scene of a crime is concerned with is whether a DX'er thousands of miles away can hear him!

However, under the right conditions, VHF/UHF transmissions can travel hundreds, and even thousands, of miles. Listeners in California have monitored the transmissions of the New York transit system, and the audio from European television stations is occasionally audible when the 11-year sunspot cycle is at its peak.

Most scanners are capable of monitoring the following frequency ranges: 29–30 MHz (10-meter Amateur), 30–50 MHz (VHF Low), 50–54 MHz (6-meter Amateur), 118–136 MHZ (VHF/AM-Aircraft), 136–144 MHz (Military Land Mobile), 144–148 MHz (2-meter Amateur), 148–174 MHz (VHF High), 406–420 MHz (Federal government Land Mobile), 421–450 MHz (70-Centimeter Amateur), 450–470 MHz (UHF), and 470–512 MHz (UHF ("T").

One of the newest frequency ranges is the "800-MHz band." Covering 806 to 912 MHz, it is beginning to come into wide use with police and fire departments throughout the country. Cellular telephone conversations can also be heard in this range. Unfortunately, there has been no nationwide coordination of frequency blocks, so what might be allocated as a fire frequency in Chicago might be used for cellular telephones in Salt Lake City.

In addition to the 800-MHz band, many scanner enthusiasts monitor frequencies between 216 and 400 MHz. This range includes everything from inland water communications (216–220 MHz) to amateur radio (220–225 MHz), and military aircraft AM transmissions (225–400 MHz). Although most commercially available scanners do not cover the 216 to 400 MHz range, an add-on frequency converter will allow you to monitor these ranges.

Action Is Awaiting You...

What attracts people to monitoring the Public Service band? Action, excitement, and adventure. The VHF/UHF bands are where you should listen for fast-breaking news that is available nowhere else. With a scanner, you can be at the scene of a fire at the same time, or even before, fire crews are battling the blaze.

Such was the case on November 28, 1981, in Lynn, Massachusetts. While most of the city was sleeping, scanner listeners were glued to 154.415 MHz, the Lynn fire channel, as firefighters battled the worst fire in Massachusetts history. For 16 hours, fire fighters fought a blaze that went on to destroy four city blocks in Lynn, and force the evacuation of 750 people. Damage was estimated at more than $35 million, and as many as 500 fire-fighters from as many as 45 neighboring communities were called in for assistance.

Of course, not everything you hear during a typical monitoring session will be as dramatic as a 10-alarm fire. But, on the Public Service bands, you never know. They aren't nicknamed the "Action Bands" for nothing!

Getting Started

For about $100, you can get in on the action with a simple crystal scanner. Radio Shack, Electra, Regency Electronics, Fanon-Courier, and other companies offer crystal scanners (Figs. 16-1 and 16-2). While the radios offer great value for the money, they also have a serious limitation. To change frequencies, you must manually change crystals. If you live in a small town where five to eight frequencies are in use, this inconvenience might not be a problem. However, in a major metro area, where thousands of frequencies can be in use, you'll either need a bank of crystal scanners to keep up with the action, or a synthesized programmable scanner.

Programmable scanners from Electra, Regency, Radio Shack, J.I.L., and Fanon-Courier range in price from about $200 to as much as $800. Because they do not use crystals, they provide you with the flexibility of programming thousands of different frequencies. Programmable scanners (Fig. 16-3) also offer considerably more storage capacity than their crystal-controlled counterparts. A top-of-the-line scanner can include 50 channels or more. In fact, a deluxe computer-controlled scanner from Electra, the Bearcat CP-2100 (Fig. 16-4), scans up to 200 channels.

Fig. 16-1. The Regency HX 750 hand-held crystal scanner (Courtesy Regency Electronics Inc.).

But even more significant than a programmable scanner's programmability is its ability to help you find new frequencies. With an *automatic search* feature, the scanner can search for active frequencies between limits you specify. This gives you the flexibility of using your own listening capabilities, instead of relying solely on directories, for up-to-date frequency information. Even hand-held portable models (Fig. 16-5) are available.

Scanners are found in many applications. Fig. 16-6 shows scanners in use in the ABC television booth at the Indianapolis 500 race.

Even if you don't know the frequency of your local police department, with the automatic search feature and a little detective work, you can probably find it. The first clue to solving the "phantom frequency caper" comes from your local squad cars. If the whip antenna on the patrol car is about 17 inches long, your "men in blue" are probably using a frequency somewhere around 154

Fig. 16-2. The Bearcat Five-Six hand-held crystal scanner which includes the VHF/AM aircraft band (Courtesy Uniden Corporation of America).

MHz. A longer whip means the frequency is lower; try searching in the 30- to 54-MHz range.

To help narrow down your search through the frequency ranges, Chart 16-1 includes the FCC frequency allocations for a variety of Public Service agencies. Select a range and start searching. Sooner or later, you'll find the frequency you're looking for.

What to Look for in a Scanner

The only thing to match the exploding popularity of scanners is the increasing sophistication of the radios themselves. Compared to the original crystal-controlled Bearcat scanner of 1968, today's scanners have reached twenty-first century proportions.

Chart 16-1 Frequency Ranges of Public Service Agencies* (in MHz)

POLICE	RAILROAD
37.040–37.400	160.215–161.565
39.020–39.960	452.900
42.020–42.940	472.4675–472.7875
44.620–45.060	475.4625–475.7875
45.100–45.660	**HIGHWAY MAINTENANCE**
45.700–46.020	37.920–37.960
154.665–155.130	45.680–45.840
155.190–155.370	47.020–47.400
155.415–155.790	150.995–151.130
156.210–156.240	156.105–156.240
158.730–158.790	159.105–159.195
158.850–159.210	**PRESS**
453.100–453.950	152.870–152.990
460.025–460.500	161.640–161.760
FIRE	166.250
33.440–33.980	170.150
37.100	173.225–173.375
42.280	450.050–450.950
45.880–46.500	452.975–453.000
154.130–154.445	455.050–455.950
453.050–453.750	**MOBILE TELEPHONE**
460.525–460.625	152.030–152.810
AIRCRAFT	158.490–158.700
118.000–135.975	454.025–454.975
UTILITIES	459.700–459.975
30.700–33.380	**GOVERNMENT**
37.460–37.860	37.100–37.260
47.700–48.540	39.100–39.980
150.980	45.080–45.640
153.410–153.725	46.520–46.580
158.130–158.460	46.600–47.000
451.025–451.250	154.025–154.115
451.275–451.750	154.965–155.145
471.3125–471.4125	155.715–156.240
474.3125–474.4125	162.025–162.175

MOBILE RADIO SERVICE (formerly Class "A" CB)
462.550–462.725
467.550–467.550

* Source: Scanner Association of North America F.C.C. Allocation Chart.

Electra recently announced a computer-controlled scanner that actually is an add-on peripheral for a personal computer. It comes with custom software, interfacing cable and BNC connectors, AC adapter, and special whip antenna (with cable). It's called the Bearcat CP-2100 (Fig. 16-7). Not only can it scan up to 200 frequencies, but each channel can be programmed with detailed informa-

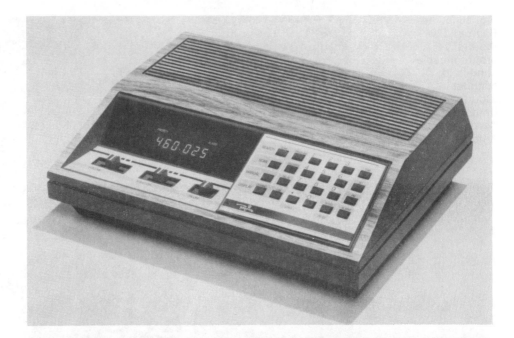

Fig. 16-3. The Regency Z-30 programmable scanner (Courtesy Regency Electronics Inc.).

Fig. 16-4. The Bearcat CompuScan™ 2100 computer-controlled scanner radio (Courtesy Uniden Corporation of America).

Fig. 16-5. The Bearcat 100— the first programmable hand-held scanner (Courtesy Uniden Corporation of America).

tion about a service, including frequency, station name, address, 10-codes, power of transmitter, and other information that you may need.

Regency Electronics, another leader in the scanner industry, now offers the MX-7000 (Fig. 16-8), a 20-channel scanner that offers continuous coverage from 27 MHz to 1.25 GHz. Not to be outmatched, J.I.L., Ltd. has entered the "ultimate scanner" sweepstakes with the SX-400 series. Through the use of optical converters, the SX-400 will continuously cover 150 kHz to 3.7 GHz. In addition, the SX-400 includes a data interface connector so it can be controlled by a computer for increased channel capacity and faster scanning.

While the Bearcat CP-2100, Regency MX-7000, and J.I.L. SX-400 represent the cutting edge in scanner technology, you might not need a scanner any more sophisticated than Al Lovell's original Bearcat scanner. Once you decide between a crystal or programmable scanner, your two biggest decisions in choosing a scanner will probably be the channel capacity and the frequency coverage

Fig. 16-6. Scanners in use in the ABC-TV broadcast booth at the Indianapolis 500 race.

Fig. 16-7. The Bearcat CompuScan™ 2100 computer-controlled scanner radio, with its software, cables, and antenna (Courtesy Uniden Corporation of America).

you need. Almost every programmable scanner covers at least VHF Low and High, UHF, and one or two of the Amateur Radio bands. You'll pay more for additional coverage. Storage capacity ranges from 10 channels to as many as 200.

When comparing various models of scanners, use the same criteria you would use to compare shortwave receivers. Sensitivity and selectivity are important. Good sensitivity will require a signal-to-noise (S/N) or signal-to-noise and distortion (SINAD) ratio of 0.5 microvolts or less. Selectivity should be at least -50 dB at 20 kHz. Once you've compared specs, compare features. Some of the most important scanner features that SWLs drool over are a priority channel, scan delay, selectable scan rate, and automatic lockout.

The *priority channel*, usually designated "Channel One," can be reserved for an important emergency frequency. While the radio is scanning for active frequencies, it automatically samples Channel One every two seconds to prevent missing important calls.

Scan delay is a useful feature for monitoring simplex communications systems, where "calls" and "answers" are on the same frequency. When activated, it adds a two-second delay on desired channels, so important conversations are never missed.

Selectable scan rate controls the number of channels per second that a scanner will sample in the scan mode. Ideally, the faster the rate, the fewer calls will be missed. Typically, scanners with this feature are capable of scanning 5 or 15 channels per second. Another way to increase the scanning cycle is to *automatically lockout* channels not of current interest. While the 24-hour National Weather Service deserves a channel in your scanner, if it wasn't locked out, your scanner would continually stop each time it swept through the scanning cycle.

A significant feature that isn't controlled on a scanner by a knob or button is "track tuning." Patented by Electra Company, "track tuning" is a function of the tuning voltage versus frequency. The radio's RF circuits are tuned directly from the tuning voltage along with the variable control oscillator to allow the scanner to peak each transmission automatically for optimum reception, regardless of frequency.

Although we discussed automatic search earlier, several scanners have a feature that takes this concept several steps further. *Automatic search/store/count* not only "finds" new frequencies between the limits you select, but it also stores them in memory and shows you how many transmissions were monitored on each frequency. *Search/store/count* is a valuable feature to have, especially during skip openings.

Now that you know what to look for, how many scanners do you need to keep track of the action? That depends. Research shows that once the "Action Band" bug bites, most listeners aren't satisfied with one scanner, they want three or more. But even three scanners aren't enough for professional news photographer Stanley Forman. He has eight scanners installed in his car. On foot, he carries three hand-held scanners, and, at home, seven more scanners are constantly in use. Incredibly, Forman says he uses all 18 radios. "I want to know what is going on, in and around Boston," he says. "Next to my camera, a scanner is my most important piece of equipment."

Forman's dedicated monitoring has paid off. Three of his photographs for the *Boston Herald-American* have won him Pulitzer Prizes, the most coveted award a photojournalist can receive, and two of those photos were the direct result of tips he heard via his scanners.

Other Equipment

One of the first accessories most listeners add is an outdoor antenna. Although the whip antenna supplied with a scanner provides adequate results for closer, more powerful transmissions, to hear the distant, low-powered stations, you'll need a bigger and better antenna system (Fig. 16-9). Companies that make scanner monitor antennas include The Antenna Specialists Co., Hustler Antennas, Shakespeare, Firestik, and Radio Shack. Expect to pay between $25.00 to $50.00.

If an outdoor antenna doesn't fit into your or your landlord's plans, don't despair. There are several ways to improve reception even without adding an outdoor antenna. Probably the simplest "trick" is to attach an extra length of wire to your scanner's whip antenna. Changing the location of your scanner can also improve reception. Still not satisfied with reception? If you have some space in your attic, install an outdoor antenna there. Mount it as high as possible. You'll be surprised at how many more of the low-powered stations you'll hear, particularly the hand-held radios used at the scene of fires and emergencies.

If you opt for a commercially made outdoor antenna, select a model that matches the ranges of frequencies you want to monitor. For good overall coverage, an omnidirectional multiband antenna (Fig. 16-10) will provide best results. Those specializing in a particular band may want to consider purchasing a directional beam antenna. During Astronaut Owen Garriott's space shuttle mission, many scanner listeners purchased 2-meter groundplane antennas that were specially cut for 145.550 MHz, Garriott's primary downlink frequency.

Listeners who'd rather design their own antenna, instead of shelling out the bucks for a commercially built one, can obtain detailed instructions from the "ARRL Antenna Book," published by the American Radio Relay League, 225

Fig. 16-9. The Antenna Specialists' MON-64 DISCAN™ multiband external antenna (Courtesy Antenna Specialists Co.).

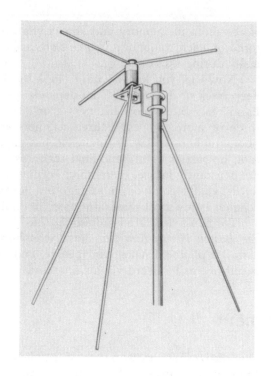

Fig. 16-10. The Antenna Specialists' MON-38 multiband external antenna (Courtesy Antenna Specialists Co.).

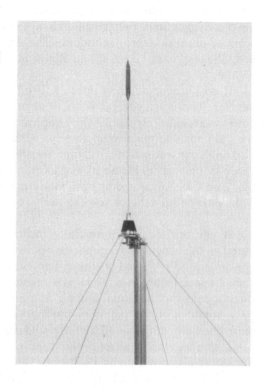

Fig. 16-11. Law enforcement officer using a mobile scanner in his patrol car.

Main Street, Newington, CT 06111. Although it is written for radio amateurs, several of the antennas listed are perfect for scanner radios, too.

Although optional goodies like preamplifiers and filters are widely used by serious scanner listeners, probably the single most important "option" you might add to your shack is a good frequency directory. One of the best directories is the *Betty Bearcat Frequency Directory*, available in both eastern and western U.S.A. editions from Betty Bearcat, Uniden® Corporation of America, 9340 Castlegate Drive, Indianapolis, IN 46256. Hollins Radio Data, P. O. Box 35022, Los Angeles, CA 90035, publishes *Police Call*, which is available in regional editions. *The Federal Frequency Directory* from Grove Enterprises, Route 1, Brasstown, NC 28902, is a good source of unclassified Federal frequency information.

What You'll Hear

Now that you know what you need to get started, let's start scanning! One of the most intriguing aspects of a scanner is not only its ability to give you news of the action—minutes or hours before you hear about it from radio or TV stations—but, also, that it can provide you with the complete version of the story.

Let's pretend a major fire is raging on the north side of Chicago. By monitoring 154.130 MHz, you may hear a dispatcher from Chicago Fireground North ordering pumpers to the scene. If you suspect the Chicago Police Department will also be involved, check 460.375 MHz, the frequency used by the north side police districts, No. 20 and No. 24. Citywide Police Channel 4, on 460.325 MHz, will probably also be active. To determine the extent of injuries that have occurred, check 460.600 MHz—the dispatch frequency for Chicago Ambulance North.

Think you're the only one keeping track of the story? Think again. Scanners are just as popular among members of the professional news media as they are among hobbyists. And while reporters extensively monitor scanners, they also make considerable use of two-way radio. So you not only can monitor the action at the scene of a story, but also the action behind the scenes at your favorite radio and TV stations or newspaper. While you're monitoring the action at the fire, why not switch to some of the Chicago news-media frequencies to see how the story will be played on tonight's news. Tune 161.640 or 161.670 MHz, and you may hear *WMAQ-TV* dispatching its reporters to the scene. To see if the Chicago Sun-Times is covering the story, switch to 173.375 MHz. It's furious. It's fast. And it's fun. That's the beauty of a scanner. With its ability to scan as many as 200 frequencies, a scanner can keep you on top of several fast-breaking stories at the same time..., without leaving your living room.

Although some police departments have begun to scramble their signals, the vast majority of police departments prefer to have the public listen in (Fig. 16-11). Besides promoting goodwill between the law enforcement community and the public sector, scanners enable citizens to take a more active role in crime prevention. In fact, one of the biggest roles of the scanner is not among hobbyists, but with members of Neighborhood Watch Programs. Fig. 16-12 shows a mobile receiver suitable for watch members. An electronic coupler which permits using the standard car antenna for mobile VHF/UHF operation is pictured in Fig. 16-13. To illustrate, in Springfield, Illinois, burglars were apprehended less than 30 minutes after a call went out over the police frequency. A Neighborhood Watch participant who monitored the call spotted the suspects' vehicle and reported it to police.

Fig. 16-12. The Regency MX 3000 mobile scanner, which features coverage of 30 channels and 6 bands (Courtesy Regency Electronics Inc.).

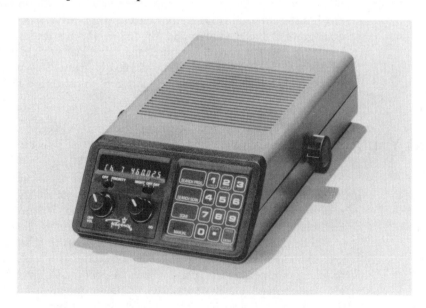

Fig. 16-13. The Antenna Specialists' MON-63 electronic coupler, which converts a standard AM/FM car antenna to cover the VHF/UHF frequencies (Courtesy Antenna Specialists Co.).

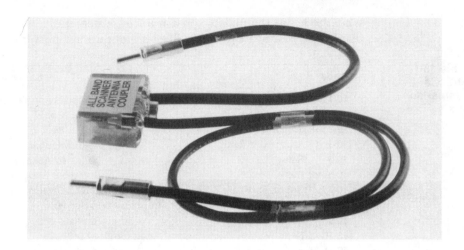

Codes, Codes, and More Codes

The more services you monitor on your scanner, the more you'll notice the use of special numeric and alphanumeric codes. These special codes are known as "10-Codes," or aural brevity codes; they're used to save air time and to convey a precise, defined meaning. The word "ten" before the code number simply alerts the listener mentally that a code is about to be spoken.

The Association of Police Communications Officers (APCO) has an official list of 10-codes that is used by many police departments. The APCO standard is by no means prevalent in all departments, however, as Table 16-1 demonstrates. Under APCO terminology, a 10-26 means "estimated arrival time," but not in Baltimore, Maryland. To a Baltimore police officer, a 10-26 is "request to change to citywide frequency." Table 16-2 gives a comparison between the APCO codes and the Baltimore police department codes.

Because 10-codes are in such wide use, collecting the different codes used by various departments and agencies throughout the country has become as popular a pastime to scanner enthusiasts as collecting QSL cards is for SWL'ers. Many Public Service agencies will provide their 10-codes just for the asking. With others, you'll have to go through about the same process of deciphering their codes as you did in finding their frequency. But that's okay—nobody said scanning was easy!

Emergencies, Disasters, and Life-saving Rescues

Listening to the nationwide emergency communications networks is like waiting for something exciting to happen. K.D. Wentzel can attest to that. He was casually monitoring 155.280 MHz, the regional hospital network frequency, when he heard the makings of a miracle. Emergency rescue teams were being dispatched to a nearby airport in preparation for a possible crash landing.

**Table 16-1.
10-Code
Comparison
Chart**

Code	APCO	Baltimore P.D.
10-1	Signal Weak	Transmission Check
10-2	Signal Good	Signal Good
10-3	Stop Transmitting	Signal Poor
10-4	Affirmative (OK)	Acknowledgement
10-5	Relay (To)	Failed to Acknowledge
10-6	Busy	Stand-by. Busy Unless Urgent
10-7	Out of Service	Out of Service, Give 10-20
10-8	In Service	In Service
10-9	Say Again	Repeat
10-10	Negative	
10-11	_____ on Duty	Meet _____ at _____
10-12	Stand By (Stop)	Not Available
10-13	Existing Conditions	Assist an Officer
10-14	Message/Information	Wagon Run
10-15	Message Delivered	Urgent Wagon Run
10-16	Reply to Message	Back Up Unit _____
10-17	En Route	Call Whatever You are Assigned to
10-18	Urgent	Go Into Whatever You Are Assigned to
10-19	(In) Contact	Return to
10-20	Location	Location
10-21	Call ( ) by Phone	Call ( ) by Phone
10-22	Disregard	Disregard
10-23	Arrived at Scene	Arrived at Scene
10-24	Assignment Completed	
10-25	Report to (Meet)	Stopping Suspicious Vehicle
10-26	Estimated Arrival Time	Request to Change to Citywide Frequency
10-27	License/Permit Information	License/Permit Information
10-28	Ownership Information	Ownership Information
10-29	Records Check	Records Check
10-30	Danger/Caution	Danger/Caution
10-31	Pick Up	Crime in Progress
10-32	_____ Units	Sufficient Units on Scene
10-33	Help Me Quick	Emergency
10-34	Time	Major Civil Disturbance in Progress

Wentzel then switched to the airport's control-tower frequency and learned the chilling details. A woman and her husband were on a pleasure flight when the husband collapsed and died at the controls, leaving her, a non-pilot, to fly the plane! Luckily for her, Wentzel wasn't the only person listening. Phil Hazel, the operator of an aircraft service, was listening to his scanner when the drama began to unfurl. Hazel took off in a small, single-engine Cessna and quickly spotted the wobbling aircraft. He established radio contact with the distraught woman and taught her how to fly right on the spot. She survived.

It takes a good ear and a knowledge of the key frequencies to be able to follow all sides of a story during an emergency rescue. One of the best frequencies to start with is 155.340 MHz, the Hospital Emergency Administration Radio System (HEAR). Every hospital uses this common working frequency to talk with their radio-equipped vehicles. Key regional hospitals also use 155.280 MHz as a "regional backbone" frequency. This frequency is often used to provide coordination between major areas or states in an emergency. Another frequency that is widely used during disaster relief operations is 154.680 MHz. Finally, during a national disaster, tune in 47.420 MHz. This channel is used principally by the American Red Cross.

The paramedic channels can also provide plenty of action. Twenty-four channels are assigned for use by hospitals and paramedics. Eight channels (with 25-kHz spacing) between 468.000 MHz and 468.175 MHz are used by paramedics to transmit to the hospital. Exactly 5 MHz below those channels are the frequencies that the hospital uses to transmit back to the field. The channels are always used in pairs and are designated "Med One" through "Med Eight."

Aircraft buffs also have a number of emergency channels which they should monitor. Aircraft Search and Rescue on 123.1 MHz deserves a channel in your scanner as does Aircraft Emergency on 121.5 MHz. Flightwatch on 122.0 MHz, while not necessarily an emergency channel, is an en-route flight advisory service used for weather forecasts and pilot reports.

Back on the ground, 146.695 MHz is used by Amateur Radio operators as a nationwide emergency frequency. During the Mt. St. Helens eruptions in 1981, 146.695 MHz provided plenty of exciting monitoring, as did 145.330, 146.920, and 146.280 MHz. And, on the waterways, the frequency to monitor is 156.800 MHz, the National Coast Guard Emergency channel. Just when you think there's nothing on the air, a faint "Mayday" may crackle through the speaker.

Although you're not permitted to divulge what you monitor on the Public Service bands, 156.800 MHz is the *one* frequency that takes exception to the rule. If you should hear an emergency call on 156.800 MHz that is not immediately answered by a local Coast Guard Station, it's up to you to take action. Call your telephone operator and tell them you have an emergency call for the Coast Guard. They will immediately connect you with the local Coast Guard search-and-rescue coordination center. If there are none in your area, the operator can hook you up to the National Coast Guard Search and Rescue Center on the East Coast.

Remember, if you hear an emergency transmission, don't be afraid to respond. You won't be the first scanner listener to come to the aid of a mariner in distress!

Highways, Railways, and Skyways

Do you wonder what time your train is due? Or, what is the condition of the morning traffic? Or, will Aunt Minnie's flight arrive on time? Your scanner can bring you the answers. As you drive to work, you can be like millions of other commuters and monitor a local radio station for traffic reports. Or, you can get up-tp-the-minute traffic reports right from the source. In New York City, 450.250 MHz will put you in the traffic copter of radio station WOR as traffic reporter/helicopter pilot George Meade comments on the traffic situation in the Big Apple. Chicago commuters can hear WGN radio's flyboy by tuning their scanners to 166.250 MHz. If you don't know the frequencies in your area for the local traffic copters, try programming your scanner to search between 161.640 and 161.750 MHz, or between 450.050 to 450.950 MHz. Also, 166.250 MHz is widely used as a remote pickup broadcast frequency.

To get another slant on road conditions, tune between 150.995 and 151.130 MHz to one of the several frequency ranges used by the highway maintenance crews (Fig. 16-14). You may learn more than the condition of the highway. If you were monitoring the Minnesota Department of Transportation on 151.055 MHz (in the Minneapolis/St. Paul area) during the major blizzard that occurred December 27–28, 1982, you would have heard everything from plow crews helping a pregnant woman get to the hospital to crews plowing snow for a heart attack victim.

Fig. 16-14. Highway maintenance crews can usually be heard around 151 MHz.

Railroad buffs will want to pay close attention to the frequencies between 160.215 and 161.565 MHz—the nation's railroad channels. There are some railroads, mostly industrials, that operate on the much higher UHF frequencies, but these are the exception and usually have no effect on the main-line railroading that you will normally encounter.

Although the larger railroads use a variety of channels, much of the activity occurs on the road channel. The road channel includes everything from conversations between the tower and dispatcher links, to crews in the train talking between the locomotive and caboose. The road channel for the Louisville and Nashville Railroad is 161.370 MHz. The Soo Line also uses 161.370 for its road channel, but the George Railroad uses 161.310 MHz.

In addition to the road channel, up to five other distinct communications channels are used by the railroads. These include the dispatcher channel, the "yard" channel, the car department channel, the "hump" channel, and the police, or special agent, channel. All of the directories described earlier contain a certain amount of railroad frequency information, but probably the most comprehensive reference book is the *Compendium of American Railroad Radio Frequencies*. It is available from Mark J. Landgraf, 3 Coralberry Circle, Albany, NY 12203.

We've already discussed which emergency aircraft frequencies to monitor. If you also want to monitor the day-to-day operations of your local airport, fasten your seat belt! Commercial airliners, private craft, control towers, and ground crews all use radio equipment in the 108–136 MHz aeronautical band. Your scanner can put you in the cockpit of a jumbo jet, or in the control tower, as air-traffic controllers provide runway assignments, weather information, and warnings about other planes.

Approach and Departure Control also provide plenty of activity. On the approach frequency, you learn everything you need to know about arriving flights. Departure controllers direct the departing flights. If Aunt Minnie is landing at Mitchell Field in Milwaukee, try monitoring the following approach control frequencies: 118.000, 118.600, 119.650, 123.800, and 126.500 MHz. To monitor the approach control frequencies at Sea-Tac International Airport in Seattle, try 119.200, 119.500, 120.400, or 123.900 MHz.

One of the best sources of frequency information for the nation's airports is available from the *fixed base operator* at any nearby airport. FBOs service planes and give flying lessons, and usually know which frequencies are in use. If they are unable or unwilling to provide you with the frequency information you need, a *sectional aeronautical chart (SAC)* for your area will list all airports and frequencies in use. Sporty's Pilot Shop, Clermont County Airport, Batavia, OH 45103, carries SACs for all areas of the country. Write to them for a free catalog.

Keep Posted on the Weather

One of the world's most comprehensive 24-hour radio systems is at your fingertips. It's called NOAA weather. NOAA stands for the National Oceanic and Atmospheric Administration. The first station was established in Rhode Island in 1960. Today, over 400 stations are audible in the United States, with more than 90% of the country's population areas covered. NOAA's most active frequency is 162.550 MHz. If that frequency isn't used in your area, try 162.400 MHz or 162.475 MHz.

Besides providing the latest weather, NOAA also provides another important service. Under a January, 1975, White House policy statement, the National Weather Service was designated as the sole government-operated radio system that is to provide direct warnings into private homes for both natural disasters and nuclear attack! To learn more about NOAA weather, write NOAA Public Affairs, Weather Radio Service, Rockville, MD 20852. They'll provide details on the frequencies and output power of all the weather stations in the country, plus a map showing exactly where they're located.

DX'ing...

Sometime during the summer months, you may be monitoring your scanner when suddenly the VHF-Low band will come alive with distant signals. Long-distance or "skip" reception doesn't occur as regularly on VHF as you would expect on the HF bands, but when it does, the DX can be phenomenal.

One of the most common forms of long-distance reception results from *tropospheric bending*. A very slow-moving high-pressure ridge is normally the most frequent cause of tropo. Many times, this ridge will stall, causing a layering in the troposphere, which traps radio signals and allows them to follow the earth's curve for hundreds of miles. Although tropo reception can occur year-round, it usually takes place in the warm-weather period from late spring to early fall, with mid-August to mid-September being the peak times. Watch for weather fronts, especially when there is a possibility of a cold front overrunning a warm front, causing a temperature inversion.

The second major type of propagation is *E-Skip*, also called "sporadic E" because of its mercurial nature. When E-Skip occurs, you can experience some incredibly long-distance DX. During the summer months, and also occasionally from mid-December through early January, the E layer of the ionosphere becomes ionized, causing signals to bounce off this "atmospheric belt" and return to the earth hundreds, even thousands, of miles away.

E-Skip does not have a mileage limit. However, when a particularly intense opening occurs, the skip distances can increase to as much as 1500 miles. There are also cases where the signal will take two hops, resulting in a signal traveling coast-to-coast. This type of skip usually is confined to the VHF low ranges, although there are documented cases of E-Skip on the high band.

The 11-year sunspot cycle also has a profound effect on VHF reception. As this cycle reaches its peak, the increasing number of eruptions (called *spots*) on our sun cause the ionosphere to reflect higher and higher frequencies from the F2 layer. F2 skip is the result and it provides the most likely way to monitor stations from different continents. During F2 skip, you can often monitor the audio from several European TV stations, including French television on 41.25 MHz and the BBC on 41.5 MHz. But, even though long-distance DX is possible on your scanner, hearing distant stations will be a much easier task than identifying their location. Fortunately, the same conditions that create distant VHF reception can have the same effect on television reception.

The television channels that correspond to your scanner's frequency ranges are: TV channels 2 to 6 which are equal to the VHF Low range, Channels 7 to 13 which are adjacent to VHF High, and the UHF TV channels which partially overlap the UHF band. So, the next time you notice a good opening on the VHF bands, switch on your TV set. If you log a TV station from Florida, there's a good chance that the transmissions you're monitoring on your scanner are coming from the Sunshine State, too!

...Who says watching Gomer Pyle reruns doesn't have its benefits!

Signals from Outer Space

It's not only possible to receive VHF signals from thousands of miles away, but also from outer space (Fig. 16-15). Every 100 minutes, satellites orbiting the earth provide us with at least five to seven minutes of VHF reception as they pass over.

Fig. 16-15. Broadcasts from orbiting satellites can be heard on the 136- to 144-MHZ band.

What can you expect to hear from an orbiting satellite? Imagine listening to a transmission from a remote jungle hospital. Or, a search-and-rescue team that is thousands of miles away. With the proper equipment, it's even possible to monitor weather-satellite facsimile broadcasts. To catch the action, you'll need a scanner that covers the 136- to 144-MHz range, the frequency range reserved for downlinking orbiting satellites. Some orbiting satellites audible in North America include *Meteor* from the USSR on 137.150 MHz and 137.300 MHz, and *Meteor 36* on 137.400 MHz. From the USA, *NOAA 6* and *NOAA 7* use 137.500 MHz and 137.600 MHz, respectively. Other U.S. satellites include *Oscar 9* on 145.825 MHz, *Oscar Phase IIIB* on 145.900 MHz, and *OSCAR 10* on 145.957 MHz.

There are also two geostationary satellites that transmit in the VHF satellite down-link band, *ATS-1* and *ATS-3*. These satellites were put into orbit several years ago by NASA as orbiting radio repeaters, for repeating *out-earth* signals from the 149.20-MHz region. Both satellites are in orbit over the U.S. and repeat signals over thousands of miles from where you may be located.

Other Space Stuff

During the 1983 mission of the STS-9, Astronaut Owen Garriott became the first astronaut to establish two-way radio contact with Amateur Radio operators on earth. Because of the success of the mission and the tremendous goodwill it generated (over 10,000 reception reports from scanner listeners were received and over 350 Hams "worked" Garriott), upcoming Space Shuttle missions may also be experimenting with two-way radio.

Even if the Space Shuttle doesn't have a Ham operator on board during a mission, you can still monitor the air-to-ground communications between the Shuttle and NASA. In the Houston, area, NASA uses 171.150 MHz as a flight information management frequency. Throughout the country, local Amateur Radio clubs rebroadcast the air-to-ground communications in the 2-meter Amateur Radio band (144 to 148 MHz). The network in California is so extensive, in fact, that the air-to-ground transmissions are audible throughout the state. With each successive mission, more and more Ham clubs are joining in. To find which Amateur Radio operators in your area are rebroadcasting the Shuttle transmissions, program your scanner to search between 146.6 MHz and 147.4 MHz, or contact your local Amateur Radio club for specific details.

More Ham Action

When Amateur Radio operators aren't rebroadcasting the NASA space program, they're busy with a host of other fascinating activities. Ham radio operators were the first to bounce signals off the moon and also make extensive use of RTTY. To get better acquainted with the Amateur Radio community, try tuning the 10-meter, 6-meter, 2-meter, or 70-centimeter bands.

One of the more fascinating activities sponsored by Ham operators throughout the country is a network that regularly meets on the airwaves to discuss relevant issues and subjects of interest to both Amateur Radio operators and scanner listeners. Sponsored by the Honeywell Amateur Radio Clubs, the North American Teleconference Radio Network presents programs covering topics as diverse as the Amateur Radio Space Program to legal action affecting the radio amateur. A recent guest speaker on the net was Arizona Senator Barry Goldwater (K7UGA). Utilizing over 180 "gateway" stations (mostly VHF repeaters, all linked together for the net), the North American Teleconference Radio Network can be heard throughout the country. In San Jose, California, 146.760 MHz is used. Florida listeners in the Ft. Lauderdale area can try 146.790 MHz. Listeners in Mobile, Alabama, should try 147.390 MHz.

For a list of scheduled broadcasts and specific frequency information, send a self-addressed, stamped envelope to Rick Whiting, 4749 Diane Drive, Minnetonka, MN 55343. If you have access to a computer and a modem, CompuServe's "Hamnet" X10 database lists all stations participating in the net.

Your Responsibilities as a Scanner Radio Owner

Although there is nothing illegal about listening to transmissions on the VHF/UHF Public Service bands, there are certain responsibilities you should be aware of. In certain states, counties, or cities, there are laws or ordinances restricting the use of scanners in vehicles, except with special permission or for certain classes of citizens, such as public safety officials, volunteer firemen, or licensed Amateur Radio operators.

When you hear of a fire, accident, or other event, it is your responsibility not to impede public safety professionals in their work. Do not go to the scene unless you have been specifically asked to do so.

Finally, it is your responsibility to help your local law enforcement officials whenever possible. If you hear of burglaries in your neighborhood, take special precautions and call the police if you spot anything suspicious. If police are looking for a stolen car, keep your eyes open, and call them if you see it.

In Closing

With the diversity of listening possibilities awaiting you, scanning can be a fascinating pastime. It's also very addictive. Once you've exposed yourself to the world of scanning, don't be surprised if you get hooked, too. The average scanner listener spends six hours per day monitoring the "Action Bands!"

For additional information on scanners and antennas, write to the following companies.

Scanner Manufacturers

1. Uniden Corporation of America, 6345 Castleway Court, Indianapolis, IN 46256.

2. Regency Electronics, Inc., 7707 Records Street, Indianapolis, IN 46226.

3. Fanon/Courier Corp., 15300 San Fernando Mission Blvd., Mission Hills, CA 91345.

4. Radio Shack, 1600 One Tandy Center, Ft. Worth, Texas 76101.

5. J.I.L.-L.A. Corporation, 17120 Edwards Road, Cerritos, CA 90701.

Antenna Manufacturers

1. The Antenna Specialists Co., 12435 Euclid Ave., Cleveland, OH 44106.

2. Hustler Antennas, 3275 N. "B" Avenue, Kissimmee, FL 32741.

3. Firestik Antennas, 2614 E. Adams, Phoenix, AZ 85034.

4. Shakespeare Antennas, P. O. Box 733, Newberry, SC 29108.

5. Radio Shack, 1600 One Tandy Center, Ft. Worth, Texas 76101.

Scanner Organizations

1. Scanner Association of North America, 240 Fencl Lane, Hillside, IL 60162. They publish the bimonthly *SCAN Magazine*, which has features devoted to all phases of radio communications as well as scanning.

2. Radio Communications Monitoring Association, P. O. Box 4563, Anaheim, CA 92803. They publish the monthly *RCMA Newsletter*, which is devoted to VHF/UHF monitoring. Also, contains outstanding columns on satellites and low-band skip transmissions.

17

UTILITY MONITORING

Mike Chabak

Mike Chabak is another 1950s-era DX'er who has kept up his strong interest ever since he began listening. Unlike most, Mike started out in utilities and has never strayed very far from that medium, nor for very long when he did. There was very little information about utilities available when Mike began, so he had to learn a lot on his own. He has been one of the pioneers in helping today's utility novice get started on the right foot.

Mike is a former editor of the "Utilities" column for the SPEEDX (Society to Preserve the Engrossing Enjoyment of DX'ing) club and was the leading inspiration behind the *SPEEDX Reference Guide to the Utilities*. His monitoring post is equipped with a number of receivers in addition to CW and RTTY capabilities. His many other interests include astronomy, hiking, military naval and aviation topics, movie trivia, and model building.

Mike works in photography and graphic arts at the University of Arizona, Tucson.

For the HF radio novice or nonutility-oriented monitor, the world of utility DX'ing is about as simple to understand as the theory of quantum mechanics. The utilities can be a formidable challenge, and not unlike a jigsaw puzzle when it first comes out of the box. There is a chaotic jumble of pieces, and one is hard pressed to determine just where to start. Our departure point for the twilight zone of shortwave monitoring begins with the basics.

The Utilities

Most of us began our shortwave interests listening to local AM broadcast stations. Mastering AM radio was simple; just turn the radio on, tune to your favorite station, and out of the speaker came news and music. The next logical step was the introduction of broadcasts made by stations in faraway countries. How we gravitated into this varied from individual to individual, but the end result was that we now had before us a shortwave receiver. We quickly found that there were several main types of transmissions—the AM broadcast band (BCB), the international shortwave broadcast stations (ISWBC), amateur radio (Ham), and Citizens band (CB). Everything seemed quite in its place, that is, until we dialed on either side of those familiar frequency bands. It is here that we encountered Morse code, Donald-Duck-like voices, chirps, pulsations, cryptic lingo, and a seemingly endless variety of odd noises. This strange world is the utilities. So you have just received your first lesson. Everything *except* BCB, ISWBC, Ham, and CB is defined as utility-type transmissions.

Basically, utilities are business communications—be they private, commercial, governmental, military, or, oh yes, clandestine. They encompass the everyday real world as it goes about its daily routine. They can be as simple as weather broadcasts, or as chair riveting as a high-seas search-and-rescue operation.

Utility stations make full use of one or more types of transmission modes. Continuous-wave (CW) Morse code, voice (AM, upper sideband, lower sideband), radioteletype (RTTY), and facsimile (Fax). Utility stations operate within one of three broad classifications, aeronautical, maritime, and point-to-point, and can either be of a mobile or fixed-site variety.

What totally sets apart utility operations from that of BCB and ISWBC is the audience. BCB and ISWBC transmissions are specifically directed to the listener. They provide news, views, entertainment, and music, with each competing for the listener's undivided attention. They want you to listen to their transmissions, and, therefore, make them as entertaining, and, at times, as controversial as possible. Ham and CB transmissions are basically public-service operations, specifically of a short- or long-range wireless-telephone communications mode. Overall, they are hobby in nature, and not within the domain of utility communications.

Utilities are, in fact, just the opposite. They are designed to be heard and utilized only by those directly involved or associated with the communication. In essence, utility communications are *private communications*. Hence, a utility monitor is an eavesdropper. An uninvited party listening to communications that are not intended for his or her ears. Sounds intriguing, doesn't it? But to quote an old proverb, one man's passion can be another man's poison, ...or to be more direct, some shortwave listeners could care less. If you prefer to be entertained with music, enjoy receiving historical and cultural lessons, or like to be on the receiving end of propaganda (preached or smoothly delivered), then stay with BCB or ISWBC stations and read no further. On the other hand, if you want to delve behind the headlines, or simply like to listen to the world and its people going about their daily business, then the utilities is where it is at for you. But be forewarned. You will be confused and frustrated in your initial attempts. Utility monitoring is a learned trait, and one must accept the fact, that as with any new endeavor, it will take both time and patience to accomplish it.

Utility communications constitute roughly 75% of all the radio usage between 0.1 hertz and 30 MHz. What you hear can be mundane and even boring. Other times you'll be intrigued and mystified. You'll laugh at some of what you hear, while others will leave you in silent shock. You're listening to the real world and, just as in real life, you'll find that your responses range the full spectrum of human emotions. Utility communicators are armchair adventurers, in the realm of reality. What you'll hear is not fiction; everything is real, happening as you hear it. The bottom line is that utility communications is the only arena where people of varying cultures, languages, and ideologies all work together. This is true, though they be thousands of miles apart in distance, and worlds apart in concepts. If a Russian freighter is sinking off the California coast, the U.S. Coast Guard won't hesitate to mount a search-and-rescue operation. When a U.S. airliner flies from New York to San Juan, the Cuban air traffic control station doesn't hesitate to render assistance to insure the airliner's safe arrival. The list can go on and on. Utilities is where the action is, but, more importantly, it is a medium that works in a unique harmony, within a very unharmonious world.

Equipment

So where do you start? First you require a general-coverage receiver. That is a shortwave rig that tunes continuously from its lowest to highest available frequency range. Shortwave receivers come in all shapes and prices. Not too long ago, the SW rig was part of the AM radio, and housed in a big wooden cabinet. A handsome work of carpentry, possibly with a 78-rpm record player, lift top, front doors, a giant speaker, and lots of vacuum tubes. World War II advanced radio into the realm of a sophisticated piece of electronics. Many SW monitors still own and use surplus receivers. Many were big and heavy, loaded with tubes, plenty of knobs and switches, and a dial-type readout. When left on for any period of time, they can double as hand warmers on cold days. Then came the transistor and diode family of semiconductors; something smaller than an aspirin tablet now did the work of a big, hot, vacuum tube. Semiconductors revolutionized the radio industry. Circuitry now occupies but a fraction of its former space. The all-tube rigs gave way to the hybrid tube/transistor models. Then came the total solid-state receiver. Still, one had to read the frequency via a slide rule or revolving plate affair. Rigs with this type of meter were called analog readout types. Next came the direct-frequency readout (digital), via mini-vacuum tubes. This, like the transistor, became more sophisticated. Today, digital readout is via mini-fluorescent tubes, or light-emitting diodes (LEDs). Advances in computer technology lead to the micro circuit. More commonly called the microchip, it too was merged with the receiver. Today, some of the most sophisticated rigs are virtually all push button.

Whether your rig is a military surplus model, or the latest state of the art, it must have the ability to tune in CW, and single sideband (both USB and LSB). Without this ability, the majority of the utility transmissions will be out of your reach. But, be it surplus or state-of-the-art, your receiver is still just a piece of electronics. Not unlike a lump of clay, it just sits there, waiting for a human mind and guiding human hands to bring it to life. A general-coverage multimode rig and some sort of antenna is your passport into the utility world. But where and when does one listen?

Activity Listings

For this, you'll need to know what the current utility activity is, and the areas of activity. A shortwave hobby club is an invaluable asset to start you on your way. There are many clubs throughout the world whose monthly bulletin includes a utility column. In the USA, we have ASWLC and SPEEDX. Both have extensive HF coverage of virtually all aspects of utility communications. Via these clubs' columns, you will discover what others have monitored and, by using the listings, you too can tune into them.

Another indispensable addition to your radio shack are the utility-oriented publications. These offer frequency, station data, and much more. You will use the publications as basic reference material, both to assist you with tips on monitoring and in providing additional required data. A list of clubs and publishers can be found in the source list given at the end of this chapter.

In front of you is your SW rig. Over there is a club bulletin and, next to it, a publication, such as the Gilfer *Confidential Frequency List*, or the Universal *Guide to Utility Stations*. Now what??? The biggest mistake a novice makes is to try and take everything in at once. Monitoring utility communications is a learned trait. All the books in the world, and a thousand dollar receiver, are just a lot of expensive junk, if you're not prepared to first listen and learn.

Each of you have interests and hobbies that can be found in the utilities. This is the place to start. Say your interest is aircraft. Check the book, or club bulletin, and find an aeronautical (aero) frequency in the 8-MHz band (assuming that you are doing this during the early evening). Now, tune to the listed frequency and remain there for five to ten minutes. Either you heard Donald Duck conversing in Duckese with his nephews or you heard nothing. Welcome to frustration....

Signals and Frequencies

You need to first understand three fundamental concepts. First, those publications list the *assigned carrier frequency*. That is, the transmitted frequency over which is sent a steady cycle audio tone. You could call this an electronic *reference* signal.

Human vocalizations are converted by the radio into electrical audio cycles, and then mixed with the steady-cycle carrier signal. This produces variables in the carrier, in relation to the pitch tones of the human voice. Allegorically, this is not unlike when one plucks a guitar string and then changes the tone, by varying the pressure and position of the fingers along the fretts. This "modulation" of the carrier manifests itself on either side of the carrier, in what are called the *sidebands*.

The sidebands are the actual "message" component of the transmission. They are identical—in effect, twins separated by the carrier. This type of transmission mode is also referred to as *double-sideband telephony*. Most simply call it amplitude modulation or AM. Double-sideband AM is a 6-kHz-wide signal (bandwidth), for its 3000-cycle audio tone, now modulated, produces two identical sidebands—3-kHz plus and 3-kHz minus the carrier frequency.

Thank of amplitude modulation as a sort of monaural stereo. The radio would represent the carrier, with the left speaker representing the lower sideband (LSB) and the right speaker representing the upper sideband (USB). Since each sideband is an exact twin of the other, only one needs to be transmitted. Likewise, the carrier signal itself (once it serves its modulating-reference purpose) need not be transmitted. Both can be filtered out, prior to transmission. This, then, constitutes a single-sideband, suppressed-carrier transmission. We refer to it as simply SSB. When we specifically indicate which sideband, it is LSB or USB.

Your receiver, when set in the single-sideband mode, generates an artificial carrier tone. When this electrical value is subtracted from the incoming modulated signal, the results are a reconstructed human voice coming out of the speaker.

All SSB voice-mode transmissions utilize a narrower bandwidth audio tone (roughly one half that of conventional AM). Since there is no carrier, you are only receiving one of its sidebands. Therefore, in conventional voice-mode USB, the signal is actually 1.5 kHz *above* the assigned carrier frequency. If you've dialed a frequency of, say, 8879.0 kHz, you've got to actually tune to its USB side, or 8880.5 kHz. (If your rig doesn't have fractional capability, you tune to 8881 kHz and nudge the bandspread/kHz/BFO control, to clarify the voice.) Unless the instruction manual of your receiver specifically states that it automatically offsets the readout to carrier, when in a SSB mode, you MUST tune to 1.5 kHz above the carrier (for USB) or 1.5 kHz below the carrier for LSB. This is a basic operating procedure, so remember it. Now all of this requires some mental arithmetic, yet there is a way for some of you to manually offset your rig and display the carrier frequency, even though you are actually on the USB side. For this, your rig must have a BFO, RIT, or fine-tuning control. If yours does not, then you will have to do the preceding arithmetic. For those that do, you may still not achieve a full offset, for many BFO/RIT controls only tune plus or minus 1 kHz. In any event, this is how to manually compensate.

Aeronautical Frequencies

Let us use the active USAF SAC frequency "Quebec," which is 6761 kHz (carrier). On this frequency, you will hear lots of coded call signs, and, at regular intervals, a transmission starting off with, "Skyking, Skyking, do not answer...". Tune your receiver so the digital display shows 6761.0 kHz (6761 for those with no decimal capability). When a transmission starts, turn the BFO/RIT/fine-tuning control until the voice is clear and normal sounding. If your control has less than 1.5-kHz capacity, the voice will be of a somewhat higher pitch than normal. Mark the position of the control, and whenever you are in the USB mode, set that control to the mark first. You'll then have a visual readout of the carrier frequency, when in the USB mode. Once you have established the carrier frequency, if the higher-pitched voice bothers you, merely nudge the "kHz" knob up frequency slightly. While playing with the BFO/RIT, etc., you may notice that even though you move it from one extreme to another, the visual readout does not change. The reason for this is that this control is not incorporated with the electrically tuned frequency-adjust circuit. There are no predictable LSB transmissions, so if you do encounter one, turn the control 180 degrees from its USB position, and adjust the "kHz" control until you receive a clear voice. At best, the resulting carrier readout should be no more than one full kilohertz off. And need I say that your rig should be in the LSB mode for this?

Whatever method you are now using, you should have heard, by now, a USB-mode utility transmission. If so, congratulations and welcome to the world of utility monitoring. But you've only got through the easy part. Now comes the real learning. Stay on the frequency you've chosen, and listen.... What you are listening for is the style and format of the communications. In two-way communications (voice, CW, RTTY), the station being called is stated first. This is followed by the caller's identity. Every radio operator has his/her own style, but follows this established radio communications format.

For a hypothetical example, let's check out a typical aeronautical communication on 8879 kHz. Full IDs are required during the initial establishment of the two-way communications, but can be and often are shortened, once the communications proceeds. Explanations of the traffic flow are given in parentheses.

"Gander, Gander, Speedbird Zero One" (airliner Speedbird 01 calling Gander air traffic control)

"Go ahead Zero One" (notice that Gander did not ID. This is common if the ATC station is working many aircraft on the frequency)

"Position 48–35 north, 45–15 west at 2305. Estimate 44 north 50 west at 0115. Flight level 380. Request flight level 400."	(message from Speedbird 01)
"Stand by Zero One"	(Gander)
"Standing by"	(Speedbird 01 acknowledging)
"Speedbird Zero One, Gander"	(Gander calling Speedbird 01)
"ATC advises Speedbird Zero One to climb to and maintain flight level 400. Report when reaching. Read back"	(Gander's message to 01)
"Understand, ATC clears Speedbird Zero One to flight level 400. Leaving 380 now. Will report when reaching."	(01 message)
"Gander"	(acknowledging 01 read back)
"Zero One"	(acknowledging Gander's acknowledgment and signing off)

Over and out is rarely used in aeronautical communications. Speedbird 01 might have said, "01 out," or even nothing at all, since Speedbird 01 doesn't have to acknowledge Gander's acknowledgment. This two-way communications is now ended and, if another aircraft is waiting, they now call Gander.

The following list contains the commercial aeronautical bands (in kHz). Check your reference material for exact frequencies, area usage, and associated ground stations.

2850–3025	5450–5680	10005–10100	17900–17970
3400–3500	5625–6685	11275–11400	21924–22000
4650–4700	8815–8965	13260–13360	

Get accustomed to the communications flow. Listen not so much for what is said, but how it is said and used. This applies to any type of two-way communications. Once you become familiar with the format flow, you can quickly jot down the logging information. This is another point to burn into your memory banks. Except for scheduled transmissions, such as weather broadcasts, *all* utility communications are catch as catch can. Many communications will be of short duration. For instance, if you tuned to 8879 kHz after the above communications, you might have heard this 10-second traffic:

"Gander, Speedbird Zero One. Flight level 400."	(Gander)

Maritime Frequencies

If you chose ships, instead, and decided to check out the maritime mobile frequencies, you may believe there is something wrong with your receiver. Commercial aeronautical voice communications are all simplex. That is, both stations transmit and receive on the same frequency. Maritime communications (ship/shore) can be a very different situation. Although some frequencies are designed for simplex operations, the majority of maritime two-way communications are duplex. Duplex means that stations use separate transmit and receive frequencies. In addition, the sidebands are 1.4 kHz, plus or minus the carrier. Much of the voice-mode maritime traffic involves telephone calls. These are known as *phone patches*. Let us assume that you are monitoring the ship side of the duplex frequency pair. You hear the Pacific Princess calling KMI (maritime coastal station San Francisco) giving a telephone number. Then there is silence, followed by the Princess acknowledging. Immediately thereafter, you hear Aunt Matilda talking to her friend Marge in Salina, Kansas. Now something is very odd. You hear Aunt Matilda perfectly, and even though it is obvious that Matilda is responding to Marge, you can't hear Marge at all. No, your rig is not going bananas, and neither are you. You are listening to only the ship side of the duplex pair. If you had two rigs (by checking a maritime ship/shore frequency listing), you could tune one rig to the ship side and one to the shore side, and then hear the entire conversation.

Below is a list of the major maritime bands (in kHz). Check your reference material for the exact frequency, duplex pairs, and shore station IDs.

SHIP SIDE	SHORE SIDE
2000–2430	2440–2850 (primarily local marine operator)
4063.0–4140.5	4357.4–4434.9
6200.0–6215.5	6506.4–6521.9
8195.0–8288.0	8718.9–8811.9
12330.0–12426.1	13100.8–13196.9
16460.0–16584.0	17232.9–17356.9
22000.0–22120.9	22596.0–22716.9

Coast Guard Frequencies

United States Coast Guard (USCG) communications have always attracted wide interest. The USCG is best known for its humanitarian search-and-rescue (SAR) efforts, and those communications can often be quite exciting to monitor. The USCG offers three types of stations: Shore, CG Cutter, and CG Aircraft. One of the most active all-night frequencies is 6506.4 kHz (carrier). On this frequency, you can hear CG shore installations working CG Cutters, U.S. Naval vessels, foreign warships, and everything else from ocean-going yachts to super tankers to research vessels. This frequency is both simplex and duplex. When the vessel is not heard on 6506.4 kHz, it is on the ship side of the duplex pair: 6200.0 kHz.

All USCG voice-mode communications are USB. 6506.4 kHz is another excellent frequency to learn about the style and flow of communications.

The USCG is quite active with voice-mode communications. The following list is just a few of the more active frequencies in use. All frequencies are the assigned carrier, and simplex. Ship/shore duplex frequencies are also included.

2182.0—International Calling and Distress

2670.0—Scheduled marine weather, notice to mariners/shipping broadcasts; also limited ship/shore communications.

3023.0—International Search and Rescue

3120.0—CG helicopter air/ground

3123.0—CG fixed wing air/ground

4428.7—CG Cutter and ship/CG Shore. Also same scheduled broadcasts as those on 2670 kHz. Duplex ship frequency is 4134.3 kHz.

5692.0—Same as 3120.0 kHz

5696.0—Same as 3123.0 kHz

6506.4—Same setup as 4428.7 kHz. Duplex frequency is 6200.0 kHz.

7836.6—USCG Northwest Pacific Loran-C network

8765.4—Same setup as 4428.7 kHz. Duplex frequency is 8241.5 kHz

8980.0—Same as 3120.0 kHz

8984.0—Same as 3123.0 kHz

11198.0—Same as 3120.0 kHz

11201.0—Same as 3123.0 kHz

12205.0—USCG Hawaiian/Central Pacific Loran-C network

13113.2—Same as 4428.7 kHz. Duplex frequency is 12342.4 kHz.

15081.0—Same as 3120.0 kHz

15084.0—Same as 3123.0 kHz

17307.3—Same as 4428.7 kHz. Duplex frequency is 16534.4 kHz.

For CW buffs, scheduled USCG marine weather broadcasts can be found on 4525.0, 9050.0, 13655.0, 16457.5, and 22472.0 kHz.

It is to be noted, that when CG Cutters are participating in a SAR mission, they can be heard on the USCG air/ground frequencies, in addition to the ship/shore channels.

USCG stations identify in one of these basic ways....

COMMSTA (name of city/location)	COMMSTA San Francisco
Radio (name), or (name) Radio	Radio Miami, or Guam Radio
(Coast Guard) Air Station (name), or (name) Air	(Coast Guard) Air Station Mobile, or Cape Cod Air

Coast Guard Group (name)	Coast Guard Group Shinnecock
(name) Loran	Iwo Jima Loran
(name) Monitor	Yokota Monitor (... indicates the Master Loran net station can also use: LORMONSTA (name))
Coast Guard Cutter (name), or Cutter (name)	Coast Guard Cutter Polar Sea, or Cutter Point Brower

USCG aircraft identify with a 4-digit number. IDs can range from: Coast Guard 1703, Rescue 1703, 1703, or (if communications are in progress) just the last two digits: 03. CG helicopters might add the word "helicopter" to the call. Only two USCG aircraft have a 2-digit call (01 and 02). Both are VIP transports and do not engage in SAR operations.

When CG shore/Cutters/Ships are operating in CW or RTTY modes, only the station call letters are used as an identification.

Unfortunately, space does not allow a full rundown of the USCG frequencies, facilities, and cutter/aircraft call letters. More in-depth information on these aspects can be found in Gilfers *Confidential Frequency List*. Also, a very comprehensive USCG coverage, including mailing addresses, can be found in the *Speedx Reference Guide to the Utilities* (SRGU).

USAF Frequencies

Back in the 1950s, as many old-timers will tell you, USAF bases, worldwide, conducted their own air traffic control operations on the HF frequencies. This was eventually streamlined down to 14 locations, called the Command Control Network. These stations used the base name followed by the word AIRWAYS. Airways stations were the military air traffic control network, and provided many military buffs ample opportunity to monitor worldwide USAF/USN flights. Except for Strategic Air Command (SAC) operations, Military Airways air/ground communications was the most widely heard USAF network. Unfortunately for the utility monitor, in late 1983, the USAF relinquished its military ATC responsibilities. These were handed over to the commercial aeronautical ATC network. All USAF Airways stations then became part of the USAF global communications network. Today, about all you hear on this network is air/ground radio checks, phone patches, and other communications of a U.S. defense nature. If conditions warrant, they will handle ATC-type operations, but, normally, these operations are referred to the civil ATC frequency. Airways stations can still be heard on HF, however. The most active frequencies are (carrier and USB mode):

4746	6750	8993	11236	13244	18019
5688	6753	9014	11239	15015	18023
5710	8964	11176	11246	15031	

```
6715   8967   11179   13201   17972
6738   8989   11182   13215   18002
```

The fourteen Airways stations are:

Albrook, Panama	Incirlik, Turkey
Andersen, Guam	Lajes, Azores
Ascension, Ascension Island	MacDill, Florida
Clark, Philippines	McClellan, California
Croughton, England	Scott, Illinois
Elmendorf, Alaska	Thule, Greenland
Hickam, Hawaii	Yokota, Japan

SAC as previously mentioned, is quite active, but all communications systems use tactical call signs to denote air and ground stations (Banjo 15, Avon Control, Night Star, and so on). Although intriguing, it takes a utility monitor skilled in military communications to make any sense of these transmissions. The "Skyking" broadcasts are part of the "Giant Talk" network. These are coded messages to U.S. nuclear deterrent forces. The more active SAC frequencies are 4725, 9027, 11243, 17975, 6761, 9057, and 15041 kHz, plus a host of lesser used frequencies.

Mystic Star

This is the identification for the worldwide Andrews Air Force Base VIP network. These communications involve flights made by high-ranking U.S. governmental and military officials. The aircraft IDs have a "SAM" prefix (Special Air Mission). The most famous of the VIP aircraft is, of course, "Air Force ONE." This call sign is used by the aircraft when carrying the President of the United States. Andrews Air Force Base, just outside of Washington, D.C., acts as the ground communications/phone patch interface for the Mystic Star network. Andrews AFB is also the home for the VIP fleet of aircraft assigned to the 89th Military Airlift Group (MAG) of the USAF Military Airlift Command (MAC). The aircraft are primarily military versions of the Boeing 707, the McDonnell-Douglas DC-9, and the Grumman Gulfstream 3 (which will eventually replace the Lockheed Jet Stars).

Tactical call signs are assigned to each VIP, be that the President, the Secretary of State, or so forth. Each administration uses tactical calls unique to its own personnel. Andrews AFB maintains communications with the VIP flights, even though they may be half a world away. This is accomplished (when required) via satellite relay through the area USAF communications station. Communications is SSB and can be either USB or LSB; they utilize a wide variety of frequencies, which are often referred to by a FOX identification number. These FOX identifications are considered classified. When enough of their frequency relationships find their way into publication, the frequencies will not change, but the FOX identifications will. Best place to check for current activity is via a SW-club utility column.

I might add that even though the President communicates with parties on the ground, you will not hear these communications (or to be more precise, be able to understand them). All of his communications are sent via secure voice methods.

United States Navy

Once, the U.S. Navy shore stations used colorful tactical call signs, such as: Orange Juice, T-Bone, Back Lash, and Dunkirk. They were static (unchanging) and, after a period of time, every long-range USN COMMSTA was identified. Then the USN adopted the everchanging alpha/numeric call sign system (Whiskey 3 Bravo, 4 Uniform, etc.). It is now rare to hear a word-type call sign on the USN circuits. A major exception is *Raspberry*. This is a prefix used to denote a US Naval Air Station; i.e., Raspberry Miramar. The Naval Air Stations use their base name. Aircraft carriers, when on the Raspberry net, may also use a word call, such as: Grey Eagle, War Chief, etc. The primary Raspberry frequency is 6723 kHz USB.

Just as with SAC communications, most USN traffic is intriguing, but it takes a dyed-in-the-wool buff to make sense of them. Some of the more active USN voice frequencies are:

3095	5430	8972	11252
3130	5716.5	9002	13182.5
3265	6697	9006	15067
4700	6705	9256	17985
4730	6720	11191	18009

U.S. Navy ships also use the alpha/numeric call-sign system, but there are instances when clear voice is used. This is primarily during phone patches, via a 2-MHz marine operator, and using naval-base harbor communications frequencies of: 2150, 2434, 2716, and 2836 kHz.

Another way to monitor USN vessels is via the MARS AFLOAT network. (MARS is the abbreviation for Military Affiliate Radio System, originally a system of military Ham operators.) This is primarily a radio phone-patch network used to link personnel at sea with their loved ones at home. USN vessels never use a clear vessel identification in MARS communications. All are assigned a MARS call sign, which begins with NNN0-, followed by three letters, which is the ID of a particular ship. MARS AFLOAT call-letter assignments and operational frequencies can be found in several publications (such as the SRGU). In addition, they often show up in the loggings presented in the SW clubs.

Propagation

Some of you might ask, why single sideband and not good old AM for voice communications? AM requires a 3-way split of the transmitter power—to the carrier, and the upper and lower sidebands. For all practical purposes, a 10-KW AM signal is really three 3.33-KW signals. By contrast, a USB transmission has its LSB mirror image filtered out, and the carrier is suppressed. Hence, the entire output power goes into just the USB portion of the frequency.

Other advantages are apparent when we examine signal propagation. All long-range HF signals propagate via bouncing off of the upper atmospheric ion layers. The F2 layer is the most important for long-range communications. With the signal beamed just above the horizon, a single hop (transmitter via F2 to ground) can span 2000 to 2500 miles. The F2 layer actually refracts or bends back the radio wave. In theory, the F2 layer is a mirror, but, in actuality, it is composed of a myriad of tiny electrically charged particles. These are all in motion, respective to one another. When sunspot/solar-flare activity is high, this excites the ions, causing the layer to be anything but a smooth refractive surface source. Under these conditions, the signal's downbounce (refractive) angle will vary in magnitude to the disturbed ion activity. This produces the well-known effect of the signal fading in and out. Transmissions going over the pole have the added factor of the pull of the intense magnetic field of the earth. These transmissions often have a "flutter" quality about them.

The AM mode of transmission is actually three frequencies, with only the sidebands carrying the actual message. Even though their frequencies are not that far apart, the absorption/refraction qualities of the frequencies are slightly different. Combine this with a very active ion layer, and one sideband can arrive at your location a few milliseconds before the other. We've all heard the results —distortion. The USB-mode signal is a single frequency, and even though the churning F2 layer can cause it to fade in and out, it will not be distorted. So the advantages of SSB versus AM are:

- Less given transmitter power requirements
- More economical, costwise
- No distortion (providing transmitter parameters are maintained)
- Uses only one half of the message carrying component. If required, the same common carrier can be used to send a separate transmission on the LSB and USB sides, simultaneously.
- The narrow bandwidth of SSB allows for closer frequency association, hence more stations per band, and less co-channel interference.

To illustrate SSB advantages, Radio Japan may have to use 100 kilowatts (100,000 watts) of power and a directional antenna beam to get a loud and clear signal into your living room. In contrast, the air traffic control station in Tokyo can do the same thing with just 2 to 5 kilowatts (2000 to 5000 watts) of power and an omnidirectional beam. Elevating the transmitter antenna produces even more dramatic contrasts. An airliner at 40,000 feet over the South China Sea can be heard equally well, even though it may be using no more than 500 watts.

Now, with all this going for voice-mode SSB utility transmissions, you should be a pro in no time. It depends on you. Patience is the key. You learn and progress through experience, at your own pace. Books and listings only assist, but there is no substitute for sitting in front of the rig and monitoring. Listen and learn. Combined with reference materials, you will slowly expand your understanding. Applying this, as you go along, builds up your knowledge base. Eventually, the jigsaw puzzle fits all together. But you will never stop learning, for that is the unique quality of utility monitoring. Even after 20 years, you will still encounter stations that you never heard before.

Frequency Allocations

You have become acquainted with some basic fundamentals of utility monitoring, along with certain voice-mode types. It must be apparent, now, that utility monitoring is anything but simple. Complex and convoluted though it may be, time and patience will sort things out. Now, let's get back to more fundamentals.

Just as with the BCB stations, the major utility classes (aero, marine, and so on) operate within established frequency areas. The North American BCB stations have but a single area of 530 to 1600 kHz. Utilities likewise have set areas, but these are segmented throughout the entire HF radio spectrum of frequencies. All stations are licensed and registered with the federal and international radio regulating bodies. As such, they, by law, must stay within the allotted frequency areas designated for their particular classification.

There are some stations that skirt around this. Let me illustrate by using 6761 kHz. This frequency is allocated to the U.S. Air Force and used specifically by SAC. All stations on this frequency use coded (tactical) call signs. You may not be able to ID the specific stations involved, but you do know that the communications on 6761 kHz are USAF/SAC.

Those who specialize in monitoring both the allocated and nonlisted tactical communication frequencies are nicknamed GMT buffs (Governmental/Military/Tactical).

Because of the nuclear deterrent mission of SAC, there are communications that are very sensitive. These transmissions use frequencies that are not listed with the FCC (Federal Communications Commission). To further foil identification attempts, many of these frequencies are outside the bands allocated for aeronautical point-to-point communications. The military, and certain government agencies, all use unlisted frequencies, to one degree or another. The Central Intelligence Agency (CIA) is a prime example. None of its HF frequencies are listed. Actually this is not a totally correct statement. Some are shown, but the authorized user given is a cover name.

National security interests transcend the rules which the average utility station must adhere to, and there is nothing illegal about these operations. It is only that the frequencies and the users are considered (because of security) as classified information. There is a class of stations which may fall into this category, but which are treated as clandestine due to the peculiar type of transmission they employ. These are the infamous *Number* stations. But we're getting ahead of ourselves. More on these later.

CW/RTTY Communications

Voice-mode communications are but a fraction of what one can hear, but, it is not to be inferred that voice communications are only a minor constitute. You could spend your entire life monitoring voice-mode utility stations, and still find it both rewarding and unending. However, there is another major type of communications mode.

CW

The oldest radio-communications medium is radio telegraphy. It is the simplest form of transmission, and was the first method employed to transmit messages. It is merely the on-and-off (make and break) transmission of the unmodulated CW, or continuous wave, signal of the carrier. Press down the key and the circuit is completed, release the key and the circuit is open. (Today, MCW, or modulated CW, is also employed.)

Morse code, or CW as it is generally referred to, is made up of only two audio pulses. A short pulse or tone is designated as a dot, and a long pulse is designated as a dash. Phonetically they are, "dit" and "dah." Combinations of "dit" and "dah" are assigned to each letter, to each number and the punctuation marks, plus an assemblage of shorthand letter combinations (CQ, for instance, meaning *calling*.) Samuel Morse perfected the code, hence the name, Morse code. Morse code has been internationally adopted as the communications medium for CW transmissions. For use on a local level, several nations use a CW code whose arrangement is not the same as that of Samuel Morse. Basically, these reflect the incompatible alphabet differences for languages such as Greek, Japanese, Chinese, Arabic, Hebrew, and Russian Cyrillic.

Code Readers

There are many publications that can instruct you in the art of mastering Morse code. Hams know it, commercial and military radio operators learned it, and you too can learn to "read" the pulsations. Thanks to the advent of microchip technology, however, one can get into CW monitoring immediately, without even knowing any Morse code. This is due to electronic devices called Morse Code readers (some are CW/RTTY types). You simply feed the audio from the receiver into them and, when properly tuned, they show, via a visual display, a translation of the Morse code signals you are monitoring. Some old-time CW buffs scoff at how easy it is for today's monitors. They toiled and practiced long hours to learn and master the Morse code. But, even though modern technology does make it easy, the electronic CW reader is only a teaching tool. When operating the reader, you will normally hear the code as well as seeing it displayed. Your mind begins to associate the sounds with the visual readout. I know of several utilities monitors who knew no code, but they bought a CW reader and, after months of reader-assisted monitoring, found that they could read (audibly) more and more of the CW traffic, as time went by. To coin a phrase, the CW reader gives you on-the-job training.

Mini-reader Audio Problems

The current crop (early 1980s) of CW/RTTY mini-readers suffer from a similar problem. They all require a healthy amount of receiver volume to properly drive the reader circuitry. It is almost impossible to operate the reader without somehow actually monitoring, via headset or external speaker, the transmission itself (near impossible from a practical standpoint). Most readers interface with the receiver, via the headset jack. As we all know, when you plug into this jack, the receiver's internal speaker (if it has one) is cut out.

Some people have solved this problem by using a dual, or "Y"-type, ¼-inch phone plug. Simply plug this device into the headset jack, and then plug both the reader interface cable and your headset (or external speaker) into the adapter. (Some readers come with ⅛-inch plugs on both ends of the interface cable, so you'll probably have to convert one of the plugs to a standard ¼-inch plug.) Another method is to interface directly (rig to reader) and plug your headset into the reader's audio output jack. In current readers, this jack will only accept the ⅛-inch plug. As it is not practical to modify your headset/speaker, most people use a ¼-inch to ⅛-inch adapter plug. But, up to this point, we've only hooked up the little beast. The real problem begins once you turn everything on.

The required volume level needed to drive the reader is enough to wake the dead. You simply cannot wear the headset without screaming and going deaf. Some operators have taped/banded the headset earcups together. Others bury the outboard speaker face down. These might do in a pinch, but, for any regular use, it is impractical (plus, there is still a good deal of audio present, even when muffled).

The answer is to attenuate the audio coming into your headset/speaker. *Attenuate?* What is that? In this instance, it means simply to electrically lower the volume before it gets to the headset/speaker. There are two ways of doing this. Most electronic supply stores carry what is called an *in-line attenuator* (it looks like a plug). These are designed to maintain usable volume levels, when tape recording via recorder jacks not designed for that purpose. Anyone who has plugged into the wrong recorder jack knows what I mean. Anyhow, these attenuators normally reduce the volume by 40 to 50 dBs. The problem is that although they mate perfectly with the reader's ⅛-inch jack, their other end only accepts (depending on model) either a RCA-type phono plug or a ⅛-inch plug. Again, you don't want to fool around with your standard plug (on the headset/

external speaker), so you'll need to purchase a ¼-inch standard-to-⅛-inch (or RCA-type) phono plug adapter. The overall cost isn't much, but you will end up with a multiplug almost 5 inches long.

All the in-line attenuator is is a resistor, in-line soldered. If you have an electrical-parts junk box, you can whip one up yourself. The required items are:

¼-inch standard phone jack

⅛-inch mini phone plug

a two-cable wire

¼-watt, 22K, resistor

soldering gun and rosin-core solder

small box to house the standard phone plug

The resistor is placed between the center terminal of the ¼-inch jack and the line coming from the center terminal (tip of plug) of the ⅛-inch plug. You may use any ¼-watt resistor from 10K to 47K—whichever you might have. The 22K resistor seems to produce about a 40-dB reduction when used with the standard 8-ohm headset. Now, if I haven't made myself perfectly clear, the attenuator is only for the headset/outboard speaker!!! DON'T attenuate the rig-to-reader interface cable!!!

Whatever attenuation method you employ, you can now listen comfortably to the audio, while operating the CW/RTTY mini-reader.

CW Users

Why fuss over an ancient transmission medium? Why? Because there are probably as many utility stations using CW as there are voice. CW/MCW has several unique advantages. CW, when properly sent, has only one decoded meaning, whereas spoken words could be misinterpreted or not understood. CW/MCW is very economical to transmit. CW is interference resistant. Often you can pick out the CW tones, even though a voice-mode station is transmitting right over it. Weak CW is more readily discernible than weak voice. By utilizing the internationally recognized Q and Z code systems, and CW shorthand, individual radio operators with dissimilar languages can make themselves perfectly understandable to one another.

"CQ CQ CQ DE JOS JOS JOS QSX 4 6 8 MHz K"

Even though transmitted by a Japanese maritime coastal station, this message is completely understandable, anywhere in the world. CW has enough advantages to keep it a viable transmission medium, now and in the future.

Maritime ship/shore stations are the major users of CW. Much of the CW coming from maritime coastal stations are markers. These are continuously repeated messages that are used to inform ships that the station is available for communications. These CW markers usually have a prologue, such as CQ or V, followed by DE, and then the station call letters. Markets can also contain QSX data; the frequencies of ship sidebands that the station is monitoring. Most markers will end with the letter K. An example of a maritime marker is:

CQ CQ CQ DE JOS JOS JOS QSX 4/6/8 MHZ K

This literally means:

"Calling all stations. This transmission is from station JOS (Nagasaki maritime coastal, Japan). I am listening for transmissions in the 4/6/8 MHz maritime ship sidebands. I am standing by for your traffic."

Some really remote places can easily be heard in North America, via CW. An example of this is: VVV DE FUX. Sometimes it comes in so loud and clear, you'd swear it was just down the block. In reality, it is from the French Naval Communications station on Reunion Island. That is a small island in the Indian Ocean, east of Madagascar. Most French Naval stations use this simple V marker, and only the last letter is changed. For example:

FUM—Papette Tahiti, French Polynesia

FUG—Noumea, New Caledonia

FUF—Fort-de-France, Martinique

FUV—Dijibouti

FUG—Castelnaudry, France.

The only major departure from this is HWN, which is the French Naval Headquarters, Paris, France.

CW markers are often repeated for hours on end. If you wish to learn to read Morse code transmissions, the maritime coastal-station marker transmissions will allow you ample time and opportunity to successfully decode the broadcasts. Of course there are some actual two-way communications, primarily on maritime duplex frequencies. Another type of broadcast the coastal stations make are the regular-scheduled marine weather forecasts. At intervals, maritime coastal stations will transmit a list of ship call letters (4-characters each, such as: GBTT A8YK, WFZK, 5MNU, and JAOT) that they are holding traffic for. These are called traffic lists and, in CW, can be identified by that title, or TFC LIST, or QTC LIST.

Morse code does have one disadvantage. This is the time required to send a message. Say to yourself, "Abraham Lincoln." This takes about one second to voice out. In contrast, when keying it in Morse code, it goes like this:

dit-dah dah-dit-dit-dit dit-dah-dit dit-dah dit-dit-dit-dit dit-dah dah-dah
dit-dah-dit-dit dit-dit dah-dit dah-dit-dah-dit dah-dah-dah dit-dah-dit-
dit dah-dit

As you can readily observe, Morse code requires considerably more time to convey the equivalent voice-mode version. CW transmissions have two limitations. The speed/accuracy of the sender versus the copy speed/accuracy of those receiving the transmission. CW can be sent at speeds in excess of 40 wpm (words per minute), but this is mainly with a machine-to-machine interface.

Utility monitors can avail themselves of the previously discussed CW mini-reader. Again, they have an audio-level problem that must be modified before they can be of practical use. Because of the high-audio driving level, weak signals often can't be successfully decoded with the reader. Raise the volume and, eventually, the background noise overpowers the reader's ability. Also, mini-readers can't discriminate between two CW transmissions on the same frequency. It will read both, resulting in a gibberish visual display. A RTTY signal on the CW frequency totally interferes with the reader's ability to decode. Some MCW transmissions have an annoyingly high carrier tone. This tone also scrambles the reader circuitry and prevents an accurate readout. Mini-readers are programmed to decode International Morse code and no other. So, in some specific instances, it will not be able to decode nonstandard Morse code.

Under normal circumstances, the CW mini-reader does an excellent job, but where the reader fails, the ear can succeed. So anyone who makes CW monitoring a major aspect of SWL'ing should learn to read Morse code by ear.

RTTY, Fax, and Other Users

Other major usages of the radio medium are *radioteletype* (RTTY) and *facsimile* (Fax).

Radioteletype

Radioteletype is the transmission of tone pulses, on one of two closely spaced frequencies. This technique is known as *Frequency Shift Keying* (FSK). RTTY is sent by typing out the messages on a modified typewriter called a teletypewriter. Standard RTTY is based on the Baudot system (named after its originator). This system consists of each letter, number, punctuation symbol, and teletypewriter function having a 5-element group. For conventional reference, these groups are made up of the binary 0s and 1s. There are a total of 32 combinations of 0 and 1, ranging from 00000 to 11111. Of course, the alphabet, numbers, and symbols will total more than 32 combinations. The way the excess is taken care of in the Baudot system is similar to that of a standard typewriter. Your typewriter has a "shift" key. When depressed, you can type capital letters and the punctuations found on the top section of certain keys. The teletypewriter has two shift keys. One is for Letter Case (ltrs) and the other for Figure Case (figs). In this way, the 32 combinations can be utilized twice, with each having a different meaning when preceded by the case shift. The actual Baudot teletypewriter code does not use all 64 possibilities. 00000 is not used in the Letters or Figures cases. The letters A and N do not have a Figure Case equivalent. The 5 teletypewriter functions (figure case, letter case, space, carriage return, and line feed) are common to both cases. By the way, in RTTY, letters are sent as capitals. There are no lowercase letters.

The use of a 5-element group allows for consistency in operation. In CW, the pulse combinations can range from a single dot or dash to five dots or dashes. This means a wide variance in the time needed to transmit a specific character. In RTTY, each character has 5 elements; hence, the character time frames are identical, since the RTTY pulses are of the same identical time length. So how is RTTY accomplished?

The binary 0 is equivalent to zero voltage values. This is known as the START or SPACE pulse. The binary 1 produces a voltage flow. This is the STOP and MARK pulse. The modern RTTY signal is produced by shifting the audio oscillator between two frequencies. This is known as AFSK. The older system shifted the carrier oscillator—FSK.

The Mark pulse has a standard audio tone of 2125 Hz. The Space audio tone is variable, depending on the shift selected. How do these pulsations make sense, when they are jumping back and forth between two frequencies? To answer this, we will examine a transmission, utilizing a paper tape.

For many of the RTTY marker-type transmissions, a perforated paper tape is still used. This tape is ¾ inch wide and as long as the message it contains. For continuously repeated markers, the tape is connected end to end, forming an endless loop.

The tape is divided into six vertical elements. Of these, the third space is utilized for feeding the tape out of the teletypewriter and into the transmitter interface reader. It also acts as an alignment indicator for each vertical element group. When the operator presses the key for the letter U, this activates the U striker to impact the tape. The Baudot letter U has a binary equivalent of 00111. On the paper tape, it is arranged vertically, bottom to top.

The tape is then fed through a mechanical device, consisting of five spring-loaded contact probes, which are linked to levers and cams. When a probe protrudes through a punched hole, it produces a voltage circuit. This indicates a Mark pulse. Where there is no hole the probe cannot go through the tape, and zero voltage is produced. This signifies a Space pulse. A 5-segment time-sequential rotary cam interprets the voltage condition of each successive probe. So the letter U goes out as: Space, Space, Mark, Mark, Mark.

We are utilizing electronic devices to encode and decode the FSK/AFSK transmissions. Electronic logic dictates the need for a standard reference to distinguish one complete group from another. This is accomplished by adding a Start pulse (binary 0) in front of the 5-element group, and a Stop pulse (binary 1) after it. These are automatically inserted into the actual RTTY transmission. When the TTY device starts to scan the tape, it inserts the Start pulse. After the 5 elements are scanned, it adds a Stop pulse. Therefore, each Baudot group consists of 7 transmitted pulses. The first pulse is always a binary 0, and the last pulse is always a binary 1. This gives the RTTY demodulator a reference to decode the signal. Although electronically noted, the Start and Stop pulses are not shown in the readout display, or occur on the paper tape. Fig. 17-1 is a stylized representation of a paper tape containing the phrase: UTILITY MONITORING 123 321. The dimensions are somewhat exaggerated for a clearer presentation.

Normal RTTY transmissions are sent in one of three standard FSK/AFSK shifts. This simply means that the frequency shifts, between the Space and

**Fig. 17-1.
Stylized
representation
of a punched
paper tape.**

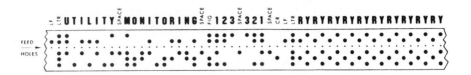

All the letters shown above the tape are there to show you the message format. They are not on the actual paper tape. This message would appear via video display or hard-copy printout as:

UTILITY MONITORING 123 321
RYRYRYRYRYRYRYRYRY

Abbreviations:

LF — line feed
CR — carriage return
LTR — letter case shift
FIG — figure case shift

Mark, have an internationally adopted spacing. These are: 170, 425, and 850 hertz. In addition, the indicated shifts are followed by a phase indicator. A small letter ''n'' indicates a normal phase, where the Mark frequency is higher than the Space frequency. A small letter ''r'' is the reverse of the above.

Another component of a RTTY transmission is its speed. Words-per-minute (wpm) have been used, but, as individual words vary in their letter lengths, a more precise method is to judge the speed of the transmission by the 5-element character groups. These groups are uniform in timeline; this method is called the *baud rate*. Ergo, a 50/425n transmission is one which is transmitted at 50 baud (roughly 66 wpm) with a 425-Hz shift and in a normal phase.

With microchip technology, the paper tape/mechanical FSK setup is rendered obsolete. Direct electronic interpretation is utilized for current state-of-the-art AFSK. Thus, electronics eliminates all moving mechanical parts, making modern AFSK a total electronic system. The older mechanical/paper tape system still has certain applications, however, and will probably be utilized for some time to come.

Obviously, RTTY is one machine talking to another. By ear alone, you could never make any sense of the FSK pulsations. For decoding purposes, one must use a RTTY demodulator. There are three basic types of demodulators. First, there is the dedicated *terminal unit (TU)*. These are high-priced, very versatile units that are capable of decoding most RTTY modes and nonstandard baud/shift notes. Some can even decode encrypted Baudot code. These units require a display unit. This is either a modified television receiver or a hard-copy printer.

Then, simpler and self-contained, there are the *mini-readers*. These have fixed baud/shift rates, but do contain a built-in billboard-type visual readout. (A billboard has a progressive left to right movement of the displayed characters.) These mini-readers will not decode ARQ, FEC, encrypted Baudot, or nonstandard baud/shift rates. (Refer to the CW section for attenuator/audio cures for the mini-reader.)

A third type of demodulator is the *computer interface* model. These utilize the computer to display the decoded RTTY. Some models allow for software add-in, to decode the more exotic RTTY modes. Like the mini-reader, it is both CW and RTTY capable.

The most important piece of equipment for RTTY monitoring is your SW receiver. A megabuck TU is useless unless your rig is "rock stable" on frequency. This is likewise true for mini-readers and computer interface models. RTTY is unforgiving when it comes to frequency stability. You are dealing with two separate transmissions, made only a few hundred hertz apart. This might not sound critical, but may I remind you that 1 kHz is 1000 hertz, and a 170 RTTY shift means that the Mark and Space pulses are only 170 hertz apart. Thus, any drifting will cause you to lose either the Mark or Space pulse, resulting in gibberish being displayed.

Many utility buffs tune the RTTY signal by ear, while watching the TU/mini-reader LED display—either blinking or bar graph. The more advanced buffs will utilize an oscilloscope to produce horizontal and vertical ellipse patterns (there are appropriate patterns for each Shift, Mark, and Space-only notation). Also, some TUs have a built-in oscilloscope or dot-matrix-type tuning screen.

A very handy device for tuning an RTTY signal is a gadget called "Blinky" (manufactured by TimeKit). It has a tinted screen that is calibrated from 1.2 to 2.3 (1200 to 2300 Hz). You simply tune your rig until the LED lamp blinks over the 2.1 (2100 Hz) position and you've correctly tuned the signal, via the Mark pulse. Offset, tuning the rig will produce two blinking LEDs. Their respective positions will indicate the RTTY shift you are tuned to. In the case of a 425 shift, if one LED is blinking over the 1.2 position, the other LED will be over the 1.7 position. This is a valuable add-on device for an RTTY monitoring setup.

What can one hear in RTTY? Virtually all types of communications, ranging from embassy, military, and governmental transmissions, to ship/shore, news services, weather, and so on. The 170 shift is used by the maritime stations. However, 425 is the most used shift. Here you will find the bulk of the news services, and a host of other transmissions. The 850 shift is primarily used by aero and marine weather stations. Just as in CW transmissions, a percentage of the RTTY transmissions are of the *marker* variety. These are used to call up other stations, inform listeners of an upcoming broadcast, or just to keep the frequency open for any possible communications.

RY, SG, and YR are very common in a marker transmission. As with the CW markers, the station sends QRA, CQ, station identification, and any other pertinent information, followed by a string of RYs, SGs, or YRs. Then, the whole process is repeated. To the ear, a RY marker is easy to distinguish. The RY portion sounds like an electronic drum roll. An example of a RTTY marker is CQ DE 6VY41 RYRYRYRYRYRYRYRYRYRYRYRYRYRYRY. SG is often used by Spanish-speaking stations. YR is Arabic and is the reverse of RY. Some military and government stations use a FOX marker. This is the phrase, "The quick brown fox jumps over the lazy dog's back." There are variances to this, but this is the basic FOX marker. Some French stations use the Le Brick marker: "Voyez le brick geant que j'examine pres du grand wharf." And endless-loop paper tape is used to send these markers. Sometimes the operator will accidentally reverse the loop and everything will come out backwards:

YRYRYRYRYRYRYRYRYRYRYR 14YV6 ED QC...

Just as with the CW markers, the RTTY types can go on and on. They are an excellent way to perfect your RTTY tuning techniques. Another source is Ham radio. 14075 to 14110 kHz is easily heard in the afternoons and early evening, with Ham operators conversing on the RTTY. Since these are mainly straightforward transmissions, they too make excellent practice lessons.

If you've ever wondered how the newspapers and TV stations get stories from around the world, it is via subscription to a wire service. These services monitor the worldwide RTTY news service (also called press service) broadcasts. So, by monitoring the news services, you can receive news that didn't show up in the newspaper or on TV. In many instances, you can receive the news before you see it on TV. There is a fly in the ointment, however. Not too long ago, RTTY monitoring was a very expensive hobby. Most news services transmitted clear broadcasts, for they realized that only a small percentage of people could afford to monitor them. Today, thanks to microchips and lower prices, anyone can do so. Some press services now encrypt or use nonstandard baud/shift rates to foil easy decoding.

In all of the utility monitoring realm, RTTY news service broadcasts is the only area in which you can get good old propaganda. If any ISWBC buffs misses being lied to, deceived, or otherwise flim-flammed, they'll feel right at home if they are into RTTY press service monitoring.

Facsimile

These are radio photo transmissions. Primarily used in weather forecasting and analysis, these Fax pictures usually depict weather charts and cloud patterns taken by orbiting weather satellites. Expensive equipment is required to decode and visualize Fax pictures. Only a handful of utility monitors are into Fax monitoring due to the expense involved. Also, there is a limited amount of interest in cloud pictures and weather maps. A Fax transmission sounds like a repeated scratchy, whirling noise, not unlike those produced by a record needle, after the end of the song, running on the inside grooves of a record.

ISWBC Feeders

These are SW broadcasts made by main studios and which are relayed to a remote station. The BBC relay on Ascension Island is one example. The transmissions are sent in SSB mode, on frequencies outside of the allocated ISWBC bands. What you hear, via the feeder, is a program before it is aired, as an official ISWBC relayed broadcast.

Point-to-Point

Once all transoceanic telephone calls went either by sea cable or HF radio. With the advent of communication satellites, many major PTP facilities switched to satellite relay. Today, there are still a few PTP networks on HF frequencies.

However, many are localized in nature, connecting remote locations to one another.

Antarctica

The very name congers up visions of cold expanses of ice and desolation. It is remote and exotic, and the quest of many a utility monitor. Most activity takes place during the brief Antarctic summer, which is roughly from September through February. Several nations have established scientific stations on the ice and the neighboring islands. Radio communications is a vital link and is extensively utilized.

The U.S. Navy provides logistical support, not only for the U.S. National Science Foundation research projects, but to other nations operating in Antarctica. During the summer, McMurdo Station is the metropolis of Antarctica. Via the U.S. Navy's airborne support, McMurdo is both a scientific station and the logistical hub. The USN uses LC-130 Hercules aircraft from the VXE-6 squadron for their flights. The aircraft voice ID with an XRAY DELTA ## call sign.

UNITED STATES

The primary United States Antarctic stations are:

- McMurdo (also IDs as MAC Center) with the call sign of NGD, and a MARS call of NNN0ICE,
- South Pole with the call sign of NPX, and a MARS call of NNN0NWB,
- Siple with the call sign of NQU, and a MARS call of NNN0ICF,
- Palmer with the call sign of NHG, and a MARS call of NNN0KMR.

Of these, only Palmer station is too far away from McMurdo for the LC-130 aircraft to reach it. Palmer is supplied by sea.

All of the stations are manned year round. After March, they have only skeleton crews/scientists, who continue their duties through the bleak Antarctic winter. Except for emergency airlift flights (when possible), they are totally cut off from the outside world, except for radio communications. Siple station is the smallest of the four, and winters over with just four personnel. During the summer, several temporary installations are used. Byrd Surface Camp and Dome Charlie are two such sites.

Air/ground communications of the U.S. facilities can be heard on 8997 kHz USB voice (primary), with 13251 kHz USB voice as the secondary frequency. Most communications are heard between 0400 to 1000 GMT. An active point-to-point network can be found on 11552 kHz USB voice from 0000 to 0400 GMT. MARS communications to the USA normally occur on 13975 kHz USB/LSB voice around 2200–0200 GMT.

AUSTRALIA

Australia has several stations in the Antarctic:

- Mawson Base—call sign VLV,

- Casey Base—call sign VNJ,
- Davis Base—call sign VLZ,
- MacQuaire Island-call sign VJM.

These stations all work one another on VNM Melbourne, Australia. The most active frequencies are: 9940, 12215, 12255, and 14415 kHz with USB voice. Try from 2300 to 0500 GMT.

GREAT BRITAIN

Great Britain also has several stations, both on the ice and on Antarctic area islands. All are part of the British Antarctic Survey (BAS). The Falkland Islands conflict made many people aware of these facilities.

- Port Stanley, Falkland Islands—call sign ZHF 88,
- Grytviken, South Georgia Island—call sign ZBH,
- Signey, South Orkley Islands—call sign ZHF 33

and on the ice:

- Halley Bay—call sign VSD,
- Rothera—call sign ZHF 45,
- Faraday (Argentine Island)—call sign ZHF 44.

The most active frequency is 9106 kHz USB voice. Try from 2300 to 0100 GMT.

RUSSIA

The USSR has five main Antarctic stations. All use CW or RTTY modes.

- Bellingshausen—call sign UGE 2,
- Molodezhnaya—call sign RUZU,
- Mirnyy—call sign UFE,
- Vostok—call sign RULE,
- Novolazerevskaya—call sign UDY.

Their active frequencies are: 8102, 10140, 13715, and 13385 kHz.

OTHER COUNTRIES

France, Belgium, New Zealand, West Germany, Poland, Japan, and South Africa each have one station. Chile and Argentina have several. Rarely have any of these stations been heard in North America.

Long Wave, VLF, and ELF

Below the AM broadcast band is a territory that many utility monitors seldom explore. For some, it is simply because their SW receiver doesn't tune below the broadcast band. For others, they've checked it out, but found little or no activity.

Today's shortwave receivers are a technological marvel, with solid-state circuitry, digital readout display, and a wide frequency spread that is electrically tuned (this is called synthesized tuning). This permits the equipment manufacturers to offer a rig that can continuously tune from the theoretical 0 Hz up to 30 MHz. Unfortunately, for the below-530-kHz buff, these modern SW rigs were designed primarily for the HF frequencies. As such, many have poor sensitivity below the broadcast band.

All is not lost, however, for there are methods and means to coax out these signals. Three conditions have a direct influence. The time, the season, the solar activity. The lower the frequency, the more prone it is to atmospheric static. Simply put, this means thunderstorm activity. The best season for long-range low-frequency monitoring is during the winter months. Any BCB buff will tell you that during the daytime, signals don't propagate very far. A 50-KW AM station may get out 500 miles. At night, that same station could be heard 2000 miles away, or more. The same holds true for below 530 kHz. When sunspot activity is at a low ebb, the ion layers are in less motion, hence more reliable long-range propagation. So the best conditions are: nighttime with darkness between you and the stations of interest; during the winter months; and with low sunspot activity.

Antennas

Long-wire antennas do work, and they work very well below the broadcast band. For optimum performance, they must be mounted outside, away from any power lines or electrical generating sources, and the long-wire must be long. Granted, a 100-foot piece of copper wire will work, but as the name implies, long wave means long wavelength frequencies. This translates into a very long piece of wire, even of one-half-wavelength dimensions.

Some very successful long-wave monitors can detect signals from nearly halfway around the globe. They do so with what is called a *Beverage* or *Wave* antenna. It is a simple thing; just a single wire, 10 to 20 feet off the ground, running in a straight line for a distance or roughtly 1500 feet. Obviously, to utilize the *Beverage* to its full capacity, you require a very big backyard, or a location in the boondocks, far from any town or city.

Long-wire antennas have drawbacks, though. Even though they have narrow directional characteristics (approximately a 4-lobe, 30° angle off of its length), they are, for all practical purposes, omnidirectional on long-wave frequencies. The long-wire antenna is also a magnet for electrical-type noises. There are ways to make a long wire more effective, however. Via electronics, we can tune, boost, and filter. At the same time, there are other types of antennas, that work just as well.

Broadcast-band DX'ers will tell you that an electrically tuned loop or ferrite bar antenna of seemingly small dimensions can outperform a 300-foot long-wire antenna. The loop and, to a lesser degree, the ferrite bar are very directional in signal pickup (actually bidirectional—180° apart). With a long-wire antenna, you've got to pray that the station you don't want to hear will fade out, if two stations are on the same frequency. With the loop/ferrite bar types, you merely turn them edge-on in the direction of the offending signal, and you null that signal out. Loop/bars are small and easily rotated. With this capability, you can either cut down the signal level by turning the antenna edge-on, or you can turn it broadside to the desired signal and obtain the maximum possible signal level. Also, their smaller physical size is far less prone to electric noise interference.

The modern SW receiver has a control called the *antenna tuner*, or *preselector*, which is used to electrically tune your antenna to match the received frequency. This is an excellent aid, providing the rig's circuitry is likewise sensitive in the long-wave (and below) frequency regions. Most general-coverage types of rigs are not. Electronics can come to the rescue, however. Be it long wire or loop/bar, all sets will benefit from a three-part electronic assist.

The first assist is circuitry designed to electrically tune your antenna to match the long-wave frequencies. Let us look at this, in terms of radio wave theory. A radio signal is an electrical charge that strikes your antenna, and runs its full length and back again. The time it takes to accomplish this is in step with the frequency wavelength. Since this electrical charge does so twice, per its wavelength, your antenna wire need only be one half the length of the wave to be tuned (or, in resonance) to that specific frequency. Thus, if you wanted to match your antenna length to say 5000 kHz, its half wavelength would be 98.4 feet. Now, if your antenna is physically longer or shorter than this length, you can peak "peak" the signal by adjusting the rig's antenna trimmer/preselector to electrically change the length of your wire and match its mathematically designated physical length for that given frequency. Sounds confusing, but the next example should make it clearer.

A long-wave frequency of, say, 150 kHz requires a half-wave antenna with a physical length of 3280 feet (talk about a BIG backyard!). Quite obviously, a SW rig designed primarily for HF frequencies won't be able to compensate for this. But an outboard circuit, specifically designed for long-wave reception, can and will do so.

The second component is *pre-amplification*. Longer wavelengths require more transmitter power to reach out the same distance. In North America, the bulk of long-wave stations are the CW Nondirectional Radio Beacon (NDB) types. Some NDBs also transmit AM-mode weather broadcasts. In any event, most only run a few hundred watts of power. The farther away that a long-wave station is from you, the weaker it will be. Hence, there is the need to amplify the signal before it is fed into your receiver. By amplifying, or boosting, the signal, however, you also boost the atmospheric and electrically generated noise.

Therefore, the third component is *filtration*. This cuts down on the QRN/QRM, while only minimally degrading the actual signal. This electrically tuning, boosting, and filtering system is more commonly known as an *active antenna*. With a system designed for long-wave frequencies, your rig will suddenly come alive with the sounds of all manner of stations.

Long-wave loop and ferrite bar antennas are available from several manufacturers. If you are into soldering, from time to time, radio magazines will run home-construction projects that are designed for long-wave applications.

Frequency Usage

What can one hear below the broadcast band? From 525 kHz down to 190 kHz is the realm of the aeronautical and maritime radio navigation beacons. These are commonly abbreviated NDB. NDBs are audible reminders of the past, when these stations constantly repeated Morse code signals, and were the primary radio-directional medium. Today, sophisticated Radar, LORAN-C, Omega and NAVSAT systems provide very precise location techniques. Although technology has outmoded the long-wave NDB, it still survives for very practical reasons. The receivers required to translate LORAN-C, Omega and NAVSAT pulsations are still very expensive, whereas the long-wave equipment is not. The NDB system itself is simple, low cost, and fully automated. This allows NDB sites to be unmanned and, when needed, located in very remote places throughout the globe. In high-density airborne traffic areas, in addition to their radio-location function, NDBs can also have a superimposed AM-mode airport/area, transcribed, weather broadcast. Even though technology is outmoding the long-wave NDB, its practicality ensures its survival for many years to come.

NDB signals consist of two or three letters, or a letter and number, sent in continuous Morse code. They act as the site indentification, and the companion carrier signal is used for the actual radio-bearing determination.

Within the 190- to 525-kHz region are other types of transmissions. 472 and 500 kHz are used for worldwide calling and distress (in the CW mode). In region 1 (the Western hemisphere), 415 to 495 kHz is reserved primarily for CW ship/shore communications. In region 2 (Europe and Africa), 148.5 to 283.5 kHz is utilized for long-wave AM-mode broadcast stations. Some of the European and African broadcast stations use upwards of 1500 KW of power. During favorable periods, many of these stations can be heard in North America.

Long-wave transmissions continue down to 30 kHz. NDBs can be found in this region. 90 to 110 kHz is reserved for LORAN-C navigation pulses. 20 and 60 kHz are set aside for standard-frequency time broadcasts. Within this area, one major user is the military. Because of the extremely wide wavelengths, voice-mode transmissions are not possible below 100 kHz.

From 30 kHz down to 10 kHz is the VLF (very low frequency) region. Almost all stations here are the military. The 10- to 50-kHz range is used by the U.S. military as the VLF component to its worldwide communications with its nuclear deterrent forces. RTTY is used, but the very long wavelengths require very slow transmissions; 5 baud (or 7.1 words-per-minute) is the norm. So even a short message would take several minutes to transmit. Omega navigation signals occur between 10.2 and 13.6 kHz.

Just for laughs, if you should ever consider constructing a half-wave antenna for 10 kHz, its length will be just over 42,600 feet, or 9 plus miles long....

Below 10 kHz is the ELF (extremely low frequency) region. This area is not allocated for any service. The U.S. Navy is experimenting with ELF for communications with its ballistic missile submarine force. ELF penetrates deeply into water; lesser so for VLF. In contrast, HF frequencies literally bounce off the surface of the water.

Scientists have monitored this "basement" of radio frequencies and have found that it is allocated for use by the Earth, itself. All manner of chirps, whistles, and other ethereal sounds have been heard. Some are associated with worldwide thunderstorm activity, others are assumed to be indications of earth crust movements. Some signals have yet to be identified. If nothing else, ELF is where the Earth, itself, sings a very mysterious song.

If you are interested in ID'ing the NDB beacons, then the Long Wave Club of America is the place to be. LWCA also publishes a NDB beacon guidebook. Their address can be found at the end of this chapter.

Other Users

We have touched upon many major types of utility users. There are others, such as accurate time signals and reports on solar/geomagnetic activity. You can hear embassy traffic, various United Nations organizations, INTERPOL, telemetry beacons from orbiting satellites, and astronomical observatory chatter. Numerous U.S. government agencies are on HF. They include the FEMA, FCC, FAA, DOI, DOE, DEA, USCS, USACE, and the FBI.

Scattered liberally throughout the frequencies are stations that cannot be readily identified. Many are military and government types, operating in what is called the *tactical mode* (coded call signs and discrete frequencies). Many stations go unnoticed due to their use of very sophisticated transmission modes. These can include voice scrambling, frequency hopping, very-narrow-shift multifrequency RTTY, and so on.

As most ISWBC buffs are aware, there are a host of CW beacons throughout the SWBC bands. These letter–letter or letter–number combinations are identifications utilized by Soviet jamming stations.

Single-Letter Beacons

There is a very puzzling group of single-letter HF beacons found throughout the HF spectrum. They are somewhat similar to the long-wave NDBs in that they continuously transmit a single letter.

The most common of these single-letter HF beacons are: K,F,U,Q. Some transmit unmodulated CW, while others, like the K beacon, are either unmodulated CW or FSK CW (simply put, FSK CW is Morse code sent in a RTTY medium). In the Speedx SRGU, a chapter was devoted to an investigation of these HF beacons. It was found that the FSK-mode "K" was routinely interrupted for several minutes, with a repeated coded numeric CW transmission. Investigation placed the FSK-K beacon to be coming from the vicinity of Khabarvosk in Asiatic USSR. It is assumed that these beacons are part of the Soviet Naval communications system. Other letter-type HF beacons are believed to come from various Soviet Naval HQ areas, and from bases outside of the USSR. In this respect, during the 1970s, a "W" beacon was well heard throughout North America on a number of frequencies. Subsequent investigation located these as coming from Cuba. For reasons unknown, the W series ceased transmitting by the late 1970s.

The FSK-K family of beacons is often heard quite well in North America. On late evenings, they can be heard on 9043 and 11155 kHz. These frequencies, plus 2844, 4006, 8012, 8158, 12250, 14478, and 14966 kHz, can be heard during local pre-dawn hours.

The "F" single-letter HF beacon family normally fades in from midnight to dawn (denoting a Pacific origin) in the frequency areas of 5305, 6800, 8646 and 10646 kHz. For more information on these mysterious HF beacons, check the Speedx SRGU.

Clandestine Stations

Many "clandestine"-type operations occur on the HF frequencies. Drug smuggling is a vast illegal enterprise, and, from time to time, utility buffs have monitored English- and Spanish-language transmissions, which were, by their conversation content, very probably drug related. Smugglers seem to prefer the frequency areas above the Ham bands, in the 7300- to 7500-kHz and 14300- to 14600-kHz areas. They are also to be found within or adjacent to the 4-, 6-, 8-, and 12-MHz maritime mobile bands. The message content is usually cryptic with, "you-know-what-I-mean" type of references, buzz word identifications, and so on.

Commercial fishing is a legal occupation, but one that is very competitive. In the past few years, some fishing fleet operations have been utilizing out-of-band (maritime mobile) frequencies for communications. They use a variety of colorful call signs, such as Kojak, Mr. T, Tomcat, Diamond Jim, and Pelican Pete. At first glance, one might assume some type of military tactical operation, but when you listen to the conversations, it becomes quite apparent that they are not military. The hallmark of the fishing fleet transmission is the liberal use of profanity and X-rated tales of a sexual nature.

Even truckers have gotten into the act, apparently not satisfied with the short-range Citizen band (CB) radio frequencies. However, fishing fleet and truckers face stiff fines if caught by the FCC. Obviously, drug smugglers could care less about the consequences of violating radio regulations.

Another type of clandestine traffic that has been monitored appears to be para-military or guerrilla motivated. These and other types of oddball transmissions can be broadly grouped as either illegal or clandestine. The *clandestine* title is usually reserved for operations which presumably orginate from within the military or governmental circles. Of all the clandestine stations, there is one group that draws the most interest and controversy. These are the enigmatic *Number Transmissions*.

Number Transmissions

Number transmissions is a catch-all designation for a group of stations that utilize a predictable presentation format, with all stations sending coded numerical or alphanumeric groups. These stations have been labeled, by some, as "SPY," although none has proven that they are of that gender. No government has acknowledged them as their own, nor do Federal or ITU listings show frequency/station identification data. Legally, they do not exist, yet as any Number buff can tell you, they are easily heard throughout North America (also in Europe and the Far East).

Number transmissions can be sorted into specific groups, for each type has an individualized procedure and format. The major types and characteristics are:

- (SY5) SS/YL 5-digit AM mode: They start on the hour with "Atencion ###, plus a group count," which is repeated for about 3 minutes. Then, the transmission goes into the 5-digit groups. The sign-off is, "Final, Final." The entire transmission is repeated several minutes later on a different frequency. The SY5 may have as many as four different transmissions going per hour. This is the most widely heard Spanish-language Number type.

- (SY4) SS/YL 4-digit AM mode: They start on the hour. The sign-on is 3 digits repeated three times and followed by 1234567890. The whole thing is repeated for almost 10 minutes. A series of 10 tone pulses and a group count is heard; then the transmission goes into the 4-digit text. Immediately, the entire 4-digit text is repeated. Sign-off is "Fin." The SY4 simultaneously keys on two frequencies, and has only one transmission per hour. This type has a distinctive pulsation that precedes and follows the actual voice transmission. It has been christened "The Walking Man" because it sounds exactly like footsteps, moving at a steady pace.

- (SY3) SS/YL 3-digit AM mode: This starts at plus 25 minutes after the hour, with 5 minutes of one-second-interval time tones. (Note: some transmissions do not have this time-tone prologue.) On the half hour, and for the next 10 minutes, the station repeats three to six of the 3-digit groups. There is no sign-off identification. It uses the same person's voice as the SY4 transmissions. So far, these signals have only been heard at 0125, 0425, 0725, and 1225 GMT. Transmission is simultaneously keyed on two frequencies.

- (GY5) GG/YL 5-digit USB mode: This uses two types of sign-on signals. One repeats two phonetic letters (such as Papa November), along with musical notes, for five minutes. The words "Achtung" and "Gruppen" follow with numerical identification and the group count. Each 5-digit group is repeated twice. The sign-off signal is "Ende." The transmissions use a simultaneously keyed method of frequency usage, with the entire message repeated approximately 30 minutes later. The second variety sends an "N" in Morse code for 5 minutes, followed by the German word "Gruppe" (group count) and 5-digits groups that are repeated twice in a row. The sign-off signal is the German word "Ende." The transmissions are simultaneously keyed. Both start on the hour.

- (EY5) EE/YL 5-digit AM/USB mode: This is apparently the sister of the GY5 transmission. It uses "N" in Morse code for 5 minutes, and then a word group, and then 5-digit groups, which are repeated twice. Sign-off signal is "End." The English voice has a German accent.

- (EY4) EE/YL 4-digit AM/USB mode: This transmission starts on the hour. Sign-on signal is a 3-digit group, plus a 1 to 0 count.

- (SY5T) SS/YL 5-digit AM mode: This also starts on the hour. The sign on uses a triple "Atencion," plus ### and a group count. The group count is not given in individual numerics, however, For example: 90 would be nueve cero. This SY5T transmission denotes it as ninety—noventa. Sign off is a triple "Final." The station sends a repeat transmission on the half hour, on a different frequency.

- (SO5) SS/OM 5-digit AM mode: This transmission uses one of the few male voices heard. It is called The Blabber by some, for the haphazard fashion of vocalizing the 5-digit text. The sign-off signal is a series of 5 zeros.

- (CW5) CW mode 5-digit: The transmission starts on hour, and the sign-on signal is a repeated segment consisting of 3 letters and 2 letters. The transmission uses only 10 letters: A, D, G, I, M, N, R, T, U, and W. The sign-off signal is a triple AR and VA.

- (CW4) CW mode 4-alphanumerics: The transmission starts at plus 25 minutes with 5 minutes of one-second-interval time tones. (As with the SY3 transmission, it sometimes does not have these tones.) On the half hour, for 10 minutes, three or four 4-alphanumeric groups are repeated. There is no sign-off signal. The transmission uses only 10 elements: A, B, D, E, N, T, U, V, 4, and 6. Morse code is a very slow 6 wpm.

- (PO) Phonetics AM/USB mode: These signals are just repeated transmissions of three phonetic letters and one numeric letter, such as: Charlie India Oscar 2. Among the frequencies used are the same as those of the GY5/EY5 signals, and the phonetic pronunciation is in English, with a German accent.

These are the basic Number transmission types heard in North America. Some are definitely related to one another. Many are heard throughout the 24-hour period, while others are heard only at selected times. The transmissions also have other common characteristics. On any given day of the week, a particular Number type will use the same frequency, during the same time period. This has been referred to as the Day-Time-Frequency pattern, abbreviated to DTF. With a little time and effort, anyone can assemble a daily schedule of Number broadcasts, not unlike that of a *TV Guide* listing. This rigid DTF trait seems most odd, considering that the concensus believes these transmissions are SPY or otherwise intelligence related.

All Number transmissions repeat the initial message. This is most logical, for part or all of the first transmission might have been missed. Logic breaks down when the same transmission is found being repeated weeks, and even months, apart. An example is the SS/YL 5-digit transmissions. Many, using the same DTF or other DTF periods, send the same exact message, from two to twelve times in succession. In the mid-1970s, this SY5 transmitted the same message, 7 days a week, for almost 180 consecutive days. Repeats of the same transmission for weeks, and even months, is another oddity that doesn't fit into the framework of what one would consider a SPY-type broadcast network.

Rigid Day-Time-Frequency patterns, a multiplexity of message repeats, straightforward transmission methods and modes, plus a static presentation format, all serve to make these Number transmissions, as predictable and obvious as possible. If they are part of cloak-and-dagger operations, then they are as secret as the presence of a full moon in the nighttime sky. But, then, there is the other side of the coin.

We know that the Spanish-language types of messages have been on the air for roughly 20 years. A 20-year, 7-days-a-week, year-to-year cycle says many things. The operation is well planned and well financed. There is a dedication of purpose. Since no one directly involved with these number transmissions has ever gone public, there is obviously very effective iron-clad security surrounding them. Federal and international radio regulating bodies have no information on them. No call signs, frequencies, station identification data is known. In other words, on paper, the Number transmissions do not exist. All of this suggests a secret operation that, in one form or another, uses the awesome clout of the government to ensure nondisclosure of the basic details. But, it is simplistic to assume that these are cloak-and-dagger operations of a type where messages are being sent to agents in the field. The obvious fact is that, to this date, not one shred of evidence to directly prove an intelligence connection has ever been uncovered.

The U.S. government is of little help in trying to solve this mystery. Of the many government agencies queried, some pretend ignorance, while others state it is not within their agency's jurisdiction, and some simply pass the buck. The same agency can say one thing to Mr. Jones and have a completely different answer for Mr. Smith, even though they both asked the same basic question. From all of this, it must be assumed that there is an agency-wide directive not to disclose valid information on any Number-type operation. This does not mean that the U.S. is directly involved, however (although the clues seem to indicate that the U.S. government is responsible for some of them). However, the U.S. government will protect from disclosure those transmissions made by foreign governments. This is all part of the diplomatic/intelligence game. We respect the confidentiality of their communications and, for very practical reasons, don't comment on their military/intelligence operations. They, in turn, keep silent about ours. In this way, the game can go on, secure in the knowledge that those people affected by the operations will never learn the details of the operations.

For a number of years, the FCC has quoted with its one liner concerning the Spanish Number transmissions: "They're from Cuba." In the light of the game plan, the validity of this reported statement must be viewed with a measure of suspicion. It is further rendered suspect since the FCC has never provided any black-and-white proof to substantiate this alleged claim. From time to time, the FCC will deny that it even stated it. This is a familiar diplomatic/intelligence game ploy. Allow yourself to be quoted as saying one thing, and then turn around and deny it.

So it is up to individuals and concerned groups to ferret out the truth. In 1983, John Demmitt, working with several European Number buffs, RDFed the German-language Number broadcasts as coming from Nauen, East Germany. This answered only part of the mystery. As Mr. Demmitt will readily acknowledge, there are German Number transmissions heard in North America that, due to the time and frequencies employed, could not possibly be coming from East Germany.

What we do have here is a genuine mystery. In time, the basic answers will be uncovered, but only through the efforts of the individuals and groups who are dedicated to solving the Numbers enigma.

For those of you who are interested in the Number transmissions, Table 17-1 is a list of frequencies and Number-type usage. Please note that this list is far from complete. It represents the most commonly encountered transmissions. Because of space, it is not possible to indicate the specific times and days used. For current information and activity, check the utility columns of shortwave hobby clubs. Because frequency readout can vary, due to receiver/individual interpretation, some of the closely aligned frequencies can be assumed to indicate only one actual frequency. Number types are in parentheses, following the frequency.

As you can observe from the Number frequency/type list, not only do the individual Number types cluster around specific frequency areas, but all types seem to more or less congregate together. Clues abound, but it requires that Number buffs be detectives. One must monitor and record data, sift and sort, research and hypothesize in order to validate or dismiss observations and ideas. If nothing else, it is a very tedious, time-consuming process on the road to the truth.

An entire book could have been devoted to this topic of Number transmissions. This chapter was merely an introduction to this very mysterious utility arena.

QSL'ing Utility Stations

A large percentage of utility stations do verify requests, thereby opening the door to some very unique confirmations.

Table 17-1. List of Number Frequencies and Number Types.

2690 (PO)	5184 (GY5)	6800 (SY5)	7846 (SY5)	9570 (SY5)	12905 (GY5)
3060 (SY5)	5225 (GY5)	6802 (SY4)	7855 (SY5)	9700 (CW5)	12950 (PO)
3080 (SY5)	5230 (PO)	6803 (SY5T)	7873 (SY5)	9830 (CW4)	13150 (PO)
3090 (SY5)	5230 (EY5)	6820 (SY5)	7885 (SY5)	9845 (SY5)	13378 (SY5)
3150 (PO)	5315 (GY5)	6825 (SY5)	7887 (SY5)	9956 (CW5)	13436 (SY5)
3218 (GY5)	5410 (GY5)	6829 (SY5)	7890 (SY5)	9967 (CW5)	13442 (EY5)
3220 (GY5)	5435 (PO)	6830 (SY5)	7905 (GY5)	9970 (SY5)	13452 (SY4)
3232 (GY5)	5440 (GY5)	6835 (SY5)	7906 (SY5)	9973 (EY5)	13485 (SY5)
3240 (SY5)	5540 (SY5)	6840 (SY5)	7910 (SY5)	10000 (CW5)	13785 (SY5)
3258 (GY5)	5550 (GY5)	6840 (CW4)	7912 (SY5)	10015 (SY5)	13787 (SY5)
3260 (EO5)	5620 (EO5)	6845 (EO5)	8065 (EY5)	10019 (SY5)	13808 (SY3)
3264 (GY5)	5643 (PO)	6850 (SY5)	8070 (CW5)	10135 (SY5)	14414 (SY5)
3370 (GY5)	5693 (GY5)	6855 (GY5)	8078 (EY4)	10177 (GY5)	14417 (SY5)
3385 (GY5)	5750 (GY5)	6860 (GY5)	8112 (SY5)	10266 (SY5)	14419 (SY5)
3416 (PO)	5750 (SY5)	6860 (CW5)	8117 (SY5)	10345 (SY5)	14421 (SY4)
3442 (SY5)	5760 (PO)	6873 (SY5)	8120 (GY5)	10426 (SY5)	14430 (SY5)
3445 (SY5)	5760 (GY5)	6875 (SY5)	8173 (GY5)	10460 (GY5)	14441 (CW4)
3808 (GY5)	5765 (SY5)	6875 (EY4)	8185 (SY5)	10460 (PO)	14586 (SY5)
3820 (GY5)	5782 (SY5)	6877 (SY5)	8216 (CW5)	10500 (EY5)	14603 (EY5)
3852 (GY5)	5808 (GY5)	6886 (SY5T)	8405 (SY5)	10500 (GY5)	14735 (SY5)
4010 (GY5)	5810 (SY4)	6887 (SY5)	8418 (SY4)	10570 (SY3)	14845 (SY5)
4015 (SY5)	5812 (SY4)	6890 (SY5)	8425 (PO)	10610 (EY4)	14886 (EY5)
4022 (GY5)	5820 (GY5)	6892 (SY5)	8924 (EY5)	10613 (EY4)	14935 (GY5)
4025 (SY5)	5850 (SY5)	6917 (SY5)	8425 (PO)	10740 (GY5)	14945 (GY5)
4030 (SY5)	5900 (SY5)	6920 (SY5)	8842 (SY5T)	10820 (PO)	14947 (GY5)
4043 (SY5)	5905 (PO)	6928 (SY5)	8870 (PO)	10823 (SY5)	15085 (GY5)
4045 (CW5)	5910 (GY5)	6997 (GY5)	8872 (SY5)	11030 (EY4)	15610 (SY5)
4046 (SY5)	5923 (SY5)	7026 (CW5)	8873 (CW5)	11034 (EY4)	15615 (SY5)
4050 (SY5)	5930 (SY5)	7320 (EO5)	8874 (SY5)	11054 (SY5)	15650 (SY4)
4055 (SY5)	5936 (SY5)	7340 (SY5)	8900 (CW5)	11108 (GY5)	16118 (SY5)
4100 (SY5)	5940 (SY5)	7342 (SY5)	8915 (SY5T)	11190 (GY5)	16450 (SY4)
4100 (CW5)	6203 (GY5)	7375 (CW5)	8925 (PO)	11215 (EY4)	16817 (CW5)
4110 (CW5)	6214 (EO5)	7375 (EY4)	8925 (EY5)	11215 (CW4)	17392 (GY5)
4305 (SY4)	6226 (SY5)	7375 (GY5)	9000 (CW5)	11350 (GY5)	17410 (EY5)
4307 (SY5)	6230 (SY5)	7384 (SY5)	9038 (CW5)	11415 (SY5)	17410 (PO)
4395 (GY5)	6236 (GY5)	7404 (PO)	9038 (SY5)	11420 (SY5)	17430 (GY5)
4455 (GY5)	6240 (EO5)	7404 (GY5)	9042 (GY5)	11432 (SY5)	17595 (PO)
4475 (SY5)	6250 (SY5T)	7410 (GY5)	9050 (SY5)	11450 (GY5)	17650 (SY5)
4500 (CW5)	6271 (PO)	7425 (SY5)	9050 (GY5)	11512 (SY5)	17968 (PO)
4545 (GY5)	6317 (SY5)	7427 (SY5)	9052 (EY5)	11518 (SY5)	18195 (GY5)
4560 (PO)	6345 (SY5)	7430 (CW4)	9072 (SY4)	11532 (SY4)	18210 (SO5)
4562 (EY5)	6373 (SY5)	7433 (CW5)	9075 (SY5)	11534 (SY5)	18737 (CW4)
4592 (GY5)	6410 (GY5)	7446 (PO)	9080 (SY5)	11540 (SY5)	19158 (SY5)
4600 (CW5)	6450 (GY5)	7446 (SO5)	9085 (SY5)	11545 (GY5)	19295 (GY5)
4615 (EY5)	6455 (GY5)	7500 (CW5)	9222 (SY4)	11555 (SY5)	20875 (SY5)
4670 (PO)	6495 (SY5)	7525 (SY5)	9264 (GY5)	11618 (GY5)	
4670 (SY4)	6500 (CW5)	7527 (SY5)	9267 (CW5)	11625 (CW4)	
4706 (SO5)	6506 (GY5)	7530 (SY4)	9267 (GY5)	11630 (SY5)	
4730 (GY5)	6510 (GY5)	7540 (SY5)	9267 (EY5)	11633 (SY5)	
4770 (GY5)	6570 (SY5)	7600 (CW5)	9325 (GY5)	11638 (SY5)	
4780 (SY5)	6572 (SY5)	7600 (SY5)	9325 (SY5)	12093 (GY5)	
4785 (SY5)	6580 (SY5)	7605 (PO)	9360 (SO5)	12135 (SY5)	
4790 (SY5)	6735 (SY5)	7625 (EY4)	9400 (SY5)	12162 (CW4)	
4825 (SY5)	6748 (SY5)	7631 (EY4)	9432 (SY5)	12164 (CW5)	
4835 (SY5)	6767 (SY5)	7660 (GY5)	9435 (EY4)	12235 (SY5)	
5015 (GY5)	6770 (SY5)	7752 (EY5)	9443 (SY5)	12312 (SY5)	
5082 (SY5)	6772 (SY5)	7752 (GY5)	9446 (SO5)	12314 (EY5)	
5093 (PO)	6774 (SY5)	7830 (SY5)	9450 (SY5)	12315 (SY5)	
5135 (CW5)	6776 (SY5)	7835 (SY3)	9450 (CW5)	12315 (GY5)	
5135 (SY5)	6777 (SY5)	7836 (SY5)	9453 (SY5)	12324 (SY5)	
5156 (GY5)	6780 (SY5)	7837 (CW5)	9462 (SY5)	12415 (EY5)	
5164 (SY5)	6783 (SY5)	7845 (CW5)	9465 (CW4)	12415 (PO)	

Some utility types have a finite lifespan. Ship transmissions are one such type. Every year, hundreds of vessels, both large and small, end up on the scrap heap. Others encounter what every sailor dreads—a calamity at sea. Whether the vessel goes peacefully to the junkyard, or reluctantly to the bottom of the ocean, it does disappear from the face of the earth. Ship QSL'ers can show you many a QSL from a now extinct ocean roamer. So uniqueness is one drawing force in utility QSL'ing.

Location is another. Where else but the utilities can you QSL the South Pole? ISWBC buffs will toil for months on end to log and verify all of the African nations. Not many of us have actually seen the supersonic Concorde airliners, much less flown in one. But any aeronautical buff can hear and QSL them. A commercial aeronautical buff can, via the air traffic control network, in two weeks or less of diligent monitoring, log all the aeronautical stations. Islands worldwide can yield to the utility buff—places like Reunion, Yap, MacQuaire, Iwo Jima, Marcus, Johnston, Diego Garcia, Cocos, Okinawa, Ponape, Raoul, Saint Paul, South Georgia, Majuro, Truk, Saipan, Madeira, Kure, Sao Tome, and a host of others.

ISWBC QSL'ers can number their verifications in the hundreds, and BCB buffs in a few thousands, but utility QSL'ers have possibilities of several hundred thousand. Fig. 17-2 shows four examples.

Fig. 17-2. Some utility station QSL cards provided by the stations themselves.

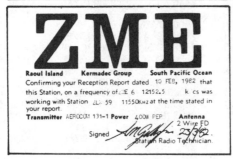

How does one get into utility QSL'ing? The first precept you must recognize is that *all utility communications are considered private communications*. FCC and ITU regulations expressly forbid any divulgence of specific details relating to these communications.

ISWBC or BCB buffs who are making the crossover to utilities will find that utility QSL'ing procedures are a very different animal from what they've been normally accustomed to. You must provide proof that you monitored the broadcast. For ISWBC/BCB'ers, this means that you have to give very specific details about the content of the program. Do this in utility QSL'ing and your chances of receiving a verification are close to zero. Why? Because you would be violating nondisclosure regulations. This is the pivotal aspect of utility QSL'ing..., you *CANNOT DISCLOSE* specific details that relate to the communications content, *not even* to those parties who participated in them.

How does one go about providing information that doesn't violate regulations, yet will confirm your reception? First, understand that radio communication QSL'ing has certain reception report information that is common to all. This information includes: date, overall time (GMT), frequency, transmission mode, SINPO data, and information on your radio equipment. The technique in utility QSL'ing is to use the "time frame" arrangement and buzz words. These buzz words state a condition:

WORKING, CALLING, PHONE PATCH VIA, RADIO CHECK, TESTING.

To illustrate, there is a hypothetical aeronautical communications breakdown from airliner NorthWest 73....

(0101–0102.5) "Honolulu this is NorthWest 73. Flight level 390.... ADENI at 0101, estimate ABSOL at 0155. Request flight level 370." "Roger 73, standby." "73 standing by."

(0105–0107) "NorthWest 73, Honolulu." "Go ahead Honolulu." "ATC advises that Northwest 73 is cleared to descend to and maintain flight level 370. Report when reaching." "Honolulu, NorthWest 73, leaving 390 for 370. Will report when reaching." "Honolulu."

(0108–0109) "Honolulu, NorthWest 73. Flight level 370." "Roger 73." "73."

(0125–0129) "Honolulu, NorthWest 73." "Go ahead 73." "Request weather, Honolulu International for my 0315 ETA."

Honolulu then transmits the weather data.

On your utility reception report, these communications are broken down as follows:

(0101–0102.5) working Honolulu.

(0105–0107) working Honolulu.

(0108–0109) working Honolulu.

(0125–0129) working Honolulu.

Another illustration, this time for a cruise ship, the Pacific Princess.... Mr. Tom Williams in Cabin 901 wishes to place a telephone call to 319-621-3984, a Mr. James Williams in Cedar Rapids, Iowa.

(1315–1317) "KMI, KMI, this is Pacific Princess." "Go ahead, Pacific Princess." "Wish to place a station to station call to, blah, blah, blah."

Finally, KMI advises the Princess that the party is on the line. The Princess radioman informs Tom to go ahead.

(1317–1335) Tom tells his father about the wonderful time he and his wife are having on their Mexican vacation cruise, and the sights seen, places visited, things bought, and so on, and so forth.

You indicate these as follows:

(1315–1317) working KMI.
(1317–1335) phone patch via KMI.

By now you should basically understand the use of buzz words. All utility stations are required to keep a radio log. Usually these notations are nothing more than the frequency, time, and who they contacted. Buzz words do not violate nondisclosure regulations, yet they do indicate the type of communications taking place. By checking the radio log, your simple data provides all that is required to confirm your reception. In any two-way communication exchange, this is the method to use. But, as with anything, there are grey areas to any procedure.

Repeated markers and test transmissions are by strict definition transmissions made in the blind, directed to no specific party. These are one-way tests, or general hailing transmissions, which are in themselves not active communications. Many stations will not object to you repeating the message in your reception report. Some, though, follow ITU regulations to the letter. Unless you know the station's policy, use a "time line" reference. An example of this is:

"A 10-second duration Morse code CQ/QSX marker, repeated 5 times per minute."... "An English and French-language voice marker, spoken by a female, 30-seconds duration, in continual repeats."

Here, again, you identify the transmission from a "time line" reference, without stating exactly what the text of the message was.

Weather broadcasts (aero/marine) are public-service types of broadcasts. These are normally scheduled and cover specific areas. Usually all you have to do is report something like this:

"A 15-minute marine weather broadcast, for the North Pacific region," or "A 10-minute aeronautical weather broadcast for locations in South America."

Even with the grey area transmissions, it is wise to follow nondisclosure principles, for radio regulations can be interpreted differently by individual stations.

It goes without saying that you will not be prosecuted for violating the regulations. The most you will lose is postage and the verification. So why all the fuss? It is your fellow utility buffs that stand to lose much more. Some utility stations, after receiving too many specific communication disclosure reports, decided not to acknowledge any further reception reports. Ergo, your lack of common sense could have contributed to the closing of a station, or even an entire network-run group, to any further QSL'ing successes. This is not just a possibility. Over the years, this has occurred a number of times. For the dedicated and responsible utility QSL buff, nothing is so frustrating as to lose a good verification, simply because a small percentage of his/her fellow DX'ers didn't follow the game plan.

Utility station radio rooms are often staffed by but a few individuals. Some stations are isolated. Often the radioman enjoys receiving a reception report, for it can break the monotony. It could be that your report was the first one the radioman ever received, or that it came from a location beyond which the radioman thought his signals could be heard. These factors usually ensure that your report will be read and checked out.

A very cordial and friendly enclosure that many include is a picture postcard of their area. It's a way of saying, "Hi there, here is where I heard you from." People are naturally curious, and showing them a place that they've never seen before can be a very positive gesture. This, in turn, could encourage them to send something (other than the QSL) back to you. I've got dozens of postcards, photos, and emblems to prove this.

Topics to Avoid

There are some other items you can include in your reception report and some topics that you must avoid. There are two NO–NOs. *NEVER* make any reference to religion or politics. You might have made the utility logging that comes once in a lifetime, but your offhand comment inadvertently insults the religious beliefs or political ideology of the individual who reads your report. Since you don't know anything about the person who will read your report, stay clear of any comments that are of a religious or political nature.

The QSL Itself

It is a very wise policy to include a prepared from card (PFC) along with your report. Utility stations are under no obligation whatsoever to even acknowledge your reception report. So make it as easy as possible for them to reply. The other factor is that many utility stations have no official QSL card/letter. The PFC should contain references to the data you seek, such as frequency, transmission mode, date, time, station identification, signature, and so forth. Since different types of utility stations require slightly different types of data responses, you would normally have several basic PFC setups.

A 4- by 6-inch index card is just the right size. There is enough room for the data you seek, plus it fits neatly into the standard #10-size mailing envelope.

You can clearly print the data, type it, or even take a copy to your local print shop and have them run off copies for you. One point I should not have to mention is that you *never fill in the data yourself*. Just send the PFC and allow the station to fill in the blank spaces.

Always include return postage. Include mint stamps if it goes to a station in your country or areas under its jurisdiction. If you have access to a stamp store that can provide mint stamps of the foreign country you are sending the card to, you can use that approach. But be sure to include enough postage. The other method in sending mail to a foreign country is to use International Reply Coupons (IRCs). These can be obtained at your local post office. Usually one or two IRCs are sufficient. The importance of including return postage is simple. You don't want to lose a verification, simply because the station doesn't have, or cannot afford, a stamp. Stations that bankroll their outgoing mail will often return, unused, your stamp or IRC.

A note about mailing to an overseas U.S. facility or warship that has an APO or FPO routing. This is in the U.S. military postal system. No matter if the location is in a foreign country, an APO/FPO routing requires only a First Class U.S. stamp for mailing, and the appropriate postcard/First Class stamp return.

Some utility stations do have their own QSL card or letter. If you prefer that, indicate in your report that they are to discard your PFC in favor of their official QSL. If you choose to go this route, then it goes without saying that you don't affix the mint stamps to your PFC; keep them separate.

Some countries prefer that reports/PFCs be in their own native language. Several publications lay out multilanguage setups that are specifically for utility station reporting. The *Speedx SRGU* has a 5-language layout, with many optional items, that can be used to configure a report/PFC to a specific type of utility station.

Mailing addresses can be found in utility-related books, magazines, and shortwave club publications.

There is virtually no limit to the types of utility stations you can obtain verifications from. Fig. 17-3 through 17-5 show representative examples of PFC returns. All are 4- by 6-inch card size that have been reduced so I could show several to the page.

Summary

This chapter on utility monitoring was not designed to make you a successful or proficient utility monitor. All the articles and books in the world cannot do that. At best, they merely inform, instruct, or act as frequency/station identification reference material. Neither does an extensive assemblage of expensive radio equipment make it any easier. In essence, they are only tools which you must master and utilize. A thousand-dollar receiver won't find a single station without your guidance. You are the key that unlocks the door to the utilities.

First, you need to have the desire to journey into the twilight world of shortwave radio. Next, you need the patience to overcome confusion and frus-

Fig. 17-3.
Samples of
prepared form-
type QSL cards
received back
from utility
stations.

tration, plus the willingness to learn and, in some cases, learn anew. You have to give it time, and that time is spent primarily in front of your shortwave receiver. There is no substitute for experience, but through it comes knowledge and understanding.

The world of utilities can either frighten you off, or present a challenge to be surmounted—for the deeper you journey into the utility world, the more diverse and complex you will find it. You may end up specializing in just one or two

Fig. 17-4. More prepared QSL cards

aspects, or you may generalize your pursuits. Whatever path you take, it will lead you into the everyday real world. You could find yourself ringside during the making of a front-page story, or you can simply be witness to the everyday occurrences that most take for granted.

All I've tried to do is to give you some helpful hints, suggestions, and gently nudge you down the path. Where and how far you go is up to you. That is the ultimate bottom line reality.

Fig. 17-5. Still
more prepared
QSL cards.

Source List

U.S. Shortwave Clubs that cater to the utility DX'er are:

- American Shortwave Listeners Club (all types of utility coverage)
  16182 Ballad Lane
  Huntington Beach, CA 92649

- Society to Preserve the Engrossing Enjoyment of DX'ing (all types)
  SPEEDX
  P. O. Box E
  Elsinore, CA 92330

- The Association of Clandestine Radio Enthusiasts (pirate radio and Numbers)
  The ACE
  P. O. Box 13225
  D. T. Station
  Minneapolis, MN 55414

- Long Wave Club of America (long- & medium-wave coverage)
  45 Wildflower Road
  Levittown, PA 19057

Some magazines to read are:

- Popular Communications (from newsstand)
- Radio Electronics (from newsstand)
- Monitoring Times (via subscription from Grove Enterprises.)

Book publishers to contact (some offer rigs and other radio accessories) are:

- Gilfer Associates, Inc.
  P. O. Box 239
  52 Park Avenue
  Park Ridge, NJ 07656

- Universal Electronics, Inc.
  4555 Groves Road Suite 3
  Columbus, OH 43227

- Grove Enterprises
  140 Dog Branch Road
  Brasstown, NC 28902

- CRB Research
  P. O. Box 56
  Commack, NY 11725

- SPEEDX
  P. O. Box E
  Elsinore, CA 92330

  The current book project for SPEEDX is the Speedx Reference Guide to the Utilities (SRGU). This is an ongoing installment-by-installment build-up of a comprehensive utility manual.

Some manufacturers of CW/RTTY equipment (mini-readers, terminal units, computer interface equipment, etc.) are:

- Kantronics
  1202 East 23rd Street
  Lawrence, KS 66044
- Digital Electronics Systems (Info-Tech)
  1633 Wisteria Court
  Englewood, FL 33533
- HAL Communications Corp.
  Box 365
  Urbana, IL 61801
- Advanced Electronic Applications (AEA)
  P. O. Box 2160
  Lynnwood, WA 98036
- MFJ Enterprises, Inc.
  Box 494
  Mississippi State, MS 39762
- Universal Amateur Radio, Inc.
  1280 Aida Drive
  Reynoldsburg, OH 43068
- TimeKit
  P. O. Box 22277
  Cleveland, OH 44122

18

WORLD RADIOTELETYPE PRESS STATIONS

Thomas P. Harrington

Tom Harrington (W8OMV) has been in radio communications for many years and has actively monitored most regions of the radio spectrum. He flew in the United States Army Air Corps during World War II, and during this period became interested in shortwave propaganda broadcasts in Europe. Later, through Amateur Radioteletype, Tom became interested in all types of radioteletype services and world news RTTY. When he is not engaged in his electronics and radio business, this interest in RTTY fills much of his time.

Tom Harrington is a Life Member of the American Radio Relay League, the Quarter Century Wireless Association, and the Experimental Aircraft Association.

In today's fast-moving world, getting knowledge of world events hours, and even days, ahead of the regular press, TV, and radio announcements holds a great fascination for many shortwave listeners. Also, the ability to receive news services from a particular part of the world has meaning for many people. Most of these news services are not reported in our usual news media and much of this news is withheld for many reasons. But, you can receive all of the world news, along with financial happenings, as they are pouring out of the many world capitals—twenty-four hours a day.

Radioteletype

Radioteletype communication has been used for many years because of its speed of transmission and its relatively simple method of producing a hard copy using the well-known Teletype™* machine. These machines are available and are now appearing on the "used" market at low prices. The strange-sounding signals which can be received with the average, general-coverage, communications receiver is fed from the audio output, or speaker, of the receiver directly into the new microprocessor units which convert the frequency shifts of the signal into intelligent copy. This copy is then shown on a video display much like a TV set (Fig. 18-1) and/or to a hard-copy printer, if desired. All that is needed in the way of equipment is:

1. A general-coverage stable receiver.
2. A terminal unit (TU).
3. A video display.
4. A hard-copy printer, if desired.

The equipment and its sources are covered in the following sections in detail, with basic recommendations.

There are hundreds of news service teletype stations operating around the clock in all parts of the world. These stations transmit on schedules, on demand, and as news develops. There are many other types of radioteletype (RTTY) stations in use by many sources. Some of these are the Army, Navy, Air Force, and the News Services of most of the countries of the world. Many of the major news services are international in scope and operate high-powered RTTY transmitters from many locations. Some of the well-known news services are United Press International (UPI), Associated Press (AP), DPA, ANSA, and TASS of the USSR.

---

* Teletype is a registered trademark of AT&T Teletype Corporation.

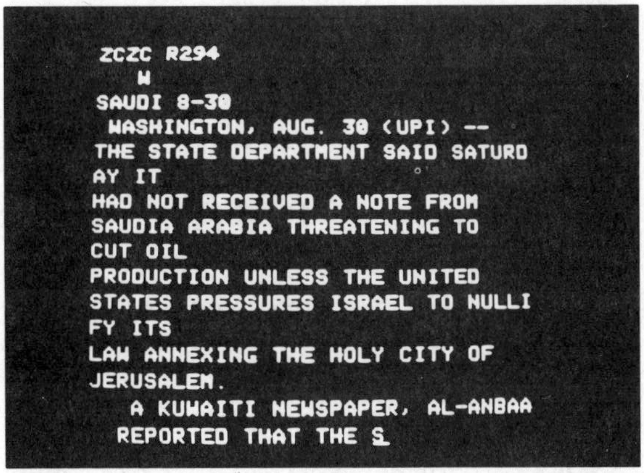

Fig. 18-1. Video
display of a
radioteletype
communication.

```
ZCZC R294
     W
SAUDI 8-30
  WASHINGTON, AUG. 30 (UPI) --
THE STATE DEPARTMENT SAID SATURD
AY IT
HAD NOT RECEIVED A NOTE FROM
SAUDIA ARABIA THREATENING TO
CUT OIL
PRODUCTION UNLESS THE UNITED
STATES PRESSURES ISRAEL TO NULLI
FY ITS
LAW ANNEXING THE HOLY CITY OF
JERUSALEM.
   A KUWAITI NEWSPAPER, AL-ANBAA
   REPORTED THAT THE S
```

World News Frequencies

After many years in amateur radio, I became interested in amateur radioteletype (RTTY). Most of this RTTY time was spent communicating with other amateur radio operators in the amateur bands. From time to time, I would receive commercial news teletypewriter stations outside of the regular amateur bands. As time went on, I became more familiar with the world news services and learned how and where to find them. During this period, I built a list of these news services and their scheduled times for transmitting. I also discovered that these news services used different frequencies for different times of the day. I found that they were sending beamed transmissions to subscribers in all parts of the world. The subscribers to these services were newspapers, radio stations, and TV stations. I also found that some of the news services were propaganda stations for their countries, i.e., TASS (U.S.S.R.), North Korea, China, etc. The listing in Table 18-1 is up-to-date, and lists only those news services that transmit in English on a regular basis.

Table 18-1. World Press Frequency List (English Transmission)

Frequency	Call	Service/Country	Information
3356.0	SUA 99	MENA Cairo, EGY	2000 E
4529.0	NFE	USN Kato Soli, GRC	VFT: 2249 E nx
4623.0	SOE 262	PAP Warsaw, POL	2000 E nx
4804.0	ISZ 48	ANSA Rome, I	1900+2110 E nx
5023.0	9KT 24	KUNA Safat, KWT	1000–1900 E nx
5097.5	JAB 35	KYODO Tokyo, J	0830–1100 E nx
5224.0	AJE	USAF Wolvey, G	VFT: 2127–0544 E nx
5460.0		VoA Tanger, MRC	75 Bd: 0000 E nx, 2300 E nx
5830.0	RWD 52	TASS Moscow, RU, URS	0430+0500+1400+1700 E nx
5849.0	SUA 79	MENA Cairo, EGY	1900+2300 E nx
5859.0	Y2V 3	ADN Berlin, DDR	1700 E nx
6418.0	IAR 26	Rome R, I	0730 E nx
6504.5	WCC	Chatham R, MA, USA	SITOR: 0024+0149+0422+0900+2329 E nx

Table 18-1
(cont.)

Frequency	Call	Service/Country	Information
6675.0		PETRA Amman, JOR	1700 E nx
6688.0	AJE	USAF Wolvey, G	VFT: 2049 E nx
6870.0	RTV 55	TASS Moscow, RU, URS	0430+0500+2000 E nx
6915.0	BAP 46	XINHUA Beijing, CHN	1630+1830 E nx
6943.0		VoA Kavalla, GRC	75 Bd: 0000+0200+0300 E nx
6960.0	LZN 3	BTA Sofia, BUL	1830 E nx
6972.0	YOG 59	AGERPRES Bucharest, ROU	1700 E nx
7460.0	LZB	BTA Sofia, BUL	1830 E nx
7520.0	BZP 57	XINHUA Beijing, CHN	0100+2030 E nx
7525.0	RTV 54	TASS Moscow, RU, URS	0430+0500 E nx
7568.0	AJE	USAF Wolvey, G	VFT: 0400+1625 E nx
7577.5	OLZ 2	CETEKA Prague, TCH	1800 E nx
7599.0	9KT 262	KUNA Safat, KWT	1000–1900 E nx
7645.0	RGE 36	TASS Moscow, RU, URS	1300+2000 E nx
7650.0	BZR 67	XINHUA Beijing, CHN	1400+1900+2030 E nx
7658.0	YZD	TANJUG Belgrade, YUG	0200+0900+1000+1100+1200+1300 +1500+1600+1730+1915 E nx
7725.0	SOH 272	PAP Warsaw, POL	0000 E nx
7756.0	SUA 34	MENA Cairo, EGY	1800 E nx
7773.5	ATU 58	INFOIND Delhi, IND	1500 E nx
7800.0	EPX 9	IRNA Tehran, IRN	1500–1730+1900-2030 E nx
7806.0	YZD 7	TANJUG Belgrade, YUG	1730+1800+1915 E nx
7842.4	CNM 20.1X	MAP Rabat, MRC	1200–1400 E nx
7960.0		IRNA Tehran, IRN	1500–1730+1900–2030 E nx
7996.0	YZD 9	TANJUG Belgrade, YUG	1730+1800+1915+2000+2100 E nx
8020.0	HME 46	KCNA Pyongyang, KRE	1800–2100 E nx
8030.0	RRQ 27	TASS Moscow, RU, URS	1300+1500+1700+1900+2100 E nx
8060.0	RAW 71	TASS Moscow, RU, URS	0403+0500+1300+1600 E nx
8062.0	IRF 80	ANSA Rome, I	1100+1900 E nx
8067.5	Y2V 7	ADN Berlin, DDR	1900+2000+2100+2200 E nx
8100.0	NGR	USN Kato Soli, GRC	VFT: 2118 E nx
8133.0	SOI 213	PAP Warsaw, POL	1700 E nx
8140.0	SLN 219	PL Habana, CUB	0800–0950 E nx
8140.0	RNN 51	TASS Moscow, RU, URS	0430+0500+1500+1600+1700+1800 E nx
8192.5	SOI 219	PAP Warsaw, POL	1600 E nx
8590.0		OANA Moscow, RU, URS	1300 E nx
8707.0	WLO	Mobile R, AL, USA	SITOR: 0620 E nx
9035.0	NGR	USN Kato Soli, GRC	VFT: 2025 E nx
9052.5	ISY 90	ANSA Rome, I	1900 E nx
9105.0	NGR	USN Kato Soli, GRC	VFT: 0703+0920+1827 E nx
9110.0	RDZ 77	TASS Moscow, RU, URS	1400+1700 E nx
9114.0	HGG 31	MTI Budapest, HNG	1700 E nx
9133.0	ZAA 6	ATA Tirana, ALB	0900–1045 E nx
9145.0	RDZ 76	TASS Moscow, RU, URS	1600 E nx
9227.0	9KT 27	KUNA Safat, KWT	1000–1900 E nx
9237.5	AJE	USAF Wolvey, G	VFT: 1620 E nx
9325.0		AUP Sydney, NSW, AUS	0530 E nx
9338.0	9KT 272	KUNA Safat, KWT	1000–1900 E nx
9349.0	GIC 29B	AP London, G	0000+0100+0200+0400+1900+2000 +2100+2300 E nx
9353.0	OLX 5	CETEKA Prague, TCH	1600+1700+1800+1900+2100 E nx
9391.0	SOJ 239	PAP Warsaw, POL	0600+1700 E nx
9395.0	HMK 21	KCNA Pyongyang, KRE	1500+1600 E nx
9396.4	FTJ 39A.G	AFP Paris, F	0330–0430+1800–1900 E nx
9402.5	ISY 94	ANSA Rome, I	1700+1800 E nx
9417.0	BZP 59	XINHUA Beijing, CHN	0200+1500+1830 E nx

Table 18-1
(cont.)

Frequency	Call	Service/Country	Information
9420.0	RMD 57	TASS Moscow, RU, URS	1800 E nx
9430.0	ZAT	ATA Tirana, ALB	1100+1500+1700–1845 E nx
9463.0	JYF 4	PETRA Amman, JOR	1700 E nx
9491.0	BZR 69	XINHUA Beijing, CHN	0100+1500+1600+1830 E nx
9797.0	YOJ 27	AGERPRES Bucharest, ROU	0900 E nx
9828.0		SPK Phnom Penh, KMP	1300 E nx
9855.0		VoA Tanger, MRC	75 Bd: 0000+2300 E nx
9950.0	YZF	TANJUG Belgrade, YUG	0430+1600+1700 E nx
10105.0	RKA 79	TASS Moscow, RU, URS	0430+0500+1500+1800 E nx
10120.0	RGI 24	TASS Moscow, RU, URS	0430+0500+2100 E nx
10153.0	9KT 281	KUNA Safat, LWT	1000–1900 E nx
10231.1		VoA Bethany, OH, USA	VFT: 75 Bd: 0100 E nx
10240.0	RGE 34	TASS Moscow, RU, URS	2000 E nx
10245.0	SOK 224	PAP Warsaw, POL	1700+2100 E nx
10258.0	RDZ 71	TASS Moscow, RU, URS	0430+0500+2000 E nx
10270.0	RKA 25	TASS Moscow, RU, URS	1300+1400+1700+1800+1900+2000 +2100 E nx
10334.0	JTD 29	MONTSAME Ulan Bator, MNG	1700 E nx
10408.0	9VF 63	ANSA Singapore, SNG	1700 E nx
10410.0		VoA Kavalla, GRC	2300 E nx
10438.0	YZB 9	TANJUG Belgrade, YUG	1600 E nx
10465.0	RKA 74	TASS Moscow, RU, URS	0430+0500+1300+1400+1500+1900 +2100 E nx
10533.0	AJE	USAF Wolvey, G	VFT: 0718–1728 E nx
10543.0	Y2V 54	ADN Berlin, DDR	1100+1500+1600+1700+1900+2000 E nx
10580.0	HMK 25	KCNA Pyongyang, KRE	0830–1000+1500–1730 E nx
10600.0	VNA 25	VNA Hanoi, VTN	1400+1500+1600 E nx
10602.5		NAN Lagos, NIG	0700+1530 E nx
10610.0	SUA 30	MENA Cairo, EGY	0700 E nx
10615.7	FTK 61H3	AFP Paris, F	0330–1430+0700–0800+1800–1900 E nx
10649.0	GIC 30B	AP London, G	0400+1500+1600+1700 E nx
10655.7	ATK 61	PTI Delhi, IND	0500 E nx
10740.0	RKA 72	APN Moscow, RU, URS	100 Bd: 1200+1400 E nx
10785.0	Y2V 43	ADN Berlin, DDR	1900+2000 E nx
10795.0	JAG 50	KYODO Tokyo, J	0720–0820 E nx
10816.0	IRH 58	ANSA Rome, I	1900 E nx
10865.0	RZA 24	TASS Moscow, RU, URS	1800 E nx
10880.0	REM 50	TASS Moscow, RU, URS	0100+1500+1600+1800 E nx
		VoA Monrovia, LBR	75 Bd: 0000+2300 E nx
10900.0		KUP Pretoria, AFS	0600 E nx
10960.0	3MA 28	CNA Taipei, TAI	0230–0330+0930–1030+1330–1500 E nx
10972.0		VoA Tanger, MRC	75 Bd: 0000+0300+0400+2300 E nx
10985.0	RCB 53	TASS Moscow, RU, URS	0430+0500 E nx
11230.0	HMF 49	KCNA Pyongyang, KRE	1500–1730 E nx
11266.0	NGR	USN Kato Soli, GRC	VFT: 0730 E nx
11291.5		KUNA Safat, KWT	1000–1900 E nx
11420.0	VNA 86	VNA Hanoi, VTN	1400 E nx
11430.0	HMN 51	KCNA Pyongyang, KRE	1000+1500+1600+1800–2100 +2130–2400 E nx
11470.0	RNK 33	TASS Moscow, RU, URS	0430+0500+1500+1600+1700+2200 E nx
11475.0	HMS 79	KCNA Pyongyang, KRE	2100+2200 E nx
11493.0	9KT 282	KUNA Safat, KWT	1000–1300+1600–1800 E nx

**Table 18-1
(cont.)**

Frequency	Call	Service/Country	Information
11494.0	SOL 249	PAP Warsaw, POL	1100 E nx
11497.0	SOL 349	PAP Warsaw, POL	0000+1400 E nx
11502.0	LZH 4	BTA Sofia, BUL	0700+1330+1830 E nx
11532.5	NGR	USN Kato Soli, GRC	VFT: 0911+1010+2143 E nx
11574.0	9KT 29	KUNA Safat, KWT	1000–1900 E nx
11575.0		APN Moscow, RU, URS	75 Bd: 1000 E nx
11680.0	BZP 51	XINHUA Beijing, CHN	0200+1600 E nx
12048.0	YZG	TANJUG Belgrade, YUG	0900+1000+1300+1700 E nx
12075.0	9KT 292	KUNA Safat, LWT	1000–1900 E nx
12085.0	RCB 55	TASS Moscow, RU, URS	0430+0500+0700+0900+1300 E nx
12108.0	IRJ 21	ANSA Rome, I	0800+1100 E nx
12128.0	IRJ 31	ANSA Rome, I	0800+1100+1500 E nx
12175.0	HMR 23	KCNA Pyongyang, KRE	0400–0730+1830–2100+2130–2300 E nx
12178.0	9KT 291	KUNA Safat, KWT	1000–1900 E nx
12186.0		JANA Tripoli, LBY	1400+1700 E nx
12212.3	YZO 7	TANJUG Belgrade, YUG	1000+1300+1500+1600 E nx
12223.5		VoA Tanger, MRC	75 Bd: 0200+2300 E nx
12250.0	RHA 41	TASS Moscow, RU, URS	0700+1600 E nx
12251.0	Y2V 32	ADN Berlin, DDR	1900 E nx
12265.0	BZR 62	XINHUA Beijing, CHN	0200+1500+1600 E nx
12275.0	JAL 82	KYODO Tokyo, J	0430–0600+0610–0640+0720–0820 +1000–1100 E nx
12280.0	GBU 32	AP London, G	0700+1100+1300+1400+1500+1600 +1700 E nx
12285.0	RKU 74	TASS Moscow, RU, URS	0430–0500 E nx
12315.0	RVW 57	TASS Moscow, RU, URS	0430+0500+0600+0800+1100+1200 +1400 E nx
12315.0	YZJ 3	TANJUG Belgrade, YUG	1300 E nx
12325.0	RDD 72	TASS Moscow, RU, URS	04300+0500 E nx
12754.0	NMA	USCG Miami, FL, USA	75 Bd: 2100 E nx
13081.0	LGJ 3	Rogaland R, NOR	SITOR: 1319+1836–0220 E nx
13083.0	WLO	Mobile R, AL, USA	SITOR: 0818+2137 E nx
13098.0	WLO	Mobile R, AL, USA	SITOR: 0340 E nx
13400.0	LZG 3	BTA Sofia, BUL	1300–1350 E nx
13410.0	RIF 38	TASS Moscow, RU, URS	0430+0500 E nx
13440.0	YZJ 5	TANJUG Belgrade, YUG	1500+1600 E nx
13462.0	VNA 16	VNA Hanoi, VTN	1500 E nx
13487.5	ISX 35	ANSA Rome, I	1900 E nx
13490.0	RCG 77	TASS Moscow, RU, URS	0430+0500+1500 E nx
13523.5	YIO 71	INA Baghdad, IRQ	1100+1200 E nx
13563.0	3MA 22	CNA Taipei, TAI	0230–0330+0930–1030+1330–1500 E nx
13580.0	HMK 25	KCNA Pyongyang, KRE	0400–0730 E nx
13597.5	OLI 2	CETEKA Prague, TCH	0600+1000+1300+1500+1630+1745 E nx
13647.5	OLI 5	CETEKA Prague, TCH	1300 E nx
13653.0	SUA 50	MENA Cairo, EGY	0700+0800+1000+1200+1800+2000 E nx
13729.7	FPN 72H3	AFP Paris, F	0700+1615–1745 E nx
13735.0	Y2V	ADN Berlin, DDR	1500+1700 E nx
13770.0		VoA Tanger, MRC	75 Bd: 0900 E nx 1200+2200+2300 E nx
13780.0	HME 28	KCNA Pyongyang, KRE	0400–0530+0830–1000+1030–2100 +1500–1730 E nx
13785.0	SON 278	PAP Warsaw, POL	1500 E nx
13793.0	SON 279	PAP Warsaw, POL	1700+2100 E nx

Table 18-1
(cont.)

Frequency	Call	Service/Country	Information
13895.0	Y2V 47	ADN Berlin, DDR	0500+0600+1000+1100+1400+1500 +1800 E nx
13920.0	RNK 39	TASS Moscow, RU, URS	0430+0500 E nx
13937.5		APN Moscow, RU, URS	75 Bd: 0900+1000 E nx
13974.0	ISX 19	ANSA Rome, I	1715+1900 E nx
13995.0		VoA Monrovia, LBR	75 Bd: 0100+0200 E nx
14362.0	SOO 236	PAP Warsaw, POL	1100+1700 E nx
14367.0	BZP 54	XINHUA Beijing, CHN	0600+0800+1100+1200+1500+1600 +1900 E nx
14362.0	YIL 71	INA Baghdad, IRQ	1100+1200+1300+1700 E nx
14417.0	NGE	USN Kato Soli, GRC	VFT: 1230-2321 E nx
14418.0	9KT 321	KUNA Safat, KWT	1000-1900 E nx
14470.0	REM 54	TASS Moscow, RU, URS	0600+0700+1200 E nx
14490.0	RNK 36	TASS Moscow, RU, URS	0430+0500+0600+1200+1400 E nx
14510.0	RIC 75	TASS Moscow, RU, URS	0500+0600+0800+1200+1300+1500 +1600+1700 E nx
14511.5		MFA Jakarta, INS	1340 msg (E nx)
14526.5		VoA La Union, Poro I., PHL	75 Bd: 0000+0100+0600 E nx
14547.5	JAL 44	KYODO Tokyo, J	0720-0820+1000-1100 E nx
14570.0	HML 61	KCNA Pyongyang, KRE	0330-0500+0800-0830+1000-1100 +1500-1730+2200-0100 E nx
14574.4	CNM 59X9	MAP Rabat, MRC	1200-1400 E nx
14630.5	ISX 46	ANSA Rome, I	1900 E nx
14632.0	YZC 2	TANJUG Belgrade, YUG	1300 E nx
14638.0	WFK 54	USIA New York, NY, USA	75 Bd: 2100 E nx
14640.0		KPL Vientiane, LAO	0900 E nx
14665.0	Y2V	ADN Berlin, DDR	1100+1200+1300+1400 E nx
14700.0	REB 24	TASS Moscow, RU, URS	0700+0800+1000+1100+1200+2000 E nx
14719.0	JAN 24	KYODO Tokyo, J	0250-0320+0440-0630 E nx
14760.0	BAT 93	XINHUA Beijing, CHN	1400+1930 E nx
	CNM 61	MAP Rabat, MRC	1200-1400 E nx
14785.0	ATP 65	INFOIND Delhi, IND	1000+1400 E nx
14800.0	Y2V 24	ADN Berlin, DDR	0700+0900+1100+1300+1400 E nx
14825.0	Y2V 25	ADN Berlin, DDR	1300+1700+1800+2100 E nx
14831.0	9KT 33	KUNA Safat, KWT	1000-1900 E nx
14880.0	RIC 72	TASS Moscow, RU, URS	0430+0500 E nx
14901.0	CLN 451	PL Habana, CUB	1900 E nx
14928.0	CLN 452	PL Habana, CUB	1400+1800 E nx
14974.0	GBW 34B	AP London, G	0400+0500+0600+0700+0900+1000 +1100+1200+1300+1400 E nx
15462.0		JANA Tripoli, LBY	0800 E nx
15480.0		APS El Djazair, ALG	1000+1100+1400+1500 E nx
15510.0	SOP 251	PAP Warsaw, POL	0600 E nx
15555.0	LZP 2	BTA Sofia, BUL	1300-1350 E nx
15575.0	REN 30	TASS Moscow, RU, URS	0430+0500 E nx
15630.0	RWM 72	APN Moscow, RU, URS	75 Bd: 0700+0900 E nx
15633.0	HMH 21	KCNA Pyongyang, KRE	0400-0530+0830-1000+1030-1200 +1500-1730 E nx
15643.0	9KT 331	KUNA Safat, KWT	1000-1900 E nx
15654.9	CNM 65.1X	MAP Rabat, MRC	1200-1400 E nx
15670.0	HGM 36	MTI Budapest, HNG	1615 E nx
15693.5	ISX 56	ANSA Rome, I	0800+1100+1500 E nx
15710.0	RWN 76	TASS Moscow, RU, URS	0600+1300+1400+1500+1800 E nx
15724.0	ISX 57	ANSA Rome, I	1500+1700+1900 E nx
15731.0		SUNA Khartoum, SDN	1600 E nx
15744.0	VNA 5	VNA Hanoi, VTN	0700 E nx

Table 18-1
(cont.)

Frequency	Call	Service/Country	Information
15780.0	RWM 71	TASS Moscow, RU, URS	0600+0800+1200 E nx
15865.0	RBK 79	TASS Moscow, RU, URS	1000 E nx
15875.0	5LA 25	VoA Monrovia, LBR	75 Bd: 0000+0100+0200+0300+0400 +2300 E nx
15897.5	OLS 4	CETEKA Prague, TCH	1500 E nx
15926.0	BZR 73	XINHUA Beijing, CHN	0600 E nx
15930.0	RBI 78	TASS Moscow, RU, URS	0430+0500 E nx
15977.0	FPP 97G	AFP Paris, F	0700+1100+1615–1745 E nx
15995.5	WER 26.45X	INFOIND N. York, NY. USA	1230+1400 E nx
15996.1	DFP 99H1	DPA Hamburg, D	1500+1600 E nx
16050.0	RCE 54	TASS Moscow, RU, URS	0700+1300 E nx
16117.5	6VK 317	PANA Dakar, SEN	0900–1200+1600–1800 E nx
16136.0	BZR 66	XINHUA Beijing, CHN	1100+1200+1300 E nx
16140.0	RGW 28	TASS Moscow, RU, URS	0600+1000+1200 E nx
16145.0	RWM 77	APN Moscow, RU, URS	100 Bd: 0700+0900+1100+1300 E nx
16150.0	RCE 59	TASS Moscow, RU, URS	0800+1500 E nx
16150.0	9VF 205	JIJI Singapore, SNG	1330 E nx
16175.0	RGW 27	APN Moscow, RU, URS	75 Bd: 1000 E nx
16210.0	SOW 221	PAP Warsaw, POL	1500 E nx
16224.0	3MA 35	CNA Taipei, TAI	0230–0330+0930–1030+1330–1500 E nx
16265.0	9VF 206	ANSA Singapore, SNG	1700 E nx
16325.0	Y2V 23	ADN Berlin, DDR	0600+1000+1100+1400+1500+1700 E nx
16343.0	YZI 4	TANJUG Belgrade, YUG	0900+1000+1300 E nx
16348.0	CLN 530	PL Habana, CUB	1800+1900+2200 E nx
16397.5	FTO 39	DIPLO Paris, F	1400 E nx
16403.0	Y2V	ADN Berlin, DDR	1100+1500 E nx
16417.0	Y2V 26	ADN Berlin, DDR	0700+0800+0900+1000+2000+2200 E nx
17002.3	NMA	USCG Miami, FL, USA	75 Bd: 2100 E nx
17049.0	UFB	Odessa R, UK, URS	2010–2023 E nx
17214.0	BZP 58	XINHUA Beijing, CHN	0700+0900+1200+1300+1400 E nx
17430.0	9VF 209	JIJI Singapore, SNG	1330 E nx
17468.0	HGO 24	MTI Budapest, HNG	0600 E nx
17492.0	SOR 249	PAP Warsaw, POL	1100 E nx
17510.0	RFD 53	TASS Moscow, RU, URS	0430+0500+0700+0800+0900+1000 +1100+1200 E nx
17525.0	OLV 3	CETEKA Prague, TCH	0600+0900 E nx
17597.5	JAQ 57	KYODO Tokyo, J	0240–0410+0430–0600+0610–0640 E nx
17950.0		KUP Pretoria, AFS	0800 E nx
18040.0	TCY 4	AA Ankara, TUR	0800 E nx
18125.0	RND 70	TASS Moscow, RU, URS	1300 E nx
18160.0	RTU 47	TASS Moscow, RU, URS	0600+0900 E nx
18192.5	CLN 603	PL Habana, CUB	2200 E nx
18220.9	CNM 76X9	MAP Rabat, MRC	1200–1400 E nx
18245.5		APN Moscow, RU, URS	75 Bd: 1100 E nx
18255.0	ATB 68	INFOIND Delhi, IND	0900+1200 E nx
18256.0	VNA 32	VNA Hanoi, VTN	0700 E nx
18260.5	9KT 346	KUNA Safat, KWT	1000–1900 E nx
18278.0	9KT 351	KUNA Safat, KWT	1000–1900 E nx
18293.0	9KT 352	KUNA Safat, KWT	1000–1900 E nx
18307.0	9KT 349	KUNA Safat, KWT	1000–1900 E nx
18385.0	RRQ 20	TASS Moscow, RU, URS	0700+0800+1000+1100+1200+1300 E nx
18440.0	RIF 32	APN Moscow, RU, URS	75 Bd: 0800 E nx

Table 18-1
(cont.)

Frequency	Call	Service/Country	Information
18496.1	CNM 80X11	MAP Rabat, MRC	1200–1400 E nx
18542.5	WFK 48	USIA New York, NY, USA	75 Bd: 2143 E nx
18548.0	GIY 38B	AP London, G	1000+1200+1300+1400+ E nx
18584.0	6MK 67	YONHAP Seoul, KOR	0730 E nx
18600.0	RBN 72	PL Moscow, RU, URS	0700+0800 E nx
18640.5	9KT 356	KUNA Safat, KWT	1000–1900 E nx
18670.7	FTS 67H3	AFP Paris, F	0815–1600 E nx
18788.0		SUNA Kartoum, SDN	0915+1630–1745 E nx
18823.5	Y2V 38A	ADN Berlin, DDR	1000+1100+1400+1500+1600+1800 E nx
18872.0	BZR 68	XINHUA Beijing, CHN	0700+0800+0900+1200+1300+1400 E nx
18985.0	OLD 2	CETEKA Prague, TCH	0800+0900+1100+1200+1600 E nx
19068.4	CNM 83X9	MAP Rabat, MRC	1200–1400 E nx
19100.3	S2M 33	BSS Dhaka, BDG	0600–0700+0900–1000+1100–1200 E nx
19114.0		MFA Jakarta, INS	0730–1750 ry or msgs or E nx
19171.1	CNM 85X11	MAP Rabat, MRC	1200–1400 E nx
19200.0		IRNA Tehran, IRN	1000–1100 E nx
19340.0	SUA 313	MENA Cairo, EGY	0700+1100 E nx
19396.0	GIW 39	GNA London, G	1330 E nx
19396.0	GIW 39A10	JIJI London, G	1200–1300 E nx
19505.0	RCD 36	PL Moscow, RU, URS	0700+0800 E nx
19525.0	OLD 4	CETEKA Prague, TCH	0900 E nx
19557.0	RFD 51	APN Moscow, RU, USR	1000 E nx
19565.0	70B 92	ANA Aden, YMS	1800 E nx
19605.0	YZJ 9	TANJUG Belgrade, YUG	0700+0900 E nx
19915.0		VoA Tanger, MRC	75 Bd: 0900+1100 E nx
20085.0	ISX 20	ANSA Rome, I	0700+1100+1515+1900 E nx
20204.0	YZJ	TANJUG Belgrade, YUG	1200+1300 E nx
20311.8	FTU 31B.G	AFP Paris, F	0700+0800+0815–1600 E nx
20327.5	6VK 221	PANA Dakar, SEN	1200–1600 E nx
20430.0	IRS 24	ANSA Rome, I	1100 E nx
20785.8	CNM 92X9	MAP Rabat, MRC	1200–1400 E nx
20910.0	9VF 232	ANSA Singapore, SNG	0645 E nx
20960.0	9VF 233	KYODO Singapore, SNG	0430–0640+1000–1100 E nx
21807.5	YOV 28	AGERPRES Bucharest, ROU	0730–0830+1100–1230 E nx
22786.0	9JA 87	ZANA Lusaka, ZMB	1000+1100 E nx

The International Telecommunications Union

Most radioteletype stations and world press stations work with and follow the frequency coordination of the International Telecommunications Union (ITU). Most countries belong to the ITU and follow their block frequency assignments and other agreed-upon operating techniques, such as call signs and standard radioteletype code systems. Most stations use the standard *Baudot* (5 level) radioteletype code. In addition to the Baudot code, there is *ASCII* (7 level), *MOORE, TOR*, and the system of limited security called *Bit-Inversion*. Some of the new terminal units now copy the commercial code known as TOR (Teleprinting Over Radio) in the ARQ and FEC modes. *SITOR* and *SPECTOR* are trademarked versions of the TOR system. (These systems have error features built into the operating mode.)

Most of the coastal marine stations transmit in TOR and, as a regular part of their schedule, carry world news to ships at sea and others who use the services. The newscasts carry AP–UPI material. However, press stations do not always use the 7-days-a-week operating schedule. Some carry a reduced weekend schedule and their transmissions are of short duration on weekends due to light world-news activity at these times.

Frequency Lists

Frequency lists have various types of information that is usually listed, such as *press station frequency* and *time of transmission*.

Press station frequency is given in kHz on all lists. These frequencies may vary, ± 2 kHz, due to the type of receiver and the sideband used to tune the RTTY station. Some news services will shift their frequencies if there is interference, or their equipment may cause a slight frequency shift. Therefore, use the stated frequency in kHz as the center spot and tune to either side of the frequency listed.

Transmission times are all given in Greenwich Mean Time (GMT), now sometimes called Universal Time Coordinated (UTC). These times will be the same the world over. For example, in any time zone, the GMT and UTC time will be the same. GMT time is based on Greenwich, England, indicated as 0 (zero) on the world time map in Fig. 18-2. This map also indicates the approximate deviation from Greenwich around the world, while Chart 18-1 gives the deviation for the time zones in the continental United States.

Station Clocks

An inexpensive, digital, 24-hour clock (Fig. 18-3) should be considered for your "shack." Many of these clocks are available either as kits or as assembled and tested units. For ease of use, they must be in the 24-hour format. One specific

Fig. 18-2. Approximate deviation of world time zones from GMT.

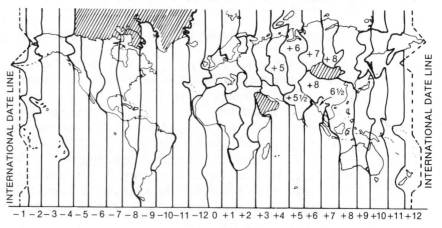

−1 −2 −3 −4 −5 −6 −7 −8 −9 −10 −11 −12 0 +1 +2 +3 +4 +5 +6 +7 +8 +9 +10 +11 +12

**Chart 18-1.
Greenwich
Mean Time—
Local Time
Zones, USA**

With this chart you can convert your local time to GMT. GMT is 4 hours ahead (+) of Eastern Daylight Time; 5 hours (+) ahead of Eastern Standard Time and Central Daylight Time; 6 hours ahead (+) of Central Standard Time and Mountain Daylight Time; 7 hours ahead (+) of Mountain Standard Time and Pacific Daylight Time; and 8 hours ahead (+) of Pacific Standard Time.

Greenwich Mean Time	EDT	EST/CDT	CST/MDT	MST/PDT	PST
0000	8 p.m.	7 p.m.	6 p.m.	5 p.m.	4 p.m.
0100	9 p.m.	8 p.m.	7 p.m.	6 p.m.	5 p.m.
0200	10 p.m.	9 p.m.	8 p.m.	7 p.m.	6 p.m.
0300	11 p.m.	10 p.m.	9 p.m.	8 p.m.	7 p.m.
0400	Midnight	11 p.m.	10 p.m.	9 p.m.	8 p.m.
0500	1 a.m.	Midnight	11 p.m.	10 p.m.	9 p.m.
0600	2 a.m.	1 a.m.	Midnight	11 p.m.	10 p.m.
0700	3 a.m.	2 a.m.	1 a.m.	Midnight	11 p.m.
0800	4 a.m.	3 a.m.	2 a.m.	1 a.m.	Midnight
0900	5 a.m.	4 a.m.	3 a.m.	2 a.m.	1 a.m.
1000	6 a.m.	5 a.m.	4 a.m.	3 a.m.	2 a.m.
1100	7 a.m.	6 a.m.	5 a.m.	4 a.m.	3 a.m.
1200	8 a.m.	7 a.m.	6 a.m.	5 a.m.	4 a.m.
1300	9 a.m.	8 a.m.	7 a.m.	6 a.m.	5 a.m.
1400	10 a.m.	9 a.m.	8 a.m.	7 a.m.	6 a.m.
1500	11 a.m.	10 a.m.	9 a.m.	8 a.m.	7 a.m.
1600	Noon	11 a.m.	10 a.m.	9 a.m.	8 a.m.
1700	1 p.m.	Noon	11 a.m.	10 a.m.	9 a.m.
1800	2 p.m.	1 p.m.	Noon	11 a.m.	10 a.m.
1900	3 p.m.	2 p.m.	1 p.m.	Noon	11 a.m.
2000	4 p.m.	3 p.m.	2 p.m.	1 p.m.	Noon
2100	5 p.m.	4 p.m.	3 p.m.	2 p.m.	1 p.m.
2200	6 p.m.	5 p.m.	4 p.m.	3 p.m.	2 p.m.
2300	7 p.m.	6 p.m.	5 p.m.	4 p.m.	3 p.m.

kit is sold by Heathkit. Your 24-hour clock should be set to GMT time not the local time.

A GMT (UTC) time check can be received on 2500, 5000, and 10,000 kHz, 24 hours a day, from WWV, the Bureau of National Standards station, Boulder, Colorado, U.S.A.

Shift and Speed

Shift

There are basically three standard frequency shifts used by most radioteletype news services. These are:

Fig. 18-3. A 24-hour digital clock.

- 170 Hz—some News Services and amateurs.

- 425 Hz—World News Services. This is the most used.

- 850 Hz—some News Services and Weather, Aero, and Military.

- Variable—many stations are now using nonstandard shifts.

The standard shift mark is the higher one. Also, it is accepted that most recent communications receivers will use the lower-sideband mode to receive RTTY. There are some communications receivers which have an RTTY position on the mode switch. The tuning of the receiver and the use of the mode switch in the upper-sideband position or the lower-sideband position will affect the frequency reading of the receiver. Experience and operation of your particular receiver will quickly tell you what slight adjustments are needed with your receiver.

Speed

Most world news service terminals are presently using 66 words per minute, which is a 50-baud rate. The following is a list of the various baud rates and speeds:

- 45 baud equals 60 words per minute.

- 50 baud equals 66/67 words per minute (most used).

- 57 baud equals 75 words per minute (seldom used).

- 74 baud equals 100 words per minute.

On some terminal units, the speed can be determined automatically. If the speed is not known, simply use the "trial" methods. First, try 66 words per minute at a 425-Hz shift in the normal position, and then other speeds until the correct speed is found and clear copy appears.

Normal or Reverse Phasing

Most foreign press services use reverse signal phasing. All terminal units have a normal/reverse switch; simply switch to one or the other to obtain copy. Phasing of a station is not listed here due to variance of equipment used.

Baudot and ASCII Code Systems

Most news services use the Baudot code. At this time, I have never copied an ASCII station other than in the amateur bands. Most of the newer terminal units, which are microprocessor based, will receive either Baudot or ASCII systems. The ASCII system was included for amateur or future use.

Press Services

The name of the news service is identified in the frequency lists with an abbreviation which is more or less standard throughout the world. For example:

- AP means Associated Press.
- UPI means United Press International.
- PTI means Press Trust of India.

Locations of Stations

Most station locations are taken from the latest International Telecommunication Union (ITU) listings. These locations can sometimes be misleading. For example, *Reuters News Agency* operated out of England and was listed as such; however, there can be other Reuters News Agencies with transmitters that are located in other parts of the world. *TASS*, the official news agency of the USSR, is also known to have transmitters located outside of their country. The same holds true for *Voice of America* (VOA) stations, which are located in various parts of the world.

Reception

It was found also that reception ratings vary from area to area and from country to country. It was decided not to list any world press services in frequency lists which did not consistently give regular reception under normal conditions. However, the seasons will affect the quality of transmissions in many areas, and all stations will be affected from time to time by unusual radio-propagation conditions. Most block frequencies that are used were chosen for optimum reception at different times of the day or night; this means that in different parts of the world, nighttime operation is affected by propagation factors which require the use of different frequencies from the daytime frequencies. Monitoring was done from my QTH in Ohio, Western North Carolina, and Southern Florida. My original frequency lists were given to other monitors on the West Coast and in the southwestern areas for confirmation and reception reports.

Other Considerations

This list in Table 18-1 covers only those world news services that transmit in English radioteletype (RTTY). Some of the stations will from time to time send their operating schedules and frequencies over the air for you to copy. *MAPS*, on its various frequencies, sends a daily schedule of the times and languages which will be transmitted, with English being one of the languages transmitted at certain times in their operating schedule.

In the original ITU work some years ago, blocks of frequencies were assigned for the news services and other RTTY stations. These general block-of-frequencies areas might well yield new stations of various types and could be considered a hunting ground for additional teletype stations of interest.

Daylight Hours
18,000 kHz to 20,800 kHz
15,480 kHz to 16,400 kHz
14,450 kHz to 15,000 kHz
Evening Hours
6760 kHz to 7400 kHz
7400 kHz to 8120 kHz
10,000 kHz to 11,160 kHz

Other blocks of frequencies where world press services might be found, along with the radioteletype (RTTY) stations, and various other types of services are:*

2300 to 2500 kHz	10,000 to 11,700 kHz
3400 to 3500 kHz	11,970 to 14,000 kHz
4000 to 4750 kHz	14,400 to 15,100 kHz

* Taken from the ITU Block of Frequencies List.

5060 to 5950 kHz	15,450 to 17,700 kHz
6200 to 7000 kHz	17,900 to 21,000 kHz
7300 to 9500 kHz	21,700 to 25,600 kHz
9775 to 10,000 kHz	26,150 to 26,950 kHz

Equipment

Antennas

An efficient, all-frequency, receiving antenna is not difficult to build. After many years of experimenting and testing many types of receiving antennas, I have found by experience, that the antenna must fill the following requirements:

- The antenna should be "cut" close to the resonant frequency to which the receiver is tuned.
- The antenna should be frequency agile and capable of covering an extremely wide frequency range.
- The antenna should be inexpensive and of a size and configuration that would be adaptable to the average person's antenna site.

Two antennas meet these requirements and give excellent performance. Of these two, the first choice is a random long-wire antenna (Fig. 18-4), which is a minimum of 50- to 60-feet (or longer) of insulated copper wire, supported on each end by an insulator, with the same type wire used as a lead-in to the receiver. Coupled to this lead-in wire should be a simple inexpensive antenna tuner, such as the one shown in Fig. 18-5. It should be able to tune the wide range of frequencies from 4 to 30 MHz. The long-wire antenna can be much longer than 60 feet, if desired; the longer the antenna wire, the more directional it becomes. Directivity will be in the direction the wire is pointing. The antenna should also be kept as straight as possible for best results.

If a suitable antenna site is not available, the second antenna choice is very close in performance to the first. This is one of the new *indoor active antennas*, such as the MFJ-1020 shown in Fig. 18-6. These units are small enough to sit on your desk, are completely self-contained, and work extremely well.

The outdoor long-wire antenna, plus a suitable antenna tuner, is my first choice. The antenna should have a switch disconnect for lightning protection that allows disconnecting of the antenna when not in use. My antenna system, in use at this time, consists of a 100-foot long-wire running east to west, and an additional 100-foot long-wire running north and south. Both are 30 feet off the ground, switchable from one to the other, and both used with a single antenna tuner. This antenna system has worked well under all conditions and covers news stations from all parts of the world at all frequencies. There are other types of antennas that work just as well. These are: verticals, dipoles cut to

Fig. 18-4. Two methods of stringing a long-wire shortwave antenna.

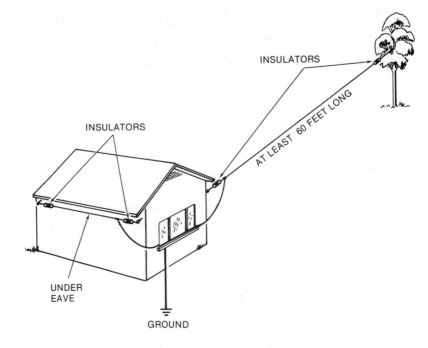

Fig. 18-5. The MFJ Model MFJ-16010 antenna tuner.

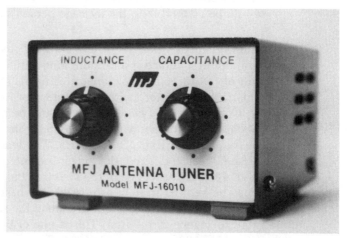

specific frequency, and several types of shortwave trap-type antennas that do not require antenna tuners. Most of the trap-type antennas require long runs in excess of 100 feet to be effective.

General-Coverage Shortwave Receivers

Requirements for the radioteletype receiver are not extreme for many of the newer, solid-state, general-coverage receivers on the market. The main require-

Fig. 18-6. The MFJ Model MFJ-1020 active antenna.

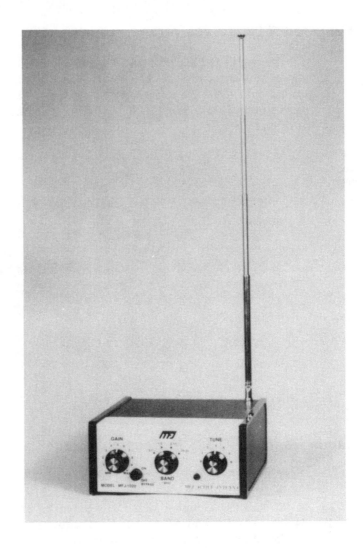

ment is frequency stability after initial warm-up and selectivity. It is possible to use older tube-type receivers if they have been proven to be stable and have the needed selectivity. Many amateur transceivers now feature general coverage receivers.

The following types of receivers have been used for radioteletype receiving with good results. There are other good receivers on the market that are not mentioned simply because I have not had the opportunity to use them in this type of service. Your local amateur radio dealer will be able to offer a suitable selection and perhaps allow you to get some hands-on experience with a receiver that will meet your needs. Fast-moving developments in solid-state electronics are producing many quality shortwave receivers at reasonable prices.

Kenwood R-2000 Receiver

The Kenwood R-2000 (Fig. 18-7) is an all-mode SSB/CW/AM/FM receiver that covers the range of 150 kHz–30 MHz in 30 bands. It is microprocessor controlled, and has a 24-hour clock, 10-memory scan, 3 IF filters, noise blanker, record output jack, S-meter. It is a very stable medium-priced RTTY receiver, with many features.

Fig. 18-7. The Kenwood Model R-2000 receiver.

Japan Radio NRD-525 Professional Receiver

The NRD-525 receiver (Fig. 18-8) covers the 20- to 34,000-kHz range, utilizing a double superheterodyne circuit with the first IF at 70.453 MHz and the second at 455 kHz. An optional VHF/UHF converter extends the coverage to include 34–60, 114–174, and 423–456 MHz. Frequency readout and receiver status are displayed on a 3-color vacuum fluorescent display. Frequency readout is to 10 Hz (e.g., 13973.250 kHz). Sensitivity is conservatively rated at 0.5 μV, 1.6–34.0 MHz (CW/SSB). Image-frequency rejection ratio is 70 dB or more. Intermediate-frequency rejection ratio is 70 dB or more and the dynamic range is 100 dB or more (IF bandwidth is 500 Hz). JRC's reputation for extreme stability continues with only $\pm$ 3-ppm drift, which makes the NRD-525 the inevitable choice for the demanding nonvoice modes, such as CW, RTTY, and Fax.

Summary

Any of the previously mentioned receivers will give excellent results. As mentioned before, there are other receivers on the market that are also in this performance class. Your local amateur radio dealer or SWL dealer can be very helpful, as most of these radio stores carry all of the needed equipment, plus the

**Fig. 18-8. The
Japan Radio
Model NRD-525
receiver.**

terminal units. Just remember that the receiver is the most important part of your RTTY system.

Many of the jazzy, military-looking, portable shortwave, or all-band, receivers sold by discount houses and mass merchandisers are not suitable for radioteletype (RTTY) work. If in doubt, seek a knowledgeable SWL hobbyist to advise you of a reliable source for this type of equipment.

Terminal Units

This unit is sometimes called a demodulator, or TU, and is the brain of the radioteletype system. The TU is connected to the audio output (speaker output or phone out) of your general-coverage receiver, which processes the audio signal into intelligent language. We will deal here with the new modern microprocessor-type (chip-type) of terminal unit, which processes the audio Frequency-Shift Keying (AFSK) signal from your receiver and then carries it to the video monitor where the information is displayed line by line. A printer for hard-copy printouts can be added to this RTTY system if desired. Most TUs now available will drive computer-type printers direct.

One audio-cable connection is all that is required from the speaker of the receiver to the audio input of the terminal unit. Fig. 18-9 shows the connection of the equipment. Each TU unit has a video output going to the video monitor through a single coaxial cable, which completes the system. The complete system hook-up is just a matter of following the simple directions furnished with each unit. No technical knowledge is required.

HAL Telereader

The HAL Model CWR-6750 Telereader (Fig. 18-10) is a compact terminal designed for reception of Baudot and ASCII RTTY signals as well as CW signals.

Fig. 18-9. Block diagram of an RTTY receiving station.

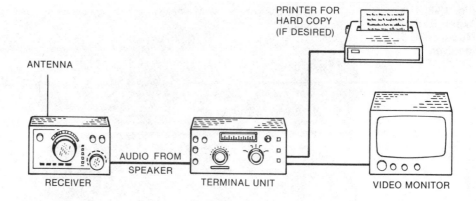

This unit includes a built-in green CRT display screen and full RTTY and CW demodulators. The small size makes it ideal for portable use. It will operate from 12-volts DC at 1.6 amps, and will drive a parallel ASCII printer. It has many other features, including all standard shifts—170, 425, and 850. A simple one-line hookup from the audio output furnishes an RTTY readout (or an RTTY printout with the optional printer).

Fig. 18-10. The HAL Model CWR-6750 Telereader.

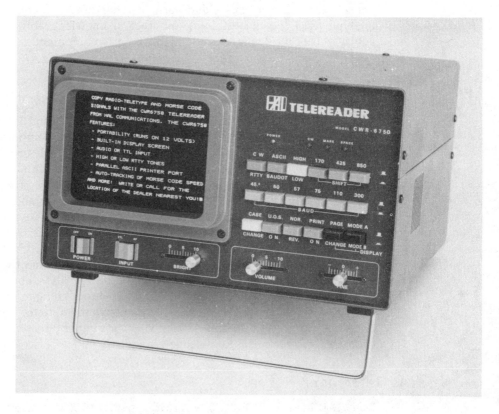

Info-Tech 200F

The Info-Tech 200F (Fig. 18-11) is an all-mode converter that will accept Morse and all standard radioteletype codes (Baudot and ASCII) from your shortwave receiver, and will convert these into a video readout on a monitor and/or a hard-copy printout on your serial printer. All the standard shifts, 170, 425, and 850 are available, plus a wide range of variable shifts. Also, all the standard RTTY speeds of 45, 50, 74, and 100 baud are present, plus the standard ASCII speeds to 300 baud. This unit is the latest state-of-the-art microprocessor/decoder TU. Its features include: scope output, full dual demodulators, ATC, clear reset, unshift on space, normal/reverse phasing switch, full limiting for weak signals, 10-step LED bar-tuning indicator, CW tuning indicator, reset/RTTY speed determination and readout, printer buffer, and wide/narrow demodulator filter selection. It has full speed conversion and code conversion for printer use and will drive all serial teletype machines. A simple connection to your general-coverage receiver and video monitor and/or printer will put the system on air in minutes.

Fig. 18-11. The Info-Tech Model 200F converter.

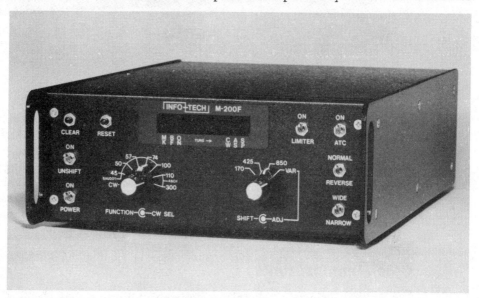

Info-Tech M-6000 Multicode Decoder

Info-Tech has once again pushed technology to the limit to offer the listener the absolute state-of-the-art in RTTY monitoring equipment. Their M-6000 (Fig. 18-12) boasts a number of new features and capabilities *never* available before to the listener. More modes, more speeds, more shifts, and more sophisticated features now let you listen to the exciting world of radioteletype via shortwave and satellite! The M-6000 is the *first* unit ever to offer the reception of *time-division multiplex*. (This code is used by diplomatic and military concerns world-wide.) The M-6000 is also the *first* unit to offer calibrated variable Baudot speeds with calibrated variable shifts, and is the *first* unit to offer complete remote control by either a terminal or computer. Thus, full system automation is now possible.

Fig. 18-12. The Info-Tech Model M-6000 multimode code receiver.

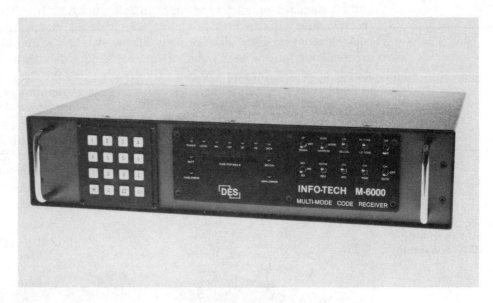

The M-6000 code receiver is the *first* unit to offer full-automatic microprocessor-controlled shift tune and selection. Press a key, and the M-6000 tunes the mark, finds the space, and displays the approximate shift. Tuning couldn't be simpler! Other features include: on-screen tuning indication, 1050- and 1800-baud ASCII, 85- and 1200-Hz shifts, 6 new ASCII shifts, parity select on ASCII, automatic gain control (3 settings), automatic threshold control, un-shift-on-space, built-in diagnostics, bit inversion (Baudot), and four video formats. Both serial and parallel printer outputs are provided.

Other Terminal Units

Both Microlog (Microlog Company, 4 Professional Drive, Gaithersburg, MD 20760) and DGM (DGM Electronics, Inc., 787 Briar Lane, Beloit, WI 53511) provide terminal units. Several systems are available from Microlog, and DGM has an RT-1100 receive terminal.

Video Monitors

The last piece of equipment in our radioteletype system (not counting the printer) is the unit which displays the final results of the signal being received. The video monitor is much the same as your black-and-white TV set which displays pictures. It processes signals through the receiver and shows them as words on the screen. The RTTY video monitor can indeed be an inexpensive portable black-and-white TV set. No conversion is necessary to accomplish this; all that is needed is a small, simple, modulator unit that is attached to the antenna terminals of the TV set. The TV set is then tuned to Channel 3 or Channel 4, whichever is clearer. These small video modulators are sold at amateur radio stores and computer stores for the same use—to convert a small black-and-white TV set into a video monitor. They cost approximately $15.00. The screen of the converted B&W TV set, or dedicated video monitor, need not be larger than 12 inches. A dedicated video monitor will produce a video quality that is superior to a converted TV set.

Hard-Copy Printers

Refer to the block diagram for the entire RTTY system in Fig. 18-9. You will see that an optional printer can be used with this system if desired. The terminal unit will drive any of the following printers and provide a permanent hard copy of the important news that you might want to keep. (*Remember*: This news is not for show; it is for your information *only*.)

- **Computer Printers**—Many new printers are available at reasonable prices. Most modern terminal units will drive these new printers. Check the printer's requirements (serial–parallel) to match the TU's printer output.

- **Model 28 Teletype™** (Baudot)—A rugged unit that is in wide use and which can be geared to most transmitting speeds used by the World Press services. Usually available at a very reasonable cost.

- **ASR-33, RO-33 Teletype™** (ASCII)—These machines can be driven directly from the Info-Tech Model 200 terminal unit at any transmission speed because of the ability of the Model 200 to do speed conversion internally—that is, Baudot code to ASCII printer, or ASCII to Baudot. The 200/6000 Info-Tech will also accept any combination of speed and code (ASCII–Baudot). They can be used with a "33"-type of printer, which is a 100-words-per-minute ASCII printer.

- **ASR-32 or RO-32 Teletype™**—These are essentially the same type of printers as the 33-types. The 32-types are Baudot systems while the 33-types are ASCII systems; both can be used with any terminal unit that will accept the printer's system of coding. That is, those able to receive Baudot code signals and use an ASCII printer, or receive an ASCII signal and accomplish a printout on a Baudot printer.

- **Other Printers**—Many other printers can be used in the RTTY system. Any modern printer that does not require handshaking (interconnection with computers) to operate will do a good job on RTTY.

Many computer and radio magazines carry ads for surplus terminal and data printers at reasonable prices. Extel, RCA, and others offer these machines. They have serial and parallel inputs with 1110-baud, Baudot or ASCII codes, and they make good RTTY printers. Good listening on RTTY!

19

AMATEUR BAND DX'ING

Mike Witkowski

Mike Witkowski has long been a "name" among the relative few listeners who actively monitor the amateur radio bands and report their results to clubs. His total contacts are up there among those at the top of the ladder. He has been involved in Amateur Band listening for a number of years and has used a wide variety of receivers in pursuing new loggings.

As noted in this chapter, Mike operates the SWL side of the national ARRL QSL bureau for Ham Band listeners—a job which takes up a great deal of time. He also publishes a Ham Band monitors' newsletter each month.

Mike is married, with two children, and is head teller at a local bank. His other interests include golf and computers.

When you think of the term "SWL," you naturally think of a person who listens to the shortwave broadcasting bands or the utility transmissions. In most situations, you are correct. But there is also a group of SWL'ers, although not a large one, who do their DX'ing on the Amateur Radio bands.

The challenge on the Ham bands is quite different from that found on shortwave broadcasts. A rare country (Fig. 19-1) may be on a particular frequency on one day and not there the next.

Fig. 19-1. A shortwave broadcast listener's dream —hearing a broadcast from Bhutan. You can do it on the Ham bands.

Receivers

A few years ago, a receiver which covered only the Ham bands was found to be far better for Ham Band listening than a general-coverage receiver. It had closer frequency readout, sensitivity, and selectivity. Selectivity was the main concern for the shortwave listener as the Ham stations are not assigned 5-kilohertz separations as are the shortwave broadcasting stations. With the coming of solid-state equipment, direct frequency readout, and so on, many of today's receivers can do it all for you, whether it's utility, shortwave broadcast, RTTY, or Ham Band DX'ing. As with any hobby, the right equipment is an essential ingredient.

Antennas

Once you have that receiver, the next step is your antenna. That super receiver will be only as good as your antenna. For Ham Band DX'ing, a dipole, vertical, or long-wire antenna will give you satisfactory results. For DX'ing the higher frequencies, a beam antenna will perform the best. A beam antenna helps you hear the weak station, while eliminating stations that are slightly off frequency and which may cause interference. A beam antenna can be aimed in the direction of the particular station that you wish to receive, while the back side of the antenna helps to depress signals coming from the opposite direction.

Long-Path Reception

In DX'ing the higher frequencies, long-path reception is possible from time to time. A beam antenna works very nicely when conditions like this exist. Long-path reception involves receiving signals from a station in the opposite direction from where you would normally hear them. If there are any openings on long-path, they will show up a little after sunrise or a little before local sunset. Examples of this would be hearing the Arab countries shortly after your sunrise or Australian stations shortly before your sunset.

With the coming of lower sunspot activity, the lower frequencies will out-perform the higher ones. To DX the lower frequencies, a good long long-wire or vertical antenna will give you maximum results. Conditions on these bands are at their best during the months of December, January, and February when static levels are at the lowest. The best time to listen on the lower frequencies is an hour or so after sunset for stations to your south or east. Look for stations to your west an hour or so before sunrise.

Reception Modes

To those who may be contemplating a try at the Ham bands, let's consider the various modes of reception and the frequencies available.

The following modes can be used in the reception of stations: SSB, CW, RTTY, slow-scan TV, and satellite. The terms, *SSB* and *CW*, should be familiar to most readers. *RTTY* is fairly new and can be copied with the various readers that are available. *Slow-scan TV* is the receiving of a TV signal on a TV monitor screen, with the signal having been sent over shortwave frequencies. *Satellite signals* can be monitored by copying the signal as the satellite passes over your part of the world.

Amateur radio transmissions occur on a number of frequency ranges:

1. 1.8 to 2 MHz is known as the 160-meter band and single-sideband transmissions, when used, are mostly in the lower sideband (LSB).
2. 3.5 to 4 MHz is the 80-meter band, with LSB again predominating.
3. 7 to 7.3 MHz is the 40-meter band and, for the most part, LSB is used. This band has quite a bit of interference as it shares its frequencies with shortwave broadcasters.
4. The 14- to 14.35-MHz area is known as 20 meters, and all SSB signals are in the upper sideband (USB).
5. 21 to 21.45 MHz is the 15-meter range, with SSB signals again in the USB.
6. The last band is 28 to 29.7 MHz, or 10 meters. The usage is again in USB. Future expansion of the Ham bands will be in the 10-, 18-, and 24-MHz areas.

Station Prefixes

When you listen on the Ham bands, you will come across quite a variety of prefixes. To know what countries these prefixes indicate can be quite confusing to a new Ham Band DX'er. A very handy and complete 12-page booklet called the *ARRL DXCC Countries List* can be of tremendous help. The booklet not only contains complete country-counting rules but it also lists ITU, CQ Zones, and the continents of each country. It also contains check-off boxes for Mixed, Phone, CW, RTTY, and Satellite reception for each band. Along with the current country list, there is also a listing of deleted countries. A prefix cross-reference (Table 19-1) is also included to help pinpoint some of the new prefixes that have been showing up. Copies of this booklet are available for $1.00 from the American Radio Relay League, 225 Main Street, Newington, CT 06111.

**Table 19-1.
Amateur Radio
Prefix Cross-
Reference**

Prefix	Country	Prefix	Country
A2	Botswana	DU	Philippine Is.
A3	Tonga Is.	EA	Spain
A4	Oman	EA6	Balearic Is.
A5	Bhutan	EA8	Canary Is.
A6	United Arab Emirates	EA9	Rio de Oro
A7	Qatar	EA9	Ceuta and Melilla
A9	Bahrein	EI	Rep. of Ireland
AP	Pakistan	EL	Liberia
BV	Formosa	EP	Iran
BY	China	ET	Ethiopia
C2	Rep. of Nauru	F	France
C3	Andorra	FB8Z	Amsterdam & St. Paul Is.
C5	The Gambia	FB8W	Crozet
C6	Bahama Isls.	FB8X	Kerguelen Is.
C9	Mozambique	FB8Y(See CE9AA-AM)	
CE	Chile	FC[1]	Corsica
CE9AA-AM,FB8Y,KC4,LA,LU-Z,OR4,UA1, UK1,VK0,VP8,ZL5,ZS1,3Y,4K,8J	Antarctica	FG	Guadeloupe
		FH	Comoro Is.
CE9AN-AZ(See VP8)		FK	New Caledonia
CE0A	Easter Is.	FL	Afars & Issas Terr.
CE0Z	Juan Fernandez	FM	Martinique
CE0X	San Felix	FO	Clipperton I.
CM,CO	Cuba	FO	Fr. Polynesia
CN	Morocco	FP	St. Pierre & Miquelon
CP	Bolivia	FR[4]	Glorioso Is.
CR3	Guiné Bissau	FR[4]	Juan de Nova, Europa
CR4	Cape Verde Is.	FR	Reunion
CR5	Principe, Sao Thome	FR	Tromelin
CR6	Angola	FS[1]	Saint Martin
CR8	Port. Timor	FW	Wallis & Futuna Is.
CR9	Macao	FY	Fr. Guiana
CT	Portugal	G	England
CT2	Azores	GC	Guernsey & Dependencies
CT3	Madeira Is.	GC	Jersey
CX	Uruguay	GD	Isle of Man
DA,DJ,DK,DL[2]	Fed. Rep. of Germany	GI	Northern Ireland
DM,DT[3]	German Democratic Rep.	GM	Scotland
		GW	Wales

1. Unofficial prefix.
2. (DA,DJ,DK,DL) Only contacts made September 17, 1973, and after, will count for this country.
3. (DM,DT) Only contacts made September 17, 1973, and after, will count for this country.

4. (FR7) Only contacts made June 25, 1960, and after, will count for this country.

**Table 19-1
(cont.)**

HA	Hungary	KX	Marshall Is.
HB	Switzerland	KZ	Canal Zone
HB0	Liechtenstein	LA,LG	Norway
HC	Ecuador	LA(See CE9AA-AM)	
HC8	Galapagos Isls.	LU	Argentina
HH	Haiti	LU-Z(See CE9AA-AM,VP8)	
HI	Dominican Rep.	LX	Luxembourg
HK	Colombia	LZ	Bulgaria
HK0	Bajo Nuevo	M1[1](See 9A)	
HK0	Malpelo I.	OA	Peru
HK0	San Andres & Providencia	OD	Lebanon
HK0(See KS4)		OE	Austria
HL,HM	Korea	OH	Finland
HP	Panama	OH0	Aland Is.
HR	Honduras	OJ0	Market
HS	Thailand	OK	Czechoslovakia
HV	Vatican	ON	Belgium
HZ,7Z	Saudi Arabia	OR4(See CE9AA-AM)	
I,IT	Italy	OX,XP	Greenland
IS	Sardinia	OY	Faroe Is.
JA,JE,JH,JR,KA	Japan	OZ	Denmark
JD,KA1[5]	Ogasawara	P2[7]	Papua New Guinea
JD,KA1[6]	Minami Torishima	PA,PD,PE,PI	Netherlands
JT	Mongolia	PJ	Neth. Antilles
JW	Svalbard	PJ	St. Maarten, Saba, St. Eustatius
JX	Jan Mayen	PY	Brazil
JY	Jordan	PY0	Fernando de Noronha
K,W	United States of America	PY0	St. Peter & St. Paul's Rocks
KA(See JA)		PY0	Trindade & Martim Vaz Is.
KA1(See JD)		PZ	Surinam
KB	Baker, Howland & American Phoenix Is.	S2	Bangladesh
KC4(See CE9AA-AM)		SK,SL,SM	Sweden
KC4	Navassa I.	SP	Poland
KC6	Eastern Caroline Is.	ST	Sudan
KC6	Western Carolina Is.	SU	Egypt
KG4	Guantanamo Bay	SV	Crete
KG6	Guam	SV	Dodecanese
KG6R,S,T	Mariana Is.	SV	Greece
KH	Hawaiian Is.	SV	Mount Athos
KH	Kure I.	TA	Turkey
KJ	Johnston I.	TF	Iceland
KL	Alaska	TG	Guatemala
KM	Midway Is.	TI	Costa Rica
KP4	Puerto Rico	TI9	Cocos I.
KP6	Kingman Reef	TJ	Cameroun
KP6	Palmyra, Jarvis Is.	TL[8]	Central African Rep.
KS4,HK0	Serrana Bank & Roncador Cay	TN[9]	Congo Rep.
KS6	American Samoa	TR[10]	Gabon Rep.
KV	Virgin Is.	TT[11]	Chad Rep.
KW	Wake I.	TU[12]	Ivory Coast
		TY[13]	Dahomey Rep.

5. (JD,KA1) Formerly Bonin and Volcano Islands.
6. (JD,KA1) Formerly Marcus Island.
7. (P2) Only contacts made September 16, 1975, and after, will count for this country.
8. (TL) Only contacts made August 13, 1960, and after, will count for this country.
9. (TN) Only contacts made August 15, 1960, and after, will count for this country.

10. (TR) Only contacts made August 17, 1960, and after, will count for this country.
11. (TT) Only contacts made August 11, 1960, and after, will count for this country.
12. (TU) Only contacts made August 7, 1960, and after, will count for this country.
13. (TY) Only contacts made August 1, 1960, and after, will count for this country.

Table 19-1
(cont.)

TZ[14] .... Mali Rep.	VQ9 .... Farquhar
UA;UK1,3,4,6;UV;UW1-6;UN1	VQ9 .... Seychelles
.... European Russian S.F.S.R.	VR1 .... Brit. Phoenix Is.
UA1,UK1 .... Franz Josef Land	VR1 .... Gilbert, Ellice Is. & Ocean I.
UA1,UK1(See CE9AA-AM)	VR3,7 .... Line Is.
UA2,UK2F .... Kaliningradsk	VR4 .... Solomon Is.
UA,UK,UV,UW9-0 .... Asiatic R.S.F.S.R.	VR6 .... Pitcairn I.
UB,UK,UT,UY5 .... Ukraine	VR7(See VR3)
UC2,UK2A/C/I/L/O/S/W .... White R.S.S.R.	VR8[16] .... Tuvalu
UD6,Uk6C/D/K .... Azerbaijan	VS5 .... Brunei
UF6,UK6F/O/Q/V .... Georgia	VS6 .... Hong Kong
UG6,UK6G .... Armenia	VS9(See 8Q)
UH8,UK8H .... Turkoman	VS9K .... Kamaran I.
UI8,UK8 .... Uzbek	VU7 .... Andaman & Nicobar Is.
UJ8,UK8J/R .... Tadzhik	VU .... India
UL7,UK7 .... Kazakh	VU .... Laccadive Is.
UM8,UK8M,N .... Kirghiz	VX9 .... Sable I.
UO5,UK50 .... Moldavia	VY0 .... St. Paul I.
UP2,UK2B/P .... Lithuania	W(See K)
UQ2,UK2G/Q .... Latvia	XE .... Mexico
UR2,UK2R/T .... Estonia	XF4 .... Revilla Gigedo
VE,VO .... Canada	XP(See OX)
VK .... Australia	XT[17] .... Voltaic Rep.
VK .... Lord Howe I.	XU .... Khmer Rep.
VK9 .... Willis I.	XV .... Vietnam
VK9 .... Christmas I.	XW .... Laos
VK9 .... Cocos I.	XZ .... Burma
VK9 .... Mellish Reef	YA .... Afghanistan
VK9 .... Norfolk I.	YB[21] .... Indonesia
VK0(See CE9AA-AM)	YI .... Iraq
VK0 .... Heard I.	YJ .... New Hebrides
VK0 .... Macquarie I.	YK .... Syria
VO(See VE)	YN .... Nicaragua
VP1 .... Belize	YO .... Rumania
VP2E[15] .... Anguilla	YS .... Salvador
VP2A[15] .... Antigua, Barbuda	YU .... Yugoslavia
VP2AV[15] .... Brit. Virgin Is.	YV .... Venezuela
VP2D[15] .... Dominica	YV0 .... Aves I.
VP2G[15] .... Grenada & Dependencies	ZA .... Albania
VP2M[15] .... Montserrat	ZB .... Gibraltar
VP2K[15] .... St. Kitts, Nevis	ZC(See 5B)
VP2L[15] .... St. Lucia	ZD7 .... St. Helena
VP2S[15] .... St. Vincent & Dependencies	ZD8 .... Ascension I.
VP5 .... Turks & Caicos Is.	ZD9 .... Tristan da Cunha & Gough I.
VP8(See CE9AA-AM)	ZE .... Rhodesia
VP8 .... Falkland, Is.	ZF .... Cayman Is.
VP8,LU-Z .... So. Georgia Is.	ZK1 .... Cook Is.
VP8,LU-Z .... So. Orkney Is.	ZK1 .... Manihiki Is.
VP8,LU-Z .... So. Sandwich Is.	ZK2 .... Niue
VP8,LU-Z,CE9AN-AZ .... So. Shetland Is.	ZL .... Auckland I. & Campbell I.
VP9 .... Bermuda Is.	ZL .... Chatham Is.
VQ9 .... Aldabra	ZL .... Kermadec Is.
VQ9 .... Chagos Is.	ZL .... New Zealand
VQ9 .... Desroches	ZL5(See CE9AA-AM)

14. (TZ) Only contacts made June 20, 1960, and after, will count for this country.

15. (VP2) For credits on QSO's made before June 1, 1958, see page 97, June 1958 QST.

16. (VR8) Only contacts made January 1, 1976, and after, will count for this country.

17. (XT) Only contacts made August 6, 1960, and after, will count for this country.

Table 19-1
(cont.)

ZM	Tokelaus		5X	Uganda
ZP	Paraguay		5Z	Kenya
ZS1,2,4,5,6	South Africa		6O	Somali Rep.
ZS1(See CE9AA-AM)			6W[20]	Senegal Rep.
ZS2	Prince Edward & Mario Is.		6Y	Jamaica
ZS3	(Namibia) Southwest Africa		7O	South Yemen
1S[1]	Spratly Is.		7P	Lesotho
3A	Monaco		7Q	Malawi
3B6,7	Agalega & St. Brandon		7X	Algeria
3B8	Mauritius		7Z(See HZ)	
3B9	Rodriguez I.		8J(See CE9AA-AM)	
3C	Equatorial Guinea		8P	Barbados
3C0	Annobon		8Q,VS9	Maldive Isls.
3D2	Fiji Is.		8R	Guyana
3D6	Swaziland		8Z4	Saudi Arabia/Iraq Neutral Zone
3V	Tunisia		9A,(M1[1])	San Marino
3X	Rep. of Guinea		9G[22]	Ghana
3Y	Bouvet		9H	Malta
3Y(See CE9AA-AM)			9J	Zambia
4K(See CE9AA-AM)			9K	Kuwait
4S	Sri Lanka		9L	Sierra Leone
4U	I.T.U. Geneva		9M2[23]	West Malaysia
4W	Yemen		9M6,8[23]	East Malaysia
4X,4Z	Israel		9N	Nepal
5A	Libya		9Q	Rep. of Zaire
5B,ZC	Cyprus		9U[24]	Burundi
5H	Tanzania		9V[25]	Singapore
5N	Nigeria		9X[24]	Rwanda
5R	Malagasy Rep.		9Y	Trinidad & Tobago
5T[18]	Mauritania			Abu Ail; Jabal at Tair
5U[19]	Niger Rep.		[26]	Geyser Reef
5V	Togo Rep.			
5W	Western Samoa			

18. (5T) Only contacts made June 20, 1960, and after, will count for this country.

19. (5U7) Only contacts made August 3, 1960, and after, will count for this country.

20. (6W8) Only contacts made June 20, 1960, and after, will count for this country.

21. (8F,YB) Only contacts made May 1, 1963, and after, will count for this country.

22. (9G1) Only contacts made March 5, 1957, and after, will count for this country.

23. (9M2,4,6,8) Only contacts made September 16, 1963, and after, will count for this country.

24. (9U5,9X5) Only contacts made July 1, 1962, and after, will count for this country.

25. (9V1) Only contacts made September 15, 1963, and before, and after August 8, 1965, will count for this country.

26. (Geyser Reef) Only contacts made after May 4, 1967, will count for this country.

Catching the Station

When listening to the Ham bands, it's easy to spot a DX station (Fig. 19-2). Most such stations automatically draw a crowd and, the rarer the station, the bigger the pileup. For the DX'er to log DX stations, he has to know the three ways to catch them. First, the most-common type of operation is hearing the station on the exact frequency that the pileup is on. DX stations will work the multitude this way only when the group will stop calling the DX station once that station has made contact. Second, a somewhat more difficult logging situation arises when the DX station cannot get through to make a contact. This is usually

**Fig. 19-2. A QSL
from a
DX'pedition.**

caused by too many stations calling the DX station. There is no break in the continuous calling so no one gets through. The DX station has no alternative but to work what is known as "split frequency." The DX station transmits on one frequency and will listen for stations calling him on a higher frequency. For example, suppose a rare station is on 14.195 MHz. He will have stations calling him that are spread out between 14.200 and 14.230 MHz. For the amateur band monitor, this means searching out the DX station's transmitting frequency. Once you've found the station, you stay on his frequency and start logging the stations he calls. Finally, the most-difficult way to log a DX station is by what is known as "Russian Roulette." This occurs when a DX station is a very rare one and the pileup is tremendous. The DX station will have the calling stations spread out over 50 kHz or more. Such an instance happened a few years ago when a DX operation occurred on Bouvet Island, Antarctica. Stations were calling from 14.200 to 14.275 MHz. What the Bouvet station did was that they tuned through the pileup, stopped at a particular frequency and worked the station calling there, and then moved on and stopped on another frequency. Although such techniques are hard for the listener to take advantage of, it can be done. Just sit on a frequency of a station calling the DX station. You may be lucky and hear the DX station coming back to the caller or callers you've been monitoring. On the Bouvet DX operation, we did this for half an hour and, luckily, were at the right place at the right time (Fig. 19-3).

We are presently on the downhill side of the 11-year sunspot cycle. With the decline in sunspot activity, the higher frequencies are of less and less use. Presently, the 10-meter band has no propagation. DX propagation on 15 meters occurs shortly after sunrise to sunset. Twenty meters is open to DX from shortly after sunrise and tends to remain open into the evening. The lower bands (40, 80, and 160 meters) are mainly nighttime DX bands. This propagation summary

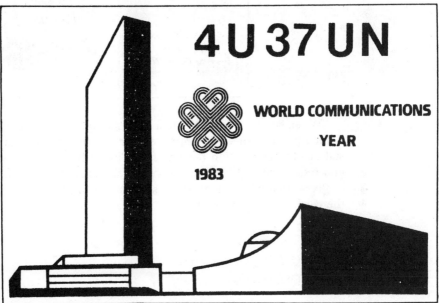

is based on early 1986 observations, and the change in the seasons, along with the changes in the 11-year sunspot cycle, will change propagation patterns.

QSL'ing

Once you have logged that special DX station, you'll want his QSL card, which means you'll have to send him your own QSL card. Your card should be as informational as you can possibly make it. The Ham station is under no obligation to send you a QSL card, so the better you make yours (Fig. 19-4), the greater your chance for a reply. Your card doesn't have to be fancy, however. If you want a card done professionally by a QSL card printer, check the Ham magazines where many of them advertise. Or perhaps you know someone locally who can print cards to your liking. Other SWL'ers have kept their cards very simple. An ordinary blank file card, with a rubber-stamped entry containing log information blanks, makes a cost-efficient card. Make sure you have your address (along with your county) somewhere on the card. One big complaint from the foreign Ham operators is that the majority of QSL cards don't have the county included on them along with the state. There are various county-based awards and your county is as important as your address.

Fig. 19-4. A sample QSL card.

When sending your QSL to a Ham station, be sure to use GMT or UTC time. A good practice that will improve your return is to list more than just one of the stations the DX station has worked. This will at least show the DX station that you indeed copied him and stayed with his signal for some time.

Now that you've logged the station and filled out your QSL, it's time to mail the card. There are three routes to go:

1. Direct.
2. QSL'ing via a manager.
3. QSL'ing via a bureau.

Direct

While you are logging the station, you may not note any information given by the station as to how to send a QSL to him. In such cases, going direct is probably the best way. Addresses of Ham stations can be found in either the U.S. or foreign callbooks. These callbooks are printed once a year by Radio Amateur Callbook Inc., 925 Sherwood Drive, Box 247, Lake Bluff, Illinois 60044. Current costs can be found in the company's regular Ham magazine advertising. The callbooks are also available in stores that sell Ham radio equipment.

In reporting to a foreign Ham operator, send your QSL card, a self-addressed envelope, and an International Reply Coupon (IRC) or two. Do not put the call letters of the foreign station on the outside of the envelope. In some countries, Hams can get into trouble with the local authorities. Also, do not enclose money, because including U.S. currency in the envelope, in the hope that it will help produce a QSL, can cause a foreign Ham to end up in jail or, at least, having to explain why he is receiving money through the mails. By not putting the call letters on the envelope, you will also help ensure that your letter will be opened only by the DX station. In some countries, there are those who are well aware that IRCs or U.S. currency is often found in certain envelopes and they are on the lookout for such opportunities. So make your letter as unobtrusive as you can.

International Reply Coupons can be purchased at your local post office. The DX station can redeem the IRCs at his post office for the equivalent return postage.

When QSL'ing direct to a stateside amateur, enclose your card and a self-addressed, stamped envelope. Then, all the Ham needs to do is fill out a QSL and mail it back to you in the envelope you've supplied.

QSL'ing Via a Manager

A lot of DX stations have QSL managers. The manager takes care of all the incoming cards for the DX station and he sees that all the incoming cards are answered. The DX station operator(s) worry only about operating; the manager handles the QSL'ing. To QSL via a manager, check the callbook or monthly lists given in the Ham magazine DX columns. QSL'ing via a manager is the fastest way to get a return QSL.

QSL'ing Via a Bureau

Receiving QSL cards via a bureau is the slowest of the three methods. Cards sent via this route seem to take forever. If you send your QSL card with no return postage, you may as well expect to receive your reply via the bureau. The reason for the long return time when it is via the bureau is that all work is done by volunteer help. If you're not in any hurry for your QSL card, then this is the plan for you. Some countries, such as the U.S.S.R., can only be QSL'ed via a bureau. Ham Station names are listed in the callbook, but not their addresses. All cards destined for Russia go through one post office box. QSL cards from Russia have been known to take anywhere from a year to 10 years, or more!

Using the ARRL Bureau

The author handles the SWL branch of the ARRL QSL Bureau, and here are some tips on how to use it. When you send that special DX station your QSL card, include a note on the card to "QSL via the bureau." To receive cards via the ARRL SWL Bureau, just arrange for us to keep a couple of Number 10, self-addressed, stamped envelopes, with your listener's "call sign" in the upper left-hand corner, on hand. When we receive some QSL cards for you, we will send them out to you in one of your envelopes. Note, however, that the ARRL SWL bureau is for *incoming* cards only, we do not send cards outgoing to other bureaus.

Types of QSLs

When chasing DX, the listener may be trying for a number of different options. One listener may be trying to verify as many different countries as he can. Others prefer to chase and confirm as many different prefixes as possible. Still others like to try for various awards. The options are there and it is up to you to decide which you enjoy the most. You may want to start with just one phase, and after you have the hang of it, decide to get into the other phases.

As stated earlier, the QSL cards you receive can vary greatly, from just a rubber-stamped index card to some very colorful cards. Half the run in waiting for your QSL is in the expectation of what the card will look like. Along with DX'ing for countries, you can also expect to see some very nice QSLs for special prefixes and special-event stations.

Clubs and Newsletters

No matter what type of DX'ing you prefer, you should always try to be well-informed on what is going on. The best way of doing this is to join a club. Presently, there are three clubs which cover listening on the Amateur Radio bands: The Association of DX Reporters (ADXR) at 7008 Plymouth Road, Baltimore, Maryland 21208; The Great Circle Shortwave Society at Box 874, Kankakee, Illinois, 60901, and "The Ham Band DX'er," which is a monthly newsletter (published by your author) covering just the Ham Bands, at 4206 Nebel Street, Stevens Point, Wisconsin 54481.

There are also various newsletters published by Hams and amateur clubs which cover Amateur Radio DX'ing. Check the Ham magazines for their addresses. When writing to any of these informational sources, please include $1.00 to help cover the cost of sending you the information and a sample copy.

Equipment

Along with staying up on the latest DX news, a few necessities will help complete your Ham monitoring shack. A logbook is your reference to what you have heard and when. Just as important is a 24-hour clock. There are various types of clocks, ranging from digital to battery-operated. When logging DX stations, be sure you use GMT time. GMT or UTC (whichever you prefer) time is the only acceptable time format in Ham Band DX'ing.

Clocks can be kept accurate by checking them regularly with WWV, the National Bureau of Standards radio station at Boulder, Colorado. If you tune to WWV at 18 minutes past the hour, you will hear the propagation forecast. It contains the solar flux, A-index, Boulder k index, and solar activity for the previous 24 hours and the prediction for the next 24 hours. A solar flux of 80 or higher will mean that the 10- and 15-meter bands are open. Indices of 10 or lower seem the best indicators of good conditions. The k index represents conditions over the past three hours, with 0–1 indicating quiet, 1–3 unsettled, and 4 or higher indicating bad news for high-frequency propagation. Keep a record of these indicators for a month and you will see the same general propagation conditions reappear just about every 27.5 days.

Summary

What does the future look like for the Ham Band listener? I think that we have just begun a new era in the electronics field. Equipment capabilities are changing fast. We have gone from tube-type receivers to solid-state equipment. Presently, we are undergoing a change to the microprocessing control to channel memories, receivers with recorder activators, and frequency-programmable timers. Not only have the receivers been upgraded, but other electronic equipment has also changed. You can take your present receiver and attach it to a "reader" that will copy CW or RTTY for you. The two most-publicized readers are the MBA reader and the Kantronics Mini-Reader. These readers are capable of displaying text or receiving Morse code, radioteletype (at 60, 67, 75, and 100 WPM), and computer ASCII (110 and 300 baud) code. You may also view on your reader such things as worldwide news bulletins, weather broadcasts, ship-to-shore messages, and Amateur Radio RTTY/CW messages. More information on readers can be found elsewhere in this book, as well as in various amateur and shortwave listening magazines, and from firms selling such equipment.

Along with the readers and the receivers has come the computer. This, in itself, has opened up more possibilities. The computer can be related to your hobby in many ways. Fig. 19-5 shows how you can use your computer to keep a log. Fig. 19-6 is a list of QSL cards received. Fig. 19-7A is a DXCC country list program printout. You merely list the call sign, country name, month, year, and one of the following letters: S, R, or C. S indicates that you have sent a QSL card, R indicates a confirmation has been received, the "C" indicates that you've been credited by the DXCC for the number of different countries you claim to have verified. When you have finished listing your DX stations, the program will tabulate the results for you (Fig. 19-7B). The printout in Fig. 19-8 shows both graphically and tabularly the maximum usable frequency. You simply input the date, longitude, latitude, and the solar flux into the program and decide whether you want a printout for each hour or prefer a graph format. The solar flux, as we noted, can also be obtained by listening to WWV at 18 minutes past the hour.

Fig. 19-5.
Computer-
printed log.

```
ENTRY NUMBER :1
DATE  :3/11/86
CONTACT STARTED :2200 UTC
CONTACT ENDED   :2300 UTC
CALLSIGN :9X5AB
NOTES :A GOOD 5/9 SIGNAL
```

Fig. 19-6.
Computer-
printed list of
QSLs received.

```
4J0BJ     JA0EIB    UK0FAP    JZ0HA
UU0IE     JA0MGR    UA0NH     UA0NT
HH0RC     VK0VK     DL0WU     ZL1ABO
ZM1ADD    JA1AEA    DL1AM     5W1AN
5W1AU     PZ1AX     PZ1AY     JA1BAL
ZK1BS     CX1CA     LU1CP     JA1CUP
9G1CY     LU1DAB    KG1FD     DL1FK
JA1FRE    VP1GFQ    OH1IT     HC1JB
UU1KAT    DU1MR     JA1QD     JA1RJU
YS1RSE    YN1TAT    JA1TRL    I1YJ
 I1ZBS    XE1ZM     ZK2AA     OK2ABU
CE2AK     HM2AO     9M2BH     OH2BH
VR2BJ     PY2BUL    DJ2BW     PY2EJ
ZL2GY     ZK2HX     HH2JT     JA2JYP
JA2KLT    VK2KU     BV2USA    VQ9AT
CALL

THERE ARE NOW 61 ENTRIES
LIST CONTAINS 366 BYTES

YOU ARE IN THE ADD MODE
```

Fig. 19-7.
Printout
obtained from
DXCC computer
program.

```
1    JT1AN      MONGOLIA   1085S
2    EA8BP      CANARY IS  1185R
3    ZS4PB      SO.AFRICA  1285C
4    T32AB      CHRISTMAS  0185R
5    TR8JD      GABON      0185R
6    JA1BL      JAPAN      0285C
7    OH2BH      FINLAND    0385C
8    XE1ZM      MEXICO     0385S
9    HH2MC      HAITI      0485R
10   CE2AK      AUSTRIA    0485C
```

(A) Printout.

```
10        COUNTRIES LISTED

10        WORKED

0         NOT WORKED

2         QSL SENT,NO RET

4         QSL SENT AND RET

4         DXCC CREDITED
```

`HIT ANY KEY TO RETURN`

(B) Tabulation.

**Fig. 19-8.
Computer
program run
showing
maximum
usable
frequency.**

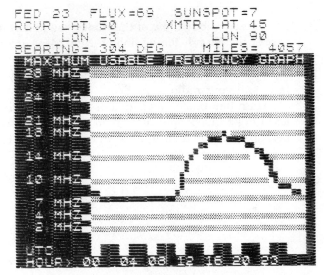

```
FEB 23   FLUX=69   SUNSPOT=7
RCVR LAT 50        XMTR LAT 45
      LON -3             LON 90
BEARING= 304 DEG       MILES= 4057
MAXIMUM USABLE FREQUENCY TABLE
UTC    MHZ         UTC    MHZ
0      7.92        1200   15.22
100    7.67        1300   16.41
200    7.47        1400   17.2
300    7.31        1500   17.57
400    7.19        1600   17.26
500    7.08        1700   16.71
600    7.01        1800   15.87
700    6.94        1900   14.61
800    7.08        2000   12.52
900    7           2100   9.08
1000   7.87        2200   8.62
1100   13.32       2300   8.23
```

```
FEB 23   FLUX=69   SUNSPOT=7
RCVR LAT 50        XMTR LAT 45
      LON -3             LON 90
BEARING= 304 DEG       MILES= 4057
MAXIMUM USABLE FREQUENCY GRAPH
28 MHZ
24 MHZ
21 MHZ
18 MHZ
14 MHZ
10 MHZ
 7 MHZ
 4 MHZ
 2 MHZ
UTC
HOUR > 00  04  08  12  16  20  23
```

```
         Propagation between
           Midwest - Europe
```

Another important way of using the computer with your radio hobby is in the creation of your own QSL cards. Fig. 19-9 shows an example of a computer-generated QSL card. You can use a printout as your QSL as all the pertinent information is listed. You can also use your regular QSL card and glue the printout onto the back of the card. A lot of QSL cards have logging information recorded on the back and by placing your computer printout on the back of the card, your card looks very professional.

Fig. 19-9. A
computer-
generated QSL
card.

```
WDX9JFT
▦▦▦▦▦▦▦▦▦▦▦▦▦▦▦▦▦
       TO RADIO-9X5AB

DATE-MAR 11,1986
                TIME-2200 UTC
BAND-14 MHZ.
                MODE-SSB
RS(T)-5/9
                QSO WITH-W1AW
RIG-YAESU FR-1015
                ANT-TA-33 JR.
▦▦▦▦▦▦▦▦▦▦▦▦▦▦▦▦▦
     I HOPE TO HEAR YOU AGAIN
     THE VERY BEST OF DX
     73S MIKE - WDX9JFT
     COMPUTIZED HAM SHACK
```

These are but a few of the things you and your computer can do together as part of the Ham listening hobby. New ideas are constantly appearing. It is now possible to copy or send CW and RTTY via your radio and your computer. In most cases, all that's needed is the unit to copy or send the material, an interface, and your radio.

Ham Band listening is one of the most fascinating forms of DX monitoring and I hope this chapter has given you a few insights into how it works and has perhaps generated an interest in exploring the Ham Bands.

20

EUROPEAN PIRATE RADIO

John Campbell, Ph.D.

John Campbell points out that he is hard to categorize, since he's been classified as a mathematician, physicist, astrophysicist, electrical engineer, and computer scientist. He is active in all of these fields, working with various universities and research institutes, currently in England (where he lives) and Sweden. His work has taken him to the far corners of the world, occasionally in "disguises" such as shown in the picture here. This, he notes, is handy when dealing with pirates who have an understandable fixation with secrecy. Although well known for his pirate-radio monitoring, his DX interests range much further, into the Indonesian and Andean stations.

He was, for some time, the "Unofficial Radio" column editor for the Danish Shortwave Clubs International bulletin, *Shortwave News*. Despite his extensive workload and travel schedule, he somehow manages to keep up with the near-weekly changes in European pirate activity.

Pirate radio on the shortwave bands needs no introductory explanations for North American listeners. In fact, pirate operations in North America are now common enough that there may be a temptation to believe that European pirate radio is just like the home-grown article except for the different accents. There are certainly similarities, but the complete DX listener needs to know more about the differences than the similarities if he is to get the most out of hunting for the European stations.

A Little History

All of the modern style of shortwave pirate broadcasting in Europe postdates the period of the offshore pirate radio ships, such as Radio Caroline, in the second half of the 1960s. Many of the people, whose involvement in home-brew pirate radio has lasted for 10 years or more, developed their taste for it by listening regularly to the offshore stations when they were high-school students, or in their early twenties. One of the attractions of the programming at that stage was that it included popular music, which was too "advanced" or daring for the official radio stations (mostly run by government monopolies) to consider playing. In some countries, particularly Denmark, there were even complaints that all modern pop music was barred from access to the official airwaves.

Quite apart from the music, the general programming of the government-run stations in the 1960s was what one might have expected from any large organization whose keywords were *monopoly* and *bureaucracy*. It is no accident that one particular genius of free-lance radio production in a certain European country (no, I'm not quite brave enough to name it) once said that his national broadcasting organization seemed to be staffed entirely by trainee pensioners. Against this background, it is easy to see why people, with an outlook modeled on the offshore stations and with some knowledge of the right technology, thought of backyard pirate radio as a way to express their opinions.

The eventual incentive to start broadcasting, for most of them, was the disappearance of almost all the pirate ships due to financial problems or hostile legislation. In one respect, the offshore pirates had served a useful purpose, by pushing governments and government broadcasters in Western Europe into the modernization of their approach to radio (and some pirate sound-alikes, such as BBC Radio 1, even unashamedly appointed former offshore pirate announcers as DJs). For many of the people who have got the do-it-yourself European pirate radio off the ground and have kept it up in the air (no pun intended), though, the push has not been effective enough.

This explains why, although some European pirate shortwave stations sound like (and are) examples of what the American DX'er Dan Ferguson has called "kids playing radio," others are run by people who are dedicated to their hobby, or to something that they see as more than a hobby: *free radio*. In Europe, "free radio" means broadcasting that is opposed in style to the bureaucratic approach. It also means free access to the airwaves for anyone who has his or her own point of view about pop music, community activities, or almost anything else, rather than a strict control over who broadcasts what by a minority or a faceless organization.

One novelty for anyone who believes that words should mean what they say is that, for free-radio freaks, politics on the air is out. The least intelligent of them get rather angry when this contradiction in the meaning of "free" radio is pointed out to them, but others say more reasonably that governments and European equivalents of the FCC do not often make the fullest use of their powers of harassment of home-brew radio operations, and that it is wise to let sleeping dogs lie. The assumption here, and quite a reasonable one in Europe, is that nothing wakes a sleeping governmental dog faster than a dose of politics. This is not to say that there are no political stations, but "free radio" pirates generally want less than nothing to do with that kind of broadcasting. The politics of mainstream free radio is rock music.

The first of the modern-style of pirates in countries that were the targets of the most intensive offshore broadcasting, particularly Britain and the Netherlands, were on the medium-wave band and on FM. Shortwave came a little later, in 1971 and early 1972, thanks mainly to the efforts of the original World Music Radio (a name which is still alive through programs on other stations, with at least one of the original presenters from the Netherlands. This was a 60-watt station with something musical for everybody, presented in more than one language, and with high-quality taped shows that were contributed weekly by DJs from several European countries. Moreover, setting another fashion for later stations, WMR showed, by being heard and verified by at least two DX'ers in Connecticut, that a small power output, backed up by careful attention to the antenna and antenna-matching, was adequate for transatlantic reception in the winter. The number of pirate-radio enthusiasts, who remember the original WMR, is continually decreasing, but the station still has an influence on how things are done. Many of the present traditions of European shortwave pirate radio are simply copies of the way that something was done first, and successfully, by WMR.

Since 1972, there has been an unbroken history of pirate broadcasting. The two mainstays of the activity over the long haul have been Britain (in first place) and the Netherlands. However, during shorter periods of time, other countries have taken the lead (for example, during the wave of German pirate activity in 1974–75, and a similar development in Ireland in late 1982 and 1983). It is one of the uncertainties of the European pirate game that forward predictions about how the scene will evolve from year to year have never been particularly reliable. Luckily, there are some steady features of the scene that North American DX'ers can continue to rely on. European shortwave piracy will be with us for a long time yet.

One of the reasons why pirates will survive is that they continue to hold a dedicated following among European SWL'ers and DX'ers. This following has continued because, even among the forgettable short-lived operations that are best described by "kids playing radio" or "Dutch modulation," or both, European pirate radio has produced a steady supply of high-quality stations which have been a pleasure to hear. For all-round excellence, the names that come to mind after World Music Radio include Radio Valentine (Germany), Radio Viking (Denmark/Sweden), ABC Europe (Netherlands), Radio Valleri (Ireland), and European Music Radio, Radio Gemini, Atlanta Radio, and ABC England (all English). Only slightly below this level of excellence, you can find stations that don't necessarily try to compete across the board with those first-line operations, but which still try to put up a grade-A performance in their chosen specializations.

One specialization that has been getting more attention recently is DX; there are stations which devote significant amounts of effort to produce programs of free-radio and pirate-radio news, with an emphasis on items of DX interest. A good example, but by no means the only example, is the English station Radio Apollo, which describes itself justifiably as "the DX'er's station." From the point of view of the North American listener, this trend represents good news because such stations are the ones most likely to respond to his wishes for specially arranged tests into North America at those times which are the best times for transatlantic reception. They also will probably arrange for special operations from those parts of Europe which count in some people's books as "rare countries."

When and Where

The most fundamental of the traditions started by World Music Radio is the tradition of the times and frequency ranges used by European pirate broadcasters. Almost all the pirate stations meet their audiences on Sunday mornings (Fig. 20-1), no earlier than 0800 GMT (0900 during the winter season), and no later than 1300 hours. When preliminary tests by WMR on frequencies like 6075 kHz had no success because of interference from the regular occupants of the 49-meter band, the station moved to 6250 kHz, almost exactly in the middle of what has since become the main European pirate band. The activity is concentrated between 6200 and 6315 kHz, although some stations occasionally venture higher than 6315, or down to 6195, in search of channels that are free from interference by other pirate or utility stations.

Fig. 20-1. Radio Britain International sends an attractive QSL sticker.

This choice of band is a good one in Europe, for two reasons that are of the "chicken and egg" variety. First, the frequencies around 6 MHz are the most reliable for all-purpose shortwave broadcasting from Europe to Europe, particularly during the daylight hours, as evidenced by the large number of 49-meter domestic European shortwave stations listed every year in the *World Radio TV Handbook*. Secondly, European manufacturers of domestic receivers, which contain a bare minimum of shortwave coverage, frequently put a general 6- to 18-MHz coverage in their products that is not much more than just an enlarged 49-meter band. This tradition of manufacture has been particularly strong in the Netherlands and Germany. Pirates started life in Europe before the equipment made and sold by local-receiver companies was outnumbered by the Asian multiband imports, so it was essential that they chose a frequency range which would attract the greatest number of potential listeners—and remember that WMR was a Dutch station.

Since late 1978, significant numbers of pirate stations (though always less than those on the main 48-meter band) have been taking advantage of the wide diffusion of multiband receivers to escape some of the 48-meter interference by moving up to the high-frequency side of the 41-meter band. The preferred range of frequencies has been roughly 7290–7400 kHz. On any Sunday morning in Europe, a typical pirate listener will divide his time between the two principal bands.

Apart from these two general traditional broadcasting ranges, stations have experimented with a wide variety of alternatives. The most successful experiments have used frequencies in the range of 11300–11590 kHz, again during the traditional Sunday broadcasting hours (although Radio East Coast Commercial ran a regular 0900–1500 service in England on Saturdays on 11400 kHz, and the Italian station IBC tried some 24-hour tests at the top of this frequency range, with both being heard in the northeastern part of the USA when they were active a few years ago), so it is rather surprising that there has been no use of the same frequency range during late 1983 and early 1984. Possibly it will be revived in the future, especially for long-distance communication.

Other experimental activities have exploited individual frequencies without trying to establish a "band." Stations have popped up, for example, on 4420, 5320, 5735, 6830, 7710 (probably the best of these choices), and 9835 kHz in the not-too-distant past. If there is any regularity in the experiments, it tends to be directed towards finding small slices of the spectrum which are interference-free and practicable for reception over long distances, as in the 1983–84 tests of the Scottish station, Radio Freedom, on 9420 kHz (perhaps with the theory that, if you can hear the BBC on 9410, then a more exotic reception ought to be possible), 15040 kHz and 15050 kHz. Both the 9- and 15-MHz efforts of Radio Freedom have been heard and verified by listeners near the East Coast of North America.

The best chance for North American reception of European pirates on their main bands will occur from roughly mid-October to mid-March, when the paths in or near darkness across the Atlantic last for the greatest lengths of time. In the middle of winter, it has been possible to hear pirates on these bands starting their regular schedules around 0900 hours, but this is a bit too late for reliable or long-lasting reception. The best period for North American listeners is 0530–0800 hours. This is now well enough understood in European pirate circles that many stations have reacted favorably to suggestions that they should consider some special broadcasts for North America during these earlier hours. Favorable reactions are not quite the same thing as actual performance, however, because it takes plenty of willpower to get out of bed in the middle of a winter night and set up a station's transmitter and antenna in the kind of isolated location that is an operator's best insurance against raids from the European cousins of the FCC. Significant numbers of stations, however, have actually made their special test broadcasts on schedule, and it seems that about half of these have been heard successfully in North America.

Publicity for special tests is given whenever possible in magazines, like *FRENDX*, or over DX-oriented programs on international stations (e.g., Radio Netherlands or Radio Canada International), though some pirate operators are still creatures of short-term impulse who don't appreciate the information that about six weeks' notice is needed if news of a test is to get into the monthly DX and SWL magazines, in time to be of value to listeners. Even with these little difficulties, though, the sources quoted here are the best ones to watch or hear if you are interested in chasing pirates on Sunday mornings.

For listeners, without special information, who wish to listen around and want to estimate their chances of hearing something unusual from Europe on any particular Sunday morning, one "early-warning system" has been a regular feature (though not present on every Sunday) of the 1980s: Radio Dublin on 6910 kHz, or some other frequency between 6908 and 6915 kHz. This station, of which more details are given below, has sometimes claimed to operate a 24-hour schedule at weekends, but it has been observed opening more realistically on Sundays, at times like 0700, 0730, and 0800 hours, carrying relays of its medium-wave outlets. During unpredictable and less-frequent periods of random enthusiasm, it has been heard to do the same thing on Saturdays. Its power varies, but it has probably not been below 150 watts in the recent past. As this is significantly higher than the normal pirate power, which tends to be between 10 and 75 watts, it suggests that if you can't hear Radio Dublin, you probably won't hear any other of the European pirates on their main bands. The catch, of course, is that you may not be hearing Radio Dublin just because it isn't on the air. But, if you're curious enough to want to use a more reliable electronic technology to find out what's going on, the telephone number is Dublin 758684.

Communication with the Stations

Another of the traditions started on shortwave by World Music Radio is one of two-way communication with the listener. Almost all of the European shortwave pirates give out their mailing addresses and respond willingly to good reception reports, especially if those reports make some constructive comments on the programming. (Recall that music rather than the technical aspects of radio communication is the primary reason why many of the stations are on the air. Return postage is needed in essentially all cases. Two International Reply

Coupons are normally enough, even from as far away as North America, but 3 IRCs (or $1.00) should put the station's management into a very good mood. The management of most of the stations seems to begin their pirate-radio lives in a good mood in any case. My own records, since 1972, show a reply rate between 95% and 96%.

Security against raids and harassment is best if the authorities can guess next to nothing about a station's location from its mailing address. Since not many operators have cooperative friends at the other end of Europe, the best general way for a station to launder its background is to arrange for mail-forwarding from one of the several general "mail drop" addresses that are maintained as a service by friends of "free radio." The situation at the time of this writing is not as clear as in the past, as some of the forwarding stations are probably on the way out because of threats of legislation against pirates, which will also discourage people running accommodation addresses (particularly in Britain). In Britain, increased pirate activity led to the drafting of a tougher new law that was due for introduction in the second half of 1984. (This law was apparently postponed to a future date.) However, for the record, it is possible to predict that at least two of the "Big Three" mail drops of 1985 will continue: P.O. Box 5, Hunstanton, Norfolk PE36 5AU, England; Postbus 41, 7700 AA Dedemsvaart, Netherlands (used almost universally by Scandinavian stations, as well as by significant numbers in other countries); and Postfach 750 925, D-2820 Bremen 75, West Germany.

In chasing European pirates, it helps to keep up with the news of what is happening among the home-brew stations. The most accessible printed source of information is *Shortwave News*, the monthly magazine of the Danish Shortwave Clubs International, which carries a detailed "Unofficial Radio" column almost every month. Up-to-date information about how to get sample copies of *Shortwave News* is given in the DSWCI advertisement in the *World Radio TV Handbook* each year. The monthly magazine with the largest number of column-inches about European pirates is *DX-Special*, which covers only pirate and utility DX'ing, but it has the possible disadvantage of being printed in German. At the time of this writing, the address of the publisher (the radio club KDKC) was Postfach 450 663, D-5000 Köln 41, West Germany. Unfortunately, the main organizer and editor has been thinking of stopping publication because of lack of help.

A promising new short bulletin, all or mostly in English, is *RadioTelex*, available from the Bremen address listed above. Finally, the best British magazine in 1985, specializing in pirates, is the weekly newsletter of Anoraks, U. K., with an address of P. O. Box 539, Blackpool, Lancashire, England. The last two of these publications have yet to show the staying power of the first two, however. Any preliminary inquiry to any of them should naturally be accompanied by the customary return postage.

Pirates in Ireland

As in many other things, Ireland is a special case when it comes to pirate radio. The first modern Irish pirate activities had the same background as in Britain, derived from the inspiration of the offshore stations and the fact that the Irish government-monopoly broadcasting organization, RTE, had developed an international reputation for stupefyingly dull programming. In fairness to the few producers of good-quality RTE programs, though, it should be said that there was never much money left to support such programs after the RTE bureaucrats had taken their share of the already small amounts of money available in the annual budgets.

Irish pirate radio was small in scale, and mostly on the medium-wave band, until the time in 1972 or early 1973, when a pirate who hauled before the Dublin courts escaped on the grounds that it was necessary to show that his transmitting equipment could be used only for transmitting before a conviction could be obtained under the existing laws. Many a potential Irish pirate quickly learned, at that point, about the virtues of using the final stages of transmitters as amplifiers for music at parties, political speeches at electoral rallies, or what have you. Strictly speaking, the legal situation is still the same in 1985, although new legislation to bring the Irish home-brew radio back under control has been promised or threatened almost continuously since 1978, and draft versions of the new laws designed to regulate broadcasting in Ireland were issued in mid-1985. But, in Irish politics, the expected never quite manages to happen.

Although it has nothing to do with shortwave, one Irish pirate story is worth telling here. As early as 1926, the Irish government promised that something would be done about a separate radio network or service for the Gaelic-speaking population in the West, but nothing came of this for almost 50 years. The *World Radio TV Handbook* now lists such a network, "Radio na Gaeltachta," but the government was embarrassed into acting on this front only because a Gaelic-language pirate with the same name moved into the gap very successfully in the early 1970s, to the point where almost no Gaelic speakers in its area of coverage bothered to continue listening to the token Gaelic programs presented on RTE.

Irish pirates first hit the shortwave bands, in 1973, in a small but determined way, with Radio Valleri on the now traditional Irish frequency of 6317 kHz taking the lead. A more significant fact for Irish pirate radio was that the famous Doctor Don Moore, and the possibly even more famous Captain Cooke (everyone is born with a title in this business), became interested in unlicensed broadcasting through having a part in the early activity. These gentlemen, quickly followed by others, reasoned that the state of the game allowed them to shift to medium wave, carry commercials, attract a wide audience, and get rich. In brief, Sunday shortwave broadcasting in Ireland soon took a back seat to the medium-wave free-for-all whose results have been summarized (now that things have settled down a little from the chaos of a few years ago when there were perhaps 65 commercial or community pirates on the air) in the *World Radio TV Handbook* (in and after 1983). However, the apparent permanence and popular support for non-RTE medium-wave broadcasting has encouraged a considerable new expansion of Sunday-morning hobby radio on the 48-meter band. And, of course, Radio Dublin (one of the founders of the medium-wave commercial fashion, and one of the three or four clear financial successes) keeps in touch with its spiritual origins by maintaining "pirates' rights of occupation" over

6910 kHz, where the only competition is from Radio Venceremos on a bad day. Radio Dublin has also started a lobbying campaign to justify its services to the image of Ireland abroad and is trying to persuade the government to give it a shortwave license if and when non-RTE broadcasting is regularized by law.

The burst of enthusiasm for shortwave piracy in Ireland in 1982 and 1983 made the pirate bands sound like an Irish monopoly on some Sundays. Some of the enthusiasm has now faded away, but at least six stations seem to have avoided withdrawal symptoms. Since Ireland is probably the best-located European country for successful broadcasts to North America, with the widest time-window for reception on Sunday mornings in winter, it may be helpful to list the stations here.

- *Capital Radio*, 6268 kHz; c/o North Street, Swords, Co. Dublin. Not a regular replier to letters.

- *Westside Radio*, 6280 kHz; 310 Collins Avenue West, Whitehall, Dublin 9. This has been by far the most regular station; a Sunday without the familiar signal on 6280 kHz is a rarity. The operator and DJ, Prince Terry (the highest rank in the Irish title stakes so far, and only likely to be upstaged if one of the Orange Lodges in the North nominates King Billy), has been in Irish pirate radio for more than 10 years. He started the first station with the name of Radio Dublin.

- *Radio Mi Amigo*, 6285 kHz; c/o North Street, Swords, Co. Dublin. Formerly on 6276 kHz. Like Westside Radio, it has the habit of relaying medium-wave commercial pirates from Dublin and neighboring towns.

- *Radio Dublin*, 6908–15 kHz; 58 Ichicore Road, Dublin 8. Among other achievements, it has been heard on these frequencies more than once in Australia and New Zealand. Claimed 800 watts output in early 1984, but these may be Irish watts.

- *Radio Ireland*, 6310 kHz; 442 Hartstown, Clonsilla, Dublin 15. Lively programming and a taste for good-quality Irish traditional music.

- *Radio Enterprise*, 6318 kHz; address same as for Radio Ireland. A Capital Radio sound-alike in pop-music programming, but, fortunately, not an imitator of Capital's letter-answering habits.

Additionally, there is an Irish history of pirate radio being present on 11463 kHz, and the crystal used by such past giants as Radio Condor International still seems to be passing from hand to hand in the Irish radio underground. It is quite likely to surface again from time to time in the transmitters of new stations.

If the predicted legislation to close off the loopholes available to Irish pirates appears at the beginning of 1986, as many people think is possible, Irish pirates voices may disappear forever from the shortwave bands—or they may just go on regardless. Both Irish history and the national character provide plenty of evidence for either one of these alternatives. I leave it as an exercise for the reader to work out which one of them it is.

Accidental Pirates and Near-Pirates

Pirate operators tend to be more casual than the official stations as far as control of harmonics is concerned. This has generally not caused interference to other broadcasters, and it has had the agreeable side-effect of making the voices of some exotic medium-wave stations available to shortwave listeners. In this field, Ireland has led the way, with such short-lived novelties as Tipperary Community Radio on 3928 kHz, ABC Radio Dublin on 4910 kHz, Community Radio Fingal on 6336 kHz, and Boyneside Radio on 6529 kHz. Replies from the first and last of these stations showed some confusion between the managers and the chief engineers as to whether the shortwave radiation was intentional or not—a truly Irish situation. But this is not only to be found in Ireland. At the end of 1983, an extremely strong second harmonic from the Liverpool-area pirate, Storeton Community Radio, was heard on 2592 kHz all over Europe and North Africa. Many listeners told the station about the signals, but the broadcasts continued even though radiation of this strength would have been easy enough to stop if the operators had regarded it as undesirable.

A mention of European harmonics is worth a place in an article such as this. Apart from questions of general interest, there are two facts which may justify it indirectly. The first is that a second harmonic of the New Jersey pirate WCBN (for "Woodbridge-Colonia Bootleg Network") was heard in England with difficulty, but still heard, on 2450 kHz in 1978. (Anyone know an address where I can send a report?) The second is that many frequencies, which are popular with medium-wave pirates in Ireland, Britain, and the Netherlands, have fourth or fifth harmonics that are in, or just above, the main 6200–6310-kHz shortwave pirate band.

A parting message about harmonics. Don't believe everything on Sunday mornings in the pirate bands with a pirate sound is a pirate. One of the BBC Radio 1 (the mindless rock channel) transmitters, on 1053 kHz, puts out a regular harmonic on 7371 kHz, though it seems to go into hiding for a few weeks whenever the BBC is told about it. And, for higher brows, there is a regular France-Culture harmonic on 6885 kHz. Finally, not everything which may sound like a harmonic is a harmonic. Vatican Radio has some very pirate-like programming on Sunday mornings on the 48-meter band, when it is not carrying the day's papal address. It was on 6210 and 6260 kHz in the past, but on 6250 kHz in 1985. BRT, the Belgian official radio, occasionally muscles in on 6225 kHz, though usually not before 1300 hours, and Trans World Radio in Monaco may have been trying to start a small religious war when it chased Vatican Radio off of 6210 kHz in favor of itself. Fortunately, the TWR programming can't possibly be mistaken for that of a pirate—unless the very occasional European religious pirates are taken into account.

On a slightly different subject, a near-pirate band has existed for some time between about 6590 and 6675 kHz. Here, there is a strange (to American ears) culture of imitation Ham Radio, with operators using imaginative call-signs and exchanging information with each other about their equipment, the weather, the mother-in-law, and so on. Part of the explanation of its existence is that Ham licensing examinations in Europe are usually stricter than those in the United States, so there is more incentive to avoid the examinations and imitate the life of a Ham operator in the easy way. The interest of the so-called Echo Charlie or EC band is that some of the musical shortwave pirates got their start through becoming EC operators, and then developed a taste for better and more adventurous things. This trend is still observable, and pirate broadcasters can ocassionally be found returning to the surroundings of the EC band after 1300 hours on Sundays to have two-way conversations when their supply of music has run out. There are occasional music-programming tests in the EC band too, although operators in this range usually give out no addresses and don't want reports. Most of the stations appear to be British, Irish, German, and Italian.

Europirate Round-up

The problem with an exhaustive listing of pirates is that it is unsuitable for any publication that is designed for as a long-life reference source, because pirate operations come and go with a speed that defeats any compiler of lists. However, it is possible to be brave and try to draw some long-term conclusions by focusing only on trends and on those stations that have shown by their past histories that they are unlikely to disappear tomorrow. The following survey is divided up into countries (except for Ireland, which has had its moment of glory already in this article), because there are marked differences between countries in the way that European pirate radio has developed.

England

Although England has been the largest single contributor to pirate radio since 1972, 1984 was not one of its good or active years. This may have been due to the fact that new legislation was being prepared for introduction in 1984 or early 1985, which would increase the penalties for pirate broadcasting, would permit searches for transmitters even when no broadcasts had taken place just before the search, would possibly remove the need for search warrants in some cases, and would possibly introduce penalties for certain kinds of help (e.g., mailing-address facilities) to pirate operators. The apparent overreaction of the government seems to be due to a rapid growth of unlicensed community-style broadcasting, mostly on FM but also on the medium-wave band, which was concentrated in areas of high unemployment (Liverpool and parts of London), and which could scarcely avoid making indirect comments on life which the government saw as criticism of its policies. A second reason for the legislation was that there were loopholes in the prevailing laws which allowed some 24-hour pirate operations to survive for up to months at a time, allowed them to carry commercials, and allowed them to take business away from the regional-franchise holders of the not-very-inspiring stations which were in the government-controlled "Independent Broadcasting Authority" (an interesting choice of name) commercial chain.

This threat to English pirate radio (in 1984 and 1985) apparently dried up the supply of new shortwave stations. Fortunately, at the time of this writing, some old favorites continue and show no signs of disappearing, even though they may come onto the air rather irregularly. A sample of the foundations of English pirate radio (as of late 1984) is as follows:

- *Radio Apollo*, "the DX'er's station," makes perhaps six to eight broadcasts annually, but these are well prepared and of high quality. There is no fixed frequency, because the operators have a large stock of crystals. Examples of frequencies are 6207, 6275, 6306, 7325, 7340, and 7373 kHz. They are very friendly to DX'ers outside Europe. The best address was uncertain in early 1984, but it seemed to be changing to the Bremen address from the Hunstanton address, both of which were mentioned earlier in this article. Power is the 30–40-watt range.

- *Atlanta Radio* established a first-rate reputation between 1978 and 1981, and has returned to the air after a period when not all its operators were in England. The return is "tentative," but this is good news, because infrequent and well-prepared broadcasts are less likely to run into problems with raids than will regular schedules. The main frequency is 6240 kHz, but 6225 and 6290 kHz are also possible. Their address is P. O. Box 319, Edenbridge, Kent, England.

- *Radio Gemini* is the oldest surviving shortwave pirate, having been founded in 1972. Programs on its two or three annual appearances are original and probably the best of any current European pirate. The Gemini history cassette, which costs about $3.50 plus postage, demonstrates this claim and is well worth having. The first Sunday in November is Gemini's birthday, and usually sees a celebratory broadcast. Its frequencies are between 6230 and 6235 kHz and its address is 10 Apsley Grange, London Road, Apsley, Hertfordshire, England. Power is a reliable 100 to 150 watts.

- *Radio Krypton*, at 6260 or 6265–68 kHz, was the longest-lived recent regular station, having existed from 1979 to 1985, with pop music enlivened by D.J. Kevin Kent, who each Sunday read his own selection of the previous week's most peculiar news items, as taken from newspapers. Krypton closed in early 1985, but there are rumors that it may still be back from time to time. During 1984, Krypton seemed to be acting as an uncle to *Spectrum World Broadcasting*, an enthusiastic operation with excellent signals, a habit of "borrowing" news broadcasts on the hour from the Independent Broadcasting Authority chain, and the best-produced pirate QSL-card that I have seen for a long time. Both stations used 7710 kHz in early 1984—a good frequency for long-distance reception. Their address was 134 Eastworth Road, Chertsey, Surrey, England.

Scotland

The English legislation on pirates applies also to Scotland, but it is unlikely to frighten away the hard-core pirate support, as there is a long history of Scottish resistance to ideas about what is good for the Scots if these ideas are invented in London. Three stations are likely to be of special interest for some time yet:

- *Weekend Music Radio*, a regular and efficient supporter of the ideal of free radio, made several tests for North America during the 1982–83 and 1983–84 winter seasons, though with less widespread success than these efforts have deserved. Nevertheless, listeners along the length of the East Coast of North America and somewhat inland have received their QSLs before now. Many frequencies have been used, among them 6235, 6240, 6285, 6305, and 7320 kHz. Their address is 42 Arran Close, Cambridge, England.

- *Radio Freedom* (Fig. 20-2) represents the strong Edinburgh nucleus of support for pirate radio. As well as the tests made on 9420 kHz, and on 15040 or 15050 kHz, mentioned above, the station favors 6235 and 6240 kHz. Their address is Dept. R, 67 Elm Row, Edinburgh EH7 4AQ, Scotland.

Fig. 20-2. Radio Freedom International of Scotland verifies with this attractive blue on yellow card.

- *Radio Stella* is less regular in its habits than the other two, but it has demonstrated a talent for survival despite various kinds of less than favorable odds. Its most popular frequencies are in the 7315–25-kHz range, and the latest of a long line of addresses is 235 Rullion Road, Penicuik, Midlothian, Scotland. In case of difficulty, it may be possible to ask Weekend Music Radio to forward mail to the chief operator.

The final word about Scotland is that several Radio Free Scotlands tend to appear at times of nationalistic excitement or elections. Most of this broadcasting is on FM, TV sound channels, or the medium-wave band, but a little of it has spilled over onto shortwave (e.g., on 6248 kHz) in the past. Some of the Radio Free Scotland activity has involved temporary changes of pirate names, just as Clark Kent happens to be known as Superman from time to time.

Wales

An exotic country indeed, look you, bach. The Welsh Language Society (motto: heb iaith, heb galog), the nationalist political party Plaid Cymru, and the compilers of the North American Shortwave Association list of radio countries would all agree on this. There has been no shortwave piracy from Wales until recently, and the scale is still small, but smallness indicates no lack of determination to continue. The Welsh dragon flag and the Welsh leek will be held aloft on the radio scene in the future by one native-born operator and one friendly visitor:

- *Llais yr Cenhinan*, or the *Voice of the Leek* to you out there who don't speak the official language of Heaven, appears two or three times annually, usually on 6265.5 kHz, but occasionally on 7373 kHz, from the central north of Wales. Power is about 30 watts. One test was made for North America in early 1984, but apparently without success. Easter Sunday is a favored fixed point on this station's calendar, as is the day following, which is a British public holiday. The Edenbridge, Kent, address given for Atlanta Radio is the latest one known. Listeners with names like Jones are excused from sending return postage.

- *Radio Pluto*, based in England, seems to come to life mainly or only at Easter, when it is "on location" at a holiday site in southern Wales. The station announces the Welsh location when it is broadcasting from there. Frequencies have included 6225 and 7350 kHz, and the address is 147 Mackie Avenue, Brighton, Sussex, England.

France

After a long period during which no shortwave pirate was positively identified as operating from French territory, a small but steady trend towards a pirate presence appeared in the early 1980s. There has been a very large amount of pirate radio in France since 1977, but all of this was on FM (defended in some cases by great and powerful friends, e.g., the present Prime Minister when he was Mayor of Lille and a pirate trade-union station there was threatened with police action). The history of FM piracy, and its subsequent partial legalization, does not seem to have had any effect on shortwave activity. Probably the longest-lived French-speaking shortwave pirate is a shadowy religious operation on 7380 kHz, which is heard occasionally in the southern half of France. It uses the name *Harmonie*, and has no address that I have been able to discover. Two more recent and less mysterious broadcasters, who also represent France on the pirate bands, are:

- *Atlantic 2000*, located in the region where the Loire river meets the Atlantic, makes regular if not frequent appearances on either 7330 or 7340 kHz. There was a period when it was using the Dedemsvaart, Netherlands, maildrop mailing address, but it now seems to prefer its original address of 75 rue Julian Grimau, 93700 Drancy, France, which is just outside Paris, on the rapid-transit line to the Charles de Gaulle airport. Power is 20 watts.

- *Radio Waves International*, the newest French station, prefers 6275 kHz, but can also use 7340 kHz. Their address is B.P. 130, 92504 Rueil-Malmaison, France. Like the probable location of the station, this address, too, is quite close to Paris.

Netherlands

The greatest amount of piracy per head of population in any European country is almost certainly Dutch. It started in the rural and relatively lightly populated eastern part of the country, where the dialect and local popular music is not encouraged by the official Dutch radio (which is based in the urbanized west). The home-brew broadcasting of the frankly "awful" local music (usually described as "Schnulzen" by the German DX magazines that mention the stations, and by "umpa-bumpa musik" in one Swedish magazine), plus the station's greetings to friends and relations has been popular around 1305–45 kHz and 1615–45 kHz for a long time. The operators have never quite mastered twentieth-century technology, and the sound quality (often atrocious) is known among pirate listeners throughout Europe as simply "Dutch modulation." Occasionally, some of the operators find 1940s-vintage army surplus transmitters and VFOs that will tune the 48-meter range, and then they cause confusion to stations all over the pirate bands, but these operations usually disappear as fast as they arrive. A number of European pirates have stickers as well as attractive QSL cards. The cards and stickers shown in Fig. 20-3 are from Holland.

A source of better-quality Dutch activity, from younger people who have a better understanding of electronics and music, has produced many useful Dutch shortwave stations in the past, but this activity has been on the wane in the 1980s because of the greatly increased popularity of FM piracy, where the high-fidelity reproduction of music is one of the attractions and the challenges. The quantity and variety of Dutch "backyard" FM is incredible, and is worth sampling at any time of the day if you ever visit the Netherlands with an FM receiver.

This does not leave many landmarks for shortwave broadcasting, but two are worth mentioning because of either their signal quality or program quality:

- *Free Radio Service Holland*, or FRS Holland, has a regular monthly program, appearing lately near 7317–20 kHz, that aims for the style and quality of World Music Radio (Fig. 20-4). Its radio news service, "FRS goes DX," at around 1015 hours, is a very good value. Its programs are also put out occasionally by Radio Delmare (see Belgium). Its location is near Maastricht, in the southeast part of the country, and the Dedemsvaart mailing address is used.

- *KBC Radio* is a station that recharges its enthusiasm with longish absences, followed by periods of activity during which its signal is of local strength over a wide area. Its claims to 1 kW of power may be exaggerated, but it certainly seems to use a power that is high for a pirate—at least

Fig. 20-3. A number of European pirates have stickers as well as cards. These are from Holland.

when it is on the air unaided at around 6228 kHz. It also has a friendly relationship with Radio Delmare, and may use Delmare's facilities from 1100–1200 on Saturdays. The latest address is: c/o GN Productions, Postbus 38, 6744 ZG Ederveen, Netherlands.

Belgium

As far as radio habits are concerned, Belgium is a little brother of its Dutch neighbor, except that it avoids one nasty habit: there is no such thing as "Belgian modulation." Most Belgian piracy is now more or less in the open, on FM, and with addresses available. The usual pop-music mixture is spiced with "language" politics in areas where both French and Dutch speakers are fighting for political and economic advantage at each other's expense. The language issue doesn't surface on the shortwave bands, but the two most representative stations are run by the Dutch language faction—who communicate with the world by doing at least half of their transmitting in English.

- *Radio Brigitte* is a fairly conventional pirate with all the Belgian national virtues, combined with an interest in coin-collecting. QSLs are often accompanied by small Belgian coins. The station can operate on 6235 kHz or 6306–10 kHz, but it has also used 6550 kHz in an attempt to avoid interference. The address is just Brigitte, Postbus 10, 7954 ZG Rouveen, Netherlands.

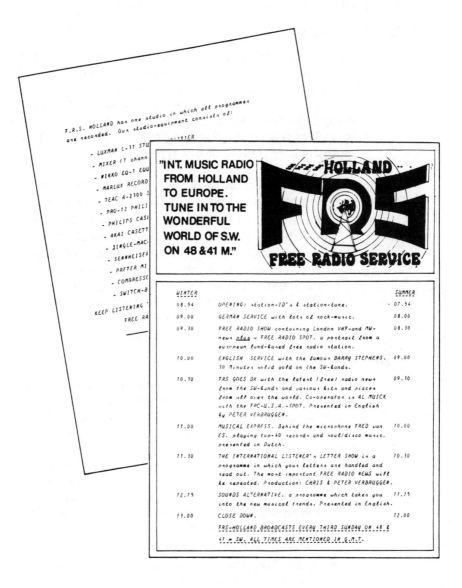

Fig. 20-4. Holland's Free Radio Service (FRS) sends a printed program schedule.

- *Radio Delmare* is a larger-scale operation, with links to FM free radio. It has such extras as a listeners' club, with membership costs of about $3.00. The regular Sunday schedule at the time of this writing was 0800–1700 hours; the frequency has always been 6206.5 kHz. Much of Delmare's style is intended to remind listeners of the earlier offshore pirate station with the same name. As mentioned earlier, Delmare provides air time to a small number of other stations. Its address is Postbus 36, B-2050 Antwerpen, Belgium.

West Germany

The German pirate wave of 1974–75 died out because of a tough and efficiently run counterwave of raids and prosecutions. Since then, although there has been plenty of overtly radical political broadcasting on FM, there has not been much to hear from Germany on the shortwave pirate bands. In order to survive, though, the few stations with ambitions for a long life have had to plan their operations and security carefully, which indicates a seriousness of purpose. Two representative stations of this type (in 1985) are:

- *Radio Batida*, on 7345, 7359, or 7364 kHz, which has a taste for soul and reggae music, a pleasing QSL, and the Bremen mailing address.
- *Radio 101*, a German station with Austrian connections, or vice versa, which may present a good propsect for North American tests, because of its frequencies (7420, 7425, 7445 kHz), its signal strength (excellent), and also, perhaps, its deliberately peculiar modulation, which cuts through interference at the price of making speech rather difficult to understand. The operator says it is FM, but that is open to debate. Its address is R.101, B.P. 2, B-4680 Gremmenich, Belgium.

Switzerland

The occasional appearance of shortwave pirates in this country seems to be a natural extension of the FM piracy present in the largest cities of the German-speaking cantons. In early 1984, Switzerland was represented by about 1½ shortwave pirates, which is an increase of roughly 100% over the normal situation in the past.

- *Radio Madison* is the "real" station, with preferred frequencies 6200 and 7400 kHz, but technical difficulties caused its late-1983 and early-1984 transmissions to be only on 6275 or 6305 kHz. Its power is 50 watts, and its schedule is biased towards fifth-Sunday operation, in months with five Sundays. The Dedemsvaart address is used.
- *Radio City* is the "½" station. News of the station has turned up occasionally in German DX magazines over a period of a few years, but I have never met anyone who has heard it. Its latest claim (in late 1984) is the planning of tests, with a power of about 1 kW, on 15030 kHz or 25920 kHz. Its address, in case anyone wants to try for a reply that will explain what the organizers are up to, is Postfach 1118, CH-8036 Zürich, Switzerland.

Italy

Here, the dividing line between pirate-style legitimate (even if unlicensed) stations and legitimate-style (always unlicensed) pirate stations is difficult to draw. The legal situation is complex, after a Constitutional Court ruling in 1976 that the Radiotelevizione Italiana monopoly was unconstitutional. The Court's ruling indicated clearly that "local" radio and TV were open to all comers, and, in effect, invited the politicians to make laws to regulate this new situation without colliding with the Constitution. Needless to say, the Italian politics of shifting coalitions has many more important priorities, and the legislators have never got around to considering radio and TV. Also, although broadcasts aimed at an international audience are not protected by the Court's first ruling, at least one subsequent case has given a station of this kind (on FM, close to the Swiss border) the benefit of the doubt. Shortwave stations have suffered from a little harassment, but nothing serious so far. It seems that shortwave operators now have only to survive through any October copy deadline at the *World Radio TV Handbook* to make it into the official-looking listing for Italy in the next year's *Handbook*.

Of the crop in the 1985 *Handbook*, the station most likely to have a long existence, and to be heard outside Europe, is "Radio Milano International," on 7295 kHz, with an address of Via Locatelli 6, I-20124 Milano, Italy. It specializes in carrying programs, disguised as "stations," from other groups without transmitters—notably Radio Victoria, "the punks on shortwave," whose address is Postfach 1214, D-6380 Bad Homburg, West Germany. Radio Victoria has also been talking of using 9810–13 kHz someday, but it is not clear whose transmitter will be involved. The Victoria organization had some internal problems in 1985 which may have closed it down, but Radio Milano continues on 7295 kHz, ID'ing under its own name.

Scandinavia

The Scandinavians have a long history of association with shortwave piracy. It is therefore regrettable to have to report that this association has been slowly running down. In early 1984, there was no regular shortwave pirate broadcasting from Scandinavia, though several stations have been dormant and may make a comeback under the right circumstances. The most interesting of these are *Sunrise Radio* and *Radio Penguin*, which used to have monthly schedules starting from 1000 hours, on frequencies around 11405–15 kHz, which were suitable for long-distance communications. Another station is *Radio Titanic*, which has used 7325 kHz, as well as frequencies in the unofficial 26-meter band. All of them used the Dedemsvaart mailing address when they were last active.

It is unlikely that the good old days of Scandinavian pirate radio, which occurred in the first half of the 1970s, will be revived. The enthusiasm is still there, but it is now mostly siphoned off by FM broadcasting, particularly on those stations in the legal Närradio network in Sweden, and to a lesser extent, in those similar possibilities in Denmark and Norway that are mentioned in the *World Radio TV Handbook*.

Eastern Europe

The nearest thing to Western pirate radio on shortwave in Eastern Europe has been in East Germany, with small collections of stations imitating the general

European pirate fashions, and using output powers of just a few watts. Halle was the center of this development between 1979 and 1981, but it stopped in mid-1981 when most of the stations were raided and various operators received sentences from 8 days to 3 months in prisons or labor camps. The stations included Sender Gigantum, Sender Delphin, and HR4 (rebroadcasting Hessischer Rundfunk from West Germany, as its name suggests). There are two reasons why similar very tentative experiments in radio may recur from time to time: East German listeners are familiar with Western Sunday pirate broadcasts, so that it is not unusual for stations to get reception reports from the East, and the knowledge of electronics (sufficient for building a transmitter) is quite easy to obtain in many countries of Eastern Europe, including East Germany, from the periodicals sold on newsstands. These periodicals are sold with the aim of making good Marxist-Leninist workers into better (i.e., more technically competent) Marxist-Leninist workers—on the workers' own time and not on the time of the State. Unfortunately for the pure theory, more human side-effects creep in occasionally!

The idea of technical education through the reading of periodicals is also common in the USSR, which can scarcely reduce the capability of "radio hooligans" to build their own transmitters. Most of the broadcasting by pirates in the USSR is on the medium-wave band, where it is said that radio hooliganism is of epidemic proportions in the largest cities. I have heard only one of these gentlemen on shortwave: on 4820 kHz, in northern Finland, in 1975. His probable location was the Arctic port of Pechenga, about 100 miles from where I was listening. To judge by the programming, the operator's most valued possession was a scratchy old LP of Bill Haley and the Comets.

Exotic Countries?

Nobody has suggested so far that any pirates are interested in using shortwave to put Liechtenstein on the map (though that's a useful idea to work on), and San Marino has (or has had) at best only one or two FM stations. Given this information, the most likely exotic European "countries" to be featured on the pirate bands are in the British Isles. Except for Wales, where piracy is established, there are three possibilities: Northern Ireland, the Channel Islands, and the Isle of Man. The first two have already had shortwave voices—most recently on Radio Liberty (6230 kHz, in 1972–73) and on Radio Channel International (6552.5 kHz in 1979, and heard at least once in the USA). One can always hope for spontaneous new broadcasters to turn up in both places, particularly because a possible nursery exists in the sporadic late-night piracy on the medium-wave band in Belfast. This is probably the explanation for the two 1984 broadcasts of Misty Mountain Radio (6310 kHz, with address at 32 Victoria Road, Salisbury, Wiltshire SPI 3NG, England), whose replies to mail were posted in Belfast. The anarchist station, Gnomes of Ulster, talked of having a working holiday in the Isle of Man early in 1973, but nothing ever came of that. However, the fact that certain shortwave pirates in less exotic parts of Britain are now strongly DX-oriented may mean that adventurous future DX'peditions are not out of the question. Any relevant information should be found, as and when it becomes available, in the Europirate news sources quoted in an earlier section of this article.

For believers in the principle of "once a country, always a country," it is worth mentioning as a curiosity that the Saarland had something that was like the political status of a country from 1947 up until the end of 1956, when the inhabitants voted to become part of Germany. More precisely, it wasn't exactly France and it wasn't exactly Germany. The German pirate Radio Black City operated from the Saarland in 1979 and 1980 on 7313 and 7382 kHz, and intended its name to be taken quite literally, in reference to the mining of coal-based industries around Saarbrücken.

Finally, real connoisseurs of microcountries can appreciate that separation of the pieces by water is not the only way in which countries can exist in disconnected pieces. Baarle-Hertog and Baarle-Nassau make up a Belgian enclave southeast of Breda that is surrounded by Dutch territory. It is about 3½ miles north of the frontier of the "real" Belgium. In this territory, Radio Tokari got off the ground briefly in 1979 on 6300 kHz, and may hit the airwaves again if the operator can be talked into doing so. The Tokari QSL with a picture of a Dutch cow in a Belgian field (or vice versa) is one of my more curious souvenirs of 12 years of listening to European pirates.

Great Pirate Stories

Most of the good stories about the semi-public disturbances caused by pirates are hoax stories of one kind or another. Transmitter hijacks on FM have been a popular staple in England, Germany, and Sweden. For example, a few years back, BBC listeners were listening to a relay station on the Isle of Wight one night, expecting their usual program of restful melodies, when they were treated instead to wild punk rock and heavy-metal music, plus demented speeches by a convincing imitation of Adi Amin. When the BBC recaptured its transmitter after almost an hour, the announcer apologized in the best decorous BBC fashion for having inadvertently broadcast the "wrong program." The secret was that the relay site normally picked up its relay feed on a different frequency from the mainland through a directional antenna, and the pirates simply set up a portable FM transmitter on that frequency, with their pre-recorded program, as close as possible to the antenna. Those responsible were not unconnected with a shortwave station that was well-known in the late 1970s. After similar hijacks in Sweden and (for radical political broadcasts) in Germany, several radio organizations were observed to be taking down the receiving antennas on relay sites and digging trenches for cables.

Guarding against transmitter hijacks is one thing, but guarding against signal hijacks is more difficult. At Easter, 1983, the new English legal medium-wave station, County Sound, in Guildford, Surrey, appeared simultaneously on its correct frequency and on 6213.9 kHz in the 48-meter pirate band. Occasional voice-overs identified the shortwave channel as the "County Sound International Service." What County Sound's chief engineer said when the authorities came looking for him is not on public record. The villains of this piece haven't been discovered, but the trick is reminiscent of the sudden appearance of the London commercial station, Capital Radio, on 6240 kHz on the first of April several years ago. After almost 48 hours, the disposable transmitter and coupled receiver on Capital's frequency were run to earth in a remote part of London's Heathrow airport. A little unofficial detective work at the time found traces of the fingerprints of a particular inventive shortwave pirate—one who, to give an obscure clue here, celebrated the sixtieth anniversary of the Russian revolution of October 1917 on one of its own programs by playing the Internationale, rendered by the Glasgow Socialist Singers, followed by "Happy birthday to you."

The Danes are equally inventive. In the early 1970s, one of the local pirates (admittedly not on shortwave at that stage) put on some good-quality programming that was much livelier than the staid (not to say stodgy) fare of Denmark's radio, and then requested all listeners who were enjoying the program to call a particular Copenhagen number and say how much they preferred the style of this "new channel" to what they had been getting on the old channels. The number, it turned out, belonged to the office of the civil servant who was in charge of such things as "tracking down illegal radio transmissions."

Stations do not always keep to their predicted schedules. Sometimes the reasons are technical faults, but, at other times, the reasons are more original. A shortwave station active in southwestern England in 1974 had a theoretical 1000–1030 Sunday schedule, which involved the erection of a dipole antenna between two tall trees on a nearby farm a short time in advance. However, the station warned its listeners across Europe that it would not come onto the air if either one of two conditions (which would make antenna building unattractive) occurred: first, if it was raining in the district (not something which a listener in Bavaria could easily check at 0930 hours), and second, if the farmer chose that particular Sunday in question to put his bull into the field. Strict adherence to the second rule has allowed the former operator to enjoy a long and healthy career later—as a BBC engineer.

The list of possible pirate stories is nowhere near exhausted yet, but I shall ration myself to just one more. Forwarding addresses are usually just that, and they don't often begin with "c/o" and somebody's name unless the management is rather innocent. A 1974–75 religious pirate with an Episcopalian outlook, backed up by some unusual (except perhaps in West Virginia and Arkansas) doctrinal frills, gave the name of a junior priest in a specific parish as its front man, in its first broadcast. As I was planning a trip shortly afterwards that would go within five miles of his address, I decided to deliver my reception report in person. On arriving there, I announced my name and business, and after a short time at the door, I was invited in. The wife of the principal priest explained that her husband's assistant seemed to be occupied with something else, but would probably not take long to finish it. From the window of the front room, I had a view of the neatly kept garden, and a brief sight of the rear of a figure dressed in priest's clothing, climbing over the side fence at high speed. After about half an hour, my hostess reported that she was very puzzled: my target was "here when you came," but was now nowhere to be found. I accepted a cup of tea and took 15 minutes more to drink it, but without any change in the situation. At that point, because of my travel schedule, all that I could do was to put my reception report on the table in an envelope and leave. However, there was a happy ending of sorts: a QSL arrived almost by return post, but the station announced a change to a more anonymous and completely different address in its next broadcast.

And to Conclude

Because of the comparatively low powers used by European pirates, transatlantic reception is almost always a case of first-rate DX. On the other hand, the output powers are not so low that the achievement is all but impossible. Many North American listeners have proved that already. The habits of pirates, and their normal choice of frequencies, means that DX'ers on or near the Atlantic coast have the best chances. Even so, other listeners are not disqualified: Radio Dublin on 6910 kHz has been heard widely, and other Irish stations with significantly lower transmitting powers plus the Dutch station, ARTO Holland, were heard in places like Texas, Missouri, and Wisconsin during a "pirate festival" of specially arranged broadcasts (from 0500 hours on 18 October 1981). Therefore, the winter season mentioned near the beginning of this article is a practicable season for the many North American DX'ers who may wish to test their skills on one of the most challenging DX frontiers that is available. The high reply rate and the friendly personal responses of the operators to reports from outside Europe are some other pluses of this DX speciality.

DX'ers to the west of the Mississippi may like to try initiating some persuasion campaigns to get the European pirates to explore the higher frequencies and, thus, increase the chances of reception over greater distances. As the Californian station, Radio North Star International, was heard in England on a frequency above 13 MHz (even though with very weak signals) in December 1981, communication in the other direction ought to be possible in this, or a slightly lower, frequency range at suitable times; e.g., throughout the year in the "26-meter band" between, say, 2000 and 2400 hours. The stations are unlikely to make the move by themselves, but North American persuasion, correspondence, and diplomacy may be able to shake loose some choice European DX in the future. I hope that this article may do something to start the process.

21

DX'ING THE USSR

Roger Legge

In 1983, **Roger Legge** celebrated his golden anniversary of involvement in shortwave radio. He first began listening in his home town of Binghampton, New York, and since that time has accumulated a large collection of pre-war verifications.

After graduation from the University of Pennsylvania in 1945, Roger went to work for the Foreign Broadcast Information Service. That was followed by 25 years with the Voice of America, working in frequency and facilities assignment and participating in international radio conferences. He retired from the VOA in 1969, but his shortwave activity has remained as high as ever.

Roger is the editor and publisher of the "USSR High Frequency Broadcast Newsletter," which specializes in sorting out Soviet broadcast schedules and locating transmitter sites. Information about the Newsletter may be obtained by sending a self-addressed, stamped envelope to Roger Legge, Box 232, McLean, VA 22101. In his spare time, Roger continues to plan and develop a combination home and monitoring post at a rural location in Virginia.

There are more transmitter sites with high-frequency broadcasting transmitters located in the Union of Soviet Socialist Republics than in any other country in the world (Fig. 21-1). There are dozens of such sites operating in the 6- to 21-MHz band and many more sites in the 3- to 5-MHz Tropical Bands, including sites that are well north of the tropical area, geographically. There are more

Fig. 21-1. Map of shortwave transmitter sites in the Soviet Union. (Courtesy *SPEEDX*.)

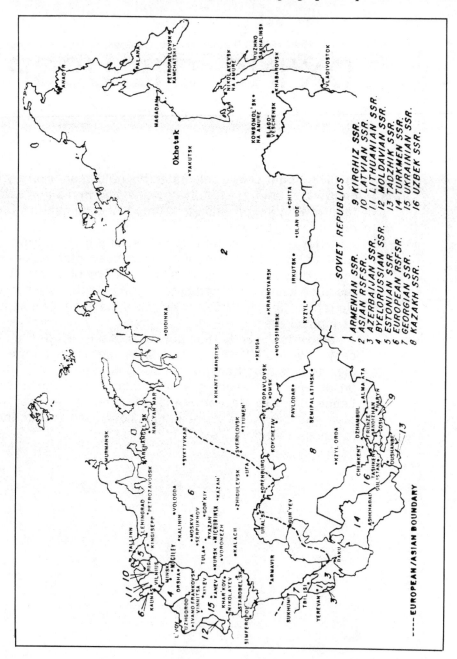

SOVIET REPUBLICS

1 ARMENIAN SSR.
2 ASIAN RSFSR.
3 AZERBAIJAN SSR.
4 BYELORUSSIAN SSR.
5 ESTONIAN SSR.
6 EUROPEAN RSFSR.
7 GEORGIAN SSR.
8 KAZAKH SSR.
9 KIRGHIZ SSR.
10 LATVIAN SSR.
11 LITHUANIAN SSR.
12 MOLDAVIAN SSR.
13 TADZHIK SSR.
14 TURKMEN SSR.
15 UKRAINIAN SSR.
16 UZBEK SSR.

---- EUROPEAN/ASIAN BOUNDARY

listed sites than actual sites, but DX'ers have learned that if you want to get QSLs from the USSR specifying sites, you have to play the game their way and use those listed sites.

The listed sites for the numerous "in-band" frequencies of USSR broadcasts are given in the quarterly "Tentative High Frequency Broadcasting Schedule" issued by the International Frequency Registration Board (IFBS) of the International Telecommunications Union in Geneva, Switzerland. Since this publication is far too expensive for purchase by most shortwave listeners, and since Radio Moscow and other USSR broadcasters follow a somewhat consistent pattern in the use of the same frequencies and listed sites in the same season from year to year, we will provide, later in the chapter, listings of the most likely seasons, frequencies, times, and services to assist you in hearing the various listed sites.

The "out of band" frequencies are not included in the IFRB publications. Sites for these frequencies, therefore, have been determined by a compilation of QSLs received in the past. However, changes from previously specified sites sometimes occur.

The locations of the actual sites have been determined by a variety of methods, including the study of satellite photographs, propagation characteristics, sign-on procedures, and so on. These sites are often different from those listed by Radio Moscow. Normally, however, they are located in the same general area as the listed sites. For example, frequencies listed at Ukrainian sites are usually in the Ukraine, but not necessarily at the same site listed. More about these actual sites later.

Radio Moscow will confirm reception reports that include program details, and will issue QSL cards specifying transmitter sites. This holds true for programs in English and some other languages, but usually not for Russian language programs. A variety of QSL card designs are used by Radio Moscow. Some of them are pictured in Fig. 21-2.

The transmitting tower at the Moscow Television Center is shown in Fig. 21-3. It also houses administrative offices, studios, transmitters, and even a restaurant. The Radio Moscow's North American Service staff of a few years ago is pictured in Fig. 21-4. The well-known Joe Adamov is on the far right. Mrs. Eugenia Stepanova, who frequently signs replys to listeners, is in the middle.

The Latin American Service of Radio Moscow has been somewhat irregular in specifying sites for reports on their Spanish and Portuguese broadcasts. One way of handling this problem is to send one or two Latin American Service reports along with a batch of World Service or North American Service reports. These will often be QSL'd with the sites specified.

Even if you know very little or none of the foreign language being used, it may still be possible to determine some program information by listening to the Radio Moscow news in English. Then listen for place and person names in the foreign language broadcast, in order to determine which news items you are hearing. A listener recently reported a QSL from Kalatch-on-Don on 15.2 MHz, the only listed frequency for this site. He used this method for obtaining program details on a broadcast in Gujarati.

Fig. 21-2. An assortment of the variety of QSL card designs received from Radio Moscow.

(A) Front view.

The following listings offer the best opportunities for verifying Radio Moscow transmitter sites. The seasons specified are "M" (March–April), "J" (May–August), "S" (September–October), and "D" (November–February). Schedules marked "M" are for March hours and will be one hour earlier in April when the USSR goes on summer time. The "S" schedule times are for September, and will be one hour later in October after the USSR returns to standard time. These time changes apply to English, Russian, and most European languages, but not to the Latin American Service and to language broadcasts to Africa and Asia.

Fig. 21-2 (cont.)

(B) Back view.

USSR High-Frequency Broadcasting

The USSR has the largest service of international and domestic high-frequency broadcasting of any country in the world. There are transmitters located throughout the USSR, many of them of 500 kilowatts. Relays are also located in Bulgaria and Cuba.

Radio Moscow (In English)

- *World Service*: On the air 24 hours a day, and transmitted to most areas of the world at approximately 6 A.M. to 12 P.M. (midnight) local time in the reception area.

Fig. 21-3. Transmitting tower at Moscow Television Center.

Fig. 21-4. Radio Moscow's North American Service staff a few years ago.

- *North American Service*: 2300–0400 GMT, October–March, and 2200–0300 GMT, April–September.

- *West North American Service*: 0400–0800 GMT, October–March, and 0300–0700 GMT, April–September.

- *Great Britain and Ireland*: 2000–2100 GMT, October–March, and 1900–2000 GMT, April–September.

- *Africa*: 1700–1800 GMT and 2000–2100 GMT.

Radio Moscow (Other Languages)

- *French to Europe and Africa* (Radio Moscou Internationale"): 0500–0700 and 1700–2200 hours during October–March, and 0400–0600 and 1600–2100 hours during April–September.

- *Spanish to South America*: 2300–0400 hours.

- *Spanish to West Indies and Central America*: 2300–0500 hours.

- *Portuguese to Brazil*: 2200–0130 hours (Radio Peace and Progress during final half hour).

Radio Moscow also transmits in numerous other languages to Europe, the Middle East, Africa, and Asia.

Russian Domestic Services

- *First Program*: Transmission is 24 hours daily, in five editions, covering the different time zones of the USSR.

- *Second Program* (also known as Mayak or "lighthouse"): Transmission is 24 hours daily, consisting mostly of light music with short periods of news on the hour and half hour.

- *Third Program*: Transmission is from 0001–2100 hours, consisting mainly of music and literary items. Only a few frequencies are used.

Russian External Service

The Russian External Service is aired on a 24-hour-a-day basis and consists of relays of the Mayak music programs and news from the First Program (at 0600, 1200, 1600, 1900, and 2200 hours; one hour earlier April through September). In addition, there are special programs as follows:

- *Radiostansiya Rodina* (or Voice of the Soviet Homeland): 0200–0300, 1400–1500, 2000–2100, and 2330–0030 hours.

- *Radiostansiya Atlantika* (or Atlantic Radio Station): 1300–1400 and 1630–1730 hours.

- *For Those at Sea*: 0030–0130 and 0715–0800 hours.

(These times are one hour earlier during April through September.)

Best reception of the Russian External Service in eastern North America is on 15.455 MHz at 1100–2000 hours, April through September, and on 15.175 MHz from 1100 to 1400 hours, year 'round. In western North America, try 15.470 MHz from 2000 to 1400 hours in April through September, and 17.720 MHz from 2300 to 0800 hours, additionally. During the period of October through March, try 17.860 MHz and 17.850 MHz from 2100 to 0530 hours.

Radio Moscow and other USSR broadcasters make frequency changes on the first Sunday of March, May, September, and November. In addition, on the first of April, when the USSR switches to summer time, many transmissions move to one hour earlier, including the Russian domestic and external services, World Service, North American Services, and many European language broadcasts.

On the first of October, when standard time is resumed, these broadcasts move back to one hour later. Some additional frequency changes take place on April 1 and October 1.

Other USSR High-Frequency Broadcasting Stations

QSL's do not specify the site, but transmitters are located at the broadcast site, except where noted herein. The stations are given individually in Tables 21-1 through 21-7.

**Table 21-1.
Radio Kiev,
Ukrainian SSR
Stations**

North American Service	East Coast	West Coast (transmitters in Soviet Far East)
	April–September	
2100–2200 Ukrainian	9.8, 11.72, 11.79 MHz	15.18, 15.405, 17.86 MHz
2300–0001 English	9.8, 11.72, 11.96 MHz	15.18, 15.405, 17.86 MHz
0200–0230 English	9.8, 11.72, 11.77 MHz	15.18, 15.405, 17.86 MHz
0230–0300 Ukrainian	9.8, 11.72, 11.77 MHz	15.18, 15.405, 17.86 MHz
	October–March	
2200–2300 Ukrainian	7.15, 7.205, 9.75 MHz	15.1, 15.24, 17.87 MHz
0030–0100 English	7.15, 7.205, 9.75 MHz	15.1, 15.24, 17.87 MHz
0300–0330 English	7.185, 7.205, 9.75 MHz	15.1, 15.24, 17.87 MHz
0330–0400 Ukrainian	7.185, 7.205, 9.75 MHz	15.1, 15,24, 17.87 MHz

To Europe	Frequencies (MHz)
	April–September
1800–1830 English	7.175, 7.320, 9.560
1830–1900 Ukrainian	7.175, 7.320, 9.560
	October–March
1900–1930 English	6.175, 7.260, 9.580
1930–2000 Ukrainian	6.175, 7.260, 9.580

**Table 21-2.
Radio Vilnius,
Lithuanian SSR
Stations**

North American Service	East Coast (transmitters in the Ukraine)	West Coast (transmitters in Far East)
	April–September	
2200–2230 English	11.72, 11.96 MHz	15.18, 15.405, 17.86 MHz
0001–0030 Lithuanian	11.72, 11.96 MHz	15.18, 15.405, 17.86 MHz
	October–March	
2300–2330 English	7.15, 9.75 MHz	15.1, 15.24, 17.87 MHz
0100–0130 Lithuanian	7.15, 9.75 MHz	15.1, 15.24, 17.87 MHz

Table 21-2 (cont.)	To Europe	Frequencies (MHz)
		April–September
	2100–2130 Lithuanian	6.100, 1.557, 1.107, 0.666
	2130–2200 English	6.100, 1.557, 1.107, 0.666
		March–October
	2200–2230 Lithuanian	6.100, 1.557, 1.107, 0.666
	2230–2300 English	6.100, 1.557, 1.107, 0.666

Table 21-3. Radio Yerevan, Armenian SSR Stations	To North America	West Coast (transmitters in Far East)
		April–September
	0230–0255 Armenian	15.18, 15.405, 17.86 MHz
	0255–0300 English	15.18, 15.405, 17.86 MHz
		October–March
	0300–0355 Armenian	15.1, 15.24, 17.87 MHz
	0355–0400 English	15.1, 15.24, 17.87 MHz
	To South America	**Frequencies**
		April–September
	2200–2300 Armenian	12.05 MHz
		October–March
	2200–2300 Armenian	9.48, 11.69 MHz
	To Europe	**Frequencies**
		April–September
	2030–2100 Armenian	15.5 MHz
		October–March
	2130–2200 Armenian	9.795 MHz

Table 21-4. Radio Minsk, Bielorussian SSR Stations	To Europe	Frequencies (MHz)
		April–September
	1700–1730 Bielorussian	7.42, 1.735, 11.86
	2030–2100 Bielorussian	6.175, 7.205, 11.735
		October–March
	1800–1830 Bielorussian	5.98, 6.01, 7.28
	2130–2200 Bielorussian	6.12, 6.185

Radio Tashkent, Uzbek SSR Stations

These stations transmit from April through September at 1200–1230 hours and 1400–1430 hours, in English, on 5.95, 9.65, 9.715, 11.785, and 15.46 MHz. Then, from October through March, from 1200–1230 hours and 1400–1430 hours, they transmit in English, on 5.945, 5.985, 9.54, 9.6, and 11.785 MHz.

Table 21-5.
Radio Riga,
Latvian SSR
Stations (Fig.
21-5)

Time	Day	Language	Frequency (MHz)
April–September (to Europe)			
0700–0730	Sun	Swedish	5.935, 0.576
0730–0825	Sun	Latvian	5.935, 0.576
2030–2100	Tu/Th/Sat	Swedish	5.935, 0.576
2030–2100	Weds	Latvian	5.935, 0.576
October–March (to Europe)			
0800–0830	Sun	Swedish	5.935, 0.576
0830–0925	Sun	Latvian	5.935, 0.576
2130–2200	Tu/Th/Sat	Swedish	5.935, 0.576
2130–2200	Weds	Latvian	5.935, 0.576

Fig. 21-5. QSL
from Radio
Riga, Latvian
SSR.

Table 21-6.
Radio Tallinn,
Estonian SSR
Stations

Time	Day	Language	Frequency (MHz)
April–September (to Europe)			
0700–0800	Sun	Finnish	5.925
0800–0830	Sun	Swedish	5.925
2005–2030		Swedish	5.925, 1.035
2035–2100		Estonian	5.925, 1.035
October–March (to Europe)			
0800–0900	Sun	Finnish	5.925
0900–0930	Sun	Swedish	5.925
2105–2135		Swedish	5.925, 1.035
2135–2200		Estonian	5.925, 1.035

Time	Language	Frequencies (MHz)
		Table 21-7. Radiostansiya Tikhiy Okean (Radio Station Pacific Ocean), Vladivostok

Table 21-7. Radiostansiya Tikhiy Okean (Radio Station Pacific Ocean), Vladivostok

Time	Language	Frequencies (MHz)
		April–September
0715–0800	Russian	9.580, 9.620, 9.810, 15.265, 15.470, 17.720, 17.860, 17.870, 21.530
1830–1930	Russian	5.015, 7.335, 9.480, 9.620, 9.770, 11.690
		October–March
0815–0900	Russian	5.940, 6.050, 7.175, 9.795, 9.810, 11.900
1930–2000	Russian	5.015, 6.035, 7.160, 7.240, 7.300, 9.780, 9.810

Radio Station Peace and Progress

Radio Station Peace and Progress (Fig. 21-6) transmits over Radio Moscow External Service transmitters.

- *Latin America*: Transmissions to Latin America are in Spanish (Radio Paz y Progresso) from 0001–0030 and 0130–0200 hours, in Portuguese from 0100–0130 hours, in Creole on Tuesday and Friday from 0030–0100 hours, and in Guarani on Tuesday and Friday from 0100–0130 hours.

- *Asia*: Transmissions to Asia in English are from 1300–1330 and from 1630–1700 hours. They also transmit in Chinese and Mongolian to Asia.

- *Africa*: Transmissions are in English from 1630–1700 hours.

Other External Services

- *Alma Ata, Kazakhstan*: Transmissions are on 5.035 and 5.915 MHz.

- *Baku, Azerbaijan*: Transmissions are on 6.110 and 6.135 MHz.

- *Dushanbe, Tadzhikistan*: Transmission is on 7.275 MHz.

Fig. 21-6. QSL from Radio Peace and Progress.

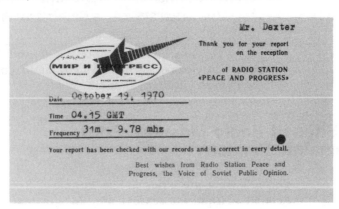

Domestic High-Frequency Broadcasts

Domestic high-frequency broadcasts are in the following frequencies and languages, and at the stated times. They are one hour earlier from April to September.

- *Riga, Latvia*: On 5.935 MHz, from 0300–2000 hours, in Russian and Latvian.
- *Tallinn, Estonia*: On 5.925 MHz, from 1600–2100 hours, in Estonian.
- *Vilnius, Lithuania*: On 6.110 MHz, from 1530–2200 hours, in Lithuanian, and on 9.710 MHz, from 0600–1500 hours, also in Lithuanian.
- *Murmansk, European RSFSR*: On 5.930 MHz, from 0200–2200, in Russian.
- *Kiev, Ukraine*: On 6.020 MHz, from 0415–1645, in Ukrainian.
- *Minsk, Bielorussia #1*: On 7.210, 9.645, 9.795, and 11.995 MHz, at 0300–2200 hours, in Russian and Bielorussian.
- *Minsk, Bielorussia #2*: On 9.645 MHz, from 0300–2200 hours, in Bielorussian.
- *Baku, Azerbaijan*: On 6.110 and 6.195 MHz from 0100–1100 hours, in Azerbaijani and Russian, and on 6.195 MHz from 0100–2400 hours in Azerbaijani.
- *Alma Ata, Kazakh #1*: On 5.970, 6.180, 9.780, and 11.950 MHz, from 0001–1930 hours in Russian and Kazakh.
- *Alma Ata, Kazakh #2*: On 5.960 and 9.505 MHz from 0001–1800 hours, in Kazakh.
- *Ashkhabad, Turkmen*: On 6.125 MHz from 0430–2200 hours, in Turkmenian and Russian.
- *Tashkent, Uzbek*: On 5.925 MHz from 0001–1900 hours, in Uzbek, Russian, and Tatar.
- *Irkutsk, ARSFSR*: On 6.090 MHz from 2100–1600, in Russian.
- *Khabarovsk, ARSFSR*: On 7.210 MHz from 2200–1900, in Russian.
- *Magadan, ARSFSR*: On 5.940, 7.320, 9.500 and 9.600 MHz from 1400–1300 hours, in Russian.
- *Yakutsk, ARSFSR*: On 7.100, 7.200, and 7.345 MHz, from 0030–0930 hours in October through March, and from 1900–1500 in April through September, in Russian and Yakut.

Listed Sites of Radio Moscow External and Domestic Service

The following are cities in the European Russian Soviet Federated Socialist Republic. Again, the seasons are "M" for March–April, "J" for May–August, "S" for September–October, and "D" for November–February.

Armavir	15.175 MHz	(J)	1600–2100 hours (French language to Europe)
	15.175 MHz	(J)	2100–2200 hours (World Service)
	9.765 MHz	(M)	2300–2400 hours (North American Service)
	9.765 MHz	(M)	0400–0600 hours (World Service)
	7.135 MHz	(D)	0230–0400 hours (North American Service)
	7.135 MHz	(D)	0400–0600 hours (World Service)
Gorky	7.185 MHz	(M)	2130–2300 hours (World Service to Middle East)
	7.175 MHz	(D)	2100–2300 hours (World Service to Central Europe)
Kalatch-on-Don	15.200 MHz	(M/J/D)	1100–1430 hours (South Asia languages)
Kalinin	21.635 MHz	(J)	1000–1130 hours (Vietnamese/Thai languages)
Kazan	11.960 MHz	(M)	1900–2300 hours (World Service and English language to Africa)
	15.110 MHz	(J)	2200–0300 hours (Portuguese and Spanish language to South America)
Kingisepp	15.320 MHz	(J)	0300–1400 hours (World Service to Middle East)
	9.760 MHz	(D)	1700–2300 hours (World Service to Europe/Africa)
	21.545 MHz	(J)	0600–1600 hours (World Service to West Africa)
Konevo	11.850 MHz	(D/M)	1100–1300 hours (Russian language to Central Europe)
Kursk	17.860 MHz	(D)	1100–1400 hours (World Service to Central Europe, including transmissions at 1000–1100 in German)
	17.745 MHz	(J)	0800–1630 hours (Russian language to the Mediterranean)
Leningrad	17.860 MHz	(J)	1600–1800 hours (French language to Africa)

	21.600 MHz	(D)	0800–1000 hours (World Service to Southeast Asia) 1000–1300 hours (Southeast Asian languages)
Michurensk	9.775 MHz		(Formerly QSL'd for this site, but more recently for Tula)
Moscow	9.685 MHz	(J)	2200–0100 hours (North American Service)
	7.390 MHz	(D)	1700–2300 hours (World Service to West Europe)
Orenburg	6.090 MHz	(D)	2100–2300 hours (World Service to Scandinavia) 1600–2100 hours (Scandinavian languages)
Riazan	5.980 MHz	(D)	1900–2200 hours (World Service to Great Britain and Ireland)
	15.140 MHz	(J)	0430–1300 hours (World Service to Southeast Asia)
	11.790 MHz	(J)	2200–0001 hours (North American Service)
Serpukhov	15.375 MHz	(J)	0500–1500 hours (World Service to Europe)
	17.730 MHz	(D)	0700–1300 hours (World Service to South and Southeast Asia)
Tula	17.765 MHz	(J)	0600–1300 hours (World Service to Southeast Asia)
	9.610 MHz	(M/S/D)	2200–0300 hours (North American Service)
Vologda	15.350 MHz	(M/D)	0400–1300 hours (World Service to Middle East)
Voronezh	9.675 MHz	(D)	2230–0500 hours (Spanish language to South America) 0600–0600 hours (World Service to Europe)
Zhigu'evsk	15.220 MHz	(D)	1000–1200 hours (World Service to South Asia)
	15.185 MHz	(J)	1100–1400 hours (South Asia languages)

Next, we have the Baltic Republics.

Estonian SSR

Tallinn	17.700 MHz	(J)	0830–1400 hours (World Service to the Middle East. Have QSL'd this site in past but present status uncertain.)

Latvian SSR

Riga	21.615 MHz	(M/S/D)	0900–1600 hours (World Service to Africa)
	7.140 MHz	(D)	0400–1100 hours (World Service to the Middle East)

Lithuanian SSR

Kaunas			(Has been listed for occasional external service use but not of late.)

Bielorussian SSR

Minsk	15.150 MHz	(D/M)	1230–1400 hours (World Service to North America)
		(J)	1000–2100 hours (World Service to North America)
	11.745 MHz	(J)	0500–0600 hours (French language to Africa)
			1030–1400 hours (World Service to Africa)
Orsha	5.920 MHz	(J)	1500–2030 hours (German language)
			2030–2100 hours (Spanish language to Spain)
			2100–2200 hours (World Service)
		(D)	0400–0600 hours (World Service to Europe)
			1500–1700 hours (World Service to Europe)
			1900–2300 hours (World Service to Europe)

Ukrainian SSR

Ivano-Frankovsk	9.665 MHz	(D)	2300–0030 hours (Spanish language to Central America)
	11.880 MHz	(J)	1900–2100 hours (Italian language)
Kharkov	11.950 MHz	(J)	1700–1800 hours (English language to Africa) 2000–2100 hours (English language to Africa) 2100–2300 hours (World Service)
	15.540 MHz	(J)	0900–1400 hours (World Service to the Middle East)
Kiev	11.770 MHz	(J)	2200–0200 hours (North American Service)
	9.710 MHz	(M)	2300–0300 hours (North American Service)
Lvov	11.750 MHz	(J)	2200–0300 hours (North American Service)
	7.150 MHz	(D)	2300–0400 hours (North American Service)
Simferopol	7.195 MHz	(D)	2300–0100 hours (North American Service)
	15.485 MHz	(M/J)	1000–1400 hours (World Service to North America)
Starobelsk	5.905 MHz	(J)	1300–2000 hours (Polish language) 2100–2200 hours (World Service)
	15.330 MHz	(J)	1400–1700 hours (World Service) 1700–1800 hours (English language to Africa) 2000–2100 hours (English language to Africa) 2100–2200 hours (World Service)
Vinnitsa	9.720 MHz	(J)	2300–0300 hours (North American Service)
	7.215 MHz	(D)	2300–0300 hours (North American Service)
	15.135 MHz	(M/S/J/D)	1100–1400 hours (World Service to North America)

These are the Caucasus Republics.

Armenian SSR

Yerevan	21.740 MHz	(J)	0700–1500 hours (World Service to Africa)
	7.130 MHz	(D)	0200–1400 hours (World Service to the Middle East)

Azerbaijan SSR

Baku	11.920 MHz	(J)	2200–0400 hours (Portuguese/Spanish languages to South America)
	15.260 MHz	(J)	0330–1400 hours (World Service to Central Europe)
	15.415 MHz	(J)	0400–0600 hours (French language to Africa)

Georgian SSR

Tbilisi	11.805 MHz	(J)	2300–0400 hours (Spanish language to Central America) 0400–0600 hours (French language to Africa)
	21.715 MHz	(J/D)	0600–1500 hours (World Service to the Middle East)

And, we have the Central Asian Republics.

Kazakh SSR

Alma Ata	11.770 MHz	(D)	0700–1300 hours (World Service to Southeast Asia)
	9.610 MHz	(D)	1900–2300 hours (Spanish language to Spain)

Kirghiz SSR

Frunze	12.075 MHz	(J)	2100–0001 hours (World Service to Australia)

	15.265 MHz	(M)	2200–0130 hours (Portuguese language to Brazil)
			0130–0400 hours (Spanish language to South America)
	11.820 MHz	(D)	0900–1300 hours (World Service to Southeast Asia)

Tadzhik SSR

Dushanbe	15.220 MHz	(J)	0030–1000 hours (World Service)
			1000–1500 hours (Southeast Asian languages)
			1500–1600 hours (World Service)
	21.585 MHz	(J)	0800–1100 hours (World Service)
			1100–1430 hours (Southeast Asian languages)
	17.730 MHz	(J)	1000–1530 hours (Southeast Asian languages)

Turkmen SSR

Ashkhabad	11.930 MHz	(J)	2300–0400 hours (Spanish language to South America)
			0400–0600 hours (French language to Africa)
	15.540 MHz	(J)	1630–1800 hours (English languages to Africa)
			1800–2000 hours (World Service to Africa)
	17.760 MHz	(J)	0500–0600 hours (French language to Africa)
			0600–0700 hours (World Service to Africa)

Uzbek SSR

Tashkent	11.785 MHz	(J/D)	0900–1000 hours (World Service)

			1000-1200 hours (Southeast Asian languages)
	15.330 MHz	(D)	1000-1300 hours (World Service to Southeast Asia)

The following sites are in the Asiatic Russian Soviet Federated Socialist Republic (North Central Asia).

Chita	15.385 MHz	(D/M)	2300-0800 hours (World Service to Southeast Asia/Australia)
	21.740 MHz	(D)	0600-1500 hours (World Service to South and Southeast Asia)
	9.625 MHz	(D)	2100-2400 hours (World Service to Australia)
Irkutsk	17.730 MHz	(J/S)	2300-0800 hours (World Service to Southeast Asia/Australia)
	17.835 MHz	(J)	2300-0900 hours (World Service to Southeast Asia)
	21.530 MHz	(J/S)	0001-0700 hours (World Service to Northeast Asia and Australia)
	21.690 MHz	(J/S)	0001-1100 hours (World Service to South Asia)
Kenga	7.440 MHz	(D)	1700-2300 hours (World Service to North America) 2300-0400 hours (North American Service)
	9.505 MHz	(M)	1700-2230 hours (World Service to North America)
	9.640 MHz	(M/S)	2200-0300 hours (North American Service)
	12.060 MHz	(J)	1630-2200 hours (World Service to North America) 2200-0300 hours (North American Service)
Krasnoiarsk	17.820 MHz	(J)	0100-0830 hours (Chinese language)
	15.315 MHz	(J)	0800-0900 hours (Chinese language) 0900-1200 hours (Korean language)

Novosibirsk	7.185 MHz	(D)	1400–1800 hours (World Service to Northeast Asia)
	21.690 MHz		0900–1100 hours (World Service to Southeast Asia)
Omsk	17.765 MHz	(D)	0200–1230 hours (World Service to South Asia)
Sverdlovsk	7.185 MHz	(M)	2200–2300 hours (World Service to North America) 2300–0300 hours (North American Service)

And, finally, the Asiatic Russian Soviet Federated Socialist Republic (Far East) lists these sites.

Blagoveschensk	9.580 MHz	(J/S/D)	0430–0700 hours (North American Service) 1300–1800 hours (World Service to west North America)
	11.790 MHz	(M)	1400–2100 hours (World Service to west North America)
Khabarovsk	7.175 MHz	(D/M)	1400–1800 hours (World Service to west North America)
	9.795 MHz	(D/M)	0530–0700 hours (Western North America)
	12.050 MHz	(J/S/D)	1830–0500 hours (World and North American Service to west North America)
	17.850 MHz	(D)	0500–0800 hours (World Service to Australia)
Komsomolsk	9.505 MHz	(D)	0530–0800 hours (West North America)
	9.655 MHz	(S)	1400–1700 hours (World Service to west North America)
	12.010 MHz	(M)	0400–0800 hours (West North America)
	15.180 MHz	(J)	0300–0700 hours (West North America)
		(S)	0300–0400 hours (West North America)
Megadan	5.930 MHz		0930–1800 hours (Mayak Service)

	7.320 MHz	(D)	1400–1300 hours (Domestic Service)
		(J)	1300–1200 hours (Domestic Service)
	9.500 MHz	(D)	1400–1300 hours (Domestic Service)
		(J)	1300–1200 hours (Domestic Service)
Nikolaevsk-Amur	9.665 MHz	(M)	0430–0700 hours (North American Service)
	9.755 MHz	(J)	0800–1130 hours (World Service to Australia)
	17.845 MHz	(J)	0830–1300 hours (Chinese language)
Okhotsk	9.600 MHz		1400–1300 hours (Russian Mayak, First Program, Radio Magadan)
Petropavlovsk-Kamchatsky	15.425 MHz	(J/S)	2200–0500 hours (West North America)
	15.455 MHz	(D)	1930–2300 hours (World Service)
			2300–0500 hours (North American Service)
	17.700 MHz	(J/S)	2000–0400 hours (North American Service)
	21.530 MHz	(D)	2300–0200 hours (North American Service)
Vladivistok	7.290 MHz	(D)	1400–2100 hours (World Service to west North America)
	11.755 MHz	(J/S/M)	0430–0700 hours (West North America)
	11.950 MHz	(D/M/J)	2330–0700 hours (North American Service feed to Petropavlovsk)
	15.130 MHz	(D)	2100–1000 hours (World Service to Australia)
		(M)	2100–1300 hours (World Service to Australia)
	21.725 MHz	(J)	0001–1000 hours (World Service to Southeast Asia)
Yakutsk	7.100 MHz	(D)	0030–0930 hours (Russian Domestic First Program)
		(J)	1900–1500 hours (Russian Domestic First Program)

	7.200 MHz	(D)	0030–0930 hours (Russian Domestic First Program)
		(J)	1900–1500 hours (Russian Domestic First Program)
	7.345 MHz	(D)	0030–0930 hours (Russian Domestic First Program)
		(J)	1900–1500 hours (Russian Domestic First Program)
Yuzhno-Sakhalinsk	7.260 MHz	(D)	0330–0800 hours (West North America)
			0800–0900 hours (Japanese language)
			0930–1300 hours (Japanese language)
	11.840 MHz		1830–2100 hours (Russian Domestic Service—low power)

Where the Transmitters Really Are

Extensive studies by my colleagues, Olle Alm, of Sweden (of propagation characteristics and transmitter scheduling of Radio Moscow facilities), and Douglas Johnson, of Washington State (of satellite photographs of the USSR), has enabled a determination to be made of the location of the actual sites of most of Radio Moscow's high-frequency broadcast transmitters.

These studies show that, in most cases, the actual site is in the same general area as the listed or QSL'd site, but not necessarily at the specific site indicated by Radio Moscow. An exception to this is that, on some frequencies, transmitters in the Ukraine are listed as Moscow, presumably due to old Moscow registration dates on those channels. Examples of this are 9.530 and 9.685 MHz. In the European RSFSR, transmitters have been located at Moscow (several sites), Kalinin, Tula, Serpukhov, and Leningrad.

A large, 16-transmitter site was found on satellite photographs of an unlisted site at Ulyanovsk. These transmitters are usually listed as Kazan, the nearest better-known city, but are also sometimes listed as Zhigulevsk (15.185 MHz), Gorky (17.755 MHz), and Orenburg (6.090 MHz).

The transmitters usually listed as Armavir are actually at a site near Krasnodar. Other listed sites are Kalatch-on-Don (15.200 MHz), Kingisepp (15.320 MHz), Konevo (11.850 MHz), Kursk (17.745 MHz), Riazan (15.140 MHz), Vologda (15.350 MHz), and Voronej (15.320 MHz).

It has also been determined that there are high-frequency broadcast transmitters in the Ukraine at Kiev, Lvov, near Simferopol, and at a site believed to be near Nikolayev. Other listed sites are Ivano-Frankovsk (9.665 MHz) and Vinnitsa (15.135 MHz). (However, the transmitter on 15.135-MHz frequency is actually in Plovdiv, Bulgaria.)

The Bulgarian relays of Radio Moscow, at Plovdiv (500 KW) and Sofia, are usually listed with the Ukraine sites. For example, 11.750 MHz is listed as Lvov, and 11.850 MHz is listed as Vinnitsa.

In the Caucasus area, the main site is Yerevan, with several transmitters at Tbilisi. Baku is occasionally listed in the IFRB publications (15.260 MHz), but the actual site is usually Yerevan.

In the Central Asia Republics, there are transmitters at Alma Ata, Dushanbe, and Tashkent (Fig. 21-7). Frunze is often listed by the IFRB, but there is no indication of high-frequency transmitters there, and the transmitters listed are actually at one of the other sites. Ashkhabad is also occasionally listed as being on 11.930 MHz (actually Alma Ata) and 17.760 MHz (actually Dushanbe).

Fig. 21-7. QSL from Radio Tashkent.

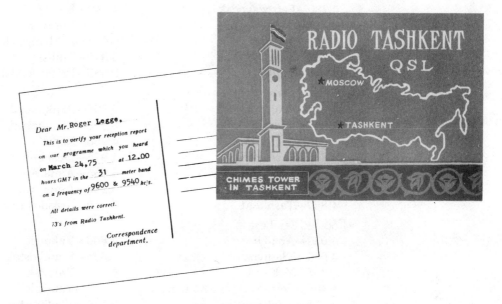

In North Central Asia, there are transmitters at Novosibirsk, Krasnoyarsk, Chita, Irkutsk, and Sverdlovsk. The Central Asia transmitter on the North American Service, listed and QSL'd as Kenga (7.440 MHz in the winter, 9.640 MHz in the spring, and 12.060 MHz in the summer), is one of the 500-kilowatt transmitters at Krasnoyarsk.

The Omsk listing on 17.765 MHz (on the winter schedule) is another one of the Krasnoyarsk 500-kilowatt transmitters.

In the Soviet Far East, transmitters have been identified at Khabarovsk (the major site), Vladivostok (including a recently activated 500-KW outlet now on 21.725 MHz and the only 500-KW transmitter in the Soviet Far East so far), Blagoveschensk, Yakutsk, Magadan, Petropavlovsk-Kamchatsky, and Lazarev—near the listed sites of Nikolaevsk-Amur and Komsomolsk-Amur.

USSR Tropical Band Stations

Many USSR local domestic service stations operate in the 4-MHz "tropical broadcasting band," which was established mainly for domestic broadcasting in tropical areas. In the case of the USSR, some of the sites are well north of the tropical area.

These outlets are most likely to be heard during the winter months when there is a path of darkness between the transmitter and the receiving location. The frequencies (in MHz) and locations of the stations listed as operating include the following:

- European USSR

4.780—Petrozavodsk	4.940—Kiev
4.785—Baku	4.990—Yerevan
4.810—Yerevan	5.015—Arkhangelsk
4.860—Kalinin	5.040—Tbilisi
4.930—Tbilisi	5.065—Petrozavodsk

- Central Asia

4.010—Frunze	4.895—Ashkhabad
4.050—Frunze	4.930—Ashkhabad
4.545—Alma Ata	4.958—Baku
4.635—Dushanbe	4.975—Dushanbe
4.795—Ulan Ude	5.035—Alma Ata
4.825—Ashkhabad	5.260—Alma Ata
4.850—Tashkent	5.290—Krasnoyarsk

- Soviet Far East

4.030—Anadyr	4.800—Yakutsk
4.040—Vladivostok	4.825—Vladivostok
4.395—Yakutsk	4.920—Yakutsk
4.485—Petropavlovsk-Kamchatsky	4.940—Yakutsk
4.610—Khabarovsk	5.015—Vladivostok

22

DOMESTIC SHORTWAVE BROADCASTING ACTIVITY IN THE ANDES

Ralph W. Perry

It is safe to say that **Ralph Perry** has turned up more first loggings of Andean stations, while pursuing them from the United States, than anyone else. "Ralphus," as he is known to his DX friends, has a passion for the Andean and Indonesian stations and has had the good fortune to visit a number of them.

Ralph has been an active DX'er for over 20 years, has edited columns and written articles for various club bulletins, including *Frendx*, *Radio Nuevo Mundo* (Japan), the *Shortwave Bulletin* (Sweden), *Fine Tuning*, and *Numero Uno*. No one I know can face an 0945 alarm ring with such eagerness.

Ralph is an international marketing executive for an Eastern hemisphere petroleum company and is currently based in Kuala Lumpur, Malaysia. He holds a BA from the University of Illinois, an MS from Northwestern, and an MBA from the Wharton School, University of Pennsylvania.

Job demands currently leave little time for DX'ing. Ralph and his wife, Fay, have two children, Mark (pictured) and Phillip.

The year is 1976 and the location is Cusco, Peru, high in the Andes of South America (Fig. 22-1). An American DX'er has wandered up and down Avenida Sol a half dozen times already, in search of Radio Tawantinsuyu, a famous Peruvian shortwave radio station. A relatively powerful outlet with a prestigious name (that of the Inca Empire itself, the "Four Quarters"), Radio Tawantinsuyu ought to occupy some grand edifice, with a gaudy neon sign out front, mammoth antenna towers in the back, and so forth. Or, at least, so thought the touring DX'er. Finally, the fellow decided to forget about signs and just look for the address of 830 Avenida Sol. But, the address wasn't posted in plain view either. Finally, out of desperation, he identified the approximate location where number 830 ought to have been located and boldly marched into the nondescript, one-story building. A receptionist at a makeshift desk just inside the foyer eyed him with a mixture of suspicion and amusement. "Tawantinsuyu?" he

Fig. 22-1. QSL card from Radiodifusora "El Sur" S.C.R.L., Cusco, Peru.

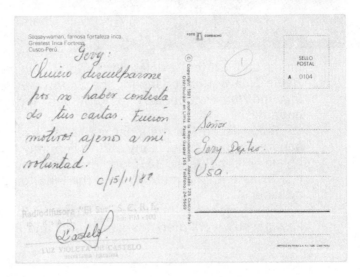

asked her? "Claro," she replied, but it had been anything but clear. So this was one of Peru's most exalted and longstanding shortwave representatives. A staff member led the visitor on a journey through the station's innards. There were rag-tag desks, the look of old wood everywhere, hand-lettered signs, dust, and an air of nonchalance throughout. A tiny, cramped "cabina," where the announcer would sit while on the air, separated by a window through which he could make eye contact with his "operador," was squeezed into the corner. And, while taking all this in with the eyes, the ears were assaulted with the *real* emisora: there it was, contradicting what the Dx'er was seeing, the sound of an intoxicatingly melancholic huayno folk-music selection, and, then, a sequence of echo-chamber commercials in the most professional style, followed by more music. The familiar sounds poured out of a speaker which allowed all at the station to monitor what was going out over the air, at any time, via Radio Tawantinsuyu. It was as if a real Peruvian station could not be seen, only heard. For, despite the sights confronting the longtime fan of the station, the sound of the programming reconfirmed the truth of the matter, that *this* was one of the best. It was as if one was tampering with magic by trying to see that which is not of the realm of the eyes, but rather of the ears. One given to mysticism might tell you that 830 Avenida Sol has nothing to do with Radio Tawantinsuyu—that Radio Tawantinsuyu is not a place. It is a sound, a mood, an experience. The DX'er took many photos of the difusora, and, if this were an authentic story of magic, he would report that none of them turned out. Well, this is my story, friends; I sent the shots to the local camera shop and every last one of them came out, crisp and colorful. So much for any possible side trips toward Twilight Zone environs....

The Tawantinsuyu story provides a handy departure point for a personal view of domestic shortwave broadcasting and broadcasters in Peru, Bolivia, and Ecuador. There is a great deal that is not understood, or at best is misunderstood, about these broadcasters. Much of this confusion stems from contradictions which seem to be embedded in the very essence of their operation. They are surrounded by an aura of ramshackle disorganization, coming one day and going the next. They beg questions as to their very viability—who can possibly make a go of it, selling peticiones or musical requests in some backwater fly-blown town in the Beni? And yet, there they are. Just when one is thoroughly convinced that there is no economic justification for commercial radio operations throughout the entire region, a verification package arrives from Oruro featuring the most extravagant pennant yet seen—all golden tassles and hand-painted designs—plus a four-color station letterhead and a complimentary album of the local musica del campo. Then, you listen to their sound on a frosty winter morning and partake of a veritable art form once again; you are surprised that any enterprise anywhere, with whatever resources imaginable, could produce such perfection. Still later that same morning, you see a New York Times' article about hunger on the Bolivian altiplano, and you are again reminded that, the region is far from rich.

For most of us who enjoy tuning the Andeans (and if you are still reading this article, it is assumed you are part of the club), it is tempting to not concern ourselves with the economic and political realities of the situation. In general, we're simply happy that there are individuals who find shortwave broadcasting a worthwhile endeavor in the Andes. It provides DX targets and we don't get worked up about why they are in the broadcasting game.

But perhaps we should. Perhaps we should be paraphrasing Butch Cassidy (who, himself, spent time in Bolivia) and be asking, in effect, "Who are those guys?" That, then, is what this chapter is about. Who, why, how? I suspect that if you have read this far, you will probably find that you care about much of this background—there is an unwritten law that upon sufficient exposure to Andean folkloric music, one begins to be converted into an "hermano por alma," a soul brother with the descendants of the Incas.

Is it all an illusion, these great sounds emerging from the midst of economic chaos, or is it real? Do roses bloom in a mound of manure? We are going to adopt the "roses theory," but with one change. It's not manure! There are compelling economic and business reasons which explain commercial broadcasting in the Andes, or, at least, will help the overall picture make more sense to an outsider. And, then, by overlaying some political, demographic, and historic/cultural factors on this base of understanding, we can arrive at a reasoned, holistic conception of just what those radio folks are doing in the thin air of Puno or La Paz.

Peru

There is a veritable explosion of shortwave activity taking place right now in Peru. There may be as many as 100 different shortwave outlets currently active, not to mention the parallel existence of a number of medium-wave stations. Interestingly, perhaps one fourth of Peru's currently active shortwave stations are less than a few years old. Shortwave broadcasting is on a major uptrend in Peru, reversing the grim forecasts of just a few years back regarding our beloved medium. We are now aware that events which were put into motion some 20 years ago, and since, have given rise to the shortwave renaissance we are now seeing in this delightful DX target area!

Everyone knows the story of Peru: the amazingly efficient (although heartless) Inca Empire, the Conquistadores, Pizzaro, the famous scene where the Inca lord, Atahualpa, is strangled (in what is current-day Cajamarca), Spanish colonial rule of the Viceroyalty, the independence struggle, and the heroes—Simon Bolivar and General San Martin. And, since the Inca Empire (Fig. 22-2) covered the breadth of the three countries under study in this chapter, ruling this vast stretch of land from, eventually, the two capital cities of Cusco and Quito, under the domain of two competing brothers, the "story of Peru" is also largely the story of Bolivia and Ecuador, although with some variations. For instance, the vast wealth of Bolivia as a mineral storehouse of tin and silver led the colonizing Spaniards to a different posture toward the region than that shown toward Peru. The exploitation was more, shall we say, rapacious. Nevertheless, a common heritage underlies the current nations of Peru, Bolivia, and Ecuador. Like sib-

Figure 22-2. Central part of Citadel with Sun Temple and Huaynapicchu Mountain in Machupicchu, Peru.

lings in an overprotective family, each has in its way struggled into the modern era of history to establish its true and unique identity. (Those seeking a detailed history of the Inca Empire are advised not to look for it in a radio hobby text! Instead, *The Conquest of Peru* by William H. Prescott, the classic work on the topic, is recommended.)

The portion of Peruvian history which directly affects today's broadcasting scene is that period dating from about 1960 to the present. Two trends have set the stage for the current situation: first, regional population growth, and second, a political shift to the left and then back again.

In 1961, approximately 10.4 million Peruvians walked the earth. At any one time, about one-fifth of the Peruvian population resides in the capital city, Lima. It is, however, the location of the remainder that is of interest. Between 1961 and 1972, the years of the official Peruvian census, the entire population grew to 13.6 million, which for the 11 years represented a compounded annual growth rate of 2.4% per annum. But the "mainline" population centers (excluding Lima), such as the departments (the equivalent of a U.S. state) of Apurimac, Cusco, Puno, Junin, Ayacucho, and Huancavelica, actually had an annual population growth of only 0.9% per year. Naturally the growth was happening elsewhere: in the north. The departments of Cajamarca, San Martin (Fig. 22-3), Amazonas, Lambayeque, and La Libertad (Fig. 22-4), for example, were growing at 2.4% per year. Thus, the population was leaving the older cities in the southern and central sierra for better opportunities in the North. The North is closer to the equator, with more moderate climates and more land available, and where new "boom" cities like Tarapoto were springing up and where, in the selva (jungle) toward the east, oil was being found. Between 1972 and 1979, the year of the next census, the nation's population grew at a hot rate of 3.5% per year—to 17.2 million. And, again, the trend was to "pump up" the population of the north.

Fig. 22-3. A bumper sticker which Radio San Martin in Tarapoto sent with their QSL.

Radio Difusora "SAN MARTIN" S.A.

UNA VOZ Y UN SONIDO PARA UN PERU MAS UNIDO

ONDA	ONDA	F. M.
MEDIA	CORTA	97.5 MHz
1130 KHz	4810 KHz	

con tu preferencia y sintonía... seremos tu eterna compañía!!!

Fig. 22-4. Card from Radio Libertad de Trujillo.

Felices Pascuas y Próspero Año 1980

Les desea Carlos Burmester B.

y Familia

LAS MEJORES 17 HORAS SON DE

RADIO LIBERTAD

**28 Años al Servicio de la Actividad
Cultural Comercial é Industrial**

POR SU MAYOR ALCANCE,
POR SU MEJOR PROGRAMACION

**Por la red más extensa de
Radiosonoras y alto parlantes en
los Mercados de
Trujillo y La Libertad.**

Por su nuevo enfoque Periodistico...

**LA VOZ DE LA CALLE
RADIO LIBERTAD de Trujillo
LA PRIMERA**

**De acuerdo al ranking de Mercados y
Tendencias, este es el veredicto de los
números:**

Dir. Zepita 450 Aptdo. 712 Telf. 231421   TRUJILLO-PERU

Suddenly, one can begin to see the business opportunity which existed for a sharp radio entrepreneur. Here was an embryonic market in the north that was starting to ripen; Peruvians who needed entertainment, news, and information about products. In short, as much as the people needed roads and telephone lines, they also needed radio. They got the roads and, to a lesser degree, the telephones. They didn't get the radio. Here's why.

Concurrent with this dramatic shift in Peru's population, which still continues to this day, the philosophy of government of the Republic also shifted. Sr. Fernando Balaunde had been elected president in July, 1963, but was removed in a military coup in October, 1968. The generals took hold of the reins and instituted what has since been dubbed "military socialism." A number of ambitious programs were instituted to redistribute the wealth and put the key industries and utilities into the hands of the general public. Besides the formation of a national oil company (Petroperu) and a telephone/telegraph national monopoly (Entelperu), a national radio system was put into effect: ENRAD (Empresa Nacional de Radio). Mainly, control of the ENRAD network was by proprietary means (the day before ENRAD, Sr. Gonzales owned Radio XX; the day after, he was an employee of the government-owned outlet, although probably remaining as manager). The state, by fiat, expropriated a string of outlets, both shortwave and medium wave, which augmented the Radio Nacional cadena. The ENRAD expropriation, at first, evidently proceeded as a semi-voluntary arrangement. Stations were invited to become controlled affiliates. Some went along, expecting that the ENRAD "Seal of Approval" for an outlet might give it a preferential status in obtaining limited spare parts. Later, however, the expropriation was by decree; also, a number of hostile takeovers of stations are known. At the same time, the generals' stance toward international trade also made an impact on the Peruvian radio system: a combination of punitive tariffs and outright bans on a variety of imports was instituted in hopes of curing the country's chronic foreign exchange problems. This meant that imported spare transmitter parts (for example: tubes and crystals for the transmitters) were largely unavailable for existing outlets. To stay on the air, the stations used methods that were equivalent to using Band-Aids®, tape, and bubble gum to patch up their equipment and keep it functioning. Needless to say, the importation of an entire transmitter from one of the well-known foreign manufacturers, like Harris in the United States, was completely out of the question. So, in a broadcasting-market sense, a pressure-cooker situation was building in the Peruvian north—a growing population was in place, advertisers were waiting, but radio transmitters were not to be had. To make matters worse, the outlets already on the air were either controlled by ENRAD or, else, they were leaving the air in a one-by-one attrition due to a lack of spare parts. By 1975, the first of the junta's leaders, Gen. Juan Velasco Alvarado, fell into ill health and was replaced by Gen. Francisco Morales Bermuda. Bermuda brought with him a rightward return toward the center and, by 1980, elections for a civilian government took place. Sr. Fernando Balaunde was elected once again (recall, he was running the show when the generals intervened in the first place). In short order, the civilian government began undoing the work of the generals, dismantling socialistic apparati like ENRAD, lifting the controls and monetary restrictions which had hobbled imports, and so on. The lid was removed from the pressure cooker.

Almost immediately, a few new stations took to the shortwaves, but the momentum had yet to build for the mass of stations to come. As with most new situations, there was a period of learning new rules, a certain span of time during which businessmen (in this case, radio entrepreneurs) were assessing the situation. When they finally realized that the door was open, things began to move.

Two factors, however, held up an immediate whole-hog multiplication of shortwave outlets on the air, especially in the north. First, by unfortunate coincidence, the Peruvian decision to lift international trade controls and let market forces do their work occurred at the same time as the worldwide recession, which was led by high U.S. interest rates and the consequently strong U.S. dollar. This drained capital from non-U.S. economies and channeled it into U.S. investments. Even had a Peruvian radio entrepreneur wanted to import a transmitter as soon as the import ban was lifted, he would have found that the price had climbed, in dollar terms, far out of reach. The second factor compounding the global recession's impact was the civilian government's decision to replace ENRAD with a "more enlightened" administration of the national radio populace.

Besides the Ministry of Transport and Communications (an engineering type unit which has historically had the role of assigning radio frequencies), the government created a new organization to minister to the political side of the radio business—the granting/cancelling of licenses. This new agency was the National Institute of Social Communications. The new bureau, feeling its way, lumbered along (it continues to work in a most somnolent fashion, needing at least two years to process a licensing request), while the stations already on the air, unable to easily acquire crystals or other needed spare parts, found themselves operating in most-unexpected sections of the shortwave radio spectrum. A most-noted specialist of the day, Sr. Juan Carlos Codina, of Lima, Peru, summed it up: "With the Peruvian radio stations, one could expect that they might use any frequency in the spectrum, with little regard to rules, licenses, and so on." Responding to the pressure of the northern Peruvian radio "void," new stations decided to risk illegality in their operations in two ways, hoping that the bureaucracy whose inefficiency helped create the mess would also be unable to enforce whatever rules which do exist. The two impunities were: (1) Operation on unassigned channels; and (2) Operation without a license.

Regarding operation on unassigned channels, Sr. Jose Carbone Castellano, in Lima, Peru, compiled the following sample list to compare official frequency assignments with the operating frequencies.

Station	Official kHz	Actual kHz
Radio Maranon, Jaen	580	1180
Radio Vision, Juanjui	1520	1545, 5360
Radio Pampas, Pampas	4875	4855
Radio Moderna, Celendin	4945	4300
Radio Imagen, Tarapoto	4970	5035
Radio Eco, Iquitos	5010	5112
Radio Huanta 2000, Huanta	5030	4746

It would be difficult to accurately proportion motives to this illegal frequency use, although a great degree of this action must be attributed to technical inadequacy, especially where the smaller outlets are concerned. The larger, smoother operations, such as Radio Eco in Iquitos (5.112 kHz) which has in the past operated on its official channel of 5.010 kHz, are obviously opting for out-of-band locations for better air quality for their listeners—away from the congestion of the usual band limits. (It has since returned to very near its assigned frequency, currently using 5011 kHz.) In such cases, one can envision a sharp broadcaster who, taking a reading of the governmental "winds," recognizes that, due to the inability of the laws' enforcers, the environment has reverted to laissez-faire. The author can state, without hesitation, that this posture by the Eco's of the world certainly benefits foreign listeners ("oyentes del exterior") as well as the local audience!

Unlicensed broadcasting is the second area where stations are skirting the rules with impunity. The sharp-eared DX'er, tuning a Peruvian which readily IDs but which does not list call letters, has probably found a station in this category. However, lack of a license seems not to be a major burden; many of these outfits readily QSL reports, have their own letterhead, and ostensibly are maintaining an official operation in all forms except for lack of official sanction. It is wrong, furthermore, to call such outlets "pirates." According to Mr. Carbone, the vast majority of these stations are those whose applications for licensing are well underway with the National Institute of Social Communications. In short, they have decided to jump the gun. Why? Under the circumstances, why not? Often a station has an economic timetable to which it must adhere, just as any commercial venture. Perhaps a bank loan underlies the new station's assets, and such a loan requires servicing, which is quite an impetus to get on the air and start selling advertising time. The plodding government bureaucracy, oblivious to and not living in a world of such economic time constraints, meanders through the licensing process at its own pace.

Sr. Carbone prepared a list of stations that have opted to broadcast before completing the licensing process and came up with:

4.900	Radio San Juan, Caraz
5.050	Radio Municipal, Cangallo
5.322	Radio Acobamba, Acobamba
5.301	Radio San Francisco
5.340	Radio Santa Cruz
6.110	Radio Yngay, Ancash
6.140	Radio Amazonas, Chachapoyas
6.296	Radio Chota
7.050	Radio Celendin
7.300	Radio Selva de Moyabamba

A letter to the author from the founder of Radio Huayabamba, in Amazonas departamento, indicated his station was not yet licensed, but it was well along the way toward being official. Evidently, in the current context, this show of good faith may be enough to fend off prosecution.

It would be incorrect to imply that tariff reductions and the lifting of the import ban on electronic equipment and parts has resulted in every potential radio-station operator writing to the Miami distributors for Sylvania tubes and whatnot! The recession of the early 1980s, combined with a history of economic frugality, encourages the newcomers to use time-tested means which, importantly, are cheaper. Many are purchasing local brand-name equipment, like Petrick, a Lima-based radio enterprise. With Petrick importing only those parts which it must, and locally assembling those it can, the result is a more affordable option. Other new stations, like our friend at Radio Huayabamba, use a familiar ploy: upgrade and refit a used amateur radio transmitter—in this case, a Johnson Viking. Still other stations rely on granddaddy's method: the home-brew transmitter. And, where the local talent isn't sufficiently able to build one's own transmitter from scrounged parts, there are always the hired guns available. One of the better-known itinerant radio-transmitter construction experts is Sr. Luis Franco, 47 years old and *blind* since 24. Senor Franco is a Ham Radio enthusiast who travels out of Trujillo in the north of Peru, and who has put a number of stations on the air, with the most recent being Radio Nuevo Continente, Cajamarca, 5.190 kHz (which blew a tube shortly after beginning its life as a shortwaver and which has since been stymied for months).

While the inefficiency of the government has created an effective substitute for a total free-market situation, another governmental deficiency has not been benevolent in its impact on the current radio scene. This deficiency is corruption. Radio Esmeralda de Huanta, a station long operating in that city (and which had its share of frequency changes, including 5.300, 3.285, 5.050 and 4.502 kHz), was ordered shut down in late 1983 (despite having passed an inspection on August 26) by the military police who are overseeing that terrorism-troubled region. The Ministry of Transport and Communications claimed to have nothing to do with the closure (and had the reprimand been based on frequency irregularities, it should have been MTC's turf). The new National System of Social Communication agency, the license-granting bureau, had pulled the plug on Radio Esmeralda. The proprietor of the victimized station claims the closure was a political favor to the new competitive station in town, Radio Huanta 2000.

With no effective enforcement, frequency "tug-o-wars" have taken place. Again, Radio Huanta 2000 was involved in one of the better-known episodes. According to Juan Carlos Codina and other monitors on the scene, Radio Huanta 2000 finally showed up on its officially appointed channel of 5.030 kHz (see earlier chart) after doodling around on 4.754 and 4.746 kHz. However, Radio Los Andes, an established squatter station (one of those in the "license-in-progress" situation), has long claimed the channel. The action erupted into a simultaneous operation, on-the-air name-calling, and an intentional co-interference squirmish. Who is right in such a case? It is tempting to mark R. Huanta 2000 as the villain but, after all, they are noted as arriving (finally!) at their government-sanctioned frequency. DX'ers' hearts must go out to the friendly Radio Los Andes, however, a station that has entertained us for nearly two years now on 5.030 kHz with smashing signals, fine folkloric programming, and even an overseas listeners' letterbox program.

These all are, of course, the types of adjustment problems that growth brings to many areas of society, whether it be roads versus new traffic congestion, water supplies versus increased population, and so forth. For these situations, good government is needed to take an effective role, as both a referee and a leader. Hopefully this is on the way in Peru, so that (to paraphrase the old cake-eating expression) we can have our Huanta 2000 and Los Andes too.

The next new-station type of occurrence, which is all too predictable, should be the emergence of a new type of station in Peru: the political clandestine. It is rather puzzling that the main anti-governmental rebel force, Sendero Luminoso (The Shining Path), has not already taken to the airwaves, in a more permanent fashion than its earlier spot takeovers of the Ayachuco-area radio stations for quick announcements to the populace at large. DX'ers should watch for this development at nearly any time now; watch for Radio Voz del Sendero Luminoso, the "Voice of the Shining Path."

One other development to watch for is the National Institute of Social Communication's promised upgrade of the Radio Nacional chain. In a message to Congress by President Belaunde in late 1982, a sweeping program of new investment was outlined for Radio Nacional, with the program halfway sounding as if much of it were in place already. The entire package, when effected, would include regional shortwave outlets at Piura, Arequipa, Cajamarca, Huarez, Huancayo, Cusco, Puno, Pucallpa, and Iquitos. The project, being conducted with assistance from West Germany, was supposed to have been completed before the end of 1983, but as of Summer, 1986, shows no signs of completion (much less initiation!).

Other Features of Peruvian Shortwave Radio

Over the years, certain aspects regarding the Peruvian domestic shortwave broadcasting industry have stood out in this DX'er's experience as being significant. Some of these observations are of obvious importance, while others may seem trivial. They are presented here in no particular order, as a smorgasbord of knowledge about Peruvian radio.

The Music

This is the irresistible bait, the addictive substance, the feature which makes Peruvian broadcasting something complete unto itself. Common to DX'er's vocabulary is the misnomer "Andean music," which is about as descriptive a term as, say, "Rocky Mountain music." It doesn't really mean anything to the folks who churn the music out, but only to a distant audience which has happened upon that insultingly simplistic term. And often, radio enthusiasts who happen to tune in a bit of "Andean music" usually haven't the slightest clue as to just what they are hearing—be it Peruvian, Bolivian, or Ecuadorian. An initial lack of knowledge about the musical forms of the Andes is by no means something to be ashamed of—after all, there are no chromosomes which genetically transmit immediate proficiency in such things. Like all worthwhile pursuits, it requires some dedication of the self and a desire to learn.

But for the dedicated DX'er, the more he hears, the more he wants. And, the more his ears understand. A proficiency, at least to the degree where you can instantly tell Peruvian from Ecuadorian from Bolivian music, is, to my mind, a necessity to enjoy DX'ing the Andes. The analogy is to a connoisseur of the arts versus a well-intentioned, but ignorant, *schlub*. While for the former, a visit to an art gallery can be an occasion for a soaring experience of the soul and intellect, for the latter, that same visit would be reduced to simply enjoying the "nice" colors. Whatever a DX'er does, he should not deprive himself of the full experience of Andean DX. Nor, by musical ignorance, should he deprive himself of a major identification aid for unidentified stations.

In truth, Peruvian music is a dichotomy, just as is Peruvian society. While Lima and Arequipa (Fig. 22-5) are dominated by the *criollo* (the pure-bred descendants of the Spanish colonists), the highlands are the realm of the *cholo* (the mixed offspring of the Spaniards and the original inhabitants). The music of the criollos is romantic, sentimental, often performed by trios on guitar and in harmony. The music of the cholos, or, to use a locational reference, the *campesinos* ("country folk"), is the form we most commonly associate with Peru. This folkloric music is a blend of the indigenous music of the Incas (and the peoples they had conquered) with the conquistadors' music. Often the agent of admixture were Jesuit, Dominican, and Franciscan missionaries, who taught campesinos hymns, formed choirs of the country folk, and so forth. (Reports have it that the colonized people *did enjoy* the music of their "captors.") While the indigenous music was pentatonic (five-note scale) and based solely on wind and percussive instruments, with limited rhythmic pattern, the conquistadoras

Fig. 22-5. Selva-Alegre park at Arequipa, Peru, showing snow-covered Mt. "Cha-Chani."

brought a musical culture that incorporated stringed instruments, syncopation and rhythmic variety, and an eight-note scale. The blending has given rise to the *musica del campo*, the folkloric music of today's Peru. The violin and saxophone only took hold sometime in the early 1900s! The most ubiquitous style is the *huayno*, which at one time was a piece of dance music but now is mainly for listening. A huayno is notable for its dolorous (i.e., mainly pentatonic scale) general tone, and its tempo which rolls on and on, often plucked out via the bass strings of a guitar, while the melody is provided by a *charango* (an Andean mandolin that is usually constructed with the help of an unwilling, and dead, armadillo, who donates his shell for the instrument's case), or a *quena* (a shrill-sounding flute of limited tones that is a true throwback to pre-Conquistador times), or violins (along with the guitars, a Spanish contribution), or voices, or any combination of all of the above. A frequently heard wind instrument is the *pinkillo*, which produces a whistle-like sound with a continuously variable pitch.

To the Western ear, the musica folklorica of Peru often recalls the music heard over Chinese stations or other outlets of the Far East. Perhaps a musicologist could explain the nature of the similarities in more objective terms; but I can say subjectively that, yes, the similarity is there. More than once DX'ers have been fooled by a weak Chinese station, coming in earlier than usual in the morning. After working the signal for a little while and hearing it fade-in better, it becomes evident what is really being heard.

The place to gain expertise in a musical style, of course, is not in a book. Like swimming, one can read all the articles and books there are about the subject, but until that first toe is dipped into the water, it's still all theoretical. Similarly, the best way to master the music of Peru is by listening to it. Because of my interest in the music, I am partial to early morning DX'ing, when D-layer absorption is all but gone and maximum-signal stability is enjoyed on the Tropical Band frequencies. But there are other ways to study this beautiful music; the major metropolitan areas in the U.S. are surprisingly fruitful places for purchasing imported Peruvian (and other) folkloric records. Over the past decade, I have put together a collection of about 60 albums, one at a time. In addition, a handful of mail-order houses specialize in Latin American music, and are also good sources. One such operation, which I can recommend, is Ipanema, Box 49452, Austin, Texas 78765. A bit pricey, but with a good selection.

The Programs

There is much programming occurring across the spectrum of Peruvian stations —enough that several serious generalizations can be made about the "typical" Peruvian formats. While not intended as an exhaustive listing, here are a few main checkpoints:

NATIONAL ANTHEM

Virtually *all* Peruvian stations, licensed or not, city or rural, commercial or state or religious, sign on and sign off with the Peruvian national anthem, "Somos Libres, Seamoslo Siempre" (We are Free, May We Always Be), written in 1821 by J. de la Torre (words) and J. B. Alcedo (melody). And it must be true patriotism, since the stations' cavalier frequency behavior (or rather, lack of same) proves they hold no great fear of the Lima radio honchos. Playing the cancion nacional at opening and close may be a regulation, but no more enforceable than, say, staying on frequency! Credit the Peruvians with loving their country. Interestingly, it seems that there exists perhaps but two or three (at most) sanctioned versions of the anthem—in fact, the most popular version (an instrumental) is heard on 50% of the outlets, if not more.

PROGRAMMING FORMAT

All Peruvian shortwave stations broadcast to the campesino audience during the very early morning hours. Whereas rural stations also devote most of the rest of their air time exclusively to the peasantry, with play lists heavy on the current huayno favorites, the big-city stations' sound, during mid-day, is more like stations playing for New York City's Hispanic community than like their sister stations in upcountry Peru. But, for the first hour or so, even the slickest of the Lima hotshots will put on its morning "farm show" and pump out a grand string of huaynos. Once the morning campo spot is over, of course, the big-city shortwave stations of Lima and Arequipa, et al, revert to form and feature rock music, ballads, radio novelas, talk shows, etc. Big-city stations will also feature dinner music programs of lush orchestrals and ballads in the mid-evening.

It is possible to further delineate the split between the urban and rural stations—and again, the split is rent by the target audience. The big-city stations run a fixed schedule, much as we North Americans would consider the norm. This is based on the fact that the urban broadcasters gain their revenue from the sale of radio time for commercials and promotional announcements. It is all amenable to scheduling. Very nice and businesslike. On the other hand, the small outlets upcountry can't plan things quite as well. Not that they wouldn't like that luxury, but the big advertisers follow the big markets—and those big paychecks aren't being handed out en masse in Huancavelica. Thus, the smaller outlets find their financial niche in the sale of air time to private citizens.

In a country where communications, as an industry, is not yet totally developed, where mails can be erratic, and telephones limited in the scope of their underlying network, radio signals can carry messages cheaply and effectively from river valley to mountaintop. When air time is sold in this form, the messages are called cumunicados. Another popular form of message is the saludo, or peticione, or dedicado, which usually takes the form of a message or greeting to Mr./Mrs. X in some other town from Mr. Y in the station's town, followed by a sometimes meaningful musical selection. One of the most popular of peticione melodies is, of course, the "Anniversary Waltz," which can usually be heard in nearly any program of dedications and messages. Naturally, since this is a prime source of revenue for the tiny outlet, the day's schedule will be adjusted to "bank" all messages in the hopper. This is the reason that during the holiday seasons, some of the smaller highland outlets (and the selva stations, too) are often heard running late into the night. Everyone has a lover, or uncle, or chum to pass holiday greetings along to. And these messages, generically known as saludos, sometimes form the bulk of some station's programming.

Back in the 1970s, for example, locals in Huancayo would come to Radio Andina and fill out small saludos forms, saying which record(s) should be played and for whom, list their message, and then pay a fee that is equivalent to a few cents per record. One saludo might consist of several records, in which case they are played consecutively (leading to those famed periods of ten minutes or more of nonstop music that is often heard late at night on these small outlets). Back then, Radio Andina's *entire* schedule consisted of saludos programming, plus a morning news bulletin! They have since found the "manna" of scheduling commercial spots, or "cunas."

The announcement of saludos can often be very stylized, and, sometimes, a saludo is combined with a time-check announcement. For example, when Radio Andina followed a 100% saludos format, they went "En la localidad de...saludamos a...deseandole muchas feliciadades de parte de (name or petitioner) con la siguente grabacion." And after the record, "En localidad de...hemos saludado a----de parte de---." Titles and messages usually preceded the records, never followed, and sometimes were omitted. In addition to musical saludos, nonmusical messages are frequently passed. All DX'ers have heard these, since they are usually preceded by an admonition of "Atencion, atencion!"

Stations dependent on peticion-based income may have some regular advertising, but these are usually clustered at the top or bottom of the hour, in advertisement "strings" of up to five minutes duration, and are often recorded messages. Often, national advertising accounts, like "Inca Cola," make up a large proportion of these canned ad strings.

SIGN-ON/SIGN-OFF TIMES

Finally, generalizations can be made about the programming in Peru, and these concern the operating schedules. While we have already seen that, upcountry, sign-off time can be a rather variable affair, sign-on times follow predictable patterns which carry useful information for the astute DX'er. The majority of Peruvian stations take to the air sometime around 6 A.M., or 1100 UTC. This seems to be the consensus start time. However, that is a norm, and there is a distribution of different sign-on times, both before and after that norm. For instance, Puno-area stations, far in the southeast, near Bolivia, usually begin broadcasting around 1000 UTC, an hour earlier than the central sierra norm. On the other hand, to the far north of the country, sign-on times begin resembling the Ecuadorian schedules, with some stations coming on at 1130 UTC. Throughout the heartland of Peru (or at dead-center), stations are noted as coming on the air from 1030 through 1100 hours. If a station is reported being heard during the local evenings, it is always worth a try to tune them in in the morning, making the appropriate sign-on time assumption as above.

The Networks

At its height, the ENRAD-Peru experiment incorporated 34 radio stations and 5 television broadcasters. This was a rather significant "bite" from the total group of broadcasters, but by no means a majority. Before ENRAD, some small chains of stations existed, and at least one of those was swallowed whole by ENRAD.

Now that ENRAD is history, the stations, which had been usurped, have been set free again, and, within Peru, we find a communications situation that is comprised of many independent stations and a number of small, often family-based, radio candenas (networks). The following is a list of the small nets known to the author; assuredly there are more:

1. *Emisoras "Cruz del Peru"* (Radio Bahia de Chimbote, Radio La Voz de Huamanga, Radio Huancavelica, Radio La Voz de Nasca, Radio Andina, Radio Mineria de la Oroya, Radio Andahuaylas, Radio La Voz del Chira, and Radio Horizonte de Tingo Maria), which is owned by the Rondinel family. ENRAD had taken two of these, Radios Andahuaylas and Andina, a pair of well-established franchises in the sierra.

2. *Organizacion Roberto Cruzado* (Radio Loreto, Radio Heroica de Trujillo, Radio Huancayo, Radio San Francisco de Piura, Radio Nuevo Mundo de Pucallpa, and Radio Atahualpa). ORC maintains its head office in the Lima suburb of San Isidro and is named for its founder/owner. No temporary casualties to ENRAD.

3. *Cadena Tawantinsuyu de Radiodifusion* (Radio Tawantinsuyu de Cusco, Radio La Hora de Cusco, Radio Amauta (Cultural) de Cusco, Radio Abancay, Radio Apurimac de Abancay, Radio Qollasuyu de Juliaca) is under the direction of the family of Sr. Raul Montesinos E., who has been primarily identified with Radio Tawantinsuyu, one of Peru's most-revered broadcasters. ENRAD snapped up only Radio La Hora from the Tawantinsuyu chain.

Also now popping up in this age of deregulation vs. governmental inefficiency, are mini-chains of unsanctioned stations! Two examples are the jointly owned Radio Moderna stations in Celendin and Cajamarca, and the jointly owned Radio Huanta 2000 and Radio Cobriza 2000.

Ecuador

One of the real problems with the unfortunate term "Andean music" is that it tends to not only imply that the different forms of Peruvian music are homogenous, it also draws a false comparison between the music of Peru and that of its two neighbors, Ecuador and Bolivia. The music of Ecuador is as different from that of Peru as the music of Mexico is from that of the U.S.A.! This is a good point of departure for a radio study of Ecuador: it is indeed separate and unique. The Peruvian evolution of radio has been dependent on a mix of political and geographic factors, past and present. The same is true in Ecuador. Both enjoy a common heritage of Inca rule, but many years have intervened since then.

The Land

Unlike Peru, however, which rambles on for some half million square miles, Ecuador is a relatively petite 104,000 square miles. As such, things tend to "hang together" somewhat better in Ecuador; this will be explained further as we proceed. While Ecuador claims the same tri-formation of geographical areas as Peru (coastal lowland area, mountains, and inland jungle east of the sierra), it all occurs in a much more compact area. Hence, "Ecuadorianidad" is a much more legitimate concept than one might initially guess. There *is* a national homogeneity which affects all aspects of life in this lush, mountainous nation— including radio communications.

The People and Their Music

Since it is the first thing one contacts when tuning in an Ecuadorian station, let's talk about the music. Whereas Peruvian music is split into criollo and cholo camps, or the "forlkoric" versus the "modern," there is a great deal more musical unity in Ecuador. When one talks about *musica Ecuadoriana* or *la musica nacional*, it is a meaningful expression. From the most distinguished business-man in Guayaquil to compesinos in a mountain hamlet, all enjoy the national music (naturally, supplementing it, perhaps, with other musical tastes) and are not embarrassed to listen to a musical form which has definite folkloric under-pinnings. While a Lima criollo would die of embarrassment rather than admit he likes huayno music, it would be very natural to hear strains of pasillos or pasacalles or sanjuanitos in the classiest of Quito restaurants. The effect is obvious, where radio is concerned: the split in programming between the big-city stations versus the rural outlets is much less, although one would rightly expect a large Quito or Guayaquil outlet to break into more rock-oriented or classical programs than would an El Puyo station, reflecting not a cultural disparity but rather the expected cosmopolitan nature of the big-city listener. Nevertheless, in general, it is more likely that you will hear the folkloric-based national music on almost all stations, a good percentage of the time. All of the prceding should not in any way be construed as a value judgement (Ecuador vs. Peru), or a statement about the relative egalitarianism of either society; rather, this is merely a comment on the amount of fusion which has taken place between the Spanish and native-Indian cultures in the two areas. In Peru, the mix has been somewhat of an oil-and-water affair, with the highbrow segment of society having chosen to stick with the oil. In Ecuador, the mix has been emulsified, and there is little for different social classes to chose from. There is just one, basic, national music.

The Ecuadorian folklorica music exhibits a fine blending of the colonial power's concepts of rhythm and stringed instrumentation with the indigenous "dolorous" melodies. The major forms of Ecuadorian music are the more-westernized *pasillo* (sounding to the non-Ecuadorian as a waltz-beat ballad, with plucked guitar), the *pasacalle* (a huayno-tempo, Indian-sounding mix of guitar and flute), and the *sanjuanito* (a syncopated, lively, and somewhat repetitive sound, due to the tendency of using a base-melody that repeats on and on in a

1-2-3, 1-2-3-4-5 rhythm). In each case, instrumentation is used to its fullest, and even the most rustic, wildest-sounding, airiest flute is used in a complex rhythmic/melodic pattern, in stark comparison to the shrill, five-noted, quena pieces which are the more traditional Peruvian huayno forms. It is not surprising that Ecuadorian music has gained a significant following in Europe; it is a sophisticated musical form displaying a raw, underlying energy. It is truly a meeting of Old and New World cultures. Again, trying to describe music on paper is akin to describing color to a sightless man, and that is not the intent here. Those interested should explore with their ears, by tuning in Ecuadorian stations, and by procuring albums from import shops. It is well worth the investment! The more you hear of this basic yet sophisticated type of music, the more amazed you will be that at one time it had sounded "wild" to your ears and indistinguishable from Peruvian folklorica music.

Population Shifts

Again, another facet of Ecuadorian radio can be best revealed by contrasting it with Peruvian music. Unlike Peru, Ecuador's population is holding steady at about 6.5 million, as it has for the past decade. Other than the expected rural-to-urban migration, which is the scourge of the Third World countries, there is little major population shift occurring, with but one exception…. The pursuit after, and the exploitation of, oil reserves in the Oriente region, in the Ecuadorian jungles to the east, has led to settlement in that area, with a consequential emergence of some "wild west" boom towns like Lago Agrio. All in all, the Ecuadorian radio market has not provided any significant new opportunities for broadcasting entrepreneurs; there has been no event akin to the opening-up of the Peruvian north. Hence, the numbers and distribution of Ecuador's shortwave stations has held remarkably constant.

It may be simplistic, but let's say it: since Ecuador is a relatively small country, there is a limit to the number of major cities it can contain. And, of these major cities, the top four of Quito (the capital), Guayaquil, Loja, and Cuenca contain an estimated 75% of the total number of shortwave transmitters in the country! This is logical. The national coverage area (the market being served by shortwave) is compact enough to allow one to choose a location in a major city, but still direct transmission to outlying areas. And the benefits of locating in a major city, with its consistency of power supply, repair expertise, parts, staff, etc., make the decision easy. One might argue that location in a large city is the "default mode"—in the absence of strong reasons to the contrary, one would virtually never choose a small-town location.

Shortwave Frequencies

Again, without the impetus of population shifts and new radio markets being formed, the status quo continues. Hence, another result is there's no out-and-out squabble for frequencies. In fact, in Ecuador, there is a hard-core assortment of shortwave broadcasters who have remained rock-steady on traditional frequencies, lending stability to the entire picture. For example, as of 25 years ago, Emisora Gran Colombia was on 4.910 kHz, Radio Nacional Espejo was on 4.680 kHz, and so forth. In fact, the situation has not been an increasingly over-crowded national radio spectrum space but, rather, a growing abundance of local channels due to the traumatic contraction in the number of Ecuadorian shortwave broadcasters over the past several decades.

Besides the tendency to operate out of the major cities, another trend is evident in Ecuadorian shortwave. That is the relative preponderance of religious stations on shortwave radio. A look at the *World Radio TV Handbook* confirms this. Why has this occurred? The relative stability of Ecuador in a political sense, compared to its many neighbors, may have been one factor in influencing missionaries to set up broadcasting in Ecuador; another factor must have been the success of the biggest "Big Brother" of them all in missionary broadcasting, HCJB in Quito. Of the perhaps 35 shortwave stations currently active in Ecua-dor, about a dozen are religiously affiliated; for instance: La Voz del Napo, Radio Bahai, HCJB, Radio Federacion Sucua, Radio Rio Amazonas, Radio Jesus del Gran Poder, Radio Luz y Vida, La Voz Sanctuario, and so forth.

Ecuadorians like to sign on at 6 A.M., 1100 UTC, although 1115 and 1130 UTC are also popular. They often use the national anthem, "Salve, O Patria!" (words by J. L. Mera, tune by A. Neumann, 1866), but, by our unofficial poll, are not as faithful to it as the Peruvians are to theirs. A typical sign-off time would be 0400 UTC, which is 11 P.M. Ecuador time, although a goodly number of HC outlets seem partial to all-night operation. Often transmitting straight through the wee hours are such frequent guests as Radio Popular Independiente de Cuenca on 4.800 kHz, or CRE on 4.656 kHz, or even Ondas Quevedenas, on 3.325 kHz.

Bolivia

As with Ecuador, and unlike Peru, Bolivia's broadcasting picture is a relatively stable one, in terms of broadcasters coming and going from the active rolls. The only real growth in shortwave broadcasting activity is occurring in the lowlands region east of the cordilleras, primarily consisting of the departments of El Beni and Santa Cruz. In the Beni area, cities of moderate size (by Bolivian standards) have spring up along the extensive river system that is comprised by all the various tributaries of the Rio Madera. (These include the rivers Mamore, Itenez, Machupo, Itonamas, Baures, Beni, and the lesser flows). The same can be said of the Santa Cruz department, although more significant road work and rail lines have improved the infrastructure in that more easterly area, giving rise to a much more substantial urban development. In fact, Santa Cruz de la Sierra, the capital of the department of the same name, is the second largest city in Bolivia (after the capital), and wholly qualifies as a bonafide "boom town." Whereas it is not unusual to read about, say, Menonite communities or massive settlements by Japanese silkworm ranchers in Santa Cruz Department, this sort of thing doesn't occur frequently in El Beni. Up there, in the dense riverine jungles,

Bolivians who call themselves "cambas" are still relying on Amateur Radio operators-turned-pro for some of the basic communications necessities—an organization known as Radio Serrano is an amalgamation of Hams in various cities of the Beni who have formed their own radiotelephone system. But, when one hears of a new station going on the air in Bolivia, it is almost always in the lowlands area. Once again, a casual chain of population growth that creates a market explains this, since, as undesirable as the jungles may sound, many Bolivians find them preferable to the frigid heights.

Clandestine Operation

To most, Bolivia is the nation of the cold, bleak Altiplano, or the towering, snow-capped Andes, or even the silver mines of the Potosi region. And that is the "Bolivia" which provides many of the most sought-after and treasured DX catches of all. Much of the shortwave broadcasting from this "traditional" (if you will) Bolivia dates back to "clandestine" outlets put on the air since the 1952 revolution by tin miner syndicates, as a way to champion the needs of that very important segment of the campesino population. As of the early 1960s, an estimated one third of the shortwave stations in Bolivia were operating outside the law, with weak outlets that were prone to spurious emissions and wandering the bands in no real pattern. Some analysts say that the mountainous configuration of the country invited such an eruption of illegals, by providing them myriad hiding places among the folds and wrinkles of the Andean terrain. One favorite operating mode for the illegals, oddly enough, was to use postal service channels, since it was assumed no other station would dare cause interference to a protected frequency! It appears most of those stations have passed on to that great radio graveyard in the sky, or else have been legitimized (the listing of call letters assumes an indication of government sanction). Some of the more familiar stations which have recently been or are today still operating, and which fall into this category, include Radio Nacional de Huanuni, Radio Cobija, Radio Pio XII, Radio Illimani (yes, despite its current role as the government mouthpiece, this one started as a clandestine!), Radio Indoamerica, Radio El Condor, Radio Grigota, Radio Altiplano, Radio Amboro, Radio Luis de Fuentes, Radio Centenario, Radio Santa Cruz, Radio Fides, Radio Sararenda, Radio 21 de Diciembre, Radio Viloco, Radio La Curz del Sur, and Radio Los Andes!

The Music

So, what does one hear on Bolivian stations? Of course, the schism between big city vs. highlands vs. lowlands again pertains. In this case, the large cities are very limited in number. Only La Paz, Santa Cruz, and Cochabamba need be considered in this category. The pervasive influence of American rock music has indeed reached down to Bolivia and La Paz itself, but during the morning time frame, when we in North America are able to tune Bolivia, roughly 0900 (earliest possible sign-on time) until perhaps 1030 UTC, even the La Paz stations are playing their morning campo music shows. During the evening hours of transmission, the La Paz stations play much the same as the Beni stations do: easy listening music, radionovelas, and romantic selections. The highlands outlets stay more consistently with the Bolivian variety of folkloric music, in the indigenous Indian language of Aymara, which features *quenas*, *tarkas* (shrill wooden flutes), and *charango* (that small mandolin, the bane of the Andean Armadillo!), the breathy *sicu-sicu*, and so forth. Perhaps as a concession to Indian immigrants to the jungles, even the Beni outlets have been heard with morning campesino music shows a la "Amanecer Andino" (which means "Andean Dawn"). This is easily the most popular program title in the Andes, and is used by a vast number of stations, closely followed (in a 3-way second-place tie) by program titles "Buenos Dias, Peru," "Buenos Dias, Ecuador," and "Buenos Dias, Bolivia." But for the vast majority of people of the Beni, the "Camba," their music is romantic, sentimental, and schmaltzy. When tuning a Beni station, except for the A.M. folklore programs, the sound is not that different from some run-of-the-mill Venezuelan outlets.

The Land and the People

The split between Bolivia's lowlands and highlands seems more pronounced than is the case in Peru, perhaps due to the greater relative disparities in wealth between the respective populations in the two countries. In Peru, while the spector of poverty is not to be downplayed, there is at least a living to be made in both the tiny mountain hamlets and the Amazon jungle villages of the Loreto Department. In Bolivia, there is an uncomfortable but viable existence in the lowlands, while there is hunger and utter exploitation in the highlands. The peasantries are not on equal footing by any means.

So perhaps it is not surprising to note that Bolivia's highland outlets and lowland outlets do act, at times, like they are different countries (and, in truth, it can be postulated that the towns on the Rio Itenez in the far northern Beni have more in common with Brazil, just across the river, than with Cochabamba or Oruro, which are far to the south and high in the mountains). Generalizing, the highland outlets are relatively good verifiers of reports. There is sensed by the DX'er, on receiving a reply, that one's report was the highlight of the day at Radio El Condor or wherever. In the Beni, on the other hand, the endless pattern of follow-up report after follow-up report is common. Blame it on the heat, the debilitating humidity of the jungle, poor roads, and a "suspect" mail service. La Paz, of course, is a world-class city and its outlets are generally very good repliers to reports.

Summary

The preceding thought-piece on domestic shortwave broadcasting in Peru, Ecuador, and Bolivia has verged on the eclectic and in no way has attempted to act as a guide of what to tune, or when, or on which frequency. That sort of information is far better obtained—especially when dealing with such ephemeral creatures as Andean reception patterns—from the numerous clubs and radio DX programs, telephone tip lines, etc. Much of the basic data is well-covered in Jerry Berg's excellent article in the 1983 edition of *World Radio TV Handbook*, "A Primer on LA DX'ing." It is hoped, however, that a somewhat better "feel" for stations in the Andes will be the result of reading this piece, and that anyone not already addicted to these marvelous outlets may be intrigued enough to crawl out of bed some morning at 0945 UTC to hear them sign on.

A number of acknowledgements should be made. Foremost, thanks to Juan Carlos Codina of Lima for his stunning research over the past year, uncovering station after station by riding the crest of the explosion in the number of Peruvian outlets; to Jack Perolo for his research on the origins of Bolivian shortwave stations; to Richard Leggett, whose classic travel articles in the Danish Shortwave Clubs International bulletins of 15 years ago helped infect the author with the Andean DX bug; to Jose Luis Carbone Castellano for his governmental research in Lima; and to my wife, Fay, who helped me incubate the "bug" years ago via travel through the region and by visits to various stations in the role of the "crazy gringo DX'er."

23

GRAYLINE DX'ING THE TROPICAL BANDS

John Tuchscherer

 **John Tushscherer** has been fascinated by radio since obtaining his first receiver in 1922. That fascination went with him into the Army in World War II, where he served with the Army Airways Communications System, based first in Ohio and, later, on various islands in the South Pacific.

John is a retired accountant, an avid reader, and a gardener. He particularly enjoys DX'ing the Tropical Bands and Latin America, and he occasionally patrols the Ham bands looking for stations located on the islands where he once served. He has an interest in propagation and antennas as well.

"Tuch," as he is sometimes called, lives with his wife, Ruth, in Neenah, Wisconsin. They have one daughter who, with her husband, is active in Ham radio.

I've spent a number of pleasant summer afternoons visiting with John and Ruth. He is one of DX'ing's real gentlemen.

If you were an astronaut out in space and could observe the planet Earth, you would note that one half of the Earth is always in shadow, while the other half is always in daylight. The line between day and night, which divides the Earth into dark and light portions, is called the *circle of illumination* or *terminator*. This circle, actually a band, is called the *grayline* in the radio hobby.

For any site on Earth, the grayline is in effect twice a day—one hour in the morning (± X minutes from sunrise), and then again in the evening at ± X minutes from sunset. Figure 23-1 illustrates the duration of the winter grayline conditions at various latitudes. Over the years, Ham operators and SWL'ers have noticed signal enhancement for propagation along the grayline. This is due to the fact that the D layer, which absorbs high-frequency signals, is disappearing at the station on the sunset side of the grayline, while it hasn't yet built up on the sunrise side. There are also other reasons that are not as yet fully understood.

Fig. 23-1. Winter duration of grayline conditions.

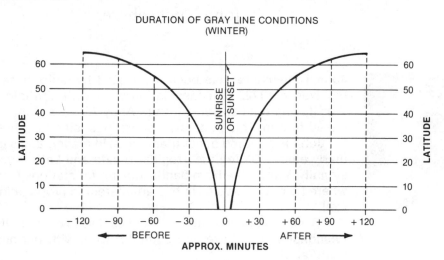

DURATION OF GRAY LINE CONDITIONS
(WINTER)

A "condition" of approximately sunrise at your location and sunset at the broadcast station, or vice versa, must exist or grayline reception will not exist. For example, if you live in New York City and hear a station in Sao Paulo, Brazil, on the first of January at 2300 GMT, you are not experiencing grayline reception since the "condition," or reception path, is simply a full darkness path.

Had you heard the same station at 2130 GMT, you still would not have had a grayline situation. Although it is approximately sunset in New York City at that hour (meeting one half of the condition), sunrise in Sao Paulo would be at 0836 GMT. Hence the path is not on the grayline. However, if you had heard an Indonesian station say in Ternate or Manado, at that time, you would have grayline reception since sunrise in those cities is at 2138 and 2151 GMT, respectively.

Even though the requisites for grayline reception exist, reception is still not assured, however. As with all shortwave propagation, the most important conditions are those that exist at the first reflection point away from the transmitter and the last reflection point before the receiver. And, *of course*, the station has to be on the air!

The annual orbit of the planet Earth around the Sun is more than 600 million miles and is directly responsible for our seasons. The axis of the Earth (that is, a line through the Earth from pole to pole) is inclined 23.5° from the axis of the sun. This means that the Earth exposes different parts of itself to the direct rays of the Sun during its yearly orbit.

The Earth's orbit is counterclockwise around the Sun, and the 23.5° inclination is outward, away from the axis of the Sun, as shown in Fig. 23-2. Therefore, the southern half of the Earth's surface is exposed to the direct rays of the sun on the right-hand wing of the orbit, and the northern half is exposed on the left-hand wing of the orbit. Note that the eccentricity of the orbit puts the Earth closer to the Sun in winter than in summer.

The limits of maximum movement (north and south) of the direct rays of the sun are known, respectively, as the *Tropic of Cancer* and the *Tropic of Capricorn*. When the sun reaches its northern zenith on 22 June, the phenomenon is known as the *summer solstice*. It reaches its southern zenith on 22 December for the *winter solstice*. When the sun crosses the equator on 21 March it is called the *vernal* or *spring equinox* and the 23 September crossing is the *autumnal* or *fall equinox*. Fig. 23-2 presents a graphic illustration of this information.

During a two-year search, I found much interest in grayline DX'ing but a paucity of information. However, I did find an excellent article by Peter Dalton, et al, in the September, 1975, issue of *CQ Magazine*. Subsequently, the same information appeared in *The Shortwave Propagation Handbook* by Jacobs and Cohen, published by Cowan Publishing Company (now CQ Publishing).

To determine where the grayline is at your sunrise or sunset time, Dalton's method requires only a world globe, a table of the inclination of the Earth's axis in relation to the sun (at about 15-day intervals—more frequently if you wish), and a few cardboard devices. For more complete details, see either of the sources just mentioned. Dalton has been grayline DX'ing since 1966, and has worked 300 countries on 80 meters, which attests to the efficiency of his method plus his skill and enthusiasm.

In 1981, a device called "The DX Edge" came on the market. The "Edge" consists of a plastic map case with twelve plastic overlays, one for each month. It is never out of date. You can easily determine the areas of the world that are in light or darkness at any time. Each plastic overlay has the shape of the grayline path shown on it. By placing the grayline on your location at your sunset or sunrise time, you can quickly determine the countries through which the grayline path passes, and see the areas which offer a good chance for DX. Since the "Edge" is designed for Hams, and because of its small size, the sites are designated by country prefixes. Perhaps a larger SWL version will become available in the future. I've made that suggestion to Mr. Anthony Japha, President of Zantec, Inc., which produces The DX Edge.

One of the best uses I have found for this product is that one can determine at a glance if there is a full darkness path to a site you may be interested in.

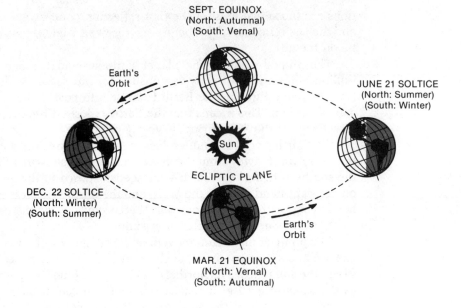

Fig. 23-2. The Earth's orbit and its ecliptic plane.

SEPT. EQUINOX
(North: Autumnal)
(South: Vernal)

Earth's Orbit

JUNE 21 SOLTICE
(North: Summer)
(South: Winter)

Sun

ECLIPTIC PLANE

DEC. 22 SOLTICE
(North: Winter)
(South: Summer)

Earth's Orbit

MAR. 21 EQUINOX
(North: Vernal)
(South: Autumnal)

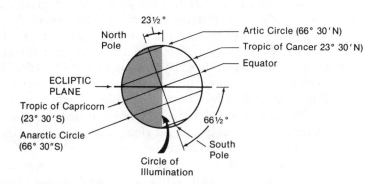

23½°

North Pole

Artic Circle (66° 30′N)

Tropic of Cancer 23° 30′N)

Equator

ECLIPTIC PLANE

Tropic of Capricorn (23° 30′S)

Anarctic Circle (66° 30″S)

66½°

South Pole

Circle of Illumination

However, its ultimate feature is that you can determine if there is a specific time of the year when your location and a station which you want to log will both lay on the grayline.

A couple of years ago, I checked the sunrise and sunset times in the Maldive Islands. At the time, they had a schedule of 1200 to 1700 GMT on 4.754 MHz. Finding that their January sunset was at 1307 and my Wisconsin sunrise was at 1327, I realized I had a good chance to hear this elusive station. I didn't hear the Maldives, but it occurred to me then that I didn't need my 16-inch world globe or any other adjunct to know the potentials for grayline DX'ing. All I really needed was a good sunrise/sunset table, which I already had.

I developed a simple method for determining the potential for grayline DX'ing. Since we want to know which sites are at sunset when it's sunrise at our location, and vice versa, I simply tabulated those conditions on a chart for each month, based on data from a comprehensive sunrise/sunset table. Since the table was based on the 15th of each month, the same date was used for local sunrise/sunset times. You can get that information from your local newspaper. Also, sunrise/sunset tables are available from many sources, including the SPEEDX club.

Table 23-1 is self-explanatory. Sunrise in Neenah, Wisconsin, occurs at 1325 GMT on 15 January. Concurrently, it is approximately sunset in Finland, Iran, Kuwait, etc. Similarly, when the sun sets at 2221 GMT in Neenah, it is approximately sunrise in Bali, Tokyo, Taiwan, etc. Charts like these can be prepared for all twelve months in just about three hours. Once prepared, the charts are valid forever (for a given location) as the slight deviations in the earth's orbit, which occur yearly, are of no significance for our purpose. The beauty of these charts is that you are readily aware of when you have the best "shot" at some of the toughies like Bhutan, Maldives, and so on.

Table 23-1. Sunrise and Sunset Times for Various Locations During January

Sunrise 1325 GMT (Neenah, WI)		Sunset 2221 GMT (Neenah, WI)	
SUNSET		*SUNRISE*	
Abu Dhabi	1351	Bali	2206
Armavir	1353	Bandjarmasin	2216
Dubai	1347	Brunei	2238
Finland	1327	Christmas Island	2241
Iran	1328	Denpasar	2214
Kuwait	1357	Jakarta	2243
Maldive Islands	1307	Japan (Tokyo)	2155
Moscow	1319	Khabarovsk	2245
Oman	1342	Philippines (Manila)	2228
Tallin	1334	Sabah (Kota Kinabalu)	2211
Tbilisi	1338	Samarinda	2219
Tula	1326	Sarawak (Miri)	2240
Yerevan	1334	Semarang	2236
		Surabaya	2234
		Taiwan	2237
		Ujung Pandang	2154
		Vladivostok	2239
		Yogyakarta	2235

Grayline DX'ing is an expecially good tool for the shortwave DX'er who is also a Ham operator. For the Ham who is interested in making "five-band Worked All Zones," targets can be contacted on 15 or 20 meters, and a schedule for 40 or 80 meters can be arranged for when the grayline potential exists.

DX'ING INDONESIA

William S. Sparks

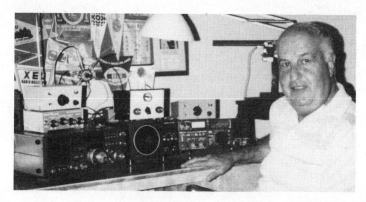

William S. Sparks' most recent DX'ing career spans over two decades. His first began in 1933 and lasted until 1941 when military commitments intervened. He began listening again some 20 years later and has been active ever since.

Bill's main DX interest is, of course, Indonesia, and he is an acknowledged expert in that area. The Pacific (where he served in World War II) and Asia are other favorites. There are several receivers in use at Bill's San Fransico home, which is situated about a mile from the Pacific Ocean. He is a member of several clubs and his name is well-known to those who read the bulletins of those organizations.

Bill Sparks is a lawyer specializing in probate, tax, and estate work. He is a great fan of the opera, and he and his wife, Winnie, enjoy traveling.

Indonesia—Its Location and Time Zones

Indonesia is one of the great archipelagos of the world. It is said to consist of over 3000 inhabited islands that are spread out along the equator between Australia and Malaysia (Fig. 24-1). There is a distance of over 5000 kilometers between Banda Aceh, on the northwestern tip of Sumatra, and Jayapura, located in Irian Jaya in the east. It is a land of many active volcanos, and even more active shortwave broadcasting stations.

While many of the islands are very small, the mainland mass is concentrated in five large islands: Sumatra, Java, the southern two-thirds of Borneo (known as Kalimantan), Sulawesi, and the western half of New Guinea (known as Irian Jaya).

The country has also been divided into three time zones as follows:

- Western: GMT + 7 hours, or *waktu Indonesia Bagian Barat* (for Sumatra, Java and Bali).
- Central: GMT + 8 hours, or *waktu Indonesia Bagian Tengah* (for Kalimantan, Sulawesi and Nusa Tenggara).
- Eastern: GMT + 9 hours, or *waktu Indonesia Bagian Timur* (for Irian Jaya and Maluku).

The DX'er is advised to learn the Indonesian names for the three time zones. Time announcements are frequently made by the stations, and they will help to identify the general location of the station. The "Indonesia Bagian" portion of the time-zone name may be eliminated, leaving the name as just *western time*, *central time*, and *eastern time;* or *waktu Barat*, *waktu Tengah*, and *waktu Timur.*

Indonesia uses the 24-hour clock in its time announcements. Experienced Indonesian DX'ers convert UTC time into local waktu Indonesia when writing their reception reports, using the 24-hour local-time format. That is more likely to be understood by the station personnel than the use of the UTC/GMT time code.

Historical Background of Indonesian Broadcasting

There has been broadcasting from Indonesia on shortwave since the early days of radio. Prior to World War II, Indonesia was under Dutch colonial rule as a part of The Netherlands East Indies. Eventually all of the broadcasting came under the control of The Netherlands Indies Broadcasting Company, Ltd. (N. I. R. O. M.). A folder sent out by N.I.R.O.M. (Fig. 24-2) to pre-war DX'ers states:

> The Nirom began broadcasting in the Netherlands Indies on March 31, 1934. Until then, only a few local stations had been in operation in the bigger cities, the cost being defrayed by contributions paid by the members of the local radio societies. These societies still exist.
>
> The Nirom started originally with four stations; in less than eighteen months, however, twenty stations were operating, while at the

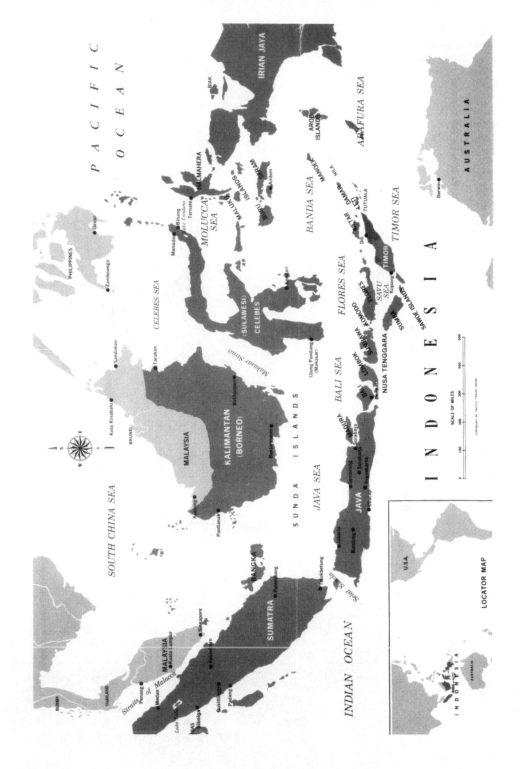

Fig. 24-1. Map of Indonesia.

Fig. 24-2. Cover of the January 1937 brochure issued by The Netherlands Indies Broadcasting Company Ltd. (N.I.R.O.M.).

The Netherlands Indies Broadcasting Company Ltd.

N. I. R. O. M.

Batavia-Java

Masts of the 10,000 Watt Transmitter Tandjong-Priok near Batavia

1-1-'37

beginning of 1937, the number of transmitters amounted to 24; namely, 5 highpower archipelago and 12 local transmitters for Western and one archipelago and 6 local transmitters for Oriental programs.

It is interesting to note that all of the stations were located on the island of Java, mainly at Batavia (now Jakarta), Surabaya, Bandung, and Solo (Surakarta). The broadcasts of "Western" programs were in Dutch. The "Oriental" programs were in various local dialects and were very-low-power transmissions. Apparently any Indonesians on the outside islands had to be DX'ers if they wanted to hear the programs.

It is worth noting that the N.I.R.O.M. stations were logged very frequently on the west coast of North America in pre-war days, and they were excellent verifiers. Many Dx'ers from those days have N.I.R.O.M. cards and folders in their QSL collections (Fig. 24-3).

World War II brought great changes to Indonesia and to Indonesian broadcasting due to the Japanese occupation and the active fighting that took place in many parts of eastern Indonesia (the Jayapura and Biak areas, in particular). Unfortunately, World War II also put an end to my early period of DX'ing. At the end of the war, Indonesians under the leadership of Sukarno declared their independence. After an armed struggle, they were finally able to attain recognition of their independence. Then came the problem of molding a nation out of a group of islands whose people spoke as many as 200 different dialects. The solution, in part, was radio. "Encourage the use of one language via radio broadcasting in that one language" was one suggestion.

Bahasa Indonesia

The "one language" (Satu Bahasa Kita) was Bahasa Indonesia, an adaption of modern Malay. Sukarno had been attemping to establish it as the official lan-

Fig. 24-3. A QSL from N.I.R.O.M. when Indonesia was the Netherlands East Indies.

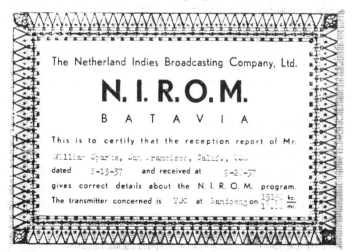

guage of Indonesia since 1928. When he came into power, he acquired in radio the tool he needed. His government took control of the radio broadcasting. The establishment of the Radio Republic Indonesia (R.R.I.) system served to educate the masses in the use of Bahasa Indonesia. This also became the means of establishing the unity of many diverse and often antagonistic groups of people.

The importance of radio and Bahasa Indonesia to the Indonesian government is illustrated by the opening line of the so-called "anthem," which is well-known to Indonesian DX'ers and is heard at the end of the 1200 and 1500 Jakarta news relays:

"Satu Nusa, Satu Bangsa, Satu Bahasa Kita," which is: One Land,
One Nation, One Language.

It was many years before the present R. R. I. system became fully established. Western New Guinea reverted to Dutch control after the Allied forces left the area. The Dutch had shortwave broadcasting stations at Biak, and Sorong, and, eventually, at Hollandia (now Jayapura). The stations were known as Radio Omroep Nieuw Guinea.

Political pressure by Indonesia eventually forced the Dutch to withdraw from Netherlands New Guinea. The United Nations moved into the area in 1963 pending final determination of the control of the area (Fig. 24-4). The U. N. took over the Hollandia station (on 6070 kHz) and broadcast its own programs until the area and the station were turned over to the Indonesian government. Thus, the shortwave stations located there became a part of the R. R. I. system in Irian Jaya. That left only the Portuguese area of eastern Timor to be acquired by Indonesia. The cultural influence of the early Portuguese traders and settlers in the islands of eastern Indonesia continues to the present day. The *kroncong*, a beautiful musical instrument that reminds many Dx'ers of the Hawaiian guitar, was introduced to the area by the Portuguese. In addition, the Portuguese retained colonial control of the eastern half of the island of Timor until forced out by the Indonesian government after heavy fighting in 1976.

The Portuguese operated a shortwave broadcasting station, Emissora de Radiodifusao di Timor (otherwise known as Radio Dili) on the 90-meter band (Fig. 24-5). This was a DX target that was eagerly sought after, as was the Falkland Islands in the 1960s and early 1970s. After Indonesia took over in 1976, it became (and is today) R. R. I. Dili. But, enough of history. Let's look at the current shortwave broadcasting situation in Indonesia.

Indonesian Propagation

The Indonesian stations are spread out over such a vast area that most of the time reception conditions will vary from different portions of the archipelago. Inasmuch as Indonesia occupies three time zones, fade-in and fade-out times will vary up to three hours, depending on the location of the stations.

DX'ers located to the east of Indonesia, such as in North and South America, are mainly interested in the evening broadcasts. This means the time period (at the transmitter) corresponding to the time from sunset until the evening sign-

Fig. 24-4. A rare card from the United Nations Temporary Executive Authority station at Hollandia.

UNITED NATIONS TEMPORY EXECUTIVE AUTHORITY
BROADCASTING SERVICES
HOLLANDIA - KOTA BARU 2.4.63 WEST NEW GUINEA – WEST IRIAN

Transmissions: 20.30 - 21.30 G.M.T.
1.30 - 5.30 G.M.T.
7.30 - 13.30 G.M.T.

Languages: Indonesian, Dutch, English.
Frequencies: 6070 kc.s - 7190 kc.s

Many thank for your report! We hope you will enjoy our programs.

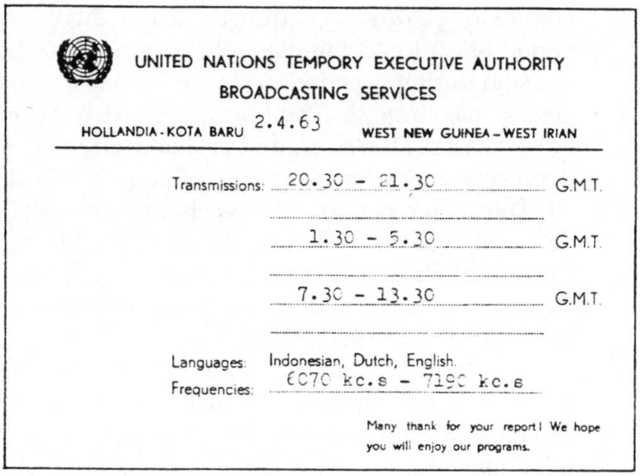

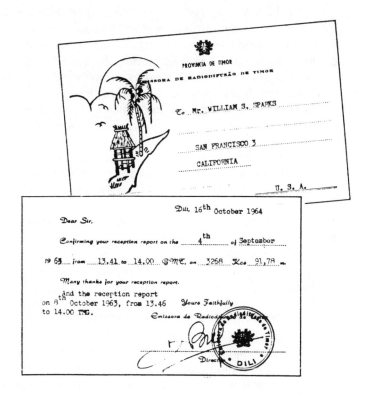

Fig. 24-5. A verification card from Dili which dates prior to Indonesia's takeover of Portuguese Timor.

PROVINCIA DE TIMOR
EMISSORA DE RADIODIFUSÃO DE TIMOR

To Mr. WILLIAM S. SPARKS
SAN FRANCISCO 3
CALIFORNIA
U. S. A.

Dili, 16th October 1964

Dear Sir,

Confirming your reception report on the 4th of September 1963 from 13.41 to 14.00 GMT, on 3268 Kcs 91,78 m.

Many thanks for your reception report.

And the reception report on 8th October 1963, from 13.46 to 14.00 TMG.

Yours Faithfully
Emissora de Radiodifusão de Timor
Director

DILI

off. The prime time to tune for Indonesian DX on the low-frequency bands, then, is the period corresponding to sunset at the transmitter site and dawn at the DX'er's location.

Tropical band reception of west to east signals will often provide short periods of peak signal strength just at dusk at the transmitter site, and for a period of 30 to 45 minutes after dawn at the DX'er's location. Thus, it is not unusual during a period of prime Indonesian reception to find two periods of peak signal strength. The first will be early, when it is sunset at the transmitter in Indonesia. The other, and probably stronger peak, will be just after dawn at the receiver location, provided, of course, that the station hasn't already signed off. During the period in between the two peaks, the signal will frequently fade down. A weak signal may be lost completely during the fade-down period. Dawn at the receiver is the witching hour when you are most likely to log that desired RPD, provided your next-door neighbor doesn't decide to use his electric razor at that time.

The Irian Jaya and Maluku stations have the earliest dusk and, therefore, the earliest fade-in time. Normally, they also sign off from two to three hours earlier than stations in Sumatra. The fade-in time for stations in Java and Sumatra tends to be about two hours later than for the Irian Jaya stations. However, the high power of RRI Ujung Pandang allows it to have an early fade-in that is well before sunset there.

A look at a sunset chart for several Indonesian transmitter locations shows some surprising differences from the conditions that one who lives a substantial distance from the equator might expect. Instead of a seasonal variation of two or more hours in the time of sunset, the variation throughout Indonesia seems to be, uniformly, about 40 minutes. However, the dates of the shortest and longest days vary from location to location depending on whether the transmitter is located north or south of the equator. As an example, Banda Aceh, located on the northwestern tip of Sumatra, has seasons corresponding to what we would expect in the northern hemisphere. On July 15, sunset is at 1154 UTC and, on November 15, it is at 1116. The stations located south of the equator differ. Surabaya has its late sunset on February 1 at 1052 UTC, and the early sunset at 1014 on June 1. Likewise, Merauke has sunset at 0903 UTC on February 1 and at 0822 on May 15.

While Indonesia is compressed into only three time zones for convenience, the actual distance between Banda Aceh in the west to Merauke in the east results in a difference in the time of sunset of almost three and a half hours. Banda Aceh is actually south of Burma, which is GMT + 6½ hours, while Banda Aceh uses GMT + 7 hours. So, when considering the fade-in time of Indonesian stations, the DX'er must consider that a variation of three and a half hours is possible; from 0822 to 1154 UTC.

European DX'ers are more concerned with the morning schedule of the Indonesian stations. They must try to find a darkness path between the sign-on time of the station and a time 30 minutes or so after dawn at the station, when fade-out will normally occur. Here again the location of the station is crucial. The Irian Jaya stations are going to fade-out three hours earlier than the Javan and Sumatran stations. Unless the station comes on before 6 A.M. local time, there is not much of a window for the DX'er before the fade-out that is caused by daylight at the station. Even so, during the winter months, some loggings of Indonesian morning broadcasts are made by dedicated DX'ers at dusk on the east coast of North America. Such loggings are made even more difficult by the presence of interference from Latin American and African stations who are also using the tropical bands.

The Stations Now on Shortwave

At the present time, shortwave broadcasting in Indonesia can be divided into four groups of stations. First, there is the international broadcaster, or Foreign Service, operated by the national government on the high frequencies. It is similar to the Foreign Service of most other governments, with programs from Jakarta in eight different languages. For schedules and frequencies, the DX'er should consult the latest edition of *World Radio TV Handbook* (WRTH). The domestic broadcasters are the normal targets for DX'ers, with a list of stations that seems unlimited in numbers.

Then, there is the Radio Republik Indonesia, or RRI, system. This also is operated by the national government with the object of providing nationwide radio coverage. Most of these stations use the tropical bands. The system is divided into five networks (Nusantaras) as follows:

Nusantara I (satu): Sumatra, with Medan as the key station.

Nusantara II (dua): Java, Bali, and Lombok-Sumbawa, where the key station is Yogyakarta.

Nusantara III (tiga): Kalimantan; the key station is Banjarmasin.

Nusantara IV (empat): Sulawesi and Timor; the key station is Ujung Pandang.

Nusantara V (lima): Irian Jaya and Maluku; the key station is Jayapura.

In addition to the five Nusantaras, RRI Jakarta has two domestic service programs—Programa Khusus and Programa Kota.

Next, there are the regional governmental stations operated by various local governmental entities, which DX'ers usually refer to, for convenience, as RPD (Radio Pemerinta Daerah) stations. These are normally either Provincial government (Propinci), District government (Kabupaten), or city or town governments (Kotamadya). There is a virtual alphabet soup of station headings for the various RPD stations. As an example, I refer the reader to their listing on page 215 of the W.R.T.H. (1984 edition). These stations (Fig. 24-6) are all low-power stations and are prime DX targets.

Finally, there are stations operated by other governmental agencies and the armed forces.

In years past, there were university stations and private commercial stations on the shortwaves. The university stations have recently left the air and may be on the medium waves. It is hoped that they will return to shortwave in the future. The private stations have all moved to the medium waves and FM on orders of the Indonesian government.

Fig. 24-6. The non-RRI broadcasters have an alphabet soup of names and abbreviations.

It should be mentioned in passing that there are also so-called "amatir" stations in operation, especially in Java on weekends. These are said to be unlicensed operations similar to the "pirate" stations now in operation in Europe and America. They are very low-power operations, sometimes heard as far away as Australia, but not elsewhere under normal circumstances.

Current Trends

In the past, many of the RRI stations broadcast on the higher frequency bands, especially on 49, 41, and 31 meters. In recent years, the tendency has been to leave the high bands and move down into the tropical bands. On the high bands, RRI Manokwari is still heard on 6188 kHz, and RRI Ambon is very irregular on 7139 kHz. RRI Biak may still be on 7211, but it is virtually impossible to hear it through the interference from Khabarvosk. RRI Jayapura sometimes can be heard on 9612.

The RRI stations mostly operate within the recognized limits of the tropical bands, or close thereto, except for Pekanbaru on 5886 kHz, Biak on 5451 kHz, Sibolga on 5257, Serui on 4607, and Dili sometimes on 3120. The tendency in recent years has been to move into the confines of the regular bands. The RPD stations, on the other hand, do most of their broadcasting out of band in the area between 2500–3200 kHz and 3400–3900 kHz.

The DX'er who wants to log Indonesians must be prepared to put up with irregularities of all kinds. The RRI stations appear to be better financed than the RPD stations, but, even so, schedule irregularities or stations dropping off the air for extended periods is common. Many stations are forced to operate on reduced power or go off the air completely pending the arrival of replacement parts. The RPDs are real "shoestring" operations and are off the air more than they are on. Tropical heat and humidity can cause all kinds of problems for radio equipment.

The Indonesian DX'er must be alert for the reports of new stations reported by Australian DX'ers, who are usually the first to identify new stations. Also, DX'ers in India and Sri Lanka do well with the Sumatra stations. The history of many RPD stations has been that they are fairly well heard when they first come on the air. After a few months, they start to fade down into the mud and eventually disappear completely. So DX'ers located far from Indonesia should try to get loggings as soon as possible or they probably never will. More and more RPD stations are leaving the shortwave for the medium waves, but there are still plenty left on the shortwaves.

RRI stations that drop off the air from time to time all seem to return to shortwave, eventually—sometimes after years of silence. The Irian Jaya RRI stations, along with Dili, all seem to suffer from wanderlust. Most of them change frequencies on a regular basis. In Indonesian DX'ing, the DX'er has to assume that any list of Indonesian stations that is over a year old is badly out of date. We all are forced to use station lists as an aid to our DX'ing. However, the DX'er concentrating on Indonesian stations is advised to keep his own list as up-to-date as possible by entering changes on it as he gets reports from DX bulletins, DX programs, etc. I strongly recommend use of the *D. S. W. C. I. Tropical Band Survey* that is published each summer by the Danish Shortwave Club International. This club also updates the Tropical Band Survey during the year for its members. The Indonesian stations listed on it are there as the result of actual loggings by DX'ers, and the stations that are no longer being heard are dropped off the list.

The best source of current Indonesian news are the Australian DX bulletins. Australian DX'ers, Geoff Cosier, David Foster, and Robert Yeo have done pioneering work in bringing to light the extensive RPD system, have visited many Indonesian stations, and carry on an extensive correspondence with the stations.

Station Schedules

Most of the Indonesian stations split their broadcasting day into three parts. There is a morning broadcast that commences between 5 A.M. and 6 A.M. local time and lasts around four hours. This is the part of the schedule that is important to European DX'ers. There is a midday program on many stations that is of little interest to foreign DX'ers because it can only be heard by the local audience. The evening schedule is the longest of the day, of up to seven hours or more, with sign-off time normally being around 11 P.M. local time, or at the end of a Jakarta relay. This is the program that is important to DX'ers in the Western Hemisphere.

Many of the stations broadcast on extended schedules during the Moslem holiday of Ramadan—up to all night long. Ramadan schedules are a great help to European DX'ers, but don't help North American DX'ers since the early morning hours in Indonesia correspond with the daylight hours in North America. At the time this article was written, a few RRI stations have been reported to be broadcasting 24 hours a day and some others have been reported as having extended their evening programs by one to two hours. This may be a temporary change, however.

A Review of the Stations Being Heard

An attempt will now be made to review the stations broadcasting from the five Nusantara areas. Please remember that it is a very volatile situation with stations constantly changing their schedules or frequencies and either dropping off the air or returning after an absence. No list can be guaranteed as 100% accurate.

Many of the RRI stations have second and even third frequencies that are very low power which are rarely heard outside of Indonesia. I will concentrate on the best frequencies. I do not know of any of the RPD stations that use more than one frequency at a time. A complete review of the RPD stations is not possible. Only those that are heard most often are included.

Nusantara I (Satu) in Sumatra

THE RRI STATIONS

1. Banda Aceh is the most westerly station in Indonesia. The two main frequencies are 3905 and 4955 kHz. Formerly, they were heard in parallel. However, at the present time, they seem to be alternate frequencies, with 3905 kHz used most of the time on the evening schedule. 3905 kHz suffers from heavy QRM from Radio New Ireland.

2. Medan normally puts in a strong signal on 4764 kHz, but the frequency has been blocked by what appears to be a strong open carrier recently, and the Radio Moscow Cuban relay on 4765 spreads out all over the place. A second frequency of 3375 kHz can be quite clear at times, although the power is lower than on 4764.

3. Pekanbaru has the highest frequency of the Sumatran stations on 5886 and usually is in the clear. At times, it drifts down as far as 5882 kHz.

4. Pangkalpinang is a very rare RRI. It was off the air for a long time but has been reported back on 3385 kHz. It is very low-power transmission on probably one of the poorest frequencies in the 90-meter band. The reported schedule in the evening is from 0900–1400 hours. Not only is there another RRI on the same frequency until after 1500 hours, but Radio Rabaul and Miri in Sarawak are also there.

5. Jambi formerly used 4927 and 3374 kHz, but the station has not been reported for a long time and presumably is not on shortwave now.

6. Palembang (Fig. 24-7) has been a regular on 4856 kHz for years and still puts out a strong signal; however, it has been bothered by heavy utility interference recently.

7. Sibolga, after a period of silence, is now heard regularly on 5257 kHz. There are reports that it has a second low-power frequency of 3241 kHz.

8. Bukittinggi now has its main transmitter on 3232 kHz with a low-power transmitter heard at times on 4910.

9. Padang (Fig. 24-8) has a potent signal on 4003 kHz, but it is subject to heavy utility QRM. This one formerly alternated between 4003 and 3960 kHz.

10. Bengkulu uses 3265 kHz, unfortunately. It is unfortunate because strong RRI Gorontalo is also on this frequency at the same time and Gorontalo dominates the airwaves in California until sign-off time after the Jakarta news at 1500 hours. The only time I have been able to log Bengkulu has been after Gorontalo signs off.

Fig. 24-7. RRI Studio Baru at Palembang.

11. Tanjung Karang can be heard on 3395 kHz; however, there has been heavy utility QRM on the frequency. A second low-power transmitter is a rare catch on 3956 kHz.

12. Tanjung Pinang recently returned to 3225 kHz after a period of silence. It now has one of the best Indonesian signals on the 90-meter band.

NON-RRI STATIONS

1. The military station Radio Angkatan Udara, located at the Medan airport, formerly was on 3367 kHz. It seems to be silent now.

2. RPDT2 Labuhanbatu has been logged outside of Indonesia and Australia during the past year. It has an unstable transmitter that is subject to drifting from one day to the next. It is now heard mainly on about 3812.5 kHz. Beware of the new Chinese C.P.B.S. station that is on 3815.

3. RPDT2 Lampung Tengah can be heard on 3511 kHz if you are lucky.

4. RPDT2 Tapanuli is a fairly new station that is reported to be on 3872 kHz.

Nusantara II (Dua) in Java, Bali, and Lombok-Sumbawa

RRI STATIONS

1. Jakarta has domestic service on several frequencies but the schedules tend to be variable. Look for it on 6045 kHz where Radio Australia dominates, on 4774 and 3277, and even down on 2332 kHz.

Fig. 24-8. A QSL
card from the
RRI station at
Padang.

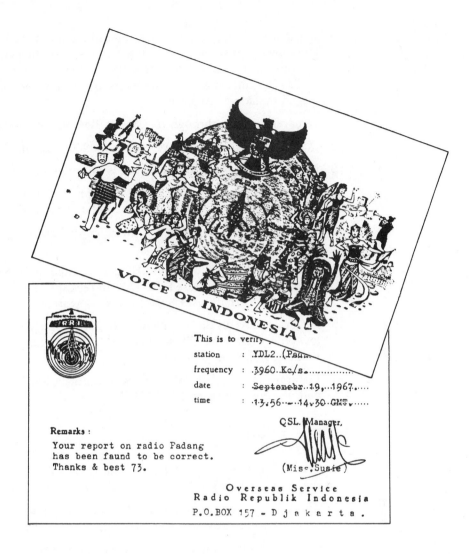

VOICE OF INDONESIA

This is to verify

station   : YDL2..(Pad...
frequency : 3960..Kc/s...............
date   : Septenebr 19, 1967......
time   : 13.56 — 14.30 GMT......

QSL. Manager.

(Miss Susie)

Remarks :
Your report on radio Padang
has been faund to be correct.
Thanks & best 73.

Overseas Service
Radio Republik Indonesia
P.O.BOX 157 - Djakarta.

2. Bandung formerly was strong on 3204 kHz but it now seems to be on very low power or not on this frequency. It has been heard on 2419 kHz several times, but not recently. This one will probably come back with a strong signal in the future.

3. Yogyakarta puts in a consistently clear signal on 5046 kHz and sometimes can be heard on 2350 kHz. It has a daily program in English at 1130 hours.

4. Surakarta has been strong and clear on 4899. It formerly used 4932 kHz and for a time alternated between the two frequencies.

5. Surabaya does well on 3975 kHz if the Ham QRM is not too strong.

6. Bogor was off the air for a long time. About a year ago, it returned with a low-power transmitter on 3960 kHz. This is another instance of two RRI stations on the same frequency. Palu is the dominant station, and there is also a Chinese station there. As far as I know, Bogor hasn't been logged outside of Australia.

7. Semarang comes through the Ham QRM on 3935 kHz at times.

8. Maidun fights constant CW QRM on 3266 kHz and is a rare catch.

9. Malang is heard on 3381 kHz but is very weak.

10. Jember has been reported in Australia on 3321 kHz but has sunk out of sight into a sea of utility QRM and splatter from Korea on 3220 kHz.

11. Sumenep on Madura Island is on 3355 kHz but subject to fierce utility QRM.

12. Denpasar on Bali is on 3945 kHz and is listed at 10 KW. I doubt that it uses anything even close to 10 KW. In recent years, it has become a rare catch, and Ternate uses 3946 kHz, so beware.

13. Singaraja, also on Bali, is about 3398 kHz, a fairly clear spot, but it uses low power. However, it is actually heard more often in my location (San Francisco) than is Denpasar.

14. Mataram on Lombok Island puts in an occasional signal on 3223 kHz and, at times, even on 2492 kHz.

NON-RRI STATIONS

1. The strongest non-RRI station on the air at this time is RKIP Surabaya, the agricultural station (Fig. 24-9). It is heard regularly on 4698 kHz.

2. Another Surabaya station is RKPDT1 Jawa Timur on 3000 kHz. A Chinese regional takes over this frequency for a portion of the year.

3. Other possibilities are RPD Stations. The majority of the RPD stations are located on Java (Figs. 24-10 and 24-11). The problem is that most of them are very low-power stations and weak. Loggings outside of Australia are very rare. The RPD station heard most often is RKPDY2 Banyuwangi (Fig. 24-11). This one is normally listed anywhere between 3502–3506 kHz. It is subject to constant drift and has been heard as low as 3498 kHz. Other stations that have been logged in the United States are RKDT2 Sukabumi on 3330 kHz; RKPKDT2 Ngawi on 2986 kHz; RSPKDT2 Gresik on 2907, but reported drifting to lower frequencies; and RDK Serang which sometimes comes in on 3102 kHz. The problem is that most of these stations are so rarely heard outside of Indonesia and Australia, that it is impossible to determine if they are still on the air at this time.

4. On Sumbawa is RKPDT2. Sumbawa is heard at times on 3775 kHz.

Fig. 24-9. A QSL
card from RKIP
in Surabaya.

RADIO KHUSUS INFORMASI PERTANIAN

Gelombang 64 M (4698 KC)

Jln. Jendral A. Yani 156 Wonocolo - Surabaya

Telex No. 380.

Tilp. D. 8285.

Kotak Post : No. 247.

MENYIARKAN SIARAN PERTANIAN/PEDESAAN

Setiap hari jam: 5.00 — 9.00 W.I.B.

12.00 — 14.00 "

17.00 — 24.00 "

(Setiap hari Kamis/malam Jum'at siaran wayang kulit semalam suntuk)

Nusantara III (Tiga) in Kalimantan

RRI STATIONS

1. Banjarmasin puts in a strong signal on 3250 kHz, but seldom is able to penetrate the utility QRM into my location. Years ago, this station also was heard well on the 49-meter band and is still listed in the *WRTH* on 5968v. However, few loggings of this outlet have been seen in recent years.

2. Samarinda does not seem to be on the air now. It also used a 49-meter-band frequency of 6134 until a couple of years ago. It came in well until Korea, on 6135 kHz, came on the air at about 1000 UTC. The Samarinda second frequency was 3294–3296 kHz. It also is silent here now, but no doubt will return to the air when repairs are made.

3. Palangkaraya has used 3325 kHz for years, but is now bothered by substantial QRM.

4. Pontianak's best transmitter is now using the unfortunate frequency of 3995 kHz, where it is subject to constant QRM from three USSR stations. In spite of the QRM, Pontianak does manage to be heard in my location several times a year. The second frequency of 3346 kHz (Fig. 24-12) is rarely heard as RRI Ternate is the dominant station on 3345 kHz.

Fig. 24-10. QSL cards from the RPD stations.

(A) Joseph Tinagari of Pimpinan RPD Tingkat II Poso.

(B) Card from RPK Blitar showing "An act of Ramayana dance."

NON-RRI STATIONS

1. RPDT2 Bulungan is a 1-KW station heard regularly on 2977 kHz, and located at Tanjungselor on the northeast coast of Kalimantan. It has been the strongest RPD in my location since it came on the air about five years ago, although it does drop off the air at times.

2. RLPDT2 Kota Baru is on 3452 kHz where it is heard on good Indonesian openings.

3. RPDT2 Tanahlaut, Pelaihari has been heard recently on about 3488 kHz —subject to drifting. This is one that changes frequencies often and may not be on 3488 for long.

Fig. 24-11. A QSL card from RKPDY2 in Banyuwangi, showing the traditional dance of Blambangan.

Nusantara IV (Empat) in Sulawesi, Timor, and Nusa Tenggara

RRI STATIONS

1. Ujung Pandang, at 4719 kHz, is a 50-KW station and has the strongest signal of any of the RRI regionals. Its high power allows it to have a very early fade-in, as early as the West Irian stations located one time zone farther east. At times, the transmitter on 4753 kHz is used as an alternate to 4719 and, on Moslem holidays, both are heard in parallel.

2. Kendari, at 4000 kHz, has an early fade-in and is usually clear.

3. Palu, on 3960 kHz, has a fairly strong signal, but is subject to heavy QRM from a Chinese regional. Also, RRI Bogor has moved onto this frequency.

4. Gorontalo, at 3265 kHz, is one of the better 90-meter Indonesian stations with an early fade-in.

5. Menado, at 3215 kHz, had a 49-meter-band transmitter on 5987 kHz several years ago that was very easy to log. This is another case of dropping the higher frequencies for the exclusive use of the tropical bands. Menado is reasonably consistent on 3215. *FLASH!* Just at the time of typing this article, I noted that RRI Menado was again using 5987.5 kHz (between 0930 and 1020 hours) in parallel with 3215 kHz. There is no sign-off announcement when the switch is pulled at 1020

hours, however. The program continues on using 3215 kHz. It will be worth watching to see if the use of 5987.5 kHz continues. The signal was excellent. (Subsequent checks revealed nothing on 5987.5, so, possibly, the transmissions were just tests.)

Fig. 24-12. Verification letter from RRI in Pontianak.

DEPARTEMEN PENERANGAN

DIREKTORAT RADIO REPUBLIK INDONESIA

STUDIO : PONTIANAK

ALAMAT : DJALAN PERWIRA — 7.   TILP. :

No.   : S.14–1–68   PONTIANAK, 10 January 19 68

HAL   :

LAMPIRAN :

> Mr. William S. Spark
> 1100 Hearst Building
> San Fransisco, California 94103
> U.S.A.

Dear sir,

Your report of reception has been received on December 10 1967 . Thanks for your pleasure to report my station. Your report is correct, and we have also received the report from : Sweden, Australia, California (Chula Vista) Finland and London.

Here by I send indentification my station ;

Callsing	: YDW2 freq. 3345 Kc/s
Type of transmitter:	GATES HY 10 / 1958.
Power output	: 10 Kw.
Antene System	: Delta macth
Operation Areal	: Regional West Kalimantan (Indonesia)
Broadcast schedule	: 06.00 – 08.15 lokal time.
	22.00 – 24.15 GMT
	12.00 – 15.15 Local time
	04.00 – 07.15 GMT
	17.00 – 23.15 Local Time
	09.00 – 15.15 GMT.

Thanking you now for this time and hoping to be able your report in the future, I am,

Your Faithfully
DIRECTOR

cc.   Miss Susie
      Radio Republik Indonesia
      P.O. Box 157
      Djakarta – Indonesia.

6. Kupang on West Timor Island uses the very poor frequency of 3385 kHz, where constant QRM from Radio Rabaul and Miri make reception almost impossible until after 1500 hours. Also, RRI Pangkalpinang is now using 3385 kHz.

7. Dili. Where is Dili today and where will it be tomorrow? This station has a history of transmitter problems since the RRI took it over from the Portuguese. For years, it was on 3120 kHz, but with such poor modulation that only the carrier could be heard for months at a time. Then it moved up to 3305 and was heard fairly well until it started to alternate between 3205 and 3120 kHz. Then, about two years ago, it started to drift all over the 90-meter band with terrible modulation. About a year ago, it was being heard on 3306//3986. Then Manokwari took over 3986 kHz and Dili was heard on 3306//2456. At the time of writing this article, I haven't been able to hear Dili on 3306 but it has been heard on 2456 kHz.

8. RRI Ende, Flores Island, is the most recent addition to the RRI system. It has been heard in Australia on 2694v, but try as I might, I haven't been able to hear it in San Francisco.

NON-RRI STATIONS

1. RPDT2 Poso is heard at times of prime Indonesian reception on 3524 kHz.

2. RKPDT2 Luwu, Palopo, has been reported at times in Australia on 3655 kHz, but not very often elsewhere.

3. RPDT2 Ngada, Bajawa, Flores Island, was heard several times fairly well on 2904 kHz.

Nusantara V (Lima) on Irian Jaya and Maluku

RRI STATIONS

1. Ambon is heard most-often now on 4845 kHz. In 1983, Ambon was moving around on the 60-meter band to 4864 and 4835 kHz, but eventually returned to 4845 kHz, where reception is subject to QRM from Brazil before 1000 UTC and QRM from Malaysia after that time. Ambon uses its old 41-meter frequency of 7139 kHz for its evening program only a few times a year. It also has a low-power transmitter on 3241 kHz, which is rarely heard. 3241 kHz is another frequency where more than one RRI is transmitting at the same time.

2. Ternate is one of the regular stations on 90 meters on 3345 kHz, with an early fade-in. It has a low-power second frequency at 3946 kHz that is difficult to log.

3. Jayapura is the key station for Irian Jaya. Unfortunately, it is a difficult catch right now. I suggest a try for 9612 kHz sometime between 0900 and 1100 hours. The 6070 frequency is hopeless due to QRM from HCJB and the powerful Chinese station. Jayapura is also available on 5044 kHz, but only on top openings into Irian Jaya.

4. Biak on 5451 kHz puts in the most consistent signal from Irian Jaya with an early fade-in.

5. Sorong has been one of the better signals on 4875v, although there are several other stations also using this frequency. In the spring of 1984, Sorong left this frequency and appeared at about 4797 kHz, with very poor modulation. The QRM has been heavy, but a definite ID was heard.

6. Manokwari is now heard in parallel on 6188 and 3986.5 kHz. Its old frequency of 3427 kHz was dropped when it took over the 80-meter frequency from Dili.

7. Serui was listed at 4606 kHz on the old Indonesian lists, but was off the air for several years. It came back on the air at 6206 kHz and, in 1983, moved back to its old frequency. It has since drifted up to about 4607.5 kHz and has extended its schedule until around 1600 hours.

8. Fak Fak was off for a while, but returned to its usual frequency of 4789 kHz recently.

9. There are three very-low-power Irian Jaya RRI stations that constantly move around the bands. I can't confirm the frequencies now reported to be in use but they are as follows:

   • Merauke reported on 3905 kHz—the same spot where you will find Banda Aceh and R. New Ireland.

   • Wamena, most-recently, was reported on 5043 kHz. It always has been an extremely difficult catch.

   • Nabire formerly was on two different spots on the 49-meter band, but it recently moved down to a reported 5055 kHz. Confirmation of this fact is virtually impossible due to QRM from the Latin American stations.

NON-RRI STATIONS

There is no non-RRI activity from the Nusantara V area that I know of.

Unidentified Indonesian Stations

There are many unidentified Indonesian stations operating in the low-frequency end of the tropical bands. These are mostly low-power RPDs or low-power second or third transmitters of RRI stations. Many of the unidentified stations eventually turn out to be RPD stations that formerly were identified and heard on higher tropical-band frequencies. Identifying these stations is a long and difficult process. DX'ers will remain forever indebted to the Australian Indonesian DXperts for their time spent both at the dials and in correspondence with Indonesian informants in an effort to identify these stations. The constant shifting of frequencies by the RPD stations makes it certain that there always will be a supply of unidentified Indonesian stations available for those DX'ers who enjoy trying to pin an ID on them.

Identification of Indonesian Stations

Like all of the domestic service stations broadcasting in the tropical bands for the benefit of a local audience, the Indonesian stations identify themselves only in their local language. This means that the DX'er must develop an ear that is accustomed to Bahasa Indonesia if he wants to identify the stations, as well as prepare a reception report, in the hope of obtaining a QSL from the stations.

Keep in mind the fact that there are at least two Indonesian stations operating at the same time on the following frequencies: 3241, 3265, 3345, 3385, 3905, 3945, and 3960 kHz. This makes accurate identification imperative. Because of the weak signals (usually accompanied by a high-noise level and static), use of a tape recorder and even an audio filter on the playback of the tape is very helpful.

The RRI stations do tend to broadcast IDs more frequently than do the RPDs, and it is easier to determine when an RRI is in the process of giving its ID. Normally, "Radio Republik Indonesia" is heard prior to the location of the station. Frequently, however, in the case of RPDs, there is nothing that will alert the DX'er prior to the mention of the location announcement.

When is the best time to look for an ID? The rule of thumb is just before and just after a Jakarta relay. Normally, there is a local ID about one to two minutes before the hour. Immediately after the ID, the well-known Song of the Coconut Islands interval signal, played on a carillon, is heard until Jakarta breaks in with the news. The DX'er will see this listed in loggings in DX bulletins as S. C. I. The correct title in Indonesian is "Rayuan Pulau Kelapa."

There will be times, however, when the local announcer doesn't have time to give his local ID prior to the Jakarta news relay. In that case, the DX'er will have to wait until the end of the relay when the local ID is again broadcast. You will be alerted to the end of a news broadcast by the playing of a patriotic song with a large chorus and orchestra. Most of us refer to the selection as an "anthem," although authorities differ as to whether Indonesia actually has an official national anthem. At any rate, even on local news programs, there is the playing of a selection that is similar to an anthem at the conclusion of the news. This is normally followed immediately by the local ID.

At other times during the regular schedule, the ID is highly irregular. I have heard them after every musical selection, but, at other times, an hour can pass with no sign of an ID. However, 30 minutes after the hour is a good time to be alert for an ID.

QSL'ing the Indonesians

QSLs from the Indonesians (Fig. 24-13) present many of the same problems DX'ers have experienced for years with the Latin American stations. Reception reports in English will be virtually useless. Some of the technicians at the stations have been educated in America, but not enough English is spoken at the stations to ensure replies to reception reports in English. It would be unusual to find anyone able to correspond in English at a small RPD station.

Fig. 24-13. QSLs from some of the many RRI stations.

This means writing your reception reports in Bahasa Indonesia. Now, the first question is "Where can I get some Indonesian forms?" Some of the DX clubs have published Indonesian reception forms. Beware of any forms prepared prior to 1972. They will have the old outdated spelling of Indonesian words. The Indonesian government officially made several substantial changes in the spelling of Bahasa Indonesia on August 17, 1972. At the time Djakarta became Jakarta, Jogjakarta became Yogyakarta, and Surabaja became Surabaya, etc. In case you are looking for an Indonesian-English dictionary, don't buy one printed prior to 1972.

It will not be easy to obtain a fully satisfactory set of forms (with vocabulary in Indonesian). The Australian Radio DX Club published its copyrighted "Indonesian Reporting Guide" in 1974. I have used it for years; however, I don't know if it is still available. Radio Netherland's publication "DX'ing Indonesia" contains an excellent set of forms to use as the basis for a personal letter to an Indonesian station; however, the vocabulary of phrases provided for use in describing the program material is not complete enough. If you can obtain a copy of both of these forms, you should be able to write an excellent reception report in Bahasa Indonesia. For those interested in obtaining a basic understanding of Indonesian, I recommend the Indonesian issue of *Teach Yourself Books* by J. B. Kwee that can be found at many book stores. It is not expensive and contains only 175 pages.

The next question is "What chance do I have for a reply after I have logged an Indonesian station and prepared a reception report in Bahasa Indonesia?" The answer is "Quite a good chance." Many Australian Indonesian DX'ers have a better reply rate from *all* of their Indonesian reports than they do from the Latin American stations. This does not mean that every RRI station is going to reply to every reception report received. Try as I might over a period of many years, I still have not been able to obtain QSLs from RRI Biak and RRI Denpasar. All of the other RRI stations that I have logged have replied (Fig. 24-13), but some required follow-up reports over a substantial period of time.

Replies from the RPD stations come in at a much lower rate, and they require follow-up reports more often than not. One of the problems discovered by some DX'ers who have visited RPD stations, while on trips to Indonesia, is that the station personnel simply cannot believe that their small low-power transmitters were actually being heard several thousand miles away. Even so, the reply rate from RPDs is high enough to make it worth the trouble of sending the reports and the follow-up reports. Any dedicated QSL collector, who has once received a friendly reply along with some of the beautiful color postcards from one of these stations, will be eagerly looking for his next logging of another station.

Opinions differ as to the value of reply postage as a necessary ingredient in getting a reply from an Indonesian station. Most of the experts believe that IRCs are of little help since few of the stations know what they are used for. I believe that most of the serious QSL collectors do try to include mint Indonesian stamps with their reports.

Program Material

The next logical question is as to what kind of program material will be heard on these stations? I think that the variety of music being heard on Indonesian stations will surprise most DX'ers, who have not heard the stations very often. The people on these islands have an extremely diverse cultural background. There is an Indian Hindu background in some of the islands; others were influenced by the early Portuguese traders and settlers. Almost all of the islands are strongly Moslem in religion due to close contacts with Arab traders in the past centuries. West Irian natives have a cultural background similar to the natives of Papua New Guinea.

All of this background results in a great variety of program material. The Islamic call to prayers is heard at all times from stations in Sumatra, Kalimantan, and Sulawesi, along with music that can only be described as typical middle-eastern style. Yet, on other stations, Christian services and hymns can be heard, especially from Irian Jaya and those islands which were influenced by the Portuguese. I have heard Portuguese spoken on both RRI Dili and Gorontalo; however, this would be unusual now.

Anyone who has looked for recordings of Indonesian music at American record stores has probably only found *Gamelan* music. From that, he would assume that only Gamelan music will be heard on the Indonesian stations. Not so! Gamelan music is only heard about 10% of the time. The music most often heard is a program of *Indonesian lagu* (Indonesian songs). Some of them are *lagu tradisionel*, the traditional Indonesian songs that are accompanied by the kroncong. Some are *lagu populer*, and, now, Indonesian pop and rock-style music is quite common. *Lagu melayu* is very popular now. It has been described as a pop version of an amalgam of Indian-style music that is combined with an Arabic or middle-eastern vocal style. A little of this goes a long way with me, but it is "the thing" in Indonesia today.

At other times, American pop music, sung by popular U. S. vocalists, or by Indonesian singers with lyrics in Bahasa Indonesia, is heard. Other stations go with *lagu hiburan* as a change of pace. This is light, easy listening, orchestral music. Hawaiian music is also very popular on these stations, and U. S. country-western music is heard often from Irian Jaya. Native drums and chanting are frequently broadcast from Irian Jaya. The *wayang kulit* (shadow puppet) plays with a Gamelan orchestra can go on for hours at a time.

One thing that you can't escape will be the *warta berita* (news broadcasts) and *komentar* (commentaries). All of the RRI and RPD stations are required to relay the main Radio Republik Indonesia news programs from Jakarta, if the station is on the air at the time of the relay. Many of the stations in the eastern part of the country sign off prior to the 1500 Jakarta news. A few of the stations will skip a Jakarta relay at times, but it is unusual for them to do so. These news relays are heard at 1200, 1300, and 1500 UTC.

The relays are listed in Indonesian as: *Warta Berita dari Program Nasional Jakarta*. At 1200 hours is Warta Berita dan Komentar (news and commentary) with Berita-berita Ekonomi dan Keuangan (news, economic and business news) at 1300, and Warta Berita dan berita-berita olaharga (news and sports) at 1500 hours. The 1300 relay tends to run until close to 1330. After five to ten minutes of news, Jakarta presents a program called "Nusantara" which involves interviews, tape recordings of political speeches, and, frequently, music. The relay is never concluded, however, until after the "anthem" is played.

Local and regional news are also broadcast by all stations, but these times are much shorter than the Jakarta relays. Normally, there is an "anthem" at the conclusion of the local news programs also.

The Anthems

What about these "anthems" heard so often at the end of the news programs? Some of the Indonesian authorities say that if Indonesia has an official national anthem, it is *Indonesia Raya*. In any event, the "news anthems" are both patriotic songs. The words to these songs are provided here in the event that your reception is clear enough to determine what you are hearing. The words are printed through the courtesy of Toshiaki Takehara, Osaka, Japan, as published in the Indonesian DX Circle in 1980.

Satu Nusa, Satu Bangsa is the "anthem" used at the end of the Jakarta relays at 1200 and 1500 hours, and expresses the "One Land, One Nation, One Language" credo of the Indonesian government. The words are:

> Satu Nusa, Satu Bangsa, Satu Bahasa Kita
> Tanah Air Pasti Jaya Untuk Selama-lamanya
> Indonesia Pusaka, Indonesia Tercinta
> Nusa Bangsa Dan Bahasa Kita Bela Bersama.

There is a different "anthem" at the end of the 1300 Jakarta relay. It is *Mars Pancasila*, otherwise known as *Garuda Pancasila*. The words are:

> Garuda Pancasila, Akulah Pendukungmu
> Patriot Proklamasi, Sedia Berkorban Untukmu
> Pancasila Dasar Negara, Rakyat Adil Makmur Sentausa,
> Pribadi Bangsaku, Ayo maju maju,
> Ayo maju maju, Ayo maju maju.

The supposedly official anthem, *Indonesia Raya*, is much longer. The words are:

> Indonesia Tanah Airku. Tanah Tumpah Darahku.
> Disanalah Aku Berdiri Jadi Pandu Ibuku.
> Indonesia Kebangsaanku Bangsa Dan Tanah Airku,
> Marilah Kita Berseru Indonesia Bersatu.
>
> Hiduplah Tanahku, Hiduplah Negeriku,
> Bangsaku Rakyatku Semuanya.
> Bangunlah Jiwanya Bangnlah Badannya
> Untuk Indonesia Raya.
> Indonesia Raya Merdeka Merdeka
> Tanahku Negeriku Yang Kucinta.
> Indonesia Raya Merdeka Hiduplah Indonesia Raya.

I have not attempted to determine which of the above anthems are used at the end of the local news programs. For those interested in acquiring a background in Indonesian music, I would recommend trying to locate a copy of the long-playing record album, *DESTO Album D505*, "Indonesia, Its Music and Its People," which has samples of various types of Gamelan, Keroncong, and Indonesian folk music, and a commentary in English. I do not know if the album is still available at record stores, however, it is worth making an effort to locate a copy. The address printed on the record cover of my copy is: Desto Records, 12 East 44th Street, New York, NY. It is sponsored by the World Federation of United Nations Association.

Love Ambon

Not everyone has a listening post in a location where it is possible to hear the sign-off of an RRI station on a regular basis. This is most unfortunate since the sign-off melody played by the RRI stations is memorable. It is *Love Ambon*, a hauntingly beautiful melody played on a kroncong in an arrangement that reminds one of the Hawaiian steel guitar. I have heard a few of the RPD stations use it at sign-off also. Recently, I noted RRI Ambon going off the air with no music at all. I hope that will not become a practice that will spread to other RRI stations.

The normal sign-off procedure is as follows: At the closing of the day's schedule, the announcer makes a short ID announcement. Then, the *Love Ambon* melody starts. After a few seconds, the announcer breaks in with his final announcement, which includes a very complete ID along with the wavelength (gelombang) in use, and the station's location. Then, *Love Ambon* continues for a minute or so until the power is cut off.

And so, what more appropriate way is there to finish a chapter on DX'ing Indonesia? *LOVE AMBON!*

ON TO THE TOP

Donald N. Jensen

No one is more qualified to write a piece about reaching the top of the ranks in shortwave broadcast DX'ing than **Don Jensen**. He has had a close acquaintence with that position for many of the 35 years that I've known him.

Since he began DX'ing in 1947, he has heard and verified close to 230 countries on the shortwave broadcast bands. His writings on various radio and shortwave topics have appeared in a number of commercial radio magazines over the years. Currently, he writes a regular column on short-wave for *Hands On Electronics*, and holds an editorial position with *Radio Database International*.

Don was the founder and the first Executive Secretary of the Association of North American Radio Clubs (ANARC). In 1969, he founded, and still continues to edit, the weekly *Numbero Uno* SWBC bulletin. He also serves on the Executive Council of the North American Shortwave Association and is chairman of the NASWA Country List Committee.

Don and his wife, Arlene, have three children. They are both journalists for the *Kenosha News* in Kenosha, Wisconsin.

How can I become a top-flight SWBC DX'er? Seldom is this question actually voiced, but it sums up the intent of a lot of specific questions actually asked by listeners who have progressed beyond the unabashed beginner stage. And, it is a good lead-off question for this chapter.

Unfortunately, there is no one magical answer, no clearcut formula for success. But one word that will be cropping up again and again is *experience*. Hearing the hard-to-hear stations is the sum and substance of serious DX'ing on the shortwave bands. And, though it may seem a flippant answer, experienced DX'ers hear the hard-to-hear stations because they're experienced!

Contrary to popular belief, top-quality equipment isn't the solution. Yes, a good receiver will be a big help. It will make serious DX'ing easier. But the important word here is "easier." An experienced DX'er with a mediocre receiver will still manage, over the long haul, to hear most of the real DX. And, a novice with a kilobuck receiver will still be asking how he can log those rare stations. There is no one so disappointed as the guy who, expecting miracles, scrapes up enough dough to buy a so-called "pro" receiver, only to find he still *isn't* bagging all those goodies!

Experience—there's that word again—is the answer! But that does not mean just "time served" in the hobby. Time is a factor, naturally, since learning takes time. But the effort to learn is critical. Experience is just the practical application of knowledge.

A top-flight DX'er? Well, when does a DX'er "arrive?" If arriving means reaching a pinnacle of proficiency, the answer is "Never!" No one can ever know all there conceivably is to know about the subject, and even the most experienced DX'ers can never stop trying to learn more. Individual definitions may vary, of course, but for our purposes, let's assume that when a DX'er is generally regarded by other DX'ers as being among the "top 10 percent," he's arrived. In other words, you're a top DX'er when others consider you one.

Or, to put it another way, if you still have to ask, you aren't. There's no magic entry point to the "charmed circle," no countries heard or verified total that proclaims it to the world. It is a matter of your reputation in DX'ing clubs as a solid, reliable, accurate, experienced listener.

It's purely a personal opinion, of course, but it seems to me that a relative beginner in the hobby, who goes all out to learn what serious DX'ing is about, could reach that vague status position in a minimum of five years. Some, of course, will take longer, and some will never reach it at all.

Spoon-feeding information is out. You've got to work very hard to learn as much as you can. Among other things, this means a great deal of reading to build a mental storehouse of knowledge which you need to be a real DX'er. When you come across a semitechnical or technical article in a DX club bulletin or magazine, do you pass it by unread or skip over it quickly because it's too hard to understand? Or, do you make a serious effort to master the subject and then use it as a take-off point for further study?

The guy who complains to editors that an article is too tough and too tedious, that he doesn't understand it, and who plaintively cries, "Why can't we have more features for the beginner?'' is the one who will only slowly, or perhaps never, progress much beyond the beginner stage!

Hard work? Definitely yes! A slow process? Probably! ''But, DX'ing is only a hobby, a pleasurable pastime,'' you say. Yes, and a casual approach to shortwave listening is perfectly acceptable if that's your thing. But excelling, even at a hobby, can be pleasurable.

Many SWL'ers simply don't care to make their hobby more than just the casual listening to some interesting overseas programs. I don't knock that approach at all. But if you've ever asked yourself the opening question: ''How can I become a top-flight SWBC DX'er?,'' you've already indicated your interest is more in the direction of DX'ing than SWL'ing. And, if it is a serious DX'er you want to be, you've got to work at it!

Concentration

Hearing rare DX stations depends on ''conditions.'' That goes without saying. But what a DX'er comes up with when ''conditions'' are right depends, to a large extent, on the storehouse of information and know-how that he has accumulated and can apply to the given situation. A good place to start is by looking at the bands themselves. Most of the rare DX will be found in the shortwave bands below about 7 MHz, the 41-, 49-, 60-, 90-, and 120-meter bands, plus the out-of-band stations. To start with, emphasize 60 meters—the best single DX band.

Concentration! Concentrate your maximum listening efforts on 60 meters. Later, you can expand your attention, in a similar way, to the other DX bands. *Hours!* Spend dozens of hours, hundreds of hours if you can, learning what there is to be heard on ''60'' under normal conditions and under unusual band conditions. Get to know that span of frequencies like the back of your hand. In order to get the rare stations, you have to be able to quickly sort out the usual from the unusual signals. When extra good conditions occur, you can't waste time on relogs of previously heard stations when you could be coming up with new catches.

Can you, at any particular time of day, cruise 60 meters and, with a fairly high degree of certainty, tentatively identify, within 15 seconds, each of the stronger signals on the band. If you can't, you're operating under a real handicap!

On a good winter afternoon opening to Africa, can you tick 'em off as you tune down the line? Lome, Togo on 5047..., Garoua, Cameroun on 5010..., Abidjan, Ivory Coast on 4940..., Chad on 4904.... If you *can* do this, the unusual signal should stand out and catch your attention, and you can spend your time logging it.

Concentration! Concentrate on learning as much as you can about shortwave propagation, if not from a ''book larnin' '' theoretical approach, then from a practical ''What can I hear?'' approach. Learn to recognize and take advantage of reception patterns.

Most beginners know you can't expect low SW frequency reception at high noon local time. It is assumed you know why this is the case. There is no black magic in DX'ing. Propagation of DX signals depends on natural phenomena. The more you know about how and why distant shortwave signals reach you, the better you will be able to determine when to tune. There is plenty of material available for study, if you only take the time and effort to seek it out.

For too long, DX'ers have neglected this field of study. In recent years, however, some have attempted a more systematic study of the propagation of weak, low-frequency, DX signals on the shortwave broadcast bands. But theories aside, careful observation and record keeping should give you a good working knowledge of practical propagation.

Do you know, for example, the time of day, and the time of year when your chances of hearing a rare station are optimum? Do you know the great circle path between your receiving location and the target DX stations? Do you know, at any particular time of year, when paths of darkness exist for these stations? Do you know when you can normally expect the 60-meter Latins to fade out and the Asians to fade in in the morning?

Do you know that your best chance to hear the home service *All India Radio* outlets may be from around mid-December to mid-January, but that Indonesian reception has much broader seasonal "window," and may, in fact, peak in the fall and again in early spring? Do you know that that rare quiet night in mid-summer can bring in some excellent African reception on 60 meters? Keep records of the excellent openings from various parts of the world. Look for patterns that are daily, seasonally, and geographically.

Geographic patterns can be broad or very selective. When you note the Peruvians coming in with exceptional strength, don't waste time with random tuning. Go after the specific unheard Peruvians you most want. Unusually powerful station signals coming from Togo? Then hunt for the other West Africans that you need!

Concentration! Concentrate on certain stations. One reason that experienced listeners manage to hear the really rare ones is, because having logged so many stations, they can concentrate efforts on a relatively few wanted targets. The less experienced, needing more stations, often take a hit-or-miss scattergun approach.

For example, when a good band opening occurs to Indonesia, the less experienced DX'er might spend his time taking a log on a Radio Republik Indonesia station, such as RRI Ujung Pandang on 4719 kHz. He "needs" it and is pleased to receive a new one. But the veteran, having logged and verified U.P. already, concentrates on trying for the rarer Indonesians—say the governmental agricultural information station, Radio Khusus Informasi Pertanian in Surabaya, on 4699 kHz. Ujunh Pandang may be a good catch—a new logging for you. But wouldn't you be happier receiving RKIP? It takes self discipline to pass up a sure catch for a chancier one, but the odds are that you will next hear RRI Ujung Pandang—and get a reportable logging—far sooner than you will get another crack at RKIP!

The choice—and, naturally, you won't always opt for the chance of hearing an ultrarare station over the sure bet logging—is tough for the less experienced listener. It means gambling on the possibility of a rare one and, for that day at least, passing by a medium-hard station already at hand. The veteran DX'er, who already has QSL'ed the easier station, doesn't have to think twice before chasing the really rare outfits when top notch openings occur. A word of caution here! Don't let the higher stakes—temporarily giving up a sure catch for a gamble—color your judgment. Apply the same identification standards you always do. The weak muddle down in the mud may be RKIP..., or it may not. Your gamble may not pay off. But if it does....

The decision as to when to play it safe and take the needed logging at hand, and when to gamble on the possible really rare logging, is easier if you have a "want" list.

Rare is the experience DX'er who doesn't have his own list of most-wanted stations. Check on what the others are hearing, particularly the DX'ers you regard as "pros." Draw up a list of about ten stations that you really want to log. In any situation where you're faced with the decision to play it safe or gamble on the rare catch, let your "want" list establish the priority. Concentrate on those stations when conditions to a certain area of the world are most favorable. If you have no luck, then go after the secondary targets. When you knock a station off, replace it on the list with one you badly want to hear.

Compile a "book" on each station on your want list. Minimum information, of course, will be the frequency and the time when the station is being received in your area. Comb DX bulletins and columns for more tips—schedules, language used, etc. Is there an especially favorable time "window?" Perhaps a normally strong utility station, which is on the same frequency, is off the air on Sundays. Can your target be heard only on the occasions after a powerful nearby station signs off? What type of programming would you expect to find? Country and western music? Outdated U.S. pop music? A French newcast? In short, get all the background data you can. It'll help *if* and *when* you actually do hear the station.

Finally, tune as often as possible for your most wanted stations. Trying for them seven times a week is better than six; six times a week is better than four. Make the odds work in your favor. Be listening when reception conditions are right.

Identification

You have followed the tips and you've tuned a shortwave signal. It's not too strong and you've troubles with interference. It may, or may not, be a real DX station—the one that you've been long trying to hear. Now you've got to identify it.

This is where the going really gets tough. It's the test that separates the men from the boys, the place where DX'ing experience really counts. But the first thing to consider, strange as it may seem, is the question of what constitutes an identification.

Identifying the station you are hearing should be an easy thing, theoretically. Either it's yes or no, go or no go! In practicality, though, station identification comes in varying degrees of certainty. Although this certainty spectrum is a continuum ranging from total certainty about a station's identity to complete uncertainty, for convenience, I'll divide it into four broad brackets.

Positive Identification

In this case, you're 100% sure of the station's identity because you heard the ID announced. However, there are other ways to positively identify a station, of course. For instance, if you hear a program called "Moscow Mailbag," you don't need an actual station identification to know that you're tuned to Radio Moscow.

Now all this is a snap when you're dealing with strong-signal stations broadcasting in English. But our topic here is the rare DX, and identification of those outlets is harder because the signals are weaker, plagued with interference, and the programs usually are in languages other than English.

Tentative (or Probable) Identification

Here, you are not certain of the identity of the station you've tuned. No definite ID has been heard, or if heard, not fully understood because of language or interference difficulties. But there are various clues in the programming that lead you to believe strongly—to a confidence level of, perhaps, 75%—that you are hearing "Station X."

Possible Identification

In this case, again, no station identification announcement has been heard or understood. There are some factors which lead you to suspect that you're hearing "Station X," but there are other aspects (in what you can make out of the programming) that could apply to other stations known to be operating on or about the same frequency at that time. It's a 50/50 proposition. Maybe it's "Station X," but there's a fair chance it might be another station. Your degree of certainty in the identification is not high.

Unidentified

Here, you either have no idea, from the announcements and program details, as to the identity—or you have a very low degree of certainty as to the identity.

Reception Reports

Broadly speaking, a reception report to a station is always justified when a station is positively ID'ed. A report is never justified with an unidentified station—if you're just guessing at its identity. In the midranges of ID certainty, even experienced DX'ers differ as to when they feel a reception report should be sent to a station. The following represents this author's viewpoint.

If you strongly believe, based on a number of factors which I'll cover later in this chapter, that you've heard "Station X," and this is coupled with some pretty convincing program details, a report may be sent to the station on the strength of tentative identification. But that report should state that you didn't positively identify the station, and should mention the reasons why you believe it to be the station reported. And, you should ask that the report be verified only if the station authority is convinced that it is correct.

Rarely should you report a station on the strength of a possible identification. A report seeking verification is justified only when you've noted some especially unique factor that could apply only to the station in question.

> *Example*: You think your station could possibly be "Station X," and during the course of your logging, a transmitter problem knocks the station off the air at 1413 GMT/UTC. It is very likely that the station could determine the accuracy of your report on the strength of this unusual detail.

The sensible rule of thumb to follow in most cases is to report only those stations which you know or strongly believe you logged. The burden should be on you, the DX'er. When in doubt, don't rely on the station's QSL to resolve those doubts.

Languages

If you haven't progressed beyond the "If it isn't in English, I can't ID it stage," you've got some basic homework ahead of you before you're ready for serious DX'ing. Minimally, you should be able to distinguish among the major languages: Spanish, Portuguese, French, German, Arabic, etc. You should be able to recognize the tonal Asian languages, such as Chinese. Foreign languages should no longer all sound alike to you. You need not be a linguist—few experienced DX'ers are—but you should be able to pick out words. If you can't, spend some months on the business of learning what languages sound like. There have been articles written on the subject in various magazines, hobby publications, and books. Seek them out and study. One of the best ways to learn is to study the Voice of America foreign language broadcasts.

Tentative Loggings

For those of you who are still with us, it should be obvious by now that a very experienced DX'er, because of his accumulated knowledge, may well be able to positively identify a station that a less experienced DX'er might not be able to identify at all, or, at best, will consider a tentative logging.

But even the best of the "pros" frequently come up short in the area of ID certainty. And, as I suggested earlier, one might choose to send a report to a station on the strength of a tentative ID. A tentative, you'll recall, is when positive identification isn't possible, but a whole series of factors surrounding the reception strongly lead the DX'er to believe he is hearing a certain station.

But, *a word of warning*! Tentatives are tricky and dangerous. Most misidentified stations that turn up as "clinkers" in the DX club bulletin lists result because the DX'er lacked information or experience, was careless, or jumped to hasty conclusions due to wishful thinking. A DX'er, anxious to hear a rare station, is easily tempted to conclude, "Well, others are hearing it, so it must be Station X that I'm hearing too." Leap to such conclusions at your own peril.

The measure of a DX'er is his reputation, especially among other experienced DX'ers. A reported rare station almost always prompts a frantic hunt by others. If you were wrong in your ID, it will be discovered quickly enough. Everyone makes mistakes now and again. It is embarrassing but not fatal to a reputation. Your error may not be challenged in print, but you can be sure that it will be noted by the more experienced DX'ers. If you later find you erred, a correction to the bulletin editor helps mitigate the mistake. Pretending it didn't happen is a mark against you.

To make a mistake occasionally is human. To be wrong regularly earns you the wrong kind of reputation. Before long, you're marked as unreliable, which is the worst of all possible curses in DX'ing. Then, no matter what rare loggings you claim, no matter how high you climb on the scoreboard, a tinge of doubt will always be attached to your reports.

Tentative IDs, to return to the subject, are the result of piecing together a number of separate factors that, in composite, convince you of the station's probable identity. The number of bits that "fit" can vary, but the more that do, the surer you can be. Too few "fits" and all you've got is a "possible," or perhaps even an "unidentified." And, the experienced DX'er has the edge since he has accumulated more know-how and background information to test for "fits."

The factors? *Frequency*, naturally. *Time*? Does your reception match the schedule? *Propagation*? Is reception possible, or likely, at the time you are hearing it? Does your reception match the *language* and *programming* known to be used? Can you, for one reason or another, eliminate from consideration *other stations* known to operate on the frequency at that time?

Other clues? You heard no country reference, but did hear a Spanish-language *commercial* for Cerveza Carta Blanca. In which country or countries of Latin America is that brand of beer distributed? Can you identify the national anthem heard at the sign-off time?

What does the *music* tell you? Can you distinguish between Mexican and Brazilian music? Between that of Peru and that of the Dominican Republic? To put a finer point to it, can you tell the difference between the music of neighboring countries—Mexican and Guatemalan, or Ecuadorian and Peruvian? What is "hi-life music?" "Gamelan?" What is a "sitar" and what does it sound like?

There are dozens of clues in the programming you hear, factors that can help you to identify a station. There is no practical way that they can be taught, except by experience. But if you're alert to the basic techniques, you can begin to file away in your mind those scraps of information that will help you in identifying stations. Naturally, none of the clues alone is sufficient to justify even a tentative identification. Enough of them, however, may make a strong enough case for a tentative ID. But, again, be careful!

Verification

The big international broadcasters of the world, those you first tuned as a beginning shortwave listener, know more than a little about those who tune in. They know hobby listeners like to collect QSL cards. They want us to listen so they go out of their way to respond with the QSLs we seek.

Rare DX stations, by their very nature, don't get many overseas reports. They usually aren't trying to reach North American audiences. Their staffs often don't have the foggiest idea what a correspondent wants when he asks for a verification. These stations, in short, are much harder to verify.

It is assumed that you know the basic report-writing technique. In your letter to the station you've heard, you include the necessary bits: the frequency; the time, date, program, and reception details; and the request for a verification. When dealing with the rarer stations, it is more important to write in the language of the station, unless you have reason to believe that someone at the station knows English. If you include mint stamps of the country to which you're writing only on rare occasions, plan to do so regularly when reporting to the lesser-heard stations. Don't expect them to pay the postage for the reply you're so anxious to receive.

It is important to get your report into the hands of someone who is known to have replied to others. Keep track of verie signers—the names of those persons who sign other DX'ers' reports. Most DX club bulletins report that sort of data in their pages.

Often your first report won't bring a response. Use periodic follow-ups, or new reports to nonrepliers. It can take a year, 5 years, or 10 years to get a verification from some rare stations. Persistence is necessary.

But beyond this, what? The serious, experienced DX'er, who is interested in QSLs, develops additional reporting techniques to get verifications from the rare, really-hard-to-QSL stations. Not infrequently, DX'ers guard the secret methods they've developed jealously. These little extra techniques are, to a degree, perishable commodities. Their usefulness may decrease with usage, so, understandably, veteran DX'ers tend to keep quiet about their best approaches. Eventually, you, too, will develop your own series of reporting techniques that work for you.

In essence, though, what most of the techniques amount to is an attempt to bring special attention to your letter, strike a note that will interest the recipient, and move him to an action he might not otherwise take—that is, reply to your letter. Anyone familiar with the field of advertising will recongnize in this the basic rules of promoting a product. It may come as a bit of a shock, but many of the same approaches used to peddle soup and soap apply to reception report writing.

In advertising, it is A-I-M-A. *Attention! Interest! Motivation! Action!* Get the attention of someone at the station. Make your letter interesting. Motivate the reader to want to respond. And, finally, hope that he actually does write that verification letter or card.

But it isn't all that coldblooded. You probably are genuinely interested in knowing more about the station, its operations and personnel, and the community in which it is located. Let that interest show in your report. Honesty doesn't have go out the window. To the contrary, candor is the most effective sales technique. It doesn't mean you must come on like gangbusters with the "hard sell." "Soft sell" is usually more effective than high pressure. But you still need a good product—the solid core of a good reception report. The more you know about your "market," however, the better you'll be able to tailor your approach to the situation.

These special reporting techniques you'll have to develop yourself. Your own personality, style, and accumulated experience will guide you if you're receptive to the basic idea. If you are rigid, if you insist on a basic master report format that you always follow, if you take the position, "If they don't like the kind of reports I write, to Hell with 'em," and if you're not willing to try, try again if your first report, or your first ten reports, fail to bring a response from a station, then you'll miss a lot of fine QSLs.

I'd be remiss, at this point, if I didn't note that some veteran DX'ers just aren't gung-ho about QSLs. They maintain that a QSL isn't always proof positive of a reception; that some stations confirm without actually checking reports for accuracy. They're right, of course.

But, for me, the quest for verifications from the rare stations I hear is enjoyable. I feel that a QSL'ed report is better than a non-QSL'ed report. It forces us to be more careful in identifying a station as we try to garner enough programming details to justify a report to that station. Plus, there is a real thrill in receiving that rare reply from Lower West Bengalistan or wherever.

To sum things up, the key to successful DX'ing is the know-how and knowledge gained through experience and an effort to learn as much as you can about the hobby and the stations you tune or want to tune.

If it has struck you that this chapter has been short on actual "how-to-do-it" tips and long on generalities, it is because there is no short course on DX'ing. That's not the way it works. All any article can do is to point you in the right direction. If this has done that... and if you carry through..., then you are, truly, on your way to the top!

Index

Dynamic range, 33, 41–42

E

Eastern Europe, pirates in, 415–416
Ecliptic plane, 470
Ecuador, 460–463
E layer, 54, 55
Electrical power, 3
Electromagnetic spectrum, 50
England, pirates in, 407–408
Equatorial regions, 55
Equipment
    grounding, 10–12
    needs, 9–14
    reviews, 47–48
    specialized, 14–17
    texts, 254–258
European pirate radio, 397–419
Europirate roundup, 407
Evening and daytime DX, 253–254
Exalted-carrier selectable sideband
    (ECSS), 44
External services
    listed sites of Radio Moscow,
        432–442
    other, 431
    Russian, 427–428

F

Facsimile, 312, 333
Fading, reducing, 44
Fall equinox, 469
Federal Frequency Directory, The, 299
Filters
    AC line, 16
    audio, 12–13
    line, 3
Five-digit stations, 230–231
F layers, 54, 55
FM, 167–168, 286
    DX, 278–280
    DX'ing, 273–286
    other forms of, 283–284
    receivers, 276–277
    subcarriers, 284–285
    translators and boosters, 284
Folcroft Radio Club, 141
Follow-up, 131–137
    letter, 134–135
    recordkeeping, 135–137

Foreign BCB DX'ing, 266–268
Four-digit stations, 231
France, pirates in, 410–411
Free Radio movement, 218–221
French, 80–82
Frequencies
    aeronautical, 316–317
    Coast Guard, 318–320
    maritime, 318
    United States Navy, 322
    USAF, 320–321
    world news, 359–365
Frequency
    allocations, 324–325
    lists, 366
    readout, 36, 38
    usage, 338–339
Fundamentals of shortwave radio
        propagation, 49–63
Furnishings for the shack, 4–9

G

Geomagnetic field, 55
GMT; *see* Greenwich Mean Time
Graveyarders, 254–258
Gray market, 46
Grayline DX'ing the Tropical Bands,
        467–471
Great Lakes Shortwave DX Club, 141
Greenwich Mean Time (GMT), 7, 367
Ground
    AC, 11, 29
    RF, 6, 11, 27–29
    systems, 27–29
    wave, 51
Grounding, equipment, 10–12
Guide to Utility Stations, 8, 314

H

Half-wave antenna, 22
Ham action, 308
Hard-copy printers, 379–380
Hauser, Glenn, 141
Headphones, 7
HF radio-wave propagation, variations
        of, 57–63
High-frequency
    broadcasting, USSR, 425–432
    broadcasts, domestic, 432
Howls and squeals, 39–42

How to
    defeat shortwave's howls and
        squeals, 39–42
    DX
        sunrise skip, 261–263
        sunset skip, 259–261
    evaluate a shortwave receiver,
        35–39

I

J

K

L

M

MORE
FROM
SAMS

☐ Radio Handbook (23rd Edition)
William I. Orr
In its 23rd edition, this is the most current reference available for radio amateurs and communications technicians/engineers. It is a complete handbook for hf, vhf, and construction theory and current practices. The book also discusses equipment and antenna design techniques and how to reduce radio interference.
ISBN: 0-672-22424-0, $24.95

☐ Commodore 64® & 128® Programs for Amateur Radio & Electronics   *Joseph J. Carr*
The electronics hobbyist, programmer, engineer, and technician will enjoy the 23 task-oriented programs for amateur radio and 19 electronics programs in this book. It contains two general categories of programs—amateur radio technology and general electronics—that will save time and simplify programming tasks when incorporated into the custom-designed software programs provided.
ISBN: 0-672-22516-6, $14.95

☐ PASCAL for Electronics and Communications   *Richard Meadows*
This book is for hobbyists, students, and engineers with no prior knowledge of computing or the PASCAL language, but it assumes a basic familiarity with math for electronics. Readers will learn to apply programming to common analytical and circuit-design problems in electronics and communication engineering through hands-on examples. Exercises and problems appear at the end of each chapter.
ISBN: 0-672-22514-X, $12.95

☐ Landmobile and Marine Radio Technical Handbook   *Edward M. Noll*
A complete atlas and study guide to two-way radio communication: private landmobile services, marine radiotelephone and radiotelegraph, marine navigation, and Citizens Band radio. Beginning with the fundamentals, this book covers everything from maintenance and installation to advanced systems and technology. It also discusses digital and microprocessor electronics, repeater stations and cellular radio, FCC licensing information, equipment testing and service, radar equipment, and satellite communications. An excellent text for radio communications courses or hobbyists.
ISBN: 0-672-22427-5, $24.95

☐ Radio Systems for Technicians
D. C. Green
This comprehensive volume examines the theory behind the broadcast and reception of radio frequencies. Discusses the principles of amplitude and frequency modulation; how to change the amplitude and frequency of a signal; how to build and use transmission lines and antennas; propagation of radio waves, RF power amplifiers; radio transmitters and receivers; and radio receiver circuits. An excellent reference for electronic technicians, radio enthusiasts, and hobbyists.
ISBN: 0-672-22464-X, $12.95

☐ Mobile Communications Design Fundamentals   *William C. Y. Lee*
This authoritative introduction to mobile communications design, including cellular radio, provides communications engineers and engineering students with an interpretation of the mobile radio environment. It presents problems related to that environment and their solutions through the choice of properly designed parameters. Well supported by illustrations and examples, this incisive book explores propagation loss, calculation of fades and methods for reducing fades, interference, frequency plans, design parameters at the base station and mobile unit, and signaling and channel access.
ISBN: 0-672-22305-8, $34.95

☐ Electronic Telephone Projects (2nd Edition)   *Anthony J. Caristi*
This book shows the reader how to create a touch-tone dialer, conferencer, computer memory, electronic ringer, and 18 other creative projects that turn a telephone into a favorite household tool. This new edition contains seven completely new projects, plus a new approach to one previous project.
ISBN: 0-672-22485-2, $10.95

☐ First Book of Modern Electronics Fun Projects   *Art Salsberg*
Novice and seasoned electronics buffs will enjoy these 20 fun and practical projects from the pages of *Modern Electronics* magazine. Electronics hobbyists are introduced to many project building areas. The necessary tools for each project accompany the step-by-step instructions, illustrations, photos, and circuit drawings.
ISBN: 0-672-22503-4, $12.95

☐ The Sams Hookup Book: Do-It-Yourself Connections for Your VCR
Howard W. Sams Engineering Staff
Here is all the information needed for simple to complex hook ups of home entertainment equipment. This step-by-step guide provides instructions to hook up a video cassette recorder to a TV, cable converter, satellite receiver, remote control, block converter, or video disk player.
ISBN: 0-672-22248-5, $4.95

☐ The Ku-Band Satellite Handbook
Mark Long
This book will be the industry standard for Ku-Band satellite technology—the future of satellite communications. Intermediate level users, technicians, and satellite professionals will learn the various aspects of the transmission and reception of video, voice, and data signals by commercial communications satellites operating on frequencies within the 11- to 12-GHZ range.
ISBN: 0-672-22522-0, $18.95

☐ The Home Satellite TV Installation and Troubleshooting Manual
Frank Baylin and Brent Gale
For the hobbyist or electronics buff, this book provides a comprehensive introduction to satellite communication theory, component operation, and the installation and troubleshooting of satellite systems— including the whys and wherefores of selecting satellite equipment. The authors are respected authorities and consultants in the satellite communication industry. If you are among the 100,000 people per month who are installing a satellite system, you'll want to have this book in your reference library.
ISBN: 0-672-22496-8, $29.95

☐ Satellites Today   *Frank Baylin*
Here are the history of satellite communications, the costs of satellite systems, system components, legal questions that have been and remain to be decided, and up-to-date coverage of the latest developments in satellites.
ISBN: 0-672-22492-5, $12.95

☐ The Hidden Signals on Satellite TV (2nd Edition)
Thomas P. Harrington and Bob Cooper
This is the authoritative guide that details satellite services available and demonstrates how to access and use such non-video signals as audio channels, news services, teletext services, and commodity and stock market reports. Don't pass up the hidden world illuminated for you by this valuable book.
ISBN: 0-672-22491-7, $19.95

☐ Introduction to Digital Communications Switching   *John P. Ronayne*
Here is a detailed introduction to the concepts and principles of communications switching and communications transmission. This technically rigorous book explores the essential topics: pulse code modulation (PCM), error sources and prevention, digital exchanges, and control. Sweeping in its scope, it discusses the present realities of the digital network, with references to the Open Systems Interconnection model (OSI), and suggests the promising future uses of digital switching.
ISBN: 0-672-22498-4, $23.95

☐ Fiber Optics Communications, Experiments, and Projects   *Waldo T. Boyd*
Another Blacksburg tutorial teaching new technology through experimentation. This book teaches light beam communication fundamentals, introduces the simple electronic devices used, and shows how to participate in transmitting and receiving voice and music by means of light traveling along slender glass fibers.
ISBN: 0-672-21834-8, $15.95

☐ A Practical Introduction to Lightwave Communications   *Forrest M. Mims III*
It is said that the most fertile copper mines in the world are located beneath the streets of New York City in a century's growth of twisted copper pair telephone lines. Today we look to other voice and data transmission technologies. This book introduces you to telecommunication by means of fiber optics and free space light transmission. All contemporary forms of lightwave communications are discussed, with working experimental circuits providing practical illustrations.
ISBN: 0-672-21976-X, $12.95

Look for these Sams Books at your local bookstore.

To order direct, call 800-428-SAMS or fill out the form below.

- -

Please send me the books whose titles and numbers I have listed below.

_____

_____

_____

_____

_____

Enclosed is a check or money order for $ _____
Include $2.50 postage and handling.
AR, CA, FL, IN, NC, NY, OH, TN, WV residents add local sales tax.

Charge my: ☐ VISA ☐ MC ☐ AE

Account No. _____   Expiration Date _____

Name *(please print)* _____

Address _____

City _____

State/Zip _____

Signature _____
(required for credit card purchases)

Mail to: Howard W. Sams & Co.
Dept. DM
4300 West 62nd Street
Indianapolis, IN 46268

DC063

SAMS™